Rebuilding Higher Education Systems Impacted by Crises:

Navigating Traumatic Events, Disasters, and More

Mustafa Kayyali
HE Higher Education Ranking, Syria

IGI Global
PUBLISHER of TIMELY KNOWLEDGE

A volume in the Advances in Higher Education and Professional Development (AHEPD) Book Series

Published in the United States of America by
IGI Global
Information Science Reference (an imprint of IGI Global)
701 E. Chocolate Avenue
Hershey PA, USA 17033
Tel: 717-533-8845
Fax: 717-533-8661
E-mail: cust@igi-global.com
Web site: http://www.igi-global.com

Library of Congress Cataloging-in-Publication Data

Names: Kayyali, Mustafa, 1990- editor.
Title: Rebuilding higher education systems impacted by crises : Edited by: navigating traumatic events, disasters, and more / Mustafa Kayyali.
Description: Hershey, PA : Information Science Reference, [2024] | Includes bibliographical references and index. | Summary: "The primary objective is to serve as a comprehensive and developing resource that improves our understanding of the numerous obstacles facing higher education systems in times of crisis while providing actionable strategies for rebuilding, renewing, and fortifying these institutions"-- Provided by publisher.
Identifiers: LCCN 2023055026 (print) | LCCN 2023055027 (ebook) | ISBN 9798369319260 (hardcover) | ISBN 9798369319277 (ebook)
Subjects: LCSH: School crisis management--Cross-cultural studies. | Universities and colleges--Administration--Cross-cultural studies. | Education, Higher--Administration--Cross-cultural studies. | Emergency management--Cross-cultural studies.
Classification: LCC LB2866.5 .R38 2024 (print) | LCC LB2866.5 (ebook) | DDC 378.1/01--dc23/eng/20231207
LC record available at https://lccn.loc.gov/2023055026
LC ebook record available at https://lccn.loc.gov/2023055027

This book is published in the IGI Global book series Advances in Higher Education and Professional Development (AHEPD) (ISSN: 2327-6983; eISSN: 2327-6991)

British Cataloguing in Publication Data
A Cataloguing in Publication record for this book is available from the British Library.

All work contributed to this book is new, previously-unpublished material. The views expressed in this book are those of the authors, but not necessarily of the publisher.

For electronic access to this publication, please contact: eresources@igi-global.com.

Advances in Higher Education and Professional Development (AHEPD) Book Series

Jared Keengwe
University of North Dakota, USA

ISSN:2327-6983
EISSN:2327-6991

Mission

As world economies continue to shift and change in response to global financial situations, job markets have begun to demand a more highly-skilled workforce. In many industries a college degree is the minimum requirement and further educational development is expected to advance. With these current trends in mind, the **Advances in Higher Education & Professional Development (AHEPD) Book Series** provides an outlet for researchers and academics to publish their research in these areas and to distribute these works to practitioners and other researchers.

AHEPD encompasses all research dealing with higher education pedagogy, development, and curriculum design, as well as all areas of professional development, regardless of focus.

Coverage

- Adult Education
- Assessment in Higher Education
- Career Training
- Coaching and Mentoring
- Continuing Professional Development
- Governance in Higher Education
- Higher Education Policy
- Pedagogy of Teaching Higher Education
- Vocational Education

IGI Global is currently accepting manuscripts for publication within this series. To submit a proposal for a volume in this series, please contact our Acquisition Editors at Acquisitions@igi-global.com or visit: http://www.igi-global.com/publish/.

The Advances in Higher Education and Professional Development (AHEPD) Book Series (ISSN 2327-6983) is published by IGI Global, 701 E. Chocolate Avenue, Hershey, PA 17033-1240, USA, www.igi-global.com. This series is composed of titles available for purchase individually; each title is edited to be contextually exclusive from any other title within the series. For pricing and ordering information please visit http://www.igi-global.com/book-series/advances-higher-education-professional-development/73681. Postmaster: Send all address changes to above address.

Titles in this Series

For a list of additional titles in this series, please visit:
http://www.igi-global.com/book-series/advances-higher-education-professional-development/73681

Global Perspectives on Micro-Learning and Micro-Credentials in Higher Education
Kizito Omona (Uganda Martyrs University, Uganda) and Modest Kayi O'dama (Uganda Martyrs University, Uganda)
Information Science Reference • © 2024 • 360pp • H/C (ISBN: 9798369303436) • US $240.00

Prioritizing Skills Development for Student Employability
Bryan Christiansen (Southern New Hampshire University, USA) and Angela M. Even (Purdue University, USA)
Information Science Reference • © 2024 • 340pp • H/C (ISBN: 9798369335710) • US $245.00

Promoting Intercultural Agility and Leadership Development at Home and Abroad for First-Year Students
Jon Stauff (South Dakota State University, USA) and Jill E. Blondin (Virginia Commonwealth University, USA)
Information Science Reference • © 2024 • 292pp • H/C (ISBN: 9781668488324) • US $245.00

Advancing Student Employability Through Higher Education
Bryan Christiansen (Southern New Hampshire University, USA) and Angela M. Even (Purdue University, USA)
Information Science Reference • © 2024 • 429pp • H/C (ISBN: 9798369305171) • US $215.00

Perspectives on Transforming Higher Education and the LGBTQIA Student Experience
Andrew Herridge (The University of Southern Mississippi, USA) and Kaity Prieto (The University of Southern Mississippi, USA)
Information Science Reference • © 2024 • 322pp • H/C (ISBN: 9781668499146) • US $280.00

Reshaping Entrepreneurial Education Within an Industry 4.0 Context
Vannie Naidoo (University of KwaZulu-Natal, South Africa) and Rahul Verma (Delhi University, India)
Business Science Reference • © 2024 • 351pp • H/C (ISBN: 9798369304099) • US $265.00

LGBTQIA Students in Higher Education Approaches to Student Identity and Policy
Kaity Prieto (The University of Southern Mississippi, USA) and Andrew Herridge (The University of Southern Mississippi, USA)
Information Science Reference • © 2024 • 291pp • H/C (ISBN: 9798369328538) • US $240.00

Facilitating Global Collaboration and Knowledge Sharing in Higher Education With Generative AI
Poshan Yu (Soochow University, China & European Business Institute, Luxembourg & Australian Studies Centre of Shanghai University, China) James Mulli (European Business Institute, Luxembourg) Zain Ali Shah Syed (Soochow University, China) and Laila Umme (Institute of Business Management, Pakistan)
Information Science Reference • © 2024 • 356pp • H/C (ISBN: 9798369304877) • US $230.00

IGI Global
PUBLISHER of TIMELY KNOWLEDGE

701 East Chocolate Avenue, Hershey, PA 17033, USA
Tel: 717-533-8845 x100 • Fax: 717-533-8661
E-Mail: cust@igi-global.com • www.igi-global.com

Editorial Advisory Board

List of Reviewers

Table of Contents

Detailed Table of Contents

This chapter examines whether post-crisis regions can benefit from quality assurance in higher education. To this end, it defines what notions of quality exist and how quality can be improved. Special attention is given to the formalized process of accreditation of study programs. From the perspective of neo-institutionalism, it is argued that quality assurance as a rule-based process requires functioning state structures. For post-crisis regions, it is therefore of crucial importance whether or not crises and traumas have an impact on the existence of the state. It is postulated that without an effective administrative and legal system, accreditation or other forms of legally binding quality assurance are not useful. Rather, low-threshold procedures based on the principle of fitness for/of purpose would be recommended in such cases.

The chapter presents the results of the impact of the two 2017 earthquakes that occurred in Mexico City and the hurricane Otis that hit Acapulco, Guerrero, Mexico. Moreover, the perception of fear of buildings collapsing during the occurrence of the two earthquakes. The case study was centered on university students of the affected cities. The main conclusions were the following: a) Overall, there is a lack of studies or reports on the extent of the impacts caused by disasters in higher educational systems in Mexico. University students are affected by the interruption of their studies and this issue has not been addressed explicitly on how to deal with it. b) The disasters covered in the chapter suggest that university students (and lecturers) exhibited prosocial behavior not only those in the affected areas but also students nationwide. c) Disasters trigger disruption to critical infrastructures such as the water supply systems, energy sector, transportation, communications systems, and d) university students exhibited fear of building collapsing during the 2017 earthquakes.

This chapter is an extensive investigation into the crucial role that non-governmental organisations (NGOs) play in facilitating the restoration and reconstruction of higher education systems following conflicts, natural disasters, and other crises. Higher education institutions usually experience major hurdles in such challenging situations. This chapter covers the many facets of the influence of NGOs on the reconstruction of higher education, spanning crucial subjects such as accessibility, quality, innovation, and sustainability. The talk thoroughly addresses the practical and ethical concerns, providing insight into the subtle dynamics of the interplay between NGOs and the recovery of higher education. This chapter stresses the relevance of non-governmental organisations (NGOs) in aiding communities that are dealing with crisis-related disturbances to their educational institutions.

This chapter describes how Wester University, a global university system, navigated the dimensions of crisis arising from recent pandemic conditions. Placing the university's strategies for sustainability in a broader context, the author considers the impact to students and institutions of threats to health and safety, economic disruption, and needed reckoning with issues of isolation, racism, and discrimination. The author reflects on the pandemic's effects on the accelerating threats to higher education's academic and business models, drawing insights from research and experience for rebuilding after crisis.

The COVID-19 pandemic affected the entire world causing every aspect of the higher education ecosystem to undergo a tipping point. The total mandatory lockdown resulted in all academic activities from learning to administrative activities coming to a standstill as institutions did not know how to go around it. Employees' job security, indefinite suspension of lectures, disruption of the university calendar, employee work-life balance, and business continuity were all operational challenges faced by universities during the pandemic. This chapter aims to establish the resilience strategies implemented by higher education institutions in South Africa and Zimbabwe as they strived to continue delivering services to stakeholders during and after the COVID-19 pandemic crisis. Digital transformation as a primary resilience strategy enabled HEIs to implement online learning and work from home for academic and non-academic staff for university business continuity.

The purpose of the research is to examine the perceptions of educators in the disaster area regarding the positive and negative aspects of distance education and their views on how the process is experienced in the field of education, in the Kahramanmaras-centered earthquake that occurred in 11 provinces of Turkey in February 2023. After determining the basic research questions for the purpose of the research, a case study design was adopted with a qualitative approach in order to obtain the expert opinions and suggestions of educators providing distance education in the disaster area during the data collection phase. The educators were interviewed using Google Forms, and they were asked to answer the semi-structured open-ended questions directed to them. The answers given by the educators were subjected to content analysis. Some of the themes obtained in this direction can be listed as follows: dynamic and updated education process, raising awareness of the people living in the region, a potential to benefit from lifelong learning, and access to true and reliable information resources.

Online course experiences of undergraduate students continue to grow due to both a pre- and mid-COVID-19 pandemic increase in online courses offered and taken at universities. In this study, students identified the benefits and challenges of an online course format and their online learning needs for achievement of learning objectives. While the majority of the students in this study had previous online course experiences, they were not confident in their ability to reach learning objectives in that course format. Students identified a lack of specific digital and life skills as well as instructor issues with communication and organization of online course content as their primary concerns. Suggested improvement of practices based on findings include: a focus on assessment and development of the digital literacy and life skills of students, clear communication of course requirements while ensuring students' realistic understanding of course commitments, and professional development opportunities for instructors that are closely reflective of students' online learning needs.

The COVID-19 pandemic has resulted in change across the world. The higher education academic workforce was largely ignored during the COVID-19 pandemic, and there has been little to no acknowledgement of the increased workload, extensive lockdowns, pivots to online learning, and the 'business as usual' expectations that were thrust upon academics during this time. In this chapter, the authors use Aristotelian ethics, the concept of human flourishing, and the United Nations Sustainable Development Goals to provide insight into how human resource in the higher education sector can be supported to recover and thrive amidst ongoing unprecedented global challenges and uncertainties. Drawing on VanderWeele's domains and pathways for human flourishing, the authors demonstrate how universities and colleges around the world can support their academic workforce to revive, thrive, and drive sustainability, productivity, economic growth, and recovery during times of challenge and uncertainty. The chapter concludes with recommendations for future directions and research.

Globally, the 21st century has been characterized by the emergence of several crises including the most recent COVID-19 pandemic. This has altered various systems and structures socially, economically, and politically. It is apparent that classical leadership approaches of response to these crises aren't a panacea anymore to navigate through these challenges hence need to explore novel, and blended transformative strategies to act as a buffer against the ever-evolving new crises. It is during the crisis that leadership becomes conspicuous and needed. It's not possible to discern when or if a crisis will occur, but education leadership could influence how the university responds to such upheavals. Drawing from a scholarly wealthy stance of leadership approaches, this discourse aims to equip higher education leadership with blended transformative strategies on how to respond to crises competently, develop practices that lessen the probability of such occurrences, and stimulate self-confidence in leadership for lasting strategies.

This chapter explores the potential for strategic organizational frameworks and applied methodological perspectives to aid workforce contexts in harnessing the transferable agency of those equipped experientially for leadership in times of unprecedented crisis. Academic research and scholarship must be needs-led not methods-driven, and as the recent pandemic demonstrated, authentic leadership had never been more important. Transitional change through the crisis of the global pandemic led to shifts to greater acknowledgement of the need for a different type of knowledge creation and replacing the contexts of validity and reliability in empirical research with those of trustworthiness and authenticity. In providing some of these organizational frameworks and methodological perspectives in an accessible manner, many can be systematically applied to the context of everyday strategic planning and institutional management settings so that tangible target outcomes are visible, achievable, and perhaps, most importantly, remain person-centered.

The purpose of this research is to investigate the effects of crises on higher education systems while examining the moderating effect of crisis management and its impacts on crises and higher education systems. In order to achieve the research objectives, the researcher collected data through an online survey. The study revealed that crises have a negative impact on higher education systems, while crisis management has a negative effect on crises and a positive impact on higher education systems. The study also found out that crisis management moderates the relationship between crises and higher education systems. Based on the results, the researcher presented recommendations for identifying and proactively preparing the next generation of leaders in education institutions to be able to enhance resilience and deal and react efficiently with any crisis that may occur.

This chapter aims to compare and contrast the experiences and responses of three universities in Africa that have faced different forms of wars and conflicts. The chapter collected data through textual analysis using a qualitative comparative case study methodology. The study compared and contrasted the experiences and reactions of three African institutions that have dealt with various types of wars and conflicts using Savo Heleta's idea of higher education in post-conflict situations. The chapter also discovered that by promoting dialogue, critical thinking, social cohesion, and respect for human rights in the participating countries, higher education has significantly contributed to post-conflict reconstruction and peacebuilding.

This chapter seeks to add to the body of knowledge on higher education in conflict-affected contexts and offer suggestions for improving the adaptability and value of higher education in these situations. As suggested in this chapter, governments should work together with a variety of stakeholders.

In the era of globalization, neoliberal free trade regards higher education as a private good that facilitates international mobility for and of higher education. Similarly, HEIs establish an economic image of overseas higher education that encourages young people to invest in global mobility to pursue higher education. In this context, following a qualitative research methodology, the chapter analyzes the (un)certainties of 18 physically mobile young Bangladeshi (Southern) students. It discusses their consciousness issues and confidence level regarding consuming higher education in Australia (North). To analyze their perceptions, they were considered global citizens (i.e., economic and moral actors in the globalized world). Based on the empirical data, the identified features of young Bangladeshi students and their areas of concern provide the scope to identify them as conscious and confident neoliberal consumers in the global education market who came to Australian universities with neoliberal aspirations.

This chapter investigates the far-reaching consequences of conflict on higher education systems in Syria, Iraq, Ukraine, and Afghanistan. It dives into the issues that these countries face, including the destruction of academic infrastructure, displacement of professors and students, curriculum changes, and gender inequities. Despite these adversities, professors and students demonstrate tremendous perseverance. International collaborations and support play an essential role in protecting education opportunities. The chapter underlines the continuing importance of education in building stability, peacemaking, and growth in war-torn communities, offering hope for a brighter future.

The conflict in Syria is almost entering its 14th year, and its negative effects have been clearly reflected in all economic, educational, health, service, and other sectors across Syria, including the education and higher education sectors, where many young people were deprived of completing their university education as a result of repeated displacement and instability. Some negative effects of the conflict on the higher education were the migration of competencies, experiences, minds, and those with advanced degrees.

This chapter explores the unstable situation of Syrian higher education in the aftermath of 13 years of ongoing conflict. As the nation grapples with the ramifications of war, its educational institutions have

Preface

In the constantly changing context of higher education, where challenges seem to arise with the relentlessness of waves crashing against the shoreline, there emerges a beacon of enlightenment regardless of the chaos - a complete book designed to illuminate how to move forward amidst the shadows of calamity. Welcome to the broad and informative world of *Rebuilding Higher Education Systems Impacted by Crises: Navigating Traumatic Events, Disasters, and More*. As we stand at the edge of uncertainties, it becomes more and more apparent that the long-term viability and flexibility of higher education systems are going through unprecedented inspection. From the seismic disruptions wrought by catastrophes of nature to the insidious spread of pandemics, from the reverberating reverberates of economic recessions to the turbulent hurricanes of political conflicts, universities and colleges around the world find themselves struggling with a wide range of challenges that threaten to erode the very pillars upon which they stand. In response to these issues, there emerges a cry for help - a call to action, a call to restructure and improve higher education systems in the wake of disaster. This call is met strongly by the pages of this groundbreaking book, which serves not only as an academic text, but as a guiding beacon for educators, administrators, legislators, and stakeholders alike. It is a monument to the resilient spirit of academia, a rallying cry for all those who share a dedication to preserving and enhancing higher education in the face of challenges.

The 21st century has borne witness to an increase in the frequency and intensity of crises facing higher education institutions. From the rapid shift to distance education forced by the COVID-19 pandemic to the ongoing consequences of climate change on university infrastructure, the problems are many and complicated. Yet, within the chaos, there is an opportunity - an opportunity to start on a transformative journey, delve deeply into the heart of these difficulties, and emerge with new solutions that pave the way for a brighter, more durable future. *Rebuilding Higher Education Systems Impacted by Crises* is not only a book, but a manifesto - a call to arms, a cry to rethink, reimagine, and rebuild higher education in a world where crises are no longer the exception but the rule. It is a framework for resilience, a plan for revitalization, and a monument to the lasting role of higher education as an inspiration for knowledge, enlightenment, and advancement for generations to come. As we commence on a journey through the corridors of higher education amid crises, we are confronted with an immense variety of chapters, each presenting a unique perspective, a distinct lens through which to view the challenges and possibilities that lie ahead.

In Chapter 1, "Resilience Strategies for Higher Education Institutions", Christopher Nyakotyo and Pedzisai Goronga lead us through a comprehensive exploration of the various resilience strategies employed by higher education institutions, emphasizing the importance of visionary leadership and transparent communication in fostering resilience.

Colleen Halupa, in Chapter 2, "Disaster Preparedness and Response in Higher Education: Applying Lessons Learned From Healthcare", draws upon lessons from the healthcare sector to craft robust emergency plans and exercises that can mitigate the impact of disasters on higher education institutions.

Stefan Handke, in Chapter 3, "Quality Assurance in Higher Education: A Fertilizer for Academic Enhancement or a Luxury of an Ideal World", delves into the notion of quality assurance in post-crisis regions, exploring the challenges and opportunities inherent in the accreditation process.

In Chapter 4, "Impact of Disasters in Higher Education: Disruption and Facility Affectations", Jaime Santos-Reyes, Galdino Santos-Reyes, and Ricardo Tejeida-Padilla present a case study on the impact of earthquakes and hurricanes on higher education institutions in Mexico, shedding light on the disruption caused to critical infrastructures.

Ahmad Kayaly, in Chapter 5, "The Impact of NGOs in Rebuilding Higher Education in the Recovery Phase", offers an extensive investigation into the role of Non-Governmental Organisations in facilitating the restoration and reconstruction of higher education systems following crises.

Elizabeth Stroble, in Chapter 6, "Sustaining a Globally Diverse Community Through a Time of Crisis: Challenges and Consequences", reflects on Wester University's strategies for sustainability amidst pandemic conditions, highlighting the impact on students and institutions.

In Chapter 7, "Exist or Exit: Resilience During and Post COVID-19 Pandemic Crisis in Higher Tertiary Institutions", Mubango Hazel, Hlanganipai Ngirande, and Khathutshelo Khashane explore resilience strategies implemented by higher education institutions in South Africa and Zimbabwe during the COVID-19 pandemic crisis.

Ebru Aksu and Hakan Kilinc, in Chapter 8, "A Case Study on the Role of Distance Education in Disaster and Crisis Management", examine the perceptions of educators in disaster areas regarding the benefits and challenges of distance education.

Chapter 9, "Learning From Chaos: The Complexity of Students' Needs in Online Learning" by Vicki Hosek and Jay Percell, delves into the evolving landscape of online learning, particularly in the context of crisis. Through a qualitative study, the authors explore the perceptions of undergraduate students regarding the benefits and challenges of online courses, shedding light on areas for improvement and suggesting strategies for enhancing the online learning experience.

In Chapter 10, "Supporting Academics to 'Revive, Thrive, and Drive' During Times of Challenge and Uncertainty" by Wendy Goff and Robert New, the focus shifts to the support mechanisms necessary for academics amidst unprecedented global challenges. Drawing on Aristotelian ethics and the United Nations Sustainable Development Goals, the authors provide insights into how higher education institutions can foster resilience and drive sustainability among their academic workforce.

Chapter 11, "Educational Leadership through the Lens of Crisis: Quest for Paradigm Shift From Classical to Transformative Leadership Approaches" by George Areba, explores transformative leadership strategies in the face of crises. By examining various leadership approaches, the chapter aims to equip higher education leaders with the tools and strategies necessary to navigate through challenges competently and stimulate lasting change.

In Chapter 12, "Tales of the Unexpected: Post-Crisis Transformational Leadership" by Catherine Hayes, the discussion centers on the role of strategic organizational frameworks and methodological perspectives in harnessing leadership during times of crisis. Through systematic application, these frameworks offer tangible outcomes and remain person-centered, guiding institutions toward resilience and authentic leadership.

Chapter 13, "Meeting the Future: Identifying and Preparing the Next Generation of Leaders for Crisis Management" by Sherihan Radi, investigates the effects of crises on higher education systems and the moderating role of crisis management. Through an online survey, the chapter explores recommendations for identifying and preparing the next generation of leaders, enhancing resilience, and efficiently managing crises in educational institutions.

In Chapter 14, "Rebuilding Higher Education Systems Impacted By Crises: The Impact of Wars and Conflicts on Higher Education" by Mulatedzi Calvin Rammbuda, the focus turns to the impact of wars and conflicts on higher education institutions in Africa. Through comparative case studies, the chapter examines experiences and responses, highlighting the importance of promoting dialogue, critical thinking, and social cohesion for post-conflict reconstruction and peacebuilding.

Chapter 15, "The Global Citizens in Global (Un)certainties: Southern Higher Education Students in the North" by Md Tariqul Islam, analyzes the experiences of Bangladeshi students pursuing higher education in Australia amidst globalization. Through qualitative research, the chapter explores their perceptions as economic and moral actors, shedding light on issues of consciousness and confidence in a globalized education market.

In Chapter 16, "The Impact of War on Higher Education: The Context of Wars in Syria, Iraq, Ukraine, and Afghanistan" by Mustafa Kayyali, the consequences of conflict on higher education systems are examined across various war-torn countries. Through a comprehensive analysis, the chapter underscores the importance of education in building stability and growth in communities affected by war, offering hope for a brighter future.

Chapter 17, "Impact of Armed Conflicts on Higher Education Northwestern Syria Context" by Mahdi Alkol, delves into the effects of the conflict in Syria on higher education, particularly in the northwestern region. Through an exploration of operational challenges and the migration of competencies, the chapter sheds light on the disruptions faced by higher education institutions amidst ongoing conflict.

Chapter 18, "Stuck in the Limbo: Syrian Higher Education After 13 Years of War" by Mustafa Kayyali, examines the unstable situation of Syrian higher education in the aftermath of prolonged conflict. Through an in-depth analysis of obstacles and alternative pathways for rebuilding, the chapter offers insights into navigating the complex processes of rebuilding and rejuvenating the industry.

Finally, in Chapter 19, "Mortality Awareness of Parents Affected by the Kahramanmaras Centred Earthquake Disaster in Turkey" by Ali Mazı, the focus shifts to the psychological impact of earthquakes on parents in Turkey. Through a survey-based approach, the chapter explores mortality awareness levels and the effects of psychological assistance on post-disaster recovery, offering valuable insights into supporting affected communities in times of crisis.

In the context of extraordinary global difficulties, the connection between higher education and disasters has become of greater significance than ever before. As we lie on the threshold of a new era defined by unpredictability and rapid change, the significance of *Rebuilding Higher Education Systems Impacted by Crises: Navigating Traumatic Events, Disasters, and More* could not be more apparent. This work emerges as a beacon of knowledge and guidance amid the turbulent seas that are modern incidents and patterns that are changing the surroundings of academia. In the past few years, we have experienced an intersection of crises that have resonated throughout higher education systems. The COVID-19 pandemic, with its nuisance force felt across all corners of the world, has thrust higher education into unknown waters, compelling HEIs to adjust swiftly to distance education, navigate difficulties with funding, and struggle with the psychological difficulties faced by students and faculty alike. The echoes of this catastrophe continue to resonate, altering the destiny of higher education for generations to come.

At the same time, the impending doom of climate change looms large, posing existential threats to university infrastructures worsening socio-economic inequalities, and spurring urgent calls for sustainability initiatives inside higher education. Natural catastrophes, political conflicts, and socio-economic upheavals further complicate the complexity facing higher education institutions, underlining the need for robust and adaptable responses in the face of adversity. Against this backdrop of crises and uncertainty, "Rebuilding Higher Education Systems Impacted By Crises" appears as an essential and valuable resource. The subject matter of adaptive tactics, preparedness for disasters, quality assurance, and involvement of stakeholders speaks directly to the critical challenges affecting higher education leaders and policymakers at this particular moment. By presenting practical insights, recommendations based on evidence, and enlightening case studies, this book offers a roadmap for navigating the stormy seas of crises and charting a course toward rebirth and restoration.

The ideas covered in this book reverberate deeply with broader socioeconomic trends and discourses impacting the prospects of higher education. From the growing focus on digital transformation and online learning to the importance of fostering inclusiveness, equity, and diversity within academic communities, the values garnered from this book transcend far beyond the bounds of academia. They speak to the more pressing importance of resilient, equitable, and resilient structures capable of weathering the storms of change and rising stronger on the other side. As we embark on this journey throughout the chapters of *Rebuilding Higher Education Systems Impacted by Crises*, let us not just work to fully understand the challenges that lie before us but also embrace the opportunities for creativity, cooperation, and positive change. In doing so, we acknowledge the perseverance of higher education and reassert its enduring dedication to knowledge, awakening, and progress in the face of challenges.

This work, *Rebuilding Higher Education Systems Impacted by Crises: Navigating Traumatic Events, Disasters, and More*, ranks as an important contribution to academia, giving an in-depth examination of the problems and opportunities faced by higher education institutions in times of crisis. Its value dwells not just in its timely assessment of critical issues but also in the breadth and breadth of its representation across multiple fields and regional scenarios. One of the major statements about this book is its multidisciplinary approach, bringing together opinions from education, disaster preparedness, leadership studies, and other related topics. By bringing together various perspectives, the book presents a comprehensive understanding of the varying characteristics of crisis and their impact on higher education systems. This interdisciplinary perspective enriches the discussion, stimulating cross-pollinating of ideas while promoting inventive solutions to challenging problems.

The importance of this book lies in its practical significance for educators, administrators, policymakers, and others involved in higher education. Each chapter presents practical insights, standards of excellence, and case studies that can inform decision-making and formulation of policies in real-world scenarios. Whether it's designing emergency plans, introducing online learning practices, or supporting the well-being of academic staff, the book gives guidelines for handling crises and building resilience. The added value of this book resides in its international perspective, with contributions from scholars as well as practitioners across numerous countries and cultural contexts. By examining crises and responses from many perspectives, the book develops cross-cultural understanding and illustrates the universality of challenges encountered by higher education institutions. This global outlook boosts the book's significance and applicability to a large popularity, overcoming geographical obstacles and stimulating collaboration on an international level.

In terms of impact, this book has an opportunity to accelerate positive change inside academia by training stakeholders with the information, tools, and strategies needed to confront crises successfully. By promoting resilience, creativity, and cooperation, the ideas acquired from this book can equip higher education institutions to not only weather the storms of calamity but also emerge with greater resilience and adaptability than before. Ultimately, the significance of this book extends beyond the field of academics, adding to the broader debate on resilience, catastrophe management, and social stability in an increasingly volatile world. Amid enormous challenges and ambiguities, *Rebuilding Higher Education Systems Impacted by Crises: Navigating Traumatic Events, Disasters, and More* emerges not just as an in-depth examination of the crises affecting higher education but also as an urgent call to action. This book is not only an observer and viewer of the turbulent circumstances that are altering academia; it is a rising call to the occasion and characterizes a road toward renewal and rebirth. The significance of this call to action is obvious as we approach an area plagued with challenges of unparalleled magnitude. The difficulties that we face are numerous and complex, requiring a response that is equally complex and adaptable.

Now is not the time for carelessness or surrender. Now is the time for determined and powerful action. The need to restore and improve higher education systems has never appeared more urgent or more persuasive. It is essential for all of us—educators, administrators, legislators, and stakeholders—to accept this call to action and devote ourselves to the process of rebuilding with unwavering recovery and determination. But what does this call to action entail? It begins with an understanding of the gravitation of the difficulties before us and a desire to address them straight away. It challenges us to recognize creative thinking, collaboration, and innovation as we go through the complexity of crises and plan our way toward rebirth. It demands that we promote fairness, inclusion, and sustainability in all our undertakings, ensuring that the advantages of higher education are available to all, regardless of origin or circumstances.

This call to action also forces us to evaluate existing paradigms and institutions within higher education, challenging the status quo and envisioning novel approaches that are more suited to resist disruptions of crisis and adapt to the evolving demands of society. It demands us to build a culture of resiliency, flexibility, and adaptation inside our institutions, empowering learners, staff, and educators to thrive in the face of hardship. Above all, this call to action is an acknowledgment of the capacity for change in education and the essential part that higher education plays in creating the future of our society. It is a call to harness the collective knowledge, innovation, and enthusiasm of academics to face the challenges of our current moment and build a brighter, more durable future for generations to come. As you take on this journey through the chapters of "Rebuilding Higher Education Systems Impacted By Crises," I encourage you to take note of this call to action and join us in the essential endeavor of rebuilding, renewing, and reimagining higher education for the challenges of today and the opportunities of tomorrow. Together, let us rise to the occasion and grasp this moment to establish a future that is worthy of our objectives and convictions.

In closing, I invite you to reflect on the tremendous significance of the journey we are going to engage in together within the chapters of *Rebuilding Higher Education Systems Impacted by Crises: Navigating Traumatic Events, Disasters, and More*. As we conclude this preface, it is essential to take a moment and contemplate the broader consequences of the ideas, insights and calls upon action contained inside this outstanding academic text. The challenges confronting higher education systems in the 21st century are numerous and complex, encompassing the areas of health, surroundings, politics, economics, and beyond. From the disruptive forces unleashed by worldwide epidemics to the profound consequences of global warming and armed conflict, higher education institutions find themselves at the intersection of crises that challenge the foundation of their resilience and adaptability.

Yet, amidst these challenges and difficulties, there emerges a beacon of hope—an understanding that crises, while challenging, also bring opportunities for growth, change, and regrowth. It is in the the crucible of hardship that the true steel of colleges and universities is made, as they rise to face the challenges of modern times with bravery, creativity, and unwavering dedication. As we get around the pages of this book, we will come across stories of resilience, creativity, and perseverance—stories of institutions and individuals who refuse to be intimidated by the enormity of the obstacles that lie before them and rather embrace the chance to reconsider and restore higher education for a more equitable, sustainable, and adaptable future.

But beyond the practical insights and practical suggestions contained inside these pages, there lurks a larger message—a call to embrace the life-changing impact of education as an instrument for positive change in the world. In the face of unpredictability and shift it is education that demonstrates how to move forward, burning the way toward a future driven by knowledge, understanding, and solidarity. As we turn to the last sentence of this preface, let us not dwell simply on the difficulties that lie ahead but also on the enormous opportunity for growth, creativity, and cooperation. Let us draw inspiration from the determination of those who have come before us and chart a road toward a better, more equitable, and more durable future for higher education as well as society as a whole. In the spirit of collaboration and shared purpose, I encourage you to join us on this journey of exploration and transformation. Together, let us dare to conceive an era wherein higher education systems emerge stronger from the ashes of the crisis, equipped to carry out their essential responsibility of nourishing knowledge, enlightenment, and development for generations to come. As we turn the page and begin our examination of *Rebuilding Higher Education Systems Impacted by Crises*, let us do so with open minds, sympathetic minds, and an unwavering dedication to building a future that is worthy of our most significant beliefs and noblest souls.

Mustafa Kayyali
HE Higher Education Ranking, Syria

Acknowledgment

Rebuilding Higher Education Systems Impacted by Crises: Navigating Traumatic Events, Disasters, and More wouldn't have been accomplished without the invaluable assistance and active involvement of several individuals who dedicated their time as well as their expertise to this project.

First and foremost, we express our deepest appreciation to IGI Global for providing us with an opportunity to bring out this academic book. Thanks must go to Jocelynn Hessler, Elizabeth Barrantes, Melissa Wagner, Casey Long, Jacob Hill, Cassandra Martin, and others at IGI Global for their unrelenting assistance and encouragement throughout the creation of this book.

We are deeply indebted to all the authors and co-authors of the chapters for having confidence in this project and their commitment to making meaningful contributions through their expertise and knowledge. Their willingness to lend a hand and share insights has enhanced this work significantly.

We also extend our heartfelt thanks to the reviewers of the book and the members of the Editorial Advisory Board for their thoughtful evaluation as well as insightful feedback, which have helped ensure the quality and validity of the chapters in question.

Particular gratitude is due to the members of the Board of Experts of HE Higher Education Ranking for their remarkable courage and dedication, which have been essential to the success of this book.

I would like to take this opportunity to express my deepest appreciation to the administration team of Azteca University for their continuous support during the development of this book.

To my beloved wife, *Noura Abduljalil*, my respected father, Abdullah Kayyali, and my three little princesses, Larin, Mira, and Celia, your love, support, and understanding have been my greatest source of inspiration and motivation.

To all those who have contributed to this project in any capacity, whether through direct involvement or steadfast support, we extend our heartfelt thanks. Your collaborative efforts contributed to making this project a reality, and we are deeply thankful for your contributions.

Chapter 1
Resilience Strategies for Higher Education Institutions

Christopher Nyakotyo
University of Zimbabwe, Zimbabwe

Pedzisai Goronga
https://orcid.org/0000-0001-9066-7498
University of Zimbabwe, Zimbabwe

ABSTRACT

This chapter explores the various resilience strategies employed by higher education institutions to navigate challenges and thrive in an ever-changing landscape. It interrogates approaches that institutions adopt to build resilience in the face of adversity. The chapter outlines the importance of resilience in higher education, emphasising the need for institutions to anticipate and respond to disruptions. It discusses challenges to building resilience in higher education institutions. Furthermore, the chapter explores the socio-political resilience of higher education institutions, addressing diversity and inclusion, student support, and community engagement. It also examines the role of leadership in fostering resilience, emphasising the significance of visionary leadership and transparent communication. Overall, the chapter provides a comprehensive understanding of the multifaceted resilience strategies that enable higher education institutions to thrive in a rapidly evolving environment.

INTRODUCTION

In an era of unprecedented challenges and disruptions, the resilience of higher education institutions has become a critical focal point for ensuring continuity, innovation, and adaptability (Dohaney et al., 2020; Almerez & Duping, 2022). This chapter looks into the significance of resilience in higher education institutions and the strategies they should employ in order to navigate the complex landscape of change and uncertainty. We are confronted with the question of how these institutions should not only survive, but also thrive in the face of multifaceted challenges. According to Bozkurt (2022), due to the rapidly evolving dynamics of global pandemics, technological advancements, and socio-economic shifts, higher

DOI: 10.4018/979-8-3693-1926-0.ch001

education institutions are compelled to reposition themselves to withstand these shocks. This implies that institutions should re-evaluate their operational frameworks, pedagogical approaches, and institutional cultures so that they are able to maintain academic quality and continuity (Dohaney et al., 2020). Institutions that have gained sufficient resilience should be able to swiftly and competently respond to crises situations in a manner that provides continuity in service provision (Bartuseviciene et al., 2021). This points to the importance of resilience in institutions of higher education so that they continue to thrive even in the midst of challenges. For higher education institutions to attain this characteristic, it is imperative that they draw resources from a diverse stakeholder base. In this light, communalism becomes critical as a grand strategy for building institutional resilience to be able to collectively weather challenges.

Defining Institutional Resilience

The concept of resilience has been conceptualised to mean organisations, systems, or individuals that are able to respond to, and recover from disruptions threatening their existence (Ortiz-de-Mandojana & Bansal, 2016; Linnenluecke, 2017). Institutional resilience refers to an institution's capability to anticipate possible risks, successfully cope with unforeseen events, and learn and adapt to changing situations (Jovanovic et al., 2022; Bento et al., 2021). The same authors add that, for system and institution adaptation, the exploration of new possibilities is required. A number of writers contend that institutional resilience is borne out of knowledge, availability of resources, and power-based relationships (Bento et al., 2021; Bartuseviciene et al., 2021). Other scholars have conceptualised institutional resilience as the ability of an institution to manage shocks and change and that it is not an isolated characteristic or feature of institutions (Linnenluecke, 2017; Bento et al., 2021). Instead, it is the product of a virtuous cycle of institutional performance which according to Linnenluecke (2017), derives from institutional efficacy which engenders trust, legitimacy and credibility. These characteristics constitute sources of resilience that act to reinforce an institution's capabilities. It must be acknowledged that institutional resilience is not a one-time response to a particular event, but rather an iterative multi-stage process from which learning and adaptation occur (Shaya et al., 2022). The concern about rebuilding higher education systems stems from the fact that such institutions play an important role in society to benefit individuals and communities. This position is supported by Bartuseviciene et al. (2021) who posit that higher education institutions strengthen social bonds, ensure employability and contribute to sustainable development hence the importance of resilience in educational institutions.

In the context of the ever-changing higher education landscape, the concept of institutional resilience has gained significant attention as higher education institutions are facing unprecedented challenges. These challenges include global economic turmoil and uncertainty, technological disruptions, and other debilitating global crises. Institutional resilience can thus be conceptualised as the capacity of a higher education institution to anticipate, adapt to, and thrive in the face of disruptive changes. Despite the shocks, institutions must maintain their core functions of teaching, research, community engagement, innovation and industrialisation that should translate to changing people's lives (Murwira, 2019; Muzira & Bondai, 2022).

According to Shava et al. (2022), resilience is not merely about bouncing back from adversity, but it is also about transforming and evolving to meet the demands of a dynamic educational, economic and social environment. This means that resilience should encompass an institution's ability to innovate, leverage human and capital resources effectively, and engender a culture of elasticity and adaptability. It should be able to regenerate itself following severe challenge while maintaining its core functions

(Muzira & Bondai, 2022; Murwira, 2019; Sriskandarajah et al., 2010). At the core of institutional resilience is the institution's ability to identify and assess potential threats and opportunities arising from the challenges confronting the institution. This should be followed by developing strategies that enable the institution to navigate through the challenges (Jovanovic et al., 2022). The approaches to building resilience should involve assembling robust governance structures, fostering strategic partnerships, and leveraging diverse sources of funding to ensure financial stability. In addition, institutional resilience can be strengthened by demonstrating a commitment to fostering a culture of collaboration, transparency, and shared responsibility among faculty, staff, students, and external stakeholders (Jovanovic et al., 2022). Taking this approach enhances a sense of collective ownership and empowerment through which institutions can successfully harness the creativity and expertise of their community to weather storms and emerge stronger. This emphasises a collective and therefore systems approach to building institutional resilience. In this digital era, technology plays a pivotal role in modernising societies hence institutional resilience should also embrace technology and pedagogical innovations, while ensuring equitable access and inclusion for all members of the academic community (Sriskandarajah et al.,2010). However, it must be noted that achieving this comes with costs hence it calls for institutions, through their leadership, to invest in digital infrastructure, professional development for faculty, and support services for students to facilitate seamless transitions to new modes of teaching and learning. In the absence of this effort, students and staff may experience mental health issues that may negatively impact their well-being (Brewer et al., 2019). Furthermore, a resilient institution must engage in strategic planning and risk management to prepare for a range of potential disruptions, from natural disasters to geopolitical shifts. Taking such a stance has the potential to minimise the impact of crises on the core mission of providing uninterrupted service (Bartuseviciene et al., 2021).

Institutional resilience is not a static state, but a dynamic process that requires on-going assessment, adaptation, and renewal (Bartuseviciene et al., 2021). It demands visionary leadership, strategic foresight, and a firm commitment to the values of academic excellence, social responsibility, and inclusivity (Bento et al., 2021; Brewer et al., 2019). Adopting a holistic approach to resilience will enable higher education institutions to position themselves as adaptive, innovative, and impactful agents of positive change in an increasingly complex and ever-changing educational space. Thus, from this discussion, resilience can be defined as the ability to bounce back from difficult experiences, adapting well in the face of adversity, trauma, tragedy, threats, or significant sources of stress that may cause mental health issues. Institutional resilience involves the utilisation of personal and external resources to withstand and overcome challenging circumstances. In a study conducted by Almerez & Duping (2022), the theme of collaboration was central to achieving institutional resilience as in the context of higher education where resilience is crucial for students to navigate the academic, social and personal challenges they encounter during their college years. Collaborations among agencies, leaders, institutions, faculty members, and students are thus critical (Almerez & Duping, 2022). In view of student mental health, it implies a wide range of considerations including institutional leadership, proactive student services, partnerships with external agencies and a comprehensive monitoring and evaluation package (West, 2018). Thus, in the context of higher education, resilience refers to the capacity of students to persevere, adapt, and thrive despite challenges that confront them so that they do not experience psychological problems such as stress and anxiety (Bozkurt, 2022) and insulate them from stressors, setbacks and other negative events (Price, 2023).

In the context of higher education, resilience involves the development of psychological resources, positive coping strategies, and social support networks that enable students to effectively manage the demands and stressors of their academic and personal lives (Almerez & Duping, 2022). By doing so, educational leaders are not in competition, but rather they are in collaboration and cooperation (Almerez & Duping, 2022). This means that, it is important to foster resilience in students to enhance their well-being, academic performance, and improve their overall success in higher education. The role of supportive learning environments, mentorship, and opportunities for skill-building in nurturing resilience within the higher education setting is emphasised (Almerez & Duping, 2022). Thus, in the context of higher education resilience encompasses the psychological, emotional, and social capacities of students, staff and faculty to effectively cope with and overcome the challenges they face (Price, 2023).

Characteristics of Resilient Higher Education Institutions

As a critical attribute, resilience enables higher education institutions to navigate challenges and adapt to changing circumstances. According to Bartuseviciene et al. (2021), there are multiple characteristics that define resilient higher education institutions, positioning them to thrive in the face of threats. Adaptability stands out as a fundamental trait, which allows institutions to respond quickly and effectively to changing needs that include students, technological advancements, and evolving job market requirements and demands (Dohaney et al., 2020). Through embracing change and innovation, resilient higher education institutions stay ahead of the demands thereby maintaining their relevance and impact as agents of change.

In addition to adaptability, a strong leadership and governance structure has been identified as pivotal for achieving resilience in higher education (Dohaney et al., 2020). This is because transparent and visionary leadership foster a culture of trust, accountability, and collaboration, which attributes are essential for canoeing the institution through turbulent times (Almerez & Duping, 2022). In order to achieve this imperative, there is need for effective communication and sound decision-making processes to bolster the institution's ability to weather challenges and remain focused on set goals (Andersson et al., 2019). However, some authors have warned against rigid hierarchical decision making in times of crises, suggesting that those with the greatest expertise with the issue at hand should be given decision-making power (Vogus & Sutcliffe, 2007; Andersson et al., 2019; Bartuseviciene et al., 2021). This is important because a diverse and inclusive community plays a vital role in building institutional resilience as it brings together varied and yet complimentary views, experiences, and expertise. This approach enriches the institution's capacity to tackle complex issues and foster creative solutions that entangle institutions from a web of challenges (Hart et al., 2009). Furthermore, financial stability and strategic resource management are indispensable for resilient higher education institutions. Prudent fiscal planning, diversified revenue streams, and efficient resource allocation enable institutions to withstand economic downturns thus safeguarding their long-term sustainability (Almerez & Duping, 2022). Similarly, in the context of technological advancements that now characterise our educational landscape, robust technological infrastructure and digital capabilities are imperative, amplifying the institution's agility and capacity to deliver quality learning experiences regardless of external disruptions (Vogus & Sutcliffe, 2007; Bartuseviciene et al., 2021).

An unwavering commitment to academic excellence and student success is another benchmark of resilient higher education institutions. If institutions prioritise pedagogical innovation, student support services, and career readiness initiatives, they are positioning themselves to equipping students with the adaptability and skills needed for responding to the rapidly evolving global marketplace. Research is fundamental to innovation and industrialisation hence a strong research and development ecosystem is needed as it has the potential to contribute to the institution's resilience by niching groundbreaking discoveries, industry partnerships, and intellectual capital, enhancing its competitive edge and societal impact (Murwira, 2019; Muzira & Bondai, 2022). It is clear that resilient higher education institutions are characterised by their adaptability, strong and visionary leadership, diversity and inclusion, financial stability, technological prowess, and a relentless focus on academic excellence and student success. It is argued that by embodying these attributes, higher education institutions will be able to endure turbulence. By embracing these imperatives, institutions will be positioning themselves as beacons of knowledge innovation, and progress in an ever-changing world (Murwira, 2019).

Importance of Resilience for Higher Education Institutions

Building institutional resilience is not a one-time event, but a long term commitment to strengthening institutions and empowering communities (Jovanovic et al., 2022). For higher education, institutional resilience is important because it helps maintain their educational mission and continue providing quality education to students in the face of threats (Shaya et al., 2022). Institutional resilience is of concern because higher education institutions are a primary vehicle for realising sustainable development and maintain sustainability. Prayag et al. (2018) ascertains that resilient institutions contribute immensely to resilient communities. Resilience is paramount as it allows higher education institutions to effectively respond and adapt to challenges in the educational landscape. The challenges can include changes in student demographics, technology advancements, shifting societal needs, or economic fluctuations (Jovanovic et al., 2022). Institutions that demonstrate resiliency possess the capacity to navigate through financial uncertainties and ensure their long-term sustainability (Brewer et al., 2019). Furthermore, according to Jovanovic et al. (2022), resilience capacitates institutions to manage budget constraints, identify and mobilise alternative revenue streams thus leading to financial stability. Institutions that are resilient are able to uphold their reputation and credibility in the midst of various pressures and crises. It has been observed that institutions that can successfully weather disruptions, such as natural disasters, leadership changes, or accreditation issues, are more likely to retain the trust and confidence of their stakeholders who in turn offer their utmost support thereby strengthening their resilience (Jovanovic et al., 2022). Student success is of utmost importance in the survival of higher education institutions hence the need to provide a supportive and flexible learning environment. It enables institutions to offer various academic and non-academic support services, adapt curriculum to meet evolving societal and industry demands, in addition to implementing innovative teaching and learning approaches. Thus, resilience is a crucial quality for higher education institutions, especially in today's rapidly changing and challenging educational landscape.

The 21st century is characterised by rapid technological advancements. Higher education institutions must be able to swiftly adapt to changes in technology, student demographics, and educational trends implying that resilient institutions must be able to respond effectively to new challenges and opportunities thus ensuring their long-term relevance and success (Sriskandarajah et al., 2010). Higher education institutions often face financial, political, and social challenges (Dohaney et al., 2020). In view of this, resilience enables them to deal with and overcome these setbacks, maintain stability, and continue providing quality education and research opportunities for their students and faculty, allowing students to learn and succeed (Reyes et al., 2022). Ultimately, resilient institutions should be able to help students navigate academic and personal challenges, leading to greater success and retention rates.

Resilient institutions are more open to innovation and experimentation in teaching methods, research initiatives, and administrative practices. This flexibility fosters creativity and a spirit of continuous improvement, leading to enhanced academic programmes and institutional effectiveness (Sriskandarajah et al., 2010). In addition, resilience fosters a sense of community and collaboration within higher education institutions. This interconnectedness allows for the sharing of resources, expertise, and best practices, ultimately benefiting the entire academic community. Resilience is essential for effective leadership and governance within higher education institutions (Goodenough et al., 2020; Shin & Kelly, 2015). Leaders who exhibit resilience can navigate complex decision-making processes, inspire confidence in other partners, and guide their institutions through periods of turbulence, uncertainty and steer them to transformation. In essence, resilience is vital for higher education institutions as it enables them to adapt to change, overcome adversity, support student success, embrace innovation, build community and collaboration, and promote effective leadership and governance (Jovanovic et al., 2022). It can be concluded that, as institutions cultivate resilience, they can thrive in an ever-evolving educational landscape and continue to fulfil their mandate.

Strategies for Building Resilience in Higher Education

In order to build institutional resilience, there is need for shared understanding of individual and collaborative networks and co-production of knowledge (Berkes, 2017). It is critical to cross-train employees throughout the organisation so that all institutional members understand what is expected of them. In addition, there is need for higher education institutions to develop a crisis management plan that clearly indicates procedures for communication, decision-making, and resource allocation (Berkes, 2017). It is thus advisable to conduct regular risk assessments to identify potential vulnerabilities and develop and deploy strategies to mitigate them. For higher education institutions, collaboration across different departments and with external partners is critical and therefore encouraged. Once specific strategies have been implemented, continuous evaluation of the effectiveness of resilience-building initiatives must be considered. It is through evaluation that institutions learn from past experiences and incorporate lessons learned into future planning. Therefore, it is imperative for higher education institutions to build community of practices (CoPs) that will enable collaborative exchange and cross pollination of ideas, teaching practices, methods, strategies and inter-institutional support during crises situations. It can be argued that, by following this trajectory, institutions can increase their ability to cope with shocks and stressors from their environment, enhance their value as a pathway to sustainable development, and maintain their educational mission.

Resilience strategies play a crucial role in fostering a positive and supportive environment within higher education institutions (Price, 2023). These strategies can help students, faculty, and staff navigate challenges and setbacks effectively. There are a number of resilience strategies that can be employed to foster institutional resilience. One such strategy involves building a strong support system and higher education institutions can do so by creating mentorship programmes, counselling services, and support groups to provide students and faculty with the necessary support and guidance during difficult times (Dohaney et al., 2020; Rossetto & Martin, 2022). If a sense of community and belonging is fostered, individuals can feel more empowered to deal with and overcome obstacles and setbacks. This is important in promoting a growth mindset in individuals. According to Reyes et al. (2022), encouraging a mindset that embraces challenges, sees failure as opportunity for growth, and believing in the power of effort can significantly enhance resilience. Thus, higher education institutions can integrate this mindset into their curriculum and institutional culture, emphasising the importance of perseverance and learning from challenges to nurture the spirit of resilience.

Furthermore, developing coping skills and emotional intelligence is vital for enhancing resilience. This may be achieved by providing training on stress management, emotional regulation, and healthy coping mechanisms thereby helping individuals navigate the pressures of higher education more effectively (Almerez & Duping, 2022). Research evidence shows that institutions can foster more resilient communities by equipping students, faculty, and staff with these skills (Jovanovic et al., 2022; Almerez & Duping, 2022). Fundamentally, fostering a culture of open communication is known to strengthen resilience by creating spaces for honest and open dialogue about mental health, personal struggles, and academic challenges (West, 2018). These initiatives can help reduce stigma and create a supportive environment. Open communication allows individuals to feel comfortable seeking help and sharing their experiences. This is made possible if individual students, staff and faculty feel a sense of belonging where they are appreciated, valued and accepted.

Another strategy to enhance institutional resilience relates to promoting self-care and well-being of institutional members. It has been argued that higher education institutions should implement wellness programmes, mindfulness activities, and initiatives that prioritise physical and mental well-being (Kamath, 2015; West, 2018). By prioritising self-care, institutions can help individuals build resilience through maintaining a healthy work-life balance and addressing their holistic needs (Chen et al., 2020). Practicing self-care enables students and staff to adapt to life changes, build strong relationships, and recover from setbacks they have experienced (Cook et al., 2022). This exposition indicates that employing resilience strategies in higher education institutions is crucial for creating a supportive and empowering environment. By building a strong support system, promoting a growth mindset, developing coping skills, fostering open communication, and prioritising well-being institutions can effectively foster resilience among students, faculty, and staff (Price, 2023). These practices will ultimately enhance their ability to overcome challenges and thrive in their institution.

Role of Students in Building Resilience in Higher Education

To be characterised as resilient, higher education institutions need to be able respond to the needs and situations of their students (Almerez & Duping, 2022). Putting students at the forefront is an indication that they important stakeholders who play a crucial role in building resilience in higher education institutions. To play their role in building institutional resilience, students should be allowed to participate in activities that promote resilience such as disaster preparedness drills, community service projects, sustainability initiatives, critical thinking, problem-solving, and adaptability skills (Almerez & Duping, 2022; West, 2018). Because of their centrality, students can provide essential feedback on institutional policies and procedures to help identify areas for improvement (Jovanovic et al., 2022). They can advocate for changes that promote institutional resilience, such as increased funding for sustainability initiatives or the development of a crisis management plan. By dialoguing with faculty and staff, students can help promote a culture of resilience and encourage the sharing of ideas as equal partners (Almerez & Duping, 2022). Thus, by allowing students to be involved in these activities they can help in building resilience in their institutions.

Resilience is a crucial attribute for success in both academic and professional spheres. In that regard, students play a significant role in cultivating a resilient culture within their academic communities and assist them overcome challenges (Howe et al., 2012). This is because of the fact that they have the power to foster a supportive and inclusive environment that encourages resilience. By way of engaging in peer support networks, mentoring programmes, and group study sessions, students can create a sense of solidarity and belonging that helps their peers navigate challenges more effectively. This is in response to the fact that the higher education in the 21st century is fast-paced, ever-changing, challenging and linked to increasingly high levels of stress for students (Ahmed & Julis, 2015; Bartuseviciene et al., 2021). In order to minimise the effects of psychological challenges, students can engage in initiatives such as mental health awareness campaigns and stress management workshops to enhance their well-being (Baxter, 2012; Reyes et al., 2015). This will contribute to a proactive approach to resilience-building within their campus communities, and enhance academic performance (Baxter, 2012).

If allowed to actively participate in the life of institutions, students can help in shaping institutional policies and practices that promote resilience. By providing feedback on academic workload, access to support services, and the overall learning environment, students can influence the development of an educational system that acknowledges and supports the diverse needs of its members (Almerez & Duping, 2022). In view of this submission, it is thus prudent to empower students by giving them a voice in decision-making processes as a way of fostering a sense of ownership and accountability, which are essential elements of resilience. Students are a critical constituency that should not be regarded as mere recipients of resilience-building efforts within higher education institutions, but should be active agents of change. Their involvement in creating a nurturing and empowering environment, advocating for supportive policies, and fostering a culture of collaboration is fundamental to the overall resilience of the institution (Bartuseviciene et al., 2021). Given these contributions that students can make, it is vital to recognise, harness and leverage their invaluable participation and contribution in the collective endeavour of building resilience of their institutions.

Role of Faculty and Staff in Building Resilience in Higher Education

In the same way as students are important in fostering institutional resilience, staff and faculty also play an equally fundamental role in building resilience. To show their importance, there is need for staff and faculty to collaborate with each other as well as external stakeholders to share knowledge and ideas (Jovanovic et al., 2022; Almerez & Duping, 2022). Staff and faculty are important in providing guidance and direction in resilience-building initiatives by leading by example, advocating for change and providing leadership. The issue of innovation is central to resilience building hence staff and faculty should spearhead in innovativeness and continuous improvement by creating a culture that values continuous learning, experimentation, risk-taking and learning from mistakes (Murwira, 2019; Muzira & Bondai, 2022; Tagwira, 2018). Staff and faculty can provide opportunities that help employees build skills required to innovate (Murwira, 2019). All these activities should culminate in staff and faculty continuously evaluating the effectiveness of their resilience-building efforts and incorporate lessons learnt into future planning. Thus, the involvement of staff and faculty goes a long way in rebuilding higher education systems impacted by crises.

Faculty and staff play a huge role in building resilience in higher education institutions as they are responsible for fostering a supportive and empowering environment for students. They are not only responsible for delivering academic content, but also play a significant role in shaping the overall student educational experience in institutions (Bartuseviciene et al., 2021). Of importance is the fact that faculty and staff serve as mentors and role models for students. As faculty and staff demonstrate resilience behaviour in the face of challenges, they serve as powerful role models for students to emulate (Rossetto & Martin, 2022). It has been documented that, when faculty and staff exhibit resilience traits, they inadvertently communicate to students that setbacks are a part of life and can be overcome with commitment, determination and perseverance (Price, 2023). This modelling of resilience can inspire students to develop their own resilience skills, which is essential for navigating academic and personal challenges they encounter in their higher education journeys.

Furthermore, faculty and staff are in pole position to provide students with the necessary support and guidance to help them develop resilience (Reyes et al., 2022). Whether it is through one-on-one interactions, classroom discussions, or extracurricular activities, faculty and staff can actively engage students by providing encouragement, advice, and resources to help them build their resilience. This support can be demonstrated by giving academic assistance, mental health resources or simply being available for them when they need help (West, 2018). By being accustomed to students' needs and providing the right kind of support, faculty and staff can help students develop the skills and mindset necessary to bounce back from adversity and thrive in the face of challenges.

Faculty and staff can contribute to building resilience in higher education institutions by designing and implementing curricular and co-curricular activities that promote resilience (West, 2018). This can be done by incorporating resilience-focused topics into the curriculum, organising workshops on stress management, as well as offering extracurricular programmes that encourage teamwork and problem-solving (Everly et al., 2013). Taken together, these activities can contribute to fostering resilience in students. Thus, by intentionally integrating resilience-building activities into the academic and social fabric of the institution, faculty and staff can create an environment that nurtures and reinforces students' ability to adapt and grow in the face of adversity (Jovanovic et al 2022; Everly et al., 2013). Creating a culture of empathy and understanding within higher education institutions can be instigated by faculty and staff thus demonstrating their leadership prowess and guidance. By understanding the diverse experiences and challenges that students bring to college or university, faculty and staff can create an inclusive and supportive environment that acknowledges and respects the different ways in which students navigate their academic and personal journeys (Sutcliffe & Vogus, 2003; Bartuseviciene et al.,2021). This inclusive approach helps to communicate to students that are respected, valued and understood, which in turn can bolster their resilience by providing them with a sense of belonging and acceptance.

Additionally, faculty and staff can contribute to building resilience by fostering a sense of community and connectedness within the higher education institution (Jovanovic et al., 2022). This can be done by creating opportunities for students to connect with each other, form supportive relationships, and engage in collaborative learning experiences that can help cultivate a sense of belonging and mutual support. By fostering a strong sense of community, faculty and staff can help students develop a network of peers and mentors that provide encouragement and assistance during challenging times, thereby enhancing their resilience (Almerez & Duping, 2022). They can contribute to building resilience by advocating for institutional policies and practices that prioritise student well-being and resilience (Bozkurt, 2022; Brewer et al., 2019). Actively participating in decision-making processes and advocating for resources and initiatives that support student mental health, academic success, and overall well-being (Clafferty & Beggs, 2014), faculty and staff can contribute to creating an institutional environment that is conducive to the development of student resilience (Brewer et al., 2019). This can involve advocating for adequate mental health services, promoting work-life balance, and creating a supportive academic environment that recognises the importance of holistic student development (Bartuseviciene et al., 2021). It is apparent from the discussion above that the role of faculty and staff in building resilience in higher education institutions is multifaceted and essential. They serve as role models, providing support and guidance, designing resilience-focused activities, fostering empathy and understanding and creating a sense of community. They also advocate for student well-being, thereby contributing to the development of resilient students who are equipped to thrive in the face of adversity. Their influence and impact extends beyond the lecture rooms, as they shape the overall culture and environment of the institution by laying the foundation for students to succeed, not only academically, but also personally and professionally.

Challenges to Building Resilience in Higher Education

In an effort to build institutional resilience, various impediments may stand in the way thus making the whole process a challenging one. Research evidence supports the view that barriers do exist that inhibits the process of building resilience in educational institutions (Dohaney et al., 2020; Kapucu & Khosa, 2013. These barriers come in different forms. Building institutional resilience requires mammoth resources. These include, among many others financial, human, and technological resources. Because there are many activities that may be taking place at the same time, higher education institutions may struggle to allocate sufficient resources to resilience-building efforts. Additionally, besides lack of resources, another threat to institutional resilience-building efforts is resistance to change involving unwillingness to adapt to changing circumstances (Dohaney et al., 2020). Lecturers and staff may be fearful about losing ownership of their educational materials especially when they are forced to migrate to online modes of teaching. Furthermore, that academics may prioritise their families and a personal life over work responsibilities is a barrier that is simultaneously beneficial to their well-being and detrimental to staffing in the institution (Seaton et al., 2012).

Building institutional resilience often requires significant changes to an institution's culture, processes, and systems. It must be noted that higher education institutions are complex organisations with multiple stakeholders and competing priorities. Thus, building resilience requires a comprehensive understanding of the institution's operations and the ability to navigate complex systems. Furthermore, building institutional resilience requires strong leadership support at all levels of the organisation. This is supported by Dohaney et al. (2020) who highlight the importance of support, community and strong leadership. This implies that, without the requisite and astute leadership and support, efforts to build institutional resilience may suffer because of lack of attention and resources. However, it should be noted that every institution needs to have a customised plan that takes into consideration its unique vulnerabilities and areas for improvement (Dohaney, et al., 2020). While challenges will always exist, it is vital to continuously make effort to build resilience to help institutions maintain their educational mission and continue to provide quality service to students and other clientele even in the midst the adversities. Dohaney and colleagues argue that, while disasters cannot be stopped from happening, efforts to minimise the vulnerability of our institutions and reduce the impacts to our people and property should always be prioritised.

Higher education institutions are unique educational entities and the challenges they may face are contextual and circumstantial, but there are some common ones that may cut across most institutions. One such common challenge is financial position of institutions (Dohaney et al., 2020). Financial constraints pose a significant threat to institutions to invest in building resilience. Budget constraints tend to restrict the ability of institutions to hire as well as retain qualified staff, invest in infrastructure and technology, and fund initiatives aimed at enhancing resilience (Bartuseviciene et al., 2021; Wildavsky, 1991). Instituting changes necessary for building resilience can be met with resistance from various stakeholders within the institution. Faculty, staff, administrators and students may be resistant to new policies, practices, or structures that disrupt established routines or challenge existing power dynamics and status quo (Berkes, 2017). Higher education institutions often have complex organisational structures, with different departments and units operating independently. The complexity may render coordination and collaboration between departments difficult and therefore hinder efforts to build resilience in a holistic and integrated manner (Almerez & Duping, 2022). Additionally, insufficient planning and preparedness can hinder an institution's ability to effectively respond to crises or mitigate the impacts of disruptions. Outdated or inadequate technology infrastructure can also limit an institution's capacity to adapt and respond quickly to changing circumstances.

Many higher education institutions have limited access to reliable internet connectivity or have outdated systems that can hinder remote learning, communication, and administrative functions of the institution (Gogoi et al., 2022). Higher education institutions often have cultures and traditions deeply rooted in the past and this can create resistance to change (Dohaney et al., 2020), making it difficult to introduce innovative teaching and learning methods, adopt emerging technologies, or implement new approaches to organisational and pedagogical practices. Furthermore, higher education institutions operate in a complex external landscape influenced by governmental policies, regulatory frameworks, and accreditation requirements. Adhering to external pressures and meeting compliance obligations can divert resources and limit the institution's ability to focus on resilience-building initiatives. In a world of multiculturalism, building resilience requires paying attention to diversity, equity, and inclusion efforts. However, addressing systemic inequities and creating inclusive environments can be challenging due to deep-rooted practices and biases within higher education institutions. Therefore, overcoming these challenges requires strong leadership, effective communication, collaborative efforts, and a willingness to embrace change and innovative solutions (Bento et al., 2021). It also necessitates strategic planning, resource mobilization and allocation, and a proactive approach to risk management. According to Bento et al. (2021), there is need to open communication channels during turbulent times. It is important to facilitate communication across all stakeholders in order to consolidate cross-scale feedback as it relates to resilience (Folke, 2006).

In their efforts to build resilience, higher education institutions face an unprecedented set of challenges. These may include the rapidly changing landscape that is compounded by the impact of global events, technological advancements, and societal shifts (Reyes et al., 2022). These have necessitated a reevaluation of traditional approaches to resilience. Building institutional resilience requires financial injection into such effort. Incidentally, higher education institutions are grappling with mounting financial pressures, including reduced public funding, fluctuating enrolment figures, and increased competition. These factors have significantly strained institutional resources, making it difficult to invest in long-term resilience-building initiatives. The just ended COVID-19 pandemic further exacerbated these financial challenges, compelling institutions to find innovative ways to sustain their operations while simultaneously investing in resilience (Bartuseviciene et al., 2012).

The rapid evolution of technology presents a significant challenge for higher education institutions. While digital transformation offers opportunities for resilience through online learning and administrative efficiency, it also demands substantial investments in infrastructure, cyber security, and human capital training (Dohaney et al., 2020). Moreover, ensuring equitable access to technology for all students and faculty members remains a pressing concern, particularly in the face of the digital divide (Floridi & Saunders, 2023). There is a digital divide between those who have access to technology and those who do not. The difference extends beyond individual, physical, and geopolitical boundaries (Floridi & Saunders, 2023). In addition, the evolving needs and expectations of students, faculty, and staff pose a serious challenge to building resilience. Demands for flexible learning modalities, mental health support, and inclusive campus environments necessitate a holistic approach to resilience that extends beyond infrastructure and operations (Clafferty & Beggs, 2014). Additionally, the dialogue surrounding diversity, equity, and inclusion requires institutions to reevaluate their policies and practices to foster a resilient and supportive community.

Higher education institutions are confronted with navigating global uncertainties and disruptions, ranging from geopolitical tensions to climate change. These external factors can have far-reaching implications on institutional resilience, requiring proactive planning and adaptability. Pandemics, natural disasters, and geopolitical conflicts underscore the imperative for higher education institutions to fortify their resilience frameworks to withstand and recover from adversities (Floridi & Saunders, 2023). The complex regulatory environment, encompassing accreditation standards, student data privacy laws, and government regulations, present a formidable challenge to building institutional resilience. Compliance with evolving mandates necessitates agile governance structures and robust risk management practices, placing additional demands on institutional leadership and resources. In light of these challenges, higher education institutions face a complex and multifaceted landscape in their pursuit for resilience. Addressing financial pressures, embracing technological advancements, responding to shifting demands, preparing for global disruptions, and navigating regulatory complexities are integral to fortifying institutional resilience. Overcoming these challenges demands a collaborative and forward-thinking approach, empowering institutions to adapt, innovate, and ultimately thrive in the face of adversity. By addressing these challenges, higher education institutions can chart a resilience path forward, ensuring the continued success and impact of higher education in an ever-changing world.

Best Practices in Building Higher Education Resilience

How can institutions build resilience in the best possible ways? By best practices is meant the working standards or ethical guidelines that provide well thought out and practical courses of action in a given situation. In the context of higher education, institutions can consider several best practice strategies for building resilience. As is known, institutions of higher education are characterised by complexity that requires the establishment of a shared understanding of roles and responsibilities (Dohaney et al., 2020). It is thus important to help ensure that everyone in the institution is working towards the same goals (Bartuseviciene et al., 2021). Because many stakeholders are involved, there is need for cross-training employees to help ensure that everyone has the skills that are needed to perform their jobs effectively, even in the face of unexpected interruptions. On-going training and broadening of skills repertoires ensures a competent workforce that is prepared with the necessary knowledge and skills to cope with disruptive events and operationalise resilience (Clafferty & Beggs, 2014; Vogus & Sutcliffe, 2007; Bartuseviciene et al., 2021). Drawing from lessons learnt from the impact of COVID-19 pandemic, building a portfolio of virtual learning opportunities can help ensure that employees have access to the training they need in order to better cope with challenging situations (Cassidy, 2015). This is important in order to stay up-to-date with the latest developments in their fields of expertise.

While it is difficult is to predict with certainty what lies ahead, it is prudent to be prepared for eventualities. Thus, it is crucial to develop a crisis management plan to help ensure that the institution is prepared to respond effectively to unexpected challenges. In addition, conducting regular risk assessments can help identify potential risks and develop strategies to deal with them. Knowledge is always in a state of flux and as such, there is need for higher education institutions to cultivate a culture of innovation and continuous improvement to help ensure that institutions are always looking for ways to improve their operations and better serve their stakeholders. It is also noble to encourage collaboration across departments and with external partners to help promote knowledge sharing. Furthermore, continuously evaluating the effectiveness of resilience-building efforts can help ensure that the institution is making progress towards its goals. Past experiences can provide a resource to learn from and can help identify what worked well and what did not and take on board those lessons into future planning. While these suggested practice strategies may not be exhaustive, higher education institutions can employ them as templates for building resilience and increase their ability to cope with stressors and environmental shocks.

Fostering resilience in higher education institutions is important for the overall well-being and success of students, faculty and staff (Clafferty & Beggs, 2014). To achieve this, several best practices can be implemented. At the core of resilience-building is the aspect of creating a supportive and inclusive environment. Resilient higher education institutions should prioritise support services integral to long-term sustainable changes (Dohaney et al., 2020). This involves promoting a culture of empathy, understanding, and open communication where individuals feel safe to express their challenges and seek assistance without fear of judgment. Resources and support are the backbone of professional academic development. Thus, staff and students need to be digitally capable and infrastructure is needed to support flexible learning (Dohaney et al., 2020). It is important to establish support networks, such as peer mentoring programmes, counselling services, and faculty support groups, in order to provide the institution with a safety net for individuals facing challenges.

Furthermore, it is also of utmost importance to integrate resilience-building activities into the curriculum and extracurricular programmes as these are known to significantly contribute to the development of resilience among students. This could include conducting workshops on stress management, mindfulness-based stress reduction strategies (Kamath, 2015; Goodenough et al., 2020, leadership training, and emotional intelligence development (Price, 2023). This is further supported by Goralnik and Marcus (2020) who say that, in higher education resilience programmes should be connected to curricula. In situations where students are equipped with practical skills and strategies to navigate challenges, they are better prepared to handle the demands of academic life and beyond. In academic activities, institutions are encouraged to involve the community and create volunteer opportunities that can help foster a sense of purpose and altruism, which are key components of resilience.

In addition, promoting self-care and mental health awareness is fundamental in nurturing resilience within higher education institutions (Bozkurt, 2022). This can be done through educational campaigns, wellness fairs, and accessible mental health resources (Clafferty & Beggs, 2014). It is important to normalise conversations around mental well-being issues and provide easy access to professional support. This can help reduce the stigma associated with seeking help, thereby helping individuals to proactively address their mental health needs. Students learn a lot from role models hence faculty and staff should be encouraged to model self-care behaviours and also prioritise their own well-being (Fletcher & Sarkar, 2016).

Fostering resilience in higher education institutions involves creating a sense of belonging and connectedness. This can be done through initiatives that celebrate diversity, equity, and inclusion, as well as by creating opportunities for meaningful engagement and collaboration (Berkes, 2017). When individuals feel valued, respected, and connected to the broader community, they are more likely to exhibit resilience behaviour in the face of challenges. By implementing these best practice options, higher education institutions can cultivate a resilient community that supports the holistic development of its members and prepares them to thrive in a rapidly changing world by expanding their knowledge and technical skills repertoires. (Wildavsky, 1991).

CONCLUSION

As the chapter that discussed the strategies for resilience in academic institutions comes to a close, it is important to emphasise the need for adopting a collaborative approach to weather the challenges. Resilience in academic institutions is not a one-size-fits-all solution, but rather a continuous process of adaptation, innovation, and support (Rossetto & Martin, 2022). In addition, since resilience relates to self- organisation, it requires collaborative networks and knowledge co-production (Berkes, 2017). Thus, it is important to create a supportive and inclusive environment where all stakeholders, including students, faculty, and staff feel valued, heard, and empowered to overcome challenges. In view of these ideas, it is recommended that higher education institutions should incorporate community-building efforts into their resilience plans (Cherrington et al., 2017). This can be done by encouraging open communication (Folke, 2006), showing empathy, and cultivating a sense of community as these can significantly contribute to resilience within academic institutions. In addition, there is need to embrace flexibility and adaptability in teaching, learning, and administrative practices. Thus, by recognising and responding to the evolving needs of the academic community, institutions can competently navigate disruptions and emerge stronger. There is need to develop assessment and evaluation strategies for online learning, develop digital literacies for academics and students, adopt quality assurance mechanisms and identify at-risk students to minimise inequalities (Bozkurt, 2022).

Resilience in institutions is steered by people hence it is important to empower students and faculty to take ownership of their learning and professional development. Providing opportunities for mentorship, skill-building, and autonomy can foster a sense of agency and resilience in academic settings. Pandemics such as COVID-19 had serious effects on all people including students who experienced mental health issues during the period (Gogoi et al., 2022). It is therefore important to emphasise and prioritise mental health and well-being initiatives within academic institutions. This can be achieved by promoting self-care, stress management, and access to mental health resources as a way of supporting the holistic resilience of their community members. As noted by Bozkurt and Sharma (2020), the time is ripe for higher education to renew, recalibrate and reposition itself in the educational landscape ultimately regenerating themselves and being born again.

Higher education institutions should encourage sharing and celebrating resilience stories within the academic community where real-life examples of resilience can be highlighted. This is in order to inspire others and create a sense of solidarity and determination in overcoming challenges. Resilience in academic institutions requires a proactive, holistic, and collaborative approach. This is what led Almerez and Duping (2022) to say that the approach of educational leaders towards academic resilience includes a way to connect and collaborate with partner agencies and some other educational leaders in order to, not just survive a particular situation, but to ceaselessly render its functions with utmost quality and efficiency. This highlights the importance of collaboration (Lambrecht et al., 2020). It is envisaged that, by adopting a holistic approach to building resilience, academic institutions can cultivate resilient communities capable of thriving in the face of adversity. This multifaceted approach will not only strengthen the fabric of the institution, but also contribute to the growth and development of its members. Resilience is not a once off response to a specific event, but rather an iterative process that requires pre and post-event preparation, learning and adaptation (Shaya et al., 2022). While making effort to achieve resilience, it must be borne in mind that resilience should not be the ultimate goal of any institution, but is necessary for organisational evolvability by which the post-event organisation will emerge improved or even completely distinct from its pre-event state (Kantur & Iseri-Say, 2012).

REFERENCES

Ahmed, Z., & Julis, S. H. (2015). Academic performance, resilience, depression, anxiety, and stress among women students. *Indian Journal of Positive Psychology*, *6*(4), 367–370.

Almerez, Q. L. G., & Duping, A. M. (2022). Challenges and responses of higher education institutions towards academic resilience. *International Journal of Research and Innovation in Social Science*, *VI*(XII), 464–472. doi:10.47772/IJRISS.2022.61225

Andersson, T., Caker, M., Tengblad, S., & Wickelgren, M. (2019). Building traits for organisational resilience through balancing organisational structures. *Scandinavian Journal of Management*, *35*(1), 36–45. doi:10.1016/j.scaman.2019.01.001

Bartuseviciene, I., Pazaver, A., & Kitada, M. (2021). Building a resilient university: Ensuring continuity-transition from face-to-face to online in the COVID-19 pandemic. *Journal of Maritime Affairs*, *20*(2), 151–172. doi:10.1007/s13437-021-00239-x

Baxter, J. (2012). Who am I and what keeps me going? Profiling the distant learning student in higher education. *International Review of Research in Open and Distance Learning*, *13*(4), 107–129. doi:10.19173/irrodl.v13i4.1283

Bento, F., Giglio-Bottino, A., Cerchiareto-Perreira, F., Forastieri de Almeida, J., & Gomes-Pereira, F. (2021). Resilience in higher education: A complex perspective to lecturers' adaptive processes in response to the COVID-19 pandemic. *Education Sciences*, *11*(9), 492–504. doi:10.3390/educsci11090492

Berkes, F. (2017). Environmental governance for anthropocene? Social ecological systems, resilience and collaborative learning. *Sustainability (Basel)*, *9*(7), 1–11. doi:10.3390/su9071232

Bozkurt, A. (2022). Resilience, adaptability, and sustainability of higher education: A systematic mapping study on the impact of the Coronavirus (COVID-19) pandemic and the transition to the new normal. *Journal of Learning for Development*, *9*(1), 1–16. doi:10.56059/jl4d.v9i1.590

Bozkurt, A., & Sharma, R. C. (2020). Education in normal, new normal and next normal: Observations from the past, insights from the present and projects for the future. *Asian Journal of Distance Education*, *15*(2), i–x. doi:10.5281/zenodo.4362664

Brewer, M. L., van Kessel, G., Sanderson, B., Naumann, F., Lane, M., & Carter, A. (2019). Resilience in higher education students: Scoping review. *Higher Education Research & Development*, *38*(6), 1105–1120. doi:10.1080/07294360.2019.1626810

Cassidy, S. (2015). Resilience building in students: The role of academic self-efficacy. *Frontiers in Psychology*, *6*, 1781. doi:10.3389/fpsyg.2015.01781 PMID:26640447

Chen, H. Y., Wang, C. W., Yang, H. F., Huang, W. T., & Fan, S. Y. (2020). The relationships between physical activity and life satisfaction and happiness among young, middle-aged and older adults. *International Journal of Environmental Research and Public Health*, *15*(4), 25–39. doi:10.29045/14784726.2021.3.5.4.25

Cherrington, S., Macaskill, A., Salmon, R., Boniface, S. M., Shep, S., & Flutey, J. (2017). Developing a pan-university professional learning community. *The International Journal for Academic Development, 23*(4), 298–311. doi:10.1080/1360144X.2017.1399271

Clafferty, E., & Beggs, B. J. (2014). *The triple C model for student retention, progression and completion matures.* Society for Research into Higher Education Conference. https://www.researchgate.net/publication/281290529

CookA.KeyteR.SprawsonI.MatharuA.MantziosM. (2022). Mental health first aid experiences: A qualitative investigation into the emotional impact of mental health first aid responsibilities. https://doi.org/doi:10.21203/rs.3.rs-1393314/v1

Dohaney, J., de Roiste, M., Salmon, R. A., & Sutherland, K. (2020). Benefits, barriers, and incentives for improved resilience to disruption in university teaching. *International Journal of Disaster Risk Reduction, 50,* 1–9. doi:10.1016/j.ijdrr.2020.101691 PMID:32509512

Everly, G. S., Smith, K. J., & Lobo, R. (2013). Resilient leadership and the organisational culture of resilience: Construct validation. *International Journal of Emergency Mental Health, 15*(2), 123–128. PMID:24558699

Fletcher, D., & Sarkar, M. (2016). Mental fortitude training: An evidence-based approach to developing psychological resilience for sustained success. *Journal of Sport Psychology in Action, 7*(3), 135–157. doi:10.1080/21520704.2016.1255496

Floridi, L., & Sanders, J. W. (2023). Artificial evil and the foundation of computer ethics. *Ethics and Information Technology, 3*(1), 55–66. doi:10.1023/A:1011440125207

Folke, C. (2006). The emergence of a perspective for social-ecological systems analyses. *Global Environmental Change, 16*(3), 253–267. doi:10.1016/j.gloenvcha.2006.04.002

Gogoi, N., Saikai, D., & Neog, P. (2021). Influence of COVID-19 pandemic on clothing interests of working women. *International Journal of Current Microbiology and Applied Sciences, 10*(01), 501–516. doi:10.20546/ijcmas.2021.1001.062

Goodenough, A. E. H., Roberts, D. M., Biggs, J. G., Derounian, A., Hart, G., & Lynch, A. (2020). A higher degree of resilience: Using psychometric testing to reveal the benefits of university internship placements. *Active Learning in Higher Education, 21*(2), 102–115. doi:10.1177/1469787417747057

Goralnik, L., & Marcus, S. (2020). Resilient learners, learning resilience: Contemplative practice in the sustainability classroom. *New Directions for Teaching and Learning, 2020*(161), 83–99. doi:10.1002/tl.20375

Hart, D., Diercks, G., & Powell, A. (2009). Exploring stakeholder engagement in impact evaluation planning in educational development work. *Evaluation, 15*(3), 285–306. doi:10.1177/1356389009105882

Howe, A., Smajdor, A., & Stokl, A. (2012). Towards an understanding of resilience and its relevance to medical training. *Medical Education, 46*(4), 349–356. doi:10.1111/j.1365-2923.2011.04188.x PMID:22429170

Jovanovic, O., Pesikan, A., & Antic, S. (2022). Resilience of higher education institutions in Serbia: A student's perspective. *Digital and Psychological Resilience, 457-461*, 457–461. Advance online publication. doi:10.46793/TIE22.457J

Kamath, S. (2015). Role of mindfulness in building resilience and emotional intelligence and improving academic adjustment among adolescents. *Indian Journal of Health and Wellbeing*, *6*(4), 414–417.

Kantur, D., & Iseri-Say, A. (2012). Organisational resilience: A conceptual integrative framework. *Journal of Management & Organization*, *18*(6), 762–773. doi:10.5172/jmo.2012.18.6.762

Kapucu, N., & Khosa, S. (2013). Disaster resiliency and culture of preparedness for university and college campuses. *Administration & Society*, *45*(1), 3–27. doi:10.1177/0095399712471626

Lambrecht, J., Lenkeit, J., Hartman, A., Ehlert, A., Knigge, M., & Sporer, N. (2020). The effect of school leadership on implementing inclusive education: How transformational and instructional leadership practices affect individualised education planning. *International Journal of Inclusive Education*. Advance online publication. doi:10.1080/13603116.2020.1752825

Linnenluecke, M. K. (2017). Resilience in business and management research: A review of influential publications and a research agenda. *International Journal of Management Reviews*, *19*(1), 4–30. doi:10.1111/ijmr.12076

Murwira, A. (2019). *Strategic plan 2019*. Ministry of Higher and Tertiary Education, Innovation, Science and Technology Development.

Muzira, D. R., & Bondai, B. (2022). Perception of educators towards the adoption of Education 5.0: A case study of State University in Zimbabwe. *East African Journal of Education and Social Sciences*, *1*(2), 43–53. doi:10.46606/eajess2020v01i02.0020

Ortiz-de-Mandojana, N., & Bansal, P. (2016). The long-term benefits of organizational resilience through sustainable business practices. *Strategic Management Journal*, *37*(8), 1615–1631. doi:10.1002/smj.2410

Prayag, G., Chowdhury, M., Spector, S., & Orchiston, C. (2018). Organisational resilience and financial performance. *Annals of Tourism Research*, *73*, 193–196. doi:10.1016/j.annals.2018.06.006

Price, R. A. (2023). A review of resilience in higher education: Toward the emerging concept of designer resilience. *Studies in Higher Education*, *48*(1), 83–99. doi:10.1080/03075079.2022.2112027

Reyes, A., Andusyszyn, M., Iwaslw, C., & Forchuk, C. (2015). Resilience in nursing education: An integrated review. *The Journal of Nursing Education*, *54*(8), 438–444. doi:10.3928/01484834-20150717-03 PMID:26230163

Reyes, E. J. D. L., Blannin, J., Cohrssen, C., & Mahat, M. (2022). Resilience of higher education Academics in the time of the 21st century pandemics: A narrative review. *Journal of Higher Education Policy and Management*, *44*(1), 39–56. doi:10.1080/1360080X.2021.1989736

Rossetto, K. R., & Martin, E. M. (2022). It's always about challenging and supporting: Communicative processes of resilience in higher education. *Communication Education*, *71*(4), 304–326. doi:10.1080/03634523.2022.2098351

Seaton, L., Seaton, P., Yarwood, J., & Ryan, M. (2012). *Preparedness for sudden change: Lessons from managing large scale disruption within a Bachelor of Nursing Community: Final Report.* Ako Aotearoa. National Centre for Tertiary Education.

Shaya, N., Abukhait, R., Madani, R., & Khattak, M. N. (2022). Organisational resilience of higher education institutions: An empirical study during COVID-19 pandemic. *Higher Education Policy*, *36*(3), 529–555. doi:10.1057/s41307-022-00272-2

Shin, Y., & Kelly, K. (2015). Resilience and decision-making strategies as predictors of career decision difficulties. *The Career Development Quarterly*, *63*(4), 291–305. doi:10.1002/cdq.12029

Sriskandarajah, N., Bawden, R., Blackmore, C., Tidball, K. G., & Wals, A. E. J. (2010). Resilience in learning systems: Case studies in university education. *Environmental Education Research*, *16*(5-6), 559–573. doi:10.1080/13504622.2010.505434

Sutcliffe, K. M., & Vogus, T. J. (2003). Organising for resilience. In J. Dutton, R. E. Quinn, & K. S. Cameron (Eds.), *Positive organisational scholarship, foundation of a new discipline* (pp. 94–110). Berrett-Koehler.

Tagwira, F. (2018). *Re-thinking education in the new dispensation in Zimbabwe.* A Presentation Paper, Mutare: Ministry of Higher Education Innovation Science Technology Development (MHTEISTD).

Vogus, T. J., & Sutcliffe, K. M. (2007). Organisational resilience towards a theory and research agenda. *IEEE International Conference on Systems, Man and Cybernetics*, 3418-3422.

West, A. (2018). The flourishing student-every tutor's guide to promoting mental health, well-being and resilience in higher education perspectives. *Perspectives*, *22*(4), 141–142. doi:10.1080/13603108.2018.1498036

Wildavsky, A. (1991). *Searching for safety*. Transaction Publishers.

Chapter 2
Disaster Preparedness and Response in Higher Education:
Applying Lessons Learned From Healthcare

Colleen Halupa
East Texas Baptist University, USA

ABSTRACT

There is a paucity of literature on disaster preparedness and response in higher education. The purpose of this chapter is to apply the lessons learned over several decades in healthcare and apply appropriate lessons to higher education institutions. Topics covered include disaster mitigation, disaster planning, disaster response, and recovery. By applying lessons from healthcare, university leaders can craft robust emergency plans and conduct effective exercises that can reduce injury, loss of life, and property damage during an actual disaster.

INTRODUCTION

Although health facilities and colleges and universities are quite different, they also have several similarities. Both have large numbers of people moving from place to place throughout a facility or multiple buildings with a larger geographic footprint. Although both facilities have records on who has appointments (in the case of health facilities), or who is to attend classes (in higher education), not everyone can be easily counted. Also, both places are likely to have external visitors, such a friends of family members, who are not easily accounted for in a disaster. In addition, higher education institutions and health facilities tend to have dense population of people in their facilities; both have people that have disabilities, although the instance of a disabled population is more common in health facilities.

There is a plethora of data on disaster preparedness in health facilities, yet there is a significant paucity of literature about disaster preparedness in higher education institutions. The majority of literature about higher education and disasters concerns student perceptions of training and preparedness (Coveleski, 2014; Izumi et al., Goddard et al., 2018; 2020; Kivunga et al., 2017; Matunhay, 2022; Mishra & Suar, 2011, Ozkazanc & Yuksel, 2016, Tan et al., 2016). Since the COVID-19 pandemic of 2020, the disas-

DOI: 10.4018/979-8-3693-1926-0.ch002

ter preparedness literature in higher education has concentrated primarily on managing pandemics of infectious disease (Izumi et al., 2020; Panganayi, 2020). Patel et al. (2020) and Tan et al. (2016) noted college students are one of the most vulnerable overlooked groups when a disaster occurs in a community because they are not permanent residents. However, there are so many more aspects of disaster preparedness that have been used and practiced in health care for decades that can easily translate to higher education institution disaster preparedness.

The Federal Emergency Management Association (2023a) defines a disaster as "the occurrence of a natural catastrophe, technological accidents or human-caused event that has resulted in severe property damage, deaths and/or multiple injuries" (p. 9). They define a large scale disaster as one that exceeds the response capability of the community and requires possible state or federal involvement. Natural catastrophes include things such as: tornadoes, severe drought, earthquakes, blizzards, hurricanes, pandemics, tsunamis, forest fires, etc. Technological accidents include things such as nuclear accidents, chemical accidents, industrial pollution, train or airplane accidents, release of biological agents from laboratories, and factory explosions or fires (Texas DEM, 2023). Human-caused events include acts of war, terrorism, biological warfare, crime, arson, civil disorder and cyber-attacks (Emergency Management, 2023). Klein and Irizarry (2023) note disasters can cause "injury, disease, illness, loss of life, destruction or property, damage to critical infrastructure and essential services" (p. 1). According to FEMA (2023b), disasters have four phases: mitigation, preparedness, response and recovery.

Disaster preparation in higher education institutions was led by the Disaster Resistant University Initiative which set aside funds for university planning for disasters in 2000 in the United States (Yamaiel, 2006). In addition, 34 CFR 668.49(d) requires that any university that has on-campus housing must maintain a fire log and fire safety report (Office of the Federal Register, 2023). However, just these items are likely not sufficient. The purpose of this proposed chapter is to present best practices in disaster preparedness and response that can be applied in higher education.

History of Disaster Preparedness and Emergency Management

Man has been attempting to deal with disasters since prehistoric times. France's Chauvet cave contains cave paintings which are thought to be 37,000 years old that depict the eruption of local volcanos (Callaway, 2016). In Genesis 6:17 in the Old Testament, God told Noah that he would bring a flood upon the earth and gave him directions on how to survive the flood and save other living creatures (Bible Gateway, n.d.) This same flood is also described in the Islamic Quran (Understanding Islam, n.d.). In 79 AD, Mount Vesuvius erupted destroying the city of Pompeii in Italy (Povoledo, 2022).

In 526 AD, Antioch was destroyed by an earthquake and buildings crumbled and burned. In 563, Geneva Switzerland was devastated by a landslide which caused a tsunami on Lake Geneva. In 1092, a tornado struck London, England. In 1257, a volcano erupted on Lombok Island and it caused cold temperatures due to ash in the air that caused a famine in Europe by preventing plant growth, which in turn, caused the death of animals such as cattle and sheep. In 1287, a massive storm destroyed dikes and changed the landscape of the Netherlands permanently; this was called St. Lucia's Flood. On January 16, 1362, an Atlantic gale flooded much of northwestern Europe. In the east in 1274 and 1281 Kublai Khan's fleets were destroyed by typhoons. There were many other disasters in medieval times (Medievalists, n.d.).

As people began to live more in town and cities, more disasters occurred. In 1665, bubonic plague decimated must of Europe (Johnson, n.d.). The Great Fire of London destroyed 13,200 houses and 87 churches (London Fire Brigade, n.d.). In 1871, The Chicago Fire left one third of the population homeless (National Geographic, n.d.).

In the United States in 1930s, the concept of emergency management, which is the mitigation of risk, began to gain traction. Emergency management was created under the auspices of the federal government. The U.S. Army Corps of Engineers build numerous flood control projects during this time. The Bureau of Public Roads was given authority to offer loans for repair of public facilities after disasters. During the 1950's, disaster preparedness centered on the potential of nuclear war and nuclear fallout. The Federal Civil Defense Administration was created to assist communities in preparation. In 1968, the National Flood Insurance Program was created for American citizens in flood zones. In 1978, emergency management functions, which had been under the control of various government agencies, were moved to the newly created Federal Emergency Management Agency (FEMA) with the director of FEMA reporting to the President. In 2001 after the attacks on the Pentagon and the World Trade Center (WTC) in New York, terrorism became the major focus of emergency management (Haddow et al., 2014). The lessons learned in the emergency response in the WTC and its subsequent collapse provided numerous hard lessons about emergency management. Environmental and man-made disasters continue to plague the world. This is what makes emergency management and disaster preparedness so critical.

Disaster Preparedness in Health Care in the U.S. and Other Countries

Disaster preparedness in health care was shaped by the U.S. military beginning in the Civil War. The Army Corps of Engineers rescued people and properly during the Mississippi Flood of 1882, the Johnstown Flood of 1889, and the San Francisco earthquake in 1906 (U.S. Army Corps of Engineers, n.d.). From the 1930s to the 1950's, hospitals participated in Civil Defense programs and prepared for mass casualties due to traditional wartime or nuclear attack. From this time, hospitals crafted disaster plans that focused on traumas and other emergencies. In addition, hospitals began coordinating with local and state health departments and other agencies (Kim, 2016). In the 1950's, emergency medical treatment through the use of emergency medical technicians (EMT) made the transition from the military to civilian life after their effective use in World War II and the Korean War. Training that was significantly influenced by former military medics was developed to create first responders for accidents and trauma (van Stralen, 2008).

In 1981, the Emergency Mobilization Preparedness Board was established to support the U.S. military in civilian emergency preparedness. This board included a working group for medical disaster preparedness. This group then established the National Disaster Medical System (NDMS) to provide medical care and evacuation for military members during war, but also to assist state and local medical resources during disasters (Lee et al., 2022). In 1997, the National Defense Authorization Act provided training for first responders to prepare for attacks from weapons of mass destruction. This training was provided by the Department of Defense and FEMA (Kim, 2016). In 2014, the Association of Healthcare Emergency Preparedness (AHEPP)was formed to refine and guide healthcare disaster planning (AHEPP, n.d.).

In 2021, Gillani et al. conducted a study of 357 members of a cross section health organizations on the preparedness of health professionals for disaster preparedness in Pakistan. One-third of the respondents at these facilities had experience in dealing with disasters; two-thirds reported knowing the risks and how disasters can impact medical and other supplies. In addition, 70.3% reported they would feel confident of their abilities in disaster response. Carmenico (2011) reported on the National Hospital Preparedness Program in the United States which includes approximately 5,350 hospitals. More than 76% met 90% or more of all measures emergency preparedness which is impressive and demonstrated that health care organizations take emergency preparedness seriously. Gao (2018) surveyed staff at 65 hospitals in the Heilongjang province of China and found 100% had disaster plans and 93% regularly reviewed this plans. They found over 95% of all hospitals had leadership groups for disaster preparedness; over 76% did drills, while 86% did simulated training exercises or table-top disaster exercises.

Disaster Preparedness in Higher Education

Disaster preparedness occurred much earlier in secondary schools in the United States. Secondary schools have regularly had scheduled fire drills and in recent years, school shooter scenarios. However, the movement towards emergency management and disaster preparedness came much later to higher education.

In 2000, the Pre-Disaster Mitigation program was authorized as part of the Stafford Act. This program was called The Disaster-Resistant University (DRU) program. Congress set aside money to be allocated to universities in order to assist them with disaster mitigation and emergency management. The first university to participate was the University of California at Berkeley which addressed earthquake hazards and physical infrastructure (Yamiel, 2006).

In 2006, Mitroff et al. identified several potential disasters that can occur on college and university campuses. This included major fires and explosions, environmental disasters, student unrest, major outbreaks of illness, terrorist attacks, violence and other major crimes. They note that universities operate numerous different businesses in addition to educating students that present risk. These include food services, retail stores, health facilities, and sporting events. In Fall of 2004, Mitroff et al. surveyed Provosts at 117 major U.S. college and universities; 36% were located in cities, 39% were in suburban areas, and 24 percent were in rural areas. Their major finding was, "surveyed colleges and universities were generally prepared only for the crises they had already experienced" (Mitroff et al., 2006, p. 65). In addition, universities did not take the lessons learned from when disasters had occurred and apply them to other disasters that might occur in the future by revising their emergency management plans. These colleges and universities had specific plans that were required by state and federal laws, but they did not have broad-based plans. Mitroff et al. recommended broad-based crisis management programs for all disasters that have the potential to occur.

In addition to emergencies that occur on college campuses, these institutions may also be called upon to provide support in community disasters. Dunlop et al. (2014) did a qualitative study on several areas of the country affected by disasters to look at the participation rates of colleges in disaster response. During Hurricane Ike and Gustav in 2008, all 18 public universities in Louisiana participated in disaster response. This included providing shelter, food, first aid, counseling medical services, as well as providing security at shelters.

The International Association of Emergency Managers (IAEM) has a caucus in the United States. The purpose of the IAEM-USA Universities and Colleges caucus is to assist in preparing faculty, students and staff for preparing, responding, and recovering from disaster. They also lobby on a national and international scale with governments and industry regarding disaster preparedness (IAEM-USA, n.d.).

In 2010, the U.S. Department of Education created the Action Guide for Emergency Management at Institutions of Higher Education. This document describes the control and security risks that are present in this environment during disasters. This document recommends the four phases of emergency management recommended by FEMA, and indicates plans must consider students, staff, faculty and visitors, including those with special needs. The Disaster Resilient Universities Network in 2020 conducted a survey of colleges and universities disaster planning and management. Only 55% of respondents reported that they have an emergency management framework, and 53% had one person or less devoted to emergency planning and response. Although most respondents had an emergency operational plans including a hazard assessment plan, 51.5% had a continuity of operations plan, 50.5% had a natural hazard mitigation plans, 47.3% had emergency management strategic plans, 42.1% had recovery plans and only 41.1% had training plans (University of Oregon, 2022). This represents a significant gap.

The evidence provided demonstrates that healthcare is much more advanced than higher education in disaster planning and preparedness, particularly in training and exercises, as well as disaster mitigation. Health care personnel are already training in triage and first aid. All of these lessons from healthcare can be applied to higher education institutions to increase their readiness for disaster response.

Mitigation

Prevention is taking action to decrease the likelihood of a disaster. Mitigation is taking actions to reduce or eliminate loss of life and property. The United States Department of Education (DOE) published a guide for emergency management at higher education institutions in 2010. Institutions of higher education are responsible for the safety of their students during normal operations and during disasters. Emergency management planning in higher education can differ somewhat from planning in other industries and organizations, but as noted previously there similarities between larger health facilities and college campuses. The DOE guide begins with outlining nine key principles in emergency management in colleges and universities. Effective emergency management:

- Begins with senior leadership
- Requires partnerships and collaboration
- Must account for a full range of possible hazards
- Should use FEMA's four phases of emergency management
- Must be comprehensive and should include provisions for staff, students, faculty, visitors and those with special needs
- Includes comprehensive planning throughout the institution
- Includes training based on prioritized threats
- Includes tabletop exercises
- Includes dissemination of information to students, staff, faculty, community partners and parents (U.S. Department of Education, 2010).

Disaster mitigation begins with university senior leadership since this leadership has the authority to devote necessary resources. A crucial part of this process is to develop community partnerships with organizations such as the police, fire department, medical services, mental health organizations and media to access their additional resources. Other community organizations can also provide additional support such as shelter and food if needed (U.S. Department of Education, 2010).

University leaders cannot plan and conduct disaster training and exercises on every possible scenario. Rather, leaders need to plan for as many disasters as possible, but conduct training and exercises on those disasters that are more likely to occur. For instance, in Oklahoma in the United States where tornadoes are common, it would be prudent to conduct disaster exercises using a tornado as a scenario.

Consequently in Japan, it would be more prudent to conduct exercises and training around earthquakes, typhoons, or tsunamis. Weather is just one aspect university leaders must consider, but there are numerous other types of disasters including violence, terrorism, biological hazards, and chemical hazards. If a university is located in a city with a nuclear reactor, it would be prudent to do evacuation and disaster exercises centered around radiation exposure. When considering all of these different types of disasters, leaders should conduct prevention/mitigation planning, preparedness exercises, response exercises and then create plans for recovery. Before a disaster occurs is the best time to think about all these issues. If a plan does not include all likely disasters and one occurs, in the chaos of the disaster important things may be forgotten that cause harm to others as well as further physical damage. In addition, a plan that only considers one member of a population (i.e. students) is likely to result in harm to others in the university population not addressed in the plan. Special consideration must be taken with those with potential physical disabilities where evacuation may take additional time, and those with mental health issues which may be exacerbated by the stress of a disaster (U.S. Department of Education, 2010).

Emergency plans for various colleges and universities may have some commonalities; however, campuses have different geographic footprints and combinations of staff, visitors and students. A one-size fits all emergency plan is unlikely to be effective in a disaster (U.S. Department of Education, 2010). In general emergency planning includes assessing physical risks such as grounds and physical facilities on the university proper, but the risks in the larger community must also be assessed.

It is not possible for a university to plan for, train for, and practice every possible scenario. Rather, primary and secondary data from the community and the campus should be used to prioritize evens that are most likely to occur. Tabletop exercises are the least intrusive and inexpensive way to test one common scenario in an emergency management plan. In tabletop exercises, multiple departments and campus populations must be represented. However, a true exercise that also tests community response and involvement is the best way to test an emergency management plan.

In addition, an emergency plan in a security office on campus is not effective. Plans should be available at various places around campus so that if one office is destroyed that houses the plan, it is also available at other locations. It should also be posted on a web site, but in many disasters internet service may be affected and the plan may not be able to access.

Risk Assessment

Risk assessment is not a one-time event that occurs when an emergency plan is created. Rather, risk assessment is ongoing, and occurs as changes in the environment occur. Risk assessment must also be conducted during the course of a disaster itself as an ongoing process. The World Health Organization (WHO) recommends that health organizations conduct a complete risk assessment every two-three years if no significant changes in the environment have occurred (2021). The same is true of colleges and universities.

WHO recommends in their Strategic Toolkit for Assessing Risks (STAR) that the following parameters be used:

- Assessing the likelihood of a disaster event
  -Almost certain (will occur in the next 12 months 95% or more of the time)
  -Very likely (will occur 70-94% of the time)

-Likely (will occur 30-69% of the time)
-Unlikely (will occur 5-29% of the time)
-Unlikely (probability less than 5%)

- Assessing the severity of a potential disaster event
  -Morbidity
  -Mortality
  -Disruption to basic needs (food, shelter, water)
  -Physical damage and associated dangers
- Assessing vulnerability to disasters
  -Located near a military base in time of war
  -Located where industrial spills or accidents could impact the campus
  -Disabled members of the population
  -Accounting for all students, and visitors
  -Community unrest (looting)
  -Possible consequences of disasters (both immediate and secondary)
  -Alternate operations will be required (such as emergency remote learning)
  -Alternate student shelter will be required for a longer period
  -Faculty and staff availability due to coping with their own homes and communities in a disaster
- Assessing changes in coping capacity or resources
  -Community partners can no longer provide support or can only provide more limited support
  -Changes in disaster team members
  -Changes in leadership in community partners (who have to be apprised of the university plan and agreements in place)
  -Environmental hazards occur in buildings on campus
  -Financial resources
  -Logistics and storage
- Assessing potential scale of a disaster
  -Just the campus
  -The local community
  -The broader community (WHO, 2021)

When risk assessment is done, a gap analysis and plan to overcome these gaps should be done and the disaster plan should be revised. The vulnerability assessment is a key component of risk assessment and should be conducted by a multi-disciplinary group from throughout the college and university, as well as key community members. Margolis Healy, in a study of multiple campuses, found that only slightly more than half of campuses had conducted a vulnerability analysis; this result was confirmed by the National Center for Public Campus Safety in 2016.

Reviewing Existing Data

There is likely to be a plethora of both primary and secondary data about the communities surrounding a college of university. Data that is helpful to review when trying to prevent and mitigate disasters include the following:

- Weather data
- Type of industrial hazards in the community that could impact the campus
- Crime data
- Earthquake/hurricane/typhoon probabilities

Other data that must be considered are the number of available community resources which would include the following:

- The number of police that may be able to respond (if the larger community is also affected, and if it is not)
- The number of available firefighters (if the larger community is also affected, and if it is not)
- Availability of special equipment such as "Jaws of Life" and ladder truck to evacuate people from the top floors of buildings
- Possible shelters that could be used in the local community as well as communities that are farther away (should the disaster be a large one).
- Resources for food, clothing, etc. for those who are affected

Community resource assessment must be done before disaster planning occurs. This is because agreements need to be made with these organizations in case of disaster and this may take a significant amount of time. Once agreements are made, a list of community agencies and governmental officials that need to be contacted in case of emergency needs to be created. Periodically the university should touch base with these community resources to ensure there have been no significant changes.

Reviewing Facilities and Grounds

Universities have limited money to devote to facilities. Some universities may have older buildings that have environmental hazards in case of a disaster such as asbestos insulation, narrow hallways that are not conducive for evacuation, and windows that cannot be opened should doors become unpassable, etc. In areas where earthquakes are common, older buildings may not have had structural modifications and enhancements that will minimize damage (U.S. Department of Education, 2010).

In places that are vulnerable to high winds, hurricanes, typhoons or tornadoes, large trees near university buildings may fall causing property damage during a disaster. In addition, a review of the grounds enables the disaster team to pick the least vulnerable areas of campus for campus member evacuation and assistance of casualties to designate as meeting and casualty collection points.

Assessing Culture and Climate

Research has shown that overall, colleges and universities do not do a good job of preparing their population for disasters. Izumi et al. (2020) reported in a sample of 150 responses from 65 universities in 29 countries that 47% of respondents reported their institution did not have an emergency management office, and 41% reported they lacked a disaster plan. In universities that had plans, 33% did not include any sort of disaster simulation or exercise plans to ensure a disaster plan actually worked.

Mishra and Suar (2011) noted that numerous studies have been done to look at people's behavior during disasters. Campus members who are trained, have participated in exercises, and are aware of emergency procedures are more likely to respond more calmly in a natural disaster. Ozkazanc and Yuksel (2015) studied disaster preparedness awareness of 164 students in Turkey. Only 1% of these students felt their disaster training in their university was sufficient, and only 26% knew who the competent authorities were in an emergency even though 55% of the students had already experienced one or more disaster incident in their lives. Matunhay (2022) in a study of 6,818 students in the Philippines confirmed Ozkanc and Yuksel's findings. Matunhay found only 5.2% knew what disaster preparedness was, and only 42.5% took any sort of action (including even packing an emergency bag), yet the same number were aware that significant risks of disaster were present. College students, like many youth, feel they are invincible; however, this attitude may increase morbidity and mortality in a disaster. This demonstrates it is important that university leaders understand the attitudes of students (who are most at risk), as well as attitudes of faculty and staff in regards to disaster preparedness and training.

According to the Healthy Minds college student survey done in 2023-2024, 44% of all students reported depression and 37% reported anxiety which was the highest rate in the 15-year history of the survey (Flannery, 2023). Disasters can exacerbate these symptoms and perception of a threat can also generate a fight or flight response. In addition, anxiety-prone people are more likely to have clinical anxiety or a mental break under stress which can significantly impair judgement. These behaviors can endanger others in a disaster since these people are more likely to engage in risky behaviors. The most important mitigation strategy for decreasing this risk is communicating clearly the nature of the risk and how to adapt to the risk (Mishra & Suar, 2011). Students are not the only university members who will be experiencing significant anxiety and stress. Faculty and staff members will experience stress worrying about their own families and homes, and visitors who are unfamiliar with the campus will also experience significant stress.

Risk Mitigation and Disaster Plan Components

The American College of Emergency Physicians (n.d.) created a hospital disaster assessment tool. Many of the items on this checklist are applicable to universities and colleges and should be considered and included as part of a disaster plan.

- Capacity (student capacity)
- Capacity (faculty and staff)
- Procedures are in place to monitor absenteeism on a daily basis (to account for everyone in a disaster)
- Location of alternate copies of the disaster plan
- Number of staff and faculty on disaster team and their names
- Percentage of population that has completed disaster preparedness training
- Number of security personnel on site at different times of the day
- Emergency power duration (if applicable), as well as the last time it was tested and any generator fuel requirements, as well as where fuel can be obtained
- Preservation of food procedures
- Sources of water
- A continuity of operation plan
- A business continuity plan
- A method is designated to track use of financial resources to deal with the disaster

- A financial plan for immediate response and recovery
- An evacuation plan
- A plan to recall personnel (staff and faculty)
- A fire prevention and response plan
- A plan for shutdown of utilities (in case usual personnel are not available)
- A hazard communication plan to release information to students, staff, faculty, parents and the press
- A security plan to limit access to the campus, in addition to a plan to control all entries and exits.
- A plan for information management to prevent loss of data
- An Incident Command method has been designated (and last time this was practiced is documented)
- There is a procedure for alternate incident commanders if the designed commander and alternates are not available.
- The university Command Center and disaster plan is fully integrated with local emergency/county planning and operations.
- An Incident Command Center Location has been established (as well as an alternate locations)
- Location of alternate communication devices (i.e. radios) is documented
- A procedure to communicate disaster information using multiple modes of communication
- Map of campus; blueprints for buildings are located with master disaster plan
- Contact information for the agencies where memorandums of understanding are in place including police, health facilities, fire department, military installations (if applicable), mental health services, and public safety
- The emergency plan (EMP) addressed four phases of emergency management: mitigation, preparedness, response and recovery.
- The EMP is accessible to everyone and has provisions for special populations (disabled, etc.).
- Designated disaster supplies are available (first aid kits, bullhorns, flashlights) and their location

Other Documentation and Checks Required

- Quarterly fire drills (and occasional active shooter drills) are conducted, evaluated and documented
- A memorandum of understanding exists with local law enforcement to provide extra security if needed
- A plan is in place to release information to the press and concerned family and friends.
- The Emergency Plan is exercised at least twice per year (for universities once per semester) including establishing security perimeters; response is evaluated and documented.
- Evacuation of dormitories and classrooms is practiced at least once a semester
- A hazard vulnerability analysis should be done and documented with emergency plan at least every 2-3 years or with significant changes
- Documented checks that exits are clearly marked in all buildings
- The financial plan has been tested (simulated) and documented (ACEP, n.d.).

A crucial concept that colleges and universities must remember is that disaster planning needs to be ongoing. This includes conducting vulnerability analyses, checks of equipment, and update of the emergency disaster plan as things change on the campus. Should a disaster occur and these things are all done, there will likely be a decrease in the loss of life, injury and property.

Disaster Teams

Every university must have a disaster team. This team at a minimum should include security, information management, physical facilities, financial services, and a member of upper leadership. In addition, representation from student affairs is critical and should also include dorm resident advisors who will be required to assist should evacuation ever happen. In addition, staff and faculty representation should also be included. A multi-disciplinary team ensures that disaster planning and mitigation will be more complete.

The size of the disaster team will depend on the size of the university. However, it is crucial that the team is not so small that university members and students are not aware of it, and not so large that it is not effective. In addition, alternates should be designated in case disaster team members are unavailable. It is preferable to have two alternates for each position to ensure continuity.

Tabletop Exercises

In her 2012 dissertation, Mary Keane Lott assessed perceptions of disaster preparedness preparation at five universities in Washington, District of Columbia. Overwhelmingly, administrators and disaster reams believed their university was prepared, but faculty, staff and students did not. Connolly (2016) noted that even if a university has plans in place, this does not mean it is prepared. All campus members must understand their role in a disaster. One way this can be done in through tabletop drills or exercises.

The purpose of tabletop emergency preparedness exercises is to test emergency policies, procedures and the emergency plan the in the most inexpensive and expedient way possible. In a tabletop exercise, a single disaster scenario is selected (Skyryabina et al., 2017). Disaster team members establish an Incident Command Center, and then notify the university community at large that a disaster has happened (letting them know this is an exercise). Depending on the nature of the scenario, the Incident Command team then simulates contacting outside agencies, and the university population at large simulates the actions that would occur with such a disaster. Tabletop exercises are frequently done by both the military and health facilities to test response. It is important that everyone take these exercises seriously and even though some items are simulated people need to be engaged and respond as if the disaster were real. These events usually last one to four hours (NYC, n.d.).

After the exercise is complete, all departments report back to disaster team members. In addition to self-reporting, observers should be assigned to evaluate key functions to provide additional feedback. Skyryabina et al. (2020) noted there tends to be a lack of methods and tools used to effectively evaluate performance in university emergency preparedness. Universities must design an evaluation plan that works best for their particular situation. Skyryabina et al. (2020) also noted that participants should also relate what they learned from the exercise, in addition to looking at what worked and what did not. This is also a crucial component in evaluating disaster plans and seeing what updates need to be made to improve performance in the future.

Live Exercises and Drills

The first type of live drill is an Incident Command Exercise. In this type of exercise, the Incident Command and Disaster Teams assembly and response is tested. Usually in this type of live exercise, one or more evaluators assess the Incident Command's execution of the plan. Evaluators usually will pull out key members of the team to simulate what can happen in an actual disaster. This is to ensure that alternate members have the opportunity to lead disaster response.

The second type of live drill is a campus-wide disaster drill which is called a field exercise (Skryabina et al., 2017). This is a full-scale drill where tasks are not simulated. Rather, the university as a whole practices response such as student evacuation, first aid, etc. at the university level only. The interaction with outside agencies is simulated.

The third type of drill is a full scale drill. This drill must be coordinated with all community partners, such as fire response and police, and a specific day must be selected by key members. The date and scenario for this exercise may be communicated in advance, or it may be unannounced. The purpose of this type of drill is to evaluate a disaster that requires a local community's emergency response on the university grounds (Skryabina et al., 2017).

A fourth type of full scale drill that is effective at some universities who have medical and nursing training is a mass casualty exercise where university students participate in patient assessment and triage. Austin et al. (2014) all discussed a university exercise with the Texas State Guard and the University of Texas in San Antonio. This scenario is a unique way to support emergency preparedness in the local community. It also can increase university preparedness since these health students would be better versed in disaster preparedness techniques such as first aid and triage should a disaster actually occur at the university.

The processes for conducting and evaluating both tabletop and live exercises is a detailed subject itself and much too complex to cover in this survey chapter. However, guidance from both the military and health facilities can be invaluable for universities who want to conduct these activities. In addition, information is also available from federal, state and local emergency management organizations in the United States, and in government emergency management organization in other countries.

The organization Readiness and Emergency Management for Schools (REMS) (2023), which is an organization for secondary education, has an emergency management toolbox that is also useful for higher education. REMS (n.d.) also notes there are numerous benefits of emergency exercises in educational institutions. This includes testing the capability of the campus, identifying gaps in emergency management plans and procedures, allowing members of the campus to participate and train which increases the effectiveness of response in an actual emergency, and allowing the university disaster team to coordinate with outside community agencies. They also note this can increase the confidence of all campus members in their response in a disaster.

Waxman et al. (2017) recommends that no-notice exercises are most effective in healthcare. The same is true in colleges and universities. There is no notice for many disasters that occur such as earthquakes, tornadoes, terrorism, large scale chemical, biological or nuclear spills, or active shooters. No-notice exercises are a better indication of true disaster response.

Disaster Plan Training and Dissemination

Tan et al. (2016) noted that university students are more vulnerable in disasters than the general public. In addition, they are often overlooked in disaster preparedness efforts within a larger community. University students in dormitories are transitional and should these residents be damaged, this may significantly interrupt their education and future plans. They note education is crucial to increase disaster preparedness among university students.

Perhaps the best way to educate students about disaster preparedness is conducting tabletop and live exercises where students are involved. This will be the most effective type of education because it is experiential and active learning.

Both students, faculty and staff should also receive information about the university disaster plan located during orientation, and periodically (at least once per year) thereafter. As part of this training, the disaster team members and alternates should be identified as well as the locations of the disaster plans throughout the university. For students and faculty/staff, evacuation procedures should also be discussed. At the department level and school level, both students, faculty and staff should be informed of the designated area they are to go in case of a disaster so they can be accounted for.

Training for disaster team members will be much more extensive and should occur more often (at least twice per year) and should include all team members and alternates. This is to ensure continuity of command and control should a disaster occur.

Disaster Response

Disaster response is implementing a disaster plan in the event of a disaster (Klein & Irizarry, 2023). In a disaster such as a hurricane, the university or college may have had advance notice. Evacuations can occur before the hurricane hits, but the university must have a contingency plan to continue operations. When a disaster occurs very quickly, such as with a tornado, university personnel are going to be the initial people on site to help the injured. The first response will be emergency medical services, but depending on the type of disaster, they may be delayed. University personnel will likely have to step up to assist others. If a university has a robust disaster response strategy, both morbidity and mortality can be decreased.

During a disaster response, incident command and control is crucial. In many universities, police and security manage disaster response (University of Oregon, 2022). However, while disaster planning, the university must create a robust system of command and control so there is not chaos should designated members who lead response are injured or unable to respond. Another problem noted by Klein and Irizarry (2023, p. 3) is the phenomena of "disaster tourism" where go to the area of the disaster intending to help. This can do more harm than good because these informal responders are not apprised of the disaster plan or in many cases, environmental hazards and risks.

Response Leadership and Command

Mahmud et al. (2020) note that leadership in a disaster differs greatly from daily university leadership. Leaders in a disaster must be able to make immediate decisions with little or no information to maximize the safety of the people affected. Leaders must be knowledgeable to the college or university's disaster protocols and must be able to effectively coordinate with university staff and students, responders and other agencies in the local community, as well as eventually potentially with state and local officials.

Just because a college or university president serves as the day-to-day leader during normal operations, this does not mean he/she is the best person to serve as Incident Commander leading a disaster team during a disaster. These duties should be delegated to campus security and a multidisciplinary disaster team. It is important that there is a duplication of disaster leaders who are well apprised of what to do in case primary leaders cannot respond. Filmer and Ranse (2013) noted that when multiple people try to take on leadership roles, the can give contradictory commands which leads to chaos. A clear line of command must also be established so someone is designated to make the final decision in difficult situations. This is why training before a disaster occurs is so critical.

Leadership in any disaster is there to help assuage panic and prevent chaos that can be even more harmful than the disaster itself. Students, staff and faculty are much more likely to follow these leaders when they have been advised of who they are in case of disaster, and when leaders take a panic-free, empathetic and confident approach to the management of the situation.

Disaster Security and Control

There are security and control risks in any disaster. Leadership must plan and provide control during a disaster for these risks. The Orthopedic Trauma Association (OTA) (n.d.) identified three areas of concern in disaster response and security. They are physical security, physical health and mental health. In regards to physical security, the evacuation and staging areas for injured students and staff should be close enough to the actual disaster to limit transportation time, but far enough away to prevent danger of additional injury. In addition, when a disaster occurs, particularly in a college or university with thousands of young people, parents, friends and other relatives are likely to come to the site. Security must be established so that family and friends do not get in the way of emergency response and in fact, do not become casualties themselves. In addition, Twigg and Mosel (2017) note disasters simulate "spontaneous responses before organizations are able to mobilize…often from outside the community (p. 443). These responders can be altruistic, or in the case of some disasters such as Hurricane Katrina, they can be harmful in the form of looters depending on the type of disaster.

In addition the OTA (n.d.) points out that not all layperson responders can safety physically participate in disaster response. If someone exerts him or herself to help another and takes a heart attack, then he or she is just becomes an additional casualty that has to be dealt with. Therefore, physical health of both responders and the injured is of paramount importance.

Last, disasters are very stressful. College of university members may see dead friends or colleagues which can result in severe mental stress or even mental breaks. Leaders must be cognizant of this and as part of disaster planning should include mental health support during disaster response for those who need it. If someone has a mental health break during a disaster response, he could injure himself or others. In addition, community members will need mental health during disaster recovery and this must also be planned for (Makuana, 2019).

Initial Disaster Crisis Communication

Communication in a disaster is critical. People will look to leadership for this communication. The 5C Model of crisis communication provides five key aspects that need to be considered when developing a communication strategy. It is important this strategy be developed before a disaster occurs and practiced to ensure it is effective. The first "C" is comprehend. This means the leader needs to have an understanding of the type of disaster and its scope. This is to ensure that communication will address the specific needs of the university or college population. The next "C" is coordinate. This ensures a consistent approach to communication. There needs to be a clearly identified leader who is responsible for communication with the university population as well as outside agencies. This ensures people do not get conflicting messages. Again, a coordination strategy must be developed and practiced prior to a disaster with disaster team members. In addition, just as with overall leadership until outside help can come, there must be duplication in case the person designated to communicate is not available to do so. The next "C" is communicate. Messages must be clear, accurate and timely to outside agencies and the community at large. The last stage is confirm. The leadership in a disaster must monitor the effectiveness of the communication, and adjust as required (Crisis Control, 2023).

The Federal Emergency Management Association (2014) notes that information during a disaster may:

- Save lives and reduce injury
- Protect property
- Change the behavior of those who are panicked and fearful
- Instills confidence that someone is in charge and the issue is being taken care of

FEMA (2014) also notes the type of emergency communication is influenced by many factors. The first is the audience which in a university initially is staff, faculty and students. However, after immediate danger has passed, this also includes parents and perhaps the community at large. The next consideration is the urgency of the message. In a disaster, communication at first can be very urgent and then level off until it is not needed anymore. The next consideration is what media will be used. Le Roux (2013) points out that during some disasters normal communication channels may fail.This includes landlines and cell phones. In order to be able to communicate with outside agencies and even within the university itself, it is recommended that two way radios be purchased for back-up communication. In addition, even items like bullhorns may be effective to get the attention of large groups of people during a disaster. FEMA (2014) also recommends that organizations craft message templates prior to a disaster. This reduces delays when the are needed and can help prevent omission of needed information.

Initial disaster communications must be concise, informational and directive in nature. They need to instill a sense of confidence and demonstrate that there is someone in charge and that person has more information than university community members do when a disaster has just occurred.

First Aid

The World Health Organization (2019) and the International Federation of Red Cross and Red Crescent Societies (IFRC) (n.d) recommend that communities educate themselves on critical topics related to disaster response. One of these critical functions is training and equipping volunteers as first responders to assist others.

Pawloski et al. (2018) reported that when immediate assistance is provided after an accident, the wounded survival rate is 85%; when the response increases to 15 minutes, the survival rate decreases to 72%, while at 20 minutes it decreases to 60%. These authors surveyed 500 students in Poland and found that among high school and university students, there was a significant impact of first aid education on student willingness to provide aid. Heard et al. (2022) confirmed this finding in 2022 by analyzing literature from 54 articles in 22 different countries.

Tan et al. (2016) in a survey of 1,764 university students in China noted students strongly desired disaster preparedness skills. This was found between genders and across numerous majors. They reported students particularly want to know rescue skills, of which first aid is a part. Labosier (2019) and Patel et al. (2020) recommend that college students maintain first aid kits for potential disasters; however, it is also important they know also how to conduct basic first aid.

Riad et al. (2013) studied a random sample of 100 secondary teachers in Abda who were provided both crisis management and first aid training. They found a training program significant improved the knowledge of teachers who were laypersons in conducting first aid. In a university that has health professions, these students can be used to conduct first aid training for disaster team members and other students. Jetten et al. (2022) noted it takes time for medical response in disasters and that initially the disaster has to be managed by the community members who are laypersons. Unaffected and slightly injured people can help more critically ensured people. Having university responders that at least have basic first aid knowledge is crucial if medical personnel are prevented from responding due to a later emergency throughout the community or because of a lack of access to the site. This can reduce casualties and further injury.

Triage

A second lesson regarding disaster response that can be gleaned from healthcare is triage. Triage is a process that is used to categorize patients based on the severity of their injuries. This term originated as a response to mass casualty situations and was sed by the military in civilian mass casualty events and wartime as early as 1792. It did not go into wide use in healthcare until the early 1900s (Robertson-Steel, 2006). The purpose of triage is to keep as many of those in danger or injured alive as possible with the limited resources available.

Djalali et al. (2011) reported that in earthquake response, patients are often evacuated without being triaged, usually by laypersons without medical training. There are two types of triage used in disaster response. The first is medical triage which will be discussed later in detail. The other type is evacuation triage. The difference between the two types of triage is that in evacuation triage, rescuers (which can be laypersons since they are present) classify patients that need to be evacuated first. This is often based on perceived medical injuries, but it can also be based on danger from physical hazards or response to mental health breaks. This process does not follow the levels of medical triage. It is essentially removing people from potential danger and staging them in a place (called a casualty collection point) away from danger so they can get medical help as quickly as possible when it arrives. This also includes assessing the approximate number of casualties to appraise first responders of population needs so additional arrangements for resources and equipment can be made. This process has been used very effectively by the military for decades when dealing with mass casualties (Koser & Suchenski, 2022).

When medical response teams arrive to a disaster, they will then re-triage using medical triage. If there are medical persons on the site of the disaster, they may also perform medical triage in addition to evacuation triage. There are five categories of triage: immediate, delayed, minimal, expectant, and dead (Frykberg, 2005). The immediate category means action needs to be taken as soon as possible to prevent loss of life or limb. This includes wounds such sucking chest wounds, or arterial blood sprays from a limb. Patients in the delayed category need medical care, but can go some time without it without the likelihood of further complications. However, patients can transition from the delayed category to immediate if something changes. People in the minimal category are the walking wounded; they may have broken arms or minor burns or abrasions. They can often assist with evacuating and helping others. Those in the expectant category are those that are unlikely to live without extensive, immediate treatment (and who likely would die even with this treatment). This includes head wounds where the brain is exposed, or third degree burns over 80% or more of the body. The last category is dead. These last two categories are usually placed further from the site where medical response is likely to occur so others can be treated first. Unfortunately in a disaster situation with limited resources, there are likely to be people that cannot be helped.

Many college campuses have medical personnel who can perform both evacuation and medical triage. However, this may not be the case, so it is important that campus members be trained in at least basic evacuation triage principles so those who need care can get it as soon as possible. Disaster team members should be training as well as at least one to two persons per building. Student volunteers should also be trained in each dormitory since disasters could impact these facilities.

Immediate Disaster Recovery

Depending on the type of disaster, a college or university may experience numerous challenges. These include housing students (particularly if dormitories were affected), and a lack of utilities such as power and water (Noboa Ramos et al., 2023). This can also make feeding students much more difficult. Most students do not have families in the local area. Their cars may have been damaged, or the roads may be impassable. Colleges need to have a response plan should a large scale disaster occur to plan for immediate recovery to ensure the basic needs of student primarily, and staff and faculty are met. This may include making arrangements with city leaders to allow students to use local shelters. If the disaster has affected the larger community, the college or university may also have to draft agreements with places that will likely be close, but out of the disaster area, to provide food for students local community resources may already be severely stretched.

In addition, the university or college must continue to try to account for all staff, student and faculty members to ensure their safety. At Tohoku University in Japan during the Great East Japan Earthquake in 2011, this process took almost three weeks for a population of over 18,000 students and 11,500 faculty and staff (Aoki & Ito, 2014). These authors also note the safety and functionality of all buildings and facilities must be checked since conditions can worsen after a disaster. If chemical or dangerous substances leak, then abatement efforts must be coordinated. In the case of Tohoku University, radiation monitoring also had to be done due to the catastrophic failure of a nuclear reactor in the area (Aoki & Ito, 2014).

As happened during the COVID-19 pandemic, if a disaster should occur, the college or university may be required to go to emergency remote learning; this may be for a short time or a long one depending on the nature of the disaster. The COVID-19 pandemic caught many university administrators by surprise. However, the plans developed and the lessons learned during the COVID-19 pandemic should allow institutions to more easily convert to emergency remote learning should it be required in the future. Beggan (2011) who evaluated disaster response after two hurricanes hit Lamar University in Texas during the first decade of the 2000's noted alternate academic calendars should also be considered.

Perhaps the most important thing universities and colleges can do is to update their disaster plans. After a college or a university has experienced a disaster, the leadership and disaster team now have a good idea of what worked well and what did not. Changes need to be made right away so these lessons can translate into more effective policy and procedure. In addition, any training deficiencies should be identified. This will not occur until the immediate disaster recovery efforts are completed, but it should not be delayed.

After immediate disaster recovery, the university will go into final disaster recovery which can take a very short time, or a long one depending on the nature of the disaster. The goal of disaster recovery is to resume normal operations as soon as possible.

CONCLUSION

The purpose of this chapter has been to discuss the lessons learned in healthcare emergency planning and disaster response that can be used in higher education. Healthcare has a long history in disaster preparedness. Disaster preparedness has only taken hold on in higher education for the last two decades; and there is a paucity of information on emergency management in higher education. Universities must ensure they institute the four phases of emergency management including mitigation, planning, response and recovery. This is necessary for the safety of the campus community and to disrupt the student educational experience as little as possible.

REFERENCES

American College of Emergency Physicians (ACEP). (n.d.). *Hospital Disaster Preparedness Self-Assessment Tool*. https://www.calhospitalprepare.org/post/hospital-disaster-preparedness-self-assessment-tool

Aoki, T., & Ito, K. (2014). What is the role of universities in disaster response, recovery, and rehabilitation? Focusing on our disaster victim identification project. *IEEE Communications Magazine*, *52*(3), 30–37. doi:10.1109/MCOM.2014.6766081

Association of Healthcare Emergency Preparedness Professionals. (n.d.). *About us*. https://www.ahepp.org/page/AboutUsTab

Austin, E. N., Bastepe-Gray, S. E., Nelson, H. W., Breitenbach, J., Ogle, K. T., Durry, A., Green, S. D., Crabtree, L. A., & Haluska, M. (2014). Pediatric mass-casualty education: Experiential learning through university-sponsored disaster simulation. *Journal of Emergency Nursing: JEN*, *40*(5), 428–433. doi:10.1016/j.jen.2014.05.015 PMID:25194652

Beggan, D. M. (2011). Disaster recovery considerations for academic institutions. *Disaster Prevention and Management*, *20*(4), 413–422. doi:10.1108/09653561111161734

Bible Gateway. (n.d.). *Genesis 6:11-20*. https://www.biblegateway.com/passage/?search=Genesis%206%3A9-9%3A17&version=NIV

Callaway, E. (2016, January 19). Cave of forgotten dreams may hold earliest painting of volcanic eruption. *Scientific American*. https://www.scientificamerican.com/article/cave-of-forgotten-dreams-may-hold-earliest-painting-of-volcanic-eruption/

Caramenico, A. (2011, May 6). *Most hospitals better prepared for disasters*. Fierce Healthcare. https://www.fiercehealthcare.com/healthcare/most-hospitals-better-prepared-for-disasters

Connolly, M. (2016). *Campus emergency preparedness: Meeting ICS and NIMS compliance*. CRC Press.

Coveleski, J. (2014). *A study of students' perceptions of natural disaster plans and emergency preparedness at a higher education institution* [Doctoral dissertation]. Florida State University.

Crisis Control. (2023, June 16). *What is the 5C model for communicating during crisis?* https://www.crises-control.com/blogs/the-5c-model-for-critical-communication/

Djalali, A., Khankeh, H., Öhlén, G., Castrén, M., & Kurland, L. (2011). Facilitators and obstacles in pre-hospital medical response to earthquakes: A qualitative study. *Scandinavian Journal of Trauma, Resuscitation and Emergency Medicine, 19*(1), 1–9. doi:10.1186/1757-7241-19-30 PMID:21575233

Dunlap, A. L., Logue, K. M., & Isakov, A. P. (2014). The engagement of academic institutions in community disaster response: A comparative analysis. *Public Health Reports, 129*(6_suppl4, Suppl. 4), 87–95. doi:10.1177/00333549141296S412 PMID:25355979

Emergency Management Monroe County, Florida. (2023). *Man-made disaster.* https://www.monroecounty-fl.gov/904/Man-Made-Disaster#:~:text=Such%20man%2Dmade%20disasters%20are,%2C%20cyber%2Dattacks%2C%20

Federal Emergency Management Association. (2014, February). *Student manual: Chapter 3 effective communication.* https://training.fema.gov/emiweb/is/is242b/student%20manual/sm_03.pdf

Federal Emergency Management Association. (2023a). *SLF 101: Guide for all hazard emergency operations planning.* https://www.fema.gov/pdf/plan/glo.pdf

Federal Emergency Management Association. (2023b). *Unit four: Emergency management in the United States.* https://training.fema.gov/emiweb/downloads/is111_unit%204.pdf

Filmer, L. B., & Ranse, J. (2013). Who is my leader? A case study from a hospital disaster scenario in a less developed country. *Australasian Emergency Nursing Journal, 16*(4), 170–174. doi:10.1016/j.aenj.2013.08.004 PMID:24199902

Flannery, M. E. (2023, March 29). *The mental health crisis on college campuses.* NEA Today. https://www.nea.org/nea-today/all-news-articles/mental-health-crisis-college-campuses

Frykberg, E. R. (2005). Triage: Principles and practice. *Scandinavian Journal of Surgery, 94*(4), 272–278. doi:10.1177/145749690509400405 PMID:16425622

Gao, L., Qunhong, W., Li, D. D., Hao, Y., Cui, Y., Mingli, J., ... Sun, H. (2018). How prepared are hospital's emergency management capacity? Factors influencing efficiency of disaster rescue. *Disaster Medicine and Public Health Preparedness, 12*(2), 176–183. doi:10.1017/dmp.2016.25 PMID:27189920

Gillani, A. H., Li, S., Akbar, J., Omer, S., Fatima, B., Ibrahim, M. I. M., & Fang, Y. (2021). How prepared are the health care professionals for disaster medicine management? An insight from Pakistan. *International Journal of Environmental Research and Public Health, 19*(1), 200. doi:10.3390/ijerph19010200 PMID:35010475

Goddard, S., Sheppard, M., Thompson, K., & Konecny, L. (2018). Disaster preparedness knowledge, beliefs, risk-perceptions, and mitigating factors of disaster preparedness behaviors of undergraduate students at a large Midwest university. *Journal of Public Health Issues and Practices, 1*(2), 115. doi:10.33790/jphip1100115

Haddow, G. D., Bullock, J. A., & Coppola, D. P. (2014). *Emergency management.* Elsevier.

Healthy Minds Network. (2023). *The Healthy Minds Study: Student survey.* https://healthymindsnetwork.org/hms/

Heard, C. L., Pearce, J. M., & Rogers, M. B. (2020). Mapping the public first-aid training landscape: A scoping review. *Disasters*, *44*(1), 205–228. doi:10.1111/disa.12406 PMID:31524986

International Association of Emergency Managers. (n.d.*) IAEM-USA Universities & Colleges Caucus.* https://www.iaem.org/groups/us-caucuses/universities-colleges

International Red Cross and Red Crescent Movement. (n.d.) *Disaster preparedness.* https://www.ifrc.org/our-work/disasters-climate-and-crises/disaster-preparedness

Izumi, T., Sukhwani, V., Surjan, A., & Shaw, R. (2021). Managing and responding to pandemics in higher educational institutions: Initial learning from COVID-19. *International Journal of Disaster Resilience in the Built Environment*, *12*(1), 51–66. doi:10.1108/IJDRBE-06-2020-0054

Jetten, W. D., Seesink, J., & Klimek, M. (2022). Prehospital triage by lay person first responders: A scoping review and proposal for a new prehospital triage tool. *Disaster Medicine and Public Health Preparedness*, *16*(3), 1242–1252. doi:10.1017/dmp.2021.19 PMID:33827731

Johnson, B. (n.d.). *The Great Plague 1665-the Black Death.* https://www.historic-uk.com/HistoryUK/HistoryofEngland/The-Great-Plague/

Kim, D. H. (2016, December). Emergency preparedness and the development of health care coalitions. *Nursing Clinics of North America, 51*(4), 545-554. doi:10.1016/j.cnur.2016.07.013

Kivunga, M. A., Mwavali, A. E., & Otuya, P. A. (2017). Assessment of higher education institutions' man-made disaster preparedness Nairobi, Kenya. *African Multidisciplinary Journal of Research, 2*(1).

Klein, T. A., & Irizarry, L. (2023, July 17). *EMS disaster response.* StatPearls. https://www.ncbi.nlm.nih.gov/books/NBK560710/

Koser, B. W., & Suchenski, M. (2022, September 19*). EMS evacuation triage.* StatPearls. https://www.ncbi.nlm.nih.gov/books/NBK537075/

Labosier, C. F. (2019). Bug out bags and first aid kits: Undergraduate college students' awareness, perceptions, preparedness, and behavior around severe weather. In Emerging Voices in Natural Hazards Research (pp. 1-25). Butterworth Heineman.

Le Roux, T. (2013). An exploration of the role of communication during the in-crisis situation. *Jamba (Potchefstroom, South Africa)*, *5*(2), 1–9. doi:10.4102/jamba.v5i2.67

Lee, C. J., Allard, R. J., Adenjii, A. A., Quintanilla, N., & Kirsch, T. D. (2022). The National Disaster Medical System and military combat readiness: A scoping review. *The Journal of Trauma and Acute Care Surgery*, (2), 136–146. doi:10.1097/TA.0000000000003703 PMID:35605023

London Fire Brigade. (n.d.) *The Great Fire of London.* https://www.london-fire.gov.uk/museum/history-and-stories/the-great-fire-of-london/

Lott, M. K. (2012). *Crisis management plans in higher education: Commonalities, attributes, and perceived effectiveness* [Doctoral Dissertation]. Gallaudet University.

Mahmud, A., Mohammad, Z., & Abdullah, K. A. (2020). Leadership in disaster management: Theory versus reality. *Journal of Clinical and Health Sciences*, *5*(1), 4–11. doi:10.24191/jchs.v5i1.9818

Makuana, N. (2019, October). Disaster and its impact on mental health: A narrative review. *Journal of Family Medicine and Primary Care*, *8*(10), 3090–3095. doi:10.4103/jfmpc.jfmpc_893_19 PMID:31742125

Margolis Healy Solutions for Safe Campuses. (2015). *Margolis Healy Campus Safety Survey.* https://www.margolishealy.com/files/resources/2015MargolisHealy_CampusSafetySurvey_1.pdf

Matunhay, L. M. (2022). Disaster preparedness and sensitivity level among higher education institution students. *International Journal of Disaster Management*, *5*(2), 79–92. doi:10.24815/ijdm.v5i2.27150

Medievalists. (n.d.). *10 natural disasters that struck the medieval world.* Medievalists.net. https://www.medievalists.net/2015/12/10-natural-disasters-that-struck-the-medieval-world/

Mishra, S., & Suar, D. (2012). Effects of anxiety, disaster education, and resources on disaster preparedness behavior. *Journal of Applied Social Psychology*, *42*(5), 1069–1087. doi:10.1111/j.1559-1816.2011.00853.x

Mitroff, I. I., Diamond, M. A., & Alpaslan, M. C. (2006). How prepared are America's colleges and universities for major crises? *Change*, *38*(1), 61–67. doi:10.3200/CHNG.38.1.61-67

National Center for Campus Public Safety. (2016). *National Higher Education Emergency Management Program Needs Assessment Report.* http://www.nccpsafety.org/news/articles/national-higher-education-emergency-management-needs-assessment

National Geographic. (n.d.) *The Chicago fire of 1871 and the great rebuilding.* https://www.london-fire.gov.uk/museum/history-and-stories/the-great-fire-of-london/

New York City Government. (n.d.). *NYC Emergency Management Tabletop Exercise Toolkit.* https://www.nyc.gov/assets/em/downloads/pdf/business/spe_facilitator_guide.pdf

Noboa-Ramos, C., Almodóvar-Díaz, Y., Fernández-Repollet, E., & Joshipura, K. (2023). Healthcare and social organizations' disaster preparedness, response, and recovery experience: Lessons learned from hurricanes Irma and Maria. *Disaster Medicine and Public Health Preparedness*, *17*, e306. doi:10.1017/dmp.2022.272 PMID:36789766

Office of the Federal Register, National Archives and Records Administration. (2023, September 23). *Title 24, Subpart B, Chapter IV,* § 668.49(d). https://www.ecfr.gov/current/title-34/subtitle-B/chapter-VI/part-668/subpart-D/section-668.49

Orthopedic Trauma Association. (n.d.). *Disaster response and security.* https://ota.org/careers-practice/disaster-management-and-emergency-preparedness/disaster-response-safety-and

Ozkazanc, S., & Yuksel, U. D. (2015). Evaluation of disaster awareness and sensitivity level of higher education students. *Procedia: Social and Behavioral Sciences*, *197*, 745–753. doi:10.1016/j.sbspro.2015.07.168

Panganayi, M. (2020). COVID-19 Lockdown and higher education. Time to look at disaster preparedness as a governance issue? *Inkanyiso*, *12*(1), 24–35. http://www.inkanyiso.uzulu.ac.za/journals/vol12i1/03%20n%2012%20%20Lockdown%20%20Panganayi%20Fin%20pdf.pdf

Patel, R. K., Kermanshachi, S., & Jahan Nipa, T. (2020, July). Establishment of a framework to measure disaster preparedness: Development of strategies to enhance disaster preparedness activities. In *Creative Construction e-Conference 2020* (pp. 22–30). Budapest University of Technology and Economics. doi:10.3311/CCC2020-051

Pawłowski, W., Lasota, D., Goniewicz, K., & Goniewicz, M. (2018). Effects of first aid training in emergency preparedness and response. *Medical Studies/Studia Medyczne, 34*(3), 259-263. doi:10.5114/ms.2018.78691

Povoledo, E. (2022, June 23). When, exactly, did Mount Vesuvius erupt. *The New York Times.* https://www.nytimes.com/2022/06/23/world/europe/pompeii-mount-vesuvius-erupt.html

Readiness and Emergency Management for Schools (REMS). (2023). *Tool box: A virtual library of practitioner created resources for planning team members and safety leaders.* https://rems.ed.gov/toolbox.aspx

Readiness and Emergency Management for Schools (REMS). (n.d.). *An effective way to practice and validate emergency operation plans.* https://rems.ed.gov/docs/EmergencyExercisesFactSheet_508C.pdf

Riad, N. A., Rabea, F. M., & Badawy, A. I. (2013). Capacity building of female governmental secondary school teachers in Abha City for crises management and first aid. *Journal of Education and Practice, 4*(23), 33-41.

Robertson-Steel, I. (2006). Evolution of triage systems. *Emergency Medicine Journal, 23*(2), 154–155. doi:10.1136/emj.2005.030270 PMID:16439754

Skryabina, E., Reedy, G., Amlot, R., Jaye, P., & Riley, P. (2017). What is the value of health emergency preparedness exercises? A scoping review study. *International Journal of Disaster Risk Reduction, 21*, 274–283. doi:10.1016/j.ijdrr.2016.12.010

Skryabina, E. A., Betts, N., Reedy, G., Riley, P., & Amlôt, R. (2020). Evaluation of emergency preparedness exercises: The design of a questionnaire to measure staff perceptions. *International Journal of Emergency Management, 16*(4), 317–346. doi:10.1504/IJEM.2020.117209

Tan, Y., Liao, X., Su, H., Li, C., Xiang, J., & Dong, Z. (2017). Disaster preparedness among university students in Guangzhou, China: Assessment of status and demand for disaster education. *Disaster Medicine and Public Health Preparedness, 11*(3), 310–317. doi:10.1017/dmp.2016.124 PMID:27666525

Texas Department of Emergency Management. (2023). *Technological hazards.* https://www.tdem.texas.gov/preparedness/technological-hazards#:~:text=Examples%20of%20technological%20hazards%20may,man%2Dmade%20incident%20or%20event

Twigg, J., & Mosel, I. (2017). Emergent groups and spontaneous volunteers in urban disaster response. *Environment and Urbanization, 29*(2), 443–458. doi:10.1177/0956247817721413

Understanding Islam. (n.d.). *Natural disasters or divine punishment?* https://free-islamic-course.org/articles-on-topical-issues/natural-disasters-divine-punishment.html

University of Oregon Institute for Resilient Organizations, Communities and Environments. (2022). *2022 DRU National Higher Education Program survey results.* https://resilient.uoregon.edu/dru/2022survey#about-survey

U.S. Army Corps of Engineers. (n.d.). *A brief history.* https://www.usace.army.mil/About/History/Brief-History-of-the-Corps/Responding-to-Natural-Disasters/#:~:text=The%20Corps'%20first%20formal%20disaster,San%20Francisco%20earthquake%20of%201906

U.S. Department of Education. (2010). *Action guide for emergency management at institutions of higher education.* Office of Safe and Drug Free Schools. https://files.eric.ed.gov/fulltext/ED515949.pdf

van Stralen, D. (2009, September 30). The origins of EMS in military medicine. *The Journal of Emergency Medical Services.* https://www.jems.com/patient-care/origins-ems-military-medicine-2/

Waxman, D. A., Chan, E. W., Pillemer, F., Smith, T. W., Abir, M., & Nelson, C. (2017). Assessing and improving hospital mass-casualty preparedness: A no-notice exercise. *Prehospital and Disaster Medicine*, *32*(6), 662–666. doi:10.1017/S1049023X17006793 PMID:28780916

World Health Organization. (2019, December 19). *Emergencies: WHO's role.* https://www.who.int/news-room/questions-and-answers/item/who-s-role-in-emergencies

World Health Organization. (2021). *Strategic toolkit for assessing risks.* https://www.who.int/publications/i/item/9789240036086

Yamaiel, A. O. (2006, August). Disaster resistant universities: In search of strategies for resilient higher education institutions. *International Journal of Mass Emergencies and Disasters*, *24*(2), 179–180. doi:10.1177/028072700602400201

Chapter 3
Quality Assurance in Higher Education:
A Fertilizer for Academic Enhancement or a Luxury of an Ideal World

Stefan Handke
https://orcid.org/0009-0002-3051-0020
HTW Dresden University of Applied Sciences, Germany

ABSTRACT

This chapter examines whether post-crisis regions can benefit from quality assurance in higher education. To this end, it defines what notions of quality exist and how quality can be improved. Special attention is given to the formalized process of accreditation of study programs. From the perspective of neo-institutionalism, it is argued that quality assurance as a rule-based process requires functioning state structures. For post-crisis regions, it is therefore of crucial importance whether or not crises and traumas have an impact on the existence of the state. It is postulated that without an effective administrative and legal system, accreditation or other forms of legally binding quality assurance are not useful. Rather, low-threshold procedures based on the principle of fitness for/of purpose would be recommended in such cases.

INTRODUCTION

In an ideal world, higher education is a driver of progress and development for humanity. Through the combination of teaching, research and practical transfer, individuals build up their competencies, which on the one hand results in individual satisfaction, but on the other hand also contributes to cultural, technological, political or social improvements. However, investing in education and training only pays off in the long run, and the intended positive impact is achieved with a time lag. Given limited resources, spending on education is competing with other, also important, financial expenditures. This is particularly the case in developing and emerging countries, but above all in all kinds of regions in

DOI: 10.4018/979-8-3693-1926-0.ch003

crisis. Countries and regions affected by natural disasters or man-made crises are usually confronted with the decision whether to focus political and social attention and resources on vital areas such as the provision of food, public security or infrastructure. Education - especially of the elementary level - is one of the most fundamental requirements for a modern society. Higher education, by contrast, has a subordinated status. Nevertheless, at a certain level of development, such education is also required to establish resilience to crises and to promote know-how in a country. Once a country has reached this stage, it is not only the mere provision and availability of higher education that is relevant, but also ensuring that it is of adequate quality.

This chapter will discuss the importance of academic higher education for states and societies in crisis from a societal perspective. In this context, a special focus is given to the function of quality assurance (QA) in higher education institutions. A critical reflection is made on whether QA should be considered as a key component of academic work in global higher education, or whether it is only a "luxury" in a highly developed educational landscape with stable structures in society.

The reasoning is presented in three steps. First, some basic perspectives on quality and quality assurance in higher education are introduced. Quality goals are almost naturally linked to the goal of being among the best compared to other universities. Thus, higher education institutions (HEIs) want to prove their *excellence*. While this aspect of competition is perfectly reasonable in the long run for strategic developments, it appears to be problematic for higher education institutions in post-crisis regions. Here, it must be about establishing and consolidating structures and processes at universities that guarantee stability and reliability. Against this background, the notion of quality should be defined as *"fitness for/of purpose"*. The emphasis is less on comparing the university's own performance with other universities in top ranks, but rather on achieving and adapting individual objectives. With recourse to practical examples of German accreditation agencies which carry out quality assurance activities in emerging and developing countries, perspectives on market effects and the development of quality cultures will be examined. This is done by contrasting the emphasis of external quality assurance in the German home market and the goals pursued by the agencies in non-domestic regions. Here, it will become clear that there is a balancing act with regard to the application and interpretation of quality standards. In a second step, QA with the instrument of accreditation is discussed as a means of institutionalization. The analytical framework of neo-institutionalism is used to illustrate which conditions must be in place to gain the benefits that accreditation can bring. In the third part, the findings are applied to quality assurance and accreditation in crisis-ridden higher education systems. As soon as there is a common understanding of quality - but also of its limitations - higher education will achieve what makes it so special: long-term, global cooperation. Quality assurance opens the doors for cooperation and promotes mutual trust. The basis for the creative and visionary work of HEIs is, first and foremost, academic freedom. This includes scientific freedom to follow ideas and interests according to one's own inclination. In addition, it means the political freedom of unhindered publication, discussion and dissemination of research results. In this sense, academic work is global and not limited by national or regional borders. From this point of view, higher education and context-appropriate QA are characterized as essential components for sustainably (re-)building crisis-ridden regions in a modern world. These insights then lead to a conclusion in which recommendations for politicians and university managers are made.

Quality and Quality Assurance in Higher Education

When people talk about the quality of academic education, very different ideas come to mind. On the one hand, high-quality universities and degree programs can be those that achieve measurable objectives in the best possible way. Such dimensions could be student enrolment rates or graduates completing their studies within a standard period of study, the level of employment of graduates directly after graduation or the number of international collaborations a university has. In the same way, the quality of a university could be measured by the amount of raised research funds or the number of Nobel Prize winners. An alternative to such quantifiable parameters, quality could be determined by individual opinions. High or low quality is the subject of a person's perception both inside and outside a university. It is the image, reputation and experience that bring about a positive or negative judgment. Based on this very elementary differentiation, concepts of quality and quality assurance can be examined in more detail regarding the higher education sector.

Notions and Concepts of Quality

Before discussing the benefits of quality assurance and its instruments in higher education, the concept of quality needs to be explained in more detail. There are already numerous publications on this subject, which mainly present a similar perspective and distinguish between the various interpretations of the concept of quality (Abbas, 2020; Duarte & Vardasca, 2023; Handke, 2023; Harvey & Green, 1993). In this respect, the definition of quality differs significantly and is either more focused on a process level or on outcome and impact. The categories of quality can be defined as value for money, as perfection and consistency, a transformative element or as excellence. The last of these concepts marks a good starting point for the issue of QA, as it is widely used in many parts of the world.

Universities regularly participate in comparative performance measurements to show what makes them better than others, or even part of the elite. Reputation and prestige play a crucial role, allowing universities to attract top researchers, students and financial supporters. In this respect, the reason for demonstrating high quality is an element of competition (Owen, 2013; Portnoi et al., 2010). The academic world lives from cooperation, but at the same time also from competition for the best ideas and concepts. Such competition concerns the research sector just as much as the race for the best students. This is why higher education institutions often participate in rankings, ratings or other benchmarking methods (Lange, 2010; Sursock, 2021). In doing so, they compete in the dimension of quality of studies and teaching. What is special about such competition is that the term "quality" is very often understood as "*excellence*". This idea of quality is at one end of a continuum of possible definitions of quality. Especially in North America and Asia, this interpretation is popular (Eaton, 2021). It also makes sense if higher education is tuition-based and students, as fee-paying customers, are to be offered the best possible product.

At the other end of the continuum is a concept of quality that is more oriented toward the common good and less focused on the value proposition of the individual. This is the notion of fitness for purpose, which is more commonly found in European countries where higher education is, for the most part, provided free of charge to students (Enders & Westerheijden, 2014). Quality as *fitness for/of purpose* is a concept that evaluates whether an organization's quality-related intentions are adequate for achieving its defined objectives and mission. It does not assess the fitness of the processes themselves in relation to external expectations.

It is not only in the context of higher education that the two ends of the spectrum of possible concepts of quality are often found in societies and countries. Rather, they generally correspond to the culture of a region, as manifested in institutionalized values, norms, and rules. In order to have a positive and encouraging impact on higher education systems, the distinction between different types of quality must also be taken into account in (external) quality assurance (Chmielecka, 2014).

Quality Assurance Models in Higher Education

As a matter of fact, quality can exist without any formal certification attesting its existence. And quality assurance as a process of maintaining a certain level of performance and continuous improvement can also exist without external assessment. Organizations are certainly capable of defining their own objectives, backing them up with strategies, implementing them and, if necessary, taking measures if goals are not achieved. Quality assurance may take the form of governmental supervision or self-regulation by professional organizations as soon as an external viewpoint is involved. Examples of such professional bodies are chambers of architects or engineering associations. Nevertheless, over the course of decades, processes have evolved that rely on peer reviews and certification procedures in order to meet quality standards in the higher education sector. Relying on this peer review process, universities and higher education systems are mainly influenced by the North American or European quality and education models (Handke, 2023). The European model will therefore be outlined below in order to draw conclusions for quality assurance in crisis regions.

The European Bologna Agreement signed in 1999 initiated the structures and processes that students experience as a natural part of their university life (The Bologna declaration of June 1999: Joint declaration of the European ministers of education., 1999). On the one hand, there is the importance of academic standards and quality criteria for universities that are obliged to comply with them in order to be in a legal position for offering degree programs. Undergraduate and graduate programs, modules, and credits have become standard features of higher education. In addition, there is now a focus on developing employability skills and promoting active citizenship. On the other hand, these criteria represent a corridor of goals for higher education institutions. Transparent assessment procedures according to widely accepted quality criteria have two major functions. First, they serve as an incentive for HEIs and their staff to improve the existing higher education system in accordance with internationally recognized standards. Second, external evaluations based on such standards provide the basis for trust that foreign partners can have in a higher education system and in individual HEIs. Only with such a basic level of trust the academic cooperation is made possible. Like that, it has become a reality to create the "European Higher Education Area" (EHEA) with a three-tier higher education system, mutual recognition of qualifications and learning periods and the introduction of a quality assurance system. In terms of QA, higher education is strongly characterized by the internal responsibility of higher education institutions for the quality of their education. The goals of quality assurance, however, go far beyond the establishment of standards, which might be a crucial aspect for post-crisis regions to consider.

The fundamental principles of the Bologna Declaration comprise autonomy and diversity. Then there are the goals of convergence and harmonization, which shall be achieved by setting standards and guidelines for teaching and academic studies at universities (Reinalda & Kulesza, 2006). These objectives are key components of the entire education policy. Thus, QA is not merely a micro-policy designed to achieve formal compliance with technical standards. Rather, the aim of European accreditation is to create a harmonized higher education area in which HEIs can bring their unique qualities about. At the same time, they should share core values and standards of academic education in order to enable the comparability of degrees and the mobility of students, teaching staff and graduates. Looking at international recognition and cooperation, the following section outlines some practical experience of cross-border QA based on the example of German QA agencies.

Cross-Border Quality Assurance According to the ESG

In particular, QA agencies from Germany have made great efforts in the last few decades to provide international accreditation and cross-border quality assurance procedures. The agencies are expanding their scope of action due to the competitive situation in the German domestic market. Moreover, in line with the Bologna Process, these organizations and the higher education system as a whole have an international orientation (Handke, 2023). While the competitive dimension cannot be explored in more detail at this point, reference should be made to the rich experience of such agencies carrying out their accreditation procedures abroad. The region of interest here is not so much EHEA, but rather the countries around it. German agencies have been actively working in countries such as Vietnam, Armenia, Jordan, Egypt, Lebanon, Syria and Iraq (EQAR, 2023).

The goal of German agencies to conduct quality assurance procedures in (post-) crisis regions is a form of empowerment. This should strengthen educational structures, civil and knowledge societies, and ultimately lead to stable state and social structures (Bromley, 2018; Handke, 2023). In assessments carried out there, the *Standards and Guidelines for Quality Assurance in the European Higher Education Area* (ESG) are usually applied as reference. The task of German and national experts from the host region is then to check at universities whether these standards are being met. This is relatively easy when it comes to formal criteria such as the modularization of a curriculum and the competence and skills-oriented nature of examinations. It is more difficult, however, when normative requirements must be checked. These include the academic freedom of universities and their teaching staff, non-discriminatory access to HEIs or the involvement of students in academic self-administration. Such criteria are rooted in the set of values of a free and democratic society, which may not be shared in every country, though.

Experience with universities abroad has shown that those who contract with German agencies aim to gain reputation through international accreditation. Quality assurance procedures aim to achieve an externally backed attribution of excellence. This is remarkable for German agencies because accreditation in the domestic market does not lead to reputation gains for German universities. Instead, quality assurance and accreditation are simply necessary obligations.

When culturally diverse QA agencies and universities meet, their goals and expectations may differ (Rosa et al., 2016). Transnational accreditation therefore has to make compromises. Although formal standards can be assessed and evaluated, criteria of a liberal quality culture can be incorporated only to a limited extent. This means that they are not elements of a substantive assessment, but rather objectives that foreign universities and social systems are encouraged to achieve. This raises the question of the validity and value of having the certificate of a German agency awarded to a university in countries such as those mentioned above.

By no means do they have the same level of value as in EHEA countries. International accreditation can only attest that formal minimum requirements for structure and organization of a university degree programme are met for the current moment. Accreditation provides only very limited information about quality policy and open and democratic principles. A long-term guarantee of reliability, which can lead to a relationship of trust, is not possible. One of the main reasons for this is that accreditation is not embedded in a system of government institutions and a legal system that would guarantee trust and accountability in all spheres of the society - including the higher education system. As a result, international accreditation is unable to provide assurance of credibility and thus fails to create trust. All it can do is pave the way for this to happen. Quality assurance can help build institutions that promote stability and reliability in post-crisis regions. The following section discusses the mechanism of such institution-building.

Institutional Perspective on Quality Assurance and Accreditation

This section explores the link between a well-functioning state with an impartial legal system and accreditation in higher education, which is seen as a crucial aspect for any (post) crisis region. It is shown that accreditation as an instrument of QA can only have a positive impact if there are government structures and a regulating regime in place which can effectively enforce sanctions. This is based on the assumption that higher education accreditation is only one element of so-called institutions in a society, which are closely tied to other institutions. Since the late 1970s, there has been an increasing discussion on the importance of institutional factors in explaining political and social processes. Both political behavior and collective decisions are shaped by institutional rules and norms of decision-making. In this notion, institutions are conceived as norms, routines as well as formal and informal rules (Mahoney & Thelen, 2010; van der Heijden & Kuhlmann, 2017). The institutionalist approach appears in research on quality assurance in higher education (Steinhardt et al., 2017) and in research on state failure and collapse (Lambach et al., 2015).

Neo-Institutional Framework

The theoretical framework of neo-institutionalism provides two important insights when studying accreditation and QA. One is that the approach addresses the rules and norms that exist in a society and a political system. The second is that the approach implies that changes in society or organizations must always take into account changes of the existing institutions. Since quality management is at its core aimed at continuous improvement, this is always connected with institutional change. Neo-institutionalism refers to a set of social science approaches that describe and explain transformation processes in organizations and societies. There are three dominant approaches: the 'rational choice' type, the 'sociological' or 'organizational' type, and the 'historical' type. The *rational choice* version attributes the creation and change of institutions to the benefit-maximizing, intentional behavior of actors. Rules are intentionally created to achieve specific goals. Institutions have a stability-building effect and create a horizon of expectations for actors. In quality assurance, for example, a legal framework may require program accreditation in order to achieve uniform program structures in a country. The *sociological perspective* explains the development of regulatory systems within organizations. Formal and informal norms arise through the interaction of actors, and institutions have an identity-forming effect. In the context of quality assurance at universities, the quality culture can be understood as an institution that leads to a formalized mission statement. The *historical approach* describes how institutions influence the actions of actors and how their interaction can lead to changes in institutions. Institutions are both identity-forming and influence action. The explanatory power of the historical approach is primarily focused on the retrospective explanation of processes of change. In university quality assurance, compliance rules can be introduced as a response to a corruption scandal. Therefore, the analytical approach considers a crisis situation as a driving force for the development of institutions. Scandals or crisis situations have the potential to undermine security, stability, and trust. The establishment of clear rules and standards can be beneficial in such cases. In the context of higher education, accreditation of programs and universities serves this goal. The following section presents accreditation as a distinct mode of quality assurance with a high level of institutionalization.

Development of Accreditation in Higher Education

Accreditation is not a recent development in the higher education sector, but rather a trust-building instrument that has been used for a long time. The Latin meaning of the verb 'accredere', which means 'to give credence to someone', is the basis for the concept of accreditation (Blauch, 1959; Scrivens, 1995). Initially, accreditation was a voluntary scheme for assessing physical study environments through peer reviews. The purpose of submitting work for external review by independent professionals in the same field is to receive constructive feedback and analysis for improvement. This feedback may include recognition of good practices and grading of achievements. However, there is no threat of consequences for insufficient results due to the voluntary nature of the review. Accreditation, which was originally a self-organized method of quality assurance and improvement in the health sector, has evolved into a tool for promoting not only disciplinary standards but also the general public interest (Scrivens, 1995). As accreditation has moved towards becoming a licensing and authorizing process, it has become less voluntary and been replaced by government regulation. It is important to note that the role of accreditation changes from a professional model of education and support to one of control when government regulation begins to use accreditation to supplement its activities.

Quality assurance by means of accreditation has a long tradition, particularly for business schools operating in the higher education sector. They are awarded certificates by specialized agencies, and these are a key element of the marketing of such schools. In this way, quality is considered primarily as excellence, which is mainly manifested in the (economic) performance of a HEI. In contrast, the compliance-oriented accreditation system of the ESG is based on the idea that universities are responsible for the quality of research, teaching and learning. This responsibility also encompasses the definition of "quality" itself within of the ESG framework.

The primary form of external accreditation is program accreditation, which assesses individual or multiple degree programs (Schwarz & Westerheijden, 2004). In this process, a university contracts a QA agency to conduct the accreditation, which culminates in the award of an accreditation seal upon successful completion. The accreditation process is carried out in several steps. The first step in the evaluation process is for the higher education institution to submit a self-evaluation report. This report should present the degree program's qualifications and curricular structure, as well as information about the institution itself. The documentation provided by the institution should demonstrate how both internal quality goals and external rules and standards are being followed within the program. The evaluation of academic and content-related standards is conducted by a group of experts who are appointed for this specific purpose. Typically, this group comprises academic and professional peers, as well as students from the area in which a degree program is focused. The experts are trained to review and assess the rules set by the agency. This is followed by an on-site visit that includes a series of meetings. The group conducts interviews with university staff, students and other stakeholders of the HEI. Following this assessment process, the expert group prepares a report that includes an evaluation. The main priority is to determine whether a HEI meets all the standards for a particular program, including issues of professional appropriateness, workload, and modularization. Additionally, the report should identify strengths and provide recommendations for improving the degree program. If compliance with the rules is established, the expert group will propose granting accreditation. If required, institutions may have additional requirements and a shorter time frame. The process concludes with the agency or an official body awarding a certificate of accreditation. Successful accreditation enables an HEI to offer degree programs, enroll students, and receive state funding. In addition, a diploma from an accredited study program is generally recognized for employment in the public sector or when a student transfers to another university to continue their studies.

Accreditation as the Institutionalization of QA

Accreditation is a formalized instrument of QA. The concept characterizes a procedure that uses transparent and verifiable assessment methods to meet one purpose: the creation of trust. This directly brings up the question on what trust in universities or accrediting organizations is based? Trust is always necessary when there is no complete information available, and decisions have to be made under uncertainty. As in human relationships, people put themselves in a situation of potential vulnerability when they make decisions on the basis of information provided by others.

For instance, individuals often depend on suggestions from acquaintances or family members. Similarly, students frequently rely on evaluations conducted by agencies that assess the quality of a university's degree programs. Accreditation serves as a formalized means of institutionalizing trust and reducing uncertainty. Once a university has obtained valid accreditation, individuals interested in pursuing higher education and other stakeholders in society can trust that the quality of education and its organization meet specific, pre-established requirements. In the event that trust is lost, it can lead to the breakdown of everyday relationships. Professional relationships can result in lawsuits and penalties due to broken promises. Disappointing or betraying trust, whether intentional or not, is subject to sanctions. Therefore, the trusted entity must make an effort to justify and maintain the trust in the relationship. Accreditation serves as a formal instrument with an intermediate function. Once an HEI is validly accredited, students and other stakeholders can trust that the program is being delivered according to specified standards, thereby reducing uncertainty. Therefore, trust is not placed in the university itself, but rather in the accreditation agency that awards the certificate or label based on an assessment.

Accrediting degree programs requires a stable legal framework in two respects. Firstly, only universities with a state license are assessed. Secondly, the validity of an institution's recognition is linked to its successful accreditation by a registered quality assurance agency. The agencies themselves must also undergo a process of recognition to issue valid accreditation for higher education institutions. The accreditation process is intentionally designed to establish trust in the higher education system while minimizing uncertainty and transaction costs for stakeholders. Accreditation represents the institutionalization in higher education according to the rational choice model.

The assessment procedure in EHEA countries is based on legal requirements set by national governments (Stensaker, 2011). Typically, these requirements mandate that higher education institutions undergo an external assessment and pass it successfully. Failure to obtain accreditation results in the inability of higher education institutions to operate. Therefore, the accreditation procedure is an administrative process that is bound to rights and obligations. Clearly, formalized, legally binding quality assurance requires a functioning governmental and administrative system. The admission of a higher education institution and the status of accredited study programs should not be based on political interests or unrelated considerations. If this happens, for example, due to corruption, nepotism, or political arbitrariness, the accreditation procedure loses all value. The following section examines higher education systems in (post)crisis regions against the background of the above-mentioned characteristics of quality assurance, accreditation, confidence building and dependence on (state) institutions.

Quality Assurance and Accreditation in Higher Education Systems impacted by Crises

Universities in crisis-ridden regions of the world often look back on a long tradition and better times. They are trying to revive their history and move forward into a better future. However, this often happens under difficult conditions, i.e. with limited resources, destroyed infrastructure or with political instability. The question is how QA works in countries affected by crises.

The Nature of Crises

Crisis situations can arise as challenges for the political system, the economy and the population of a country for various reasons. On the one hand, these are problems caused by environmental factors, such as earthquakes, floods, droughts or climate disasters. On the other hand, there are man-made crisis phenomena that arise as a result of wars, poor economic decisions or political turmoil (Shaluf et al., 2003). In the worst case, the entire order breaks down and the state collapses. According to the institutional perspective, the "ideal-type of the state as an institution characterised by monopolies on rule making, violence and taxation within a defined territory and among the population living therein. This institution finds its organisational expression in an administrative apparatus, political organs and bodies for collective decision making. It is represented by symbols and social practices that remind citizens of the existence of the political order" (Lambach et al., 2015, p. 1305). Thus, a failed or collapsed state is one, where most of the above-mentioned features are missing.

We are currently in a time of many crises, also referred as "poly-crisis" (Henig & Knight, 2023). After surviving the Covid-19 pandemic, the world is continuing to combat climate change and prevent economic upheaval. In many regions of the world, nationalism, religious and ethnic tensions, and the struggle for scarce resources are leading to conflict, resulting in forced migration, displacement, death, and disease. Whether in crisis regions such as Ukraine and the Middle East or in often-overlooked areas of conflict and crisis like Syria, Yemen, Myanmar, or Sudan. The problems and challenges in each of these cases are unique. However, what all regions have in common is that they are far from the normality of a peaceful everyday life where people's basic needs are met. Even in these special situations, routines and regular procedures develop in the region. However, they usually come in the form of informal institutions that are not part of a legal or governmental system. This comes later, as shown in the case of the former Yugoslavian states. Due to the diversity of crises, there is currently no research approach that provides a comprehensive view of issues such as quality assurance in higher education. For this reason, related analytical perspectives are used in this study.

Comparing similar situations is one way of getting started. In developing countries, quality assurance provides opportunities for the development of higher education. Universities in countries with low levels of development often have limited budgets, little experience, and know-how in establishing university education programs, as well as poor opportunities to orient university programs towards research (Lim, 1999). Public and private investment in education, especially higher education, should be prioritized.

However, education competes with other areas of political, economic and social life. These may include the development of physical infrastructure, food production, or national defense. In developing countries, material shortages are not the only challenge. Political influence can be just as detrimental as a weak academic foundation. Any kind of "[p]olitical inference makes the search for truth, especially in the social sciences, unattractive, and promotion by political or social connection discourages the pursuit of excellence in teaching and research" (Lim, 1999, p. 384).

When considering developing countries, it has been concluded that introducing a university quality management system may be futile due to conflicts between existing institutions and quality assurance requirements. If cultural, religious, or political ideas dictate a specific form of higher education, quality assurance may be imposed according to international standards but may not be effective. An example of cultural restriction is the practice in Taliban-ruled Afghanistan of granting access to universities only to men. This contradicts the principles of a liberal democracy, where access to universities should be based solely on merit and not on discriminatory factors. In this case, the university's accreditation would fail. However, it would be wrong to dismiss the issue of quality assurance for universities in crisis and post-crisis regions. While it may not be feasible to directly apply e.g. the European model to these regions, adaptation is possible and can yield positive outcomes.

Appropriate Quality Assurance in Post-Crisis Regions

The competitiveness of the higher education system is a decisive factor for post-crisis regions when embarking on the path of quality assurance. As a tool used in conjunction with higher education institutions' competition, quality assurance can be viewed as a commodity in the market. Although quality assurance may sound like a technical term at first, it is a tool of academic education and should be applied accordingly. Few would disagree that higher education should strive for quality and adhere to appropriate standards to achieve this goal. However, it is important to note that quality assurance is not a neutral concept. The goals, methods, and consequences of quality assurance are all normatively loaded.

Further, there is a lack of reliable and comparable data on the demonstrable impact of QA and accreditation on the quality of programs. This is due to the different definitions of quality and the lack of consistent measurements of quality standards. When normative parameters are included under the heading of quality, the concept becomes beyond statistical-quantitative analysis. Given a current crisis, it is necessary to determine the appropriate monetary investment required to generate an immeasurable and uncertain benefit. In other words: what is it worth to a university or a government to assess and accredit degree programs and which benefits can be expected from successful certification?

After a natural disaster, academic life should resume in the affected country or region, including quality assurance at the pre-crisis level. However, in the case of political, social, and man-made crises and traumas, institutions cannot simply continue with old traditions and procedures. Instead, they must be carefully rebuilt. The first step is to establish the basic rules of coexistence in society and define values that will lead to a constitution. After primary law, a legal system develops to regulate all areas of everyday social life. Only then are decisions made on development and quality management in higher education. A society and its political decision-makers must actively strive for the development of a high-quality higher education system. However, commitment to this goal does not guarantee immediate implementation or optimal results. But a commitment to this goal is crucial. Only after careful deliberation is it appropriate to introduce or use a comprehensive quality management system based on international models.

Where the social and political will exists to improve the higher education system, quality assurance will be able to develop its full potential. QA serves as a monitoring system for intermediate steps, motivates the achievement of goals step by step, and signals external stakeholders, such as international cooperation partners. However, it is important to adopt a sensible and realistic approach when defining objectives and QA instruments. Universities in developing and post-crisis countries should avoid setting unrealistic goals that may lead to frustration. It is not advisable to set targets that require them to achieve top positions in international rankings within a short period of time. This is because universities cannot skip stages of development and must slowly acquire competence, expertise, and reputation.

Given this context, the concept of quality as *excellence* may not be the most appropriate approach. Instead, the principle of *fitness for/of purpose* should be considered. For instance, in regions where research institutions are lacking and fundamental structures in the economy and society need to be established, a university system may not require doctoral programs. Instead, training programs should be developed to prepare professionals for the existing labor market. QA should consider the university's operating context. A simple quality management system can be implemented where each university reviews its own objectives internally, relieving external pressure and allowing for learning from mistakes without immediate sanctions. This allows for suggestions and initial criticism to improve the existing situation. In a second step, universities can seek an external perspective through advisory boards or horizontal feedback systems. It is important to consider both options for obtaining external impulses. Herbert A. Simon's concept of 'satisficing' can be a useful guideline for establishing quality assurance in post-crisis regions, as it aligns with the idea of fitness for/of purpose (1982). Satisficing is a term that combines the words 'sufficing' and 'satisfying'. Satisficing is a way of making decisions that involves trying different solutions until an acceptable one emerges. It refers to making choices which satisfy, achieving an acceptable result without wasting resources trying to find something perfect in the notion of excellence.

Once it is agreed that competition between universities is positive for increasing quality, the next steps can be taken immediately. In order to achieve their goals, HEIs must find a strategic orientation that enables them to compete for the best scientists, students, and funding. External QA can also contribute to this effort. Further, accreditation is a means of verifying legal compliance and demonstrating the reliability of a HEI. Thus, universities might go through external assessments, which objectively and critically review their compliance with rules and, if necessary, sanction non-compliance. From this point forward, collegial networks typically dissolve and competition between universities ensues, as is inherent in the pursuit of academic excellence. As a consequence, a university without certified status runs the risk of being labeled a 'degree mill' and damaging its reputation. This is the end of the cycle for higher education systems in crisis regions. In order to have a good reputation in the long term, there is a need for the gradual establishment of quality assurance measures that are appropriate to the situation.

CONCLUSION

In times of crisis, it is important to consider the future beyond the crisis. However, addressing issues of quality assurance in higher education may not be a priority in countries experiencing poverty, misery, and destruction. It is important to note that certification of excellence and perfection may only be relevant once other problems have been addressed. Quality assurance and accreditation based on the Western model may be considered a luxury that a traumatized country cannot afford. The imposition of standards of excellence in the form of corresponding rules and regulations may not be appropriate at this time. While *excellence* can be a vision, the path to achieving it must proceed in stages. Therefore, it is necessary to concentrate on stability and small successes first. Institutions should develop gradually and emerge as identity-forming norms and values of a society, according to the sociological approach. This should happen after overcoming difficult times.

Since QA is an integral part of modern higher education, post-crisis countries should for sure strive for excellence and high performance in their university systems. However, it is crucial to achieve quality in a sustainable and reasonable manner. To road to high quality is a long path and not a racing track. Therefore, policymakers and universities are recommended to start with small steps into a direction of trust creation by mutual assessments and peer learning. As soon as stable institutions are in place, the external perspective should be added with a focus on quality as *fitness of/for purpose*. Just at the last stage, quality assurance should serve as an instrument of expressing excellence on national and international levels.

The recommendation for action is based on the finding that quality assurance provides orientation and hope for a better future. This future may foster trust and enable the realization of the core ideas of agreements like the Bologna Declaration that include academic freedom, mutual recognition, and mobility in a world of peace.

REFERENCES

Abbas, J. (2020). Service quality in higher education institutions: Qualitative evidence from the students' perspectives using Maslow hierarchy of needs. *International Journal of Quality and Service Sciences, 12*(3), 371–384. doi:10.1108/IJQSS-02-2020-0016

Blauch, L. E. (1959). *Accreditation in higher education: Organized and* (L. E. Blauch, Ed.). U.S. Dep. of Health Education and Welfare.

Bromley, P., Schofer, E., & Longhofer, W. (2018). Organizing for Education: A Cross-National, Longitudinal Study of Civil Society Organizations and Education Outcomes. *Voluntas, 29*(3), 1–15. doi:10.1007/s11266-018-9979-9

Chmielecka, E. (2014). Academic Values and The Procedures of Quality Assurance. In H. Eggins (Ed.), *Drivers and Barriers to Achieving Quality in Higher Education* (pp. 43–54). SensePublishers. doi:10.1007/978-94-6209-494-9_4

Duarte, N., & Vardasca, R. (2023). Literature Review of Accreditation Systems in Higher Education. *Education Sciences, 13*(6), 582. doi:10.3390/educsci13060582

Eaton, J. S. (2021). The Role of Quality Assurance and the Values of Higher Education. In H. van't Land, A. Corcoran, & D.-C. Iancu (Eds.), *The Promise of Higher Education* (pp. 181–186). Springer International Publishing. doi:10.1007/978-3-030-67245-4_28

Enders, J., & Westerheijden, D. F. (2014). Quality assurance in the European policy arena. *Policy and Society, 33*(3), 167–176. doi:10.1016/j.polsoc.2014.09.004

EQAR. (2023, April 21). *Cross-border QA activities - EQAR.* https://www.eqar.eu/kb/cross-border-qa/drafting-cross-border-external-qa-activities/

Handke, S. (2023). *Accreditation agencies in the European higher education area: Nonprofit business models, competition and survival. New horizons in nonprofit research series*. Edward Elgar Publishing Limited. doi:10.4337/9781800881259

Harvey, L., & Green, D. (1993). Defining Quality. *Assessment & Evaluation in Higher Education, 18*(1), 9–34. doi:10.1080/0260293930180102

Henig, D., & Knight, D. M. (2023). Polycrisis: Prompts for an emerging worldview. *Anthropology Today, 39*(2), 3–6. doi:10.1111/1467-8322.12793

Lambach, D., Johais, E., & Bayer, M. (2015). Conceptualising state collapse: An institutionalist approach. *Third World Quarterly, 36*(7), 1299–1315. doi:10.1080/01436597.2015.1038338

Lange, R. (2010). Benchmarking, Rankings und Ratings. In D. Simon, A. Knie, & S. Hornbostel (Eds.), *Handbuch Wissenschaftspolitik* (pp. 322–333). VS Verlag für Sozialwissenschaften. doi:10.1007/978-3-531-91993-5_22

Lim, D. (1999). Quality Assurance in Higher Education in Developing Countries. *Assessment & Evaluation in Higher Education, 24*(4), 379–390. doi:10.1080/0260293990240402

Mahoney, J., & Thelen, K. A. (Eds.). (2010). *Explaining institutional change: Ambiguity, agency, and power*. Cambridge University Press. https://search.ebscohost.com/login.aspx?direct=true&scope=site&db=nlebk&db=nlabk&AN=304430

Owen, C. (2013). Quality Culture at What Cost? Pursuing Educational Excellence in Straightened Times. In R. Land & G. Gordon (Eds.), *International studies in higher education. Enhancing quality in higher education: International perspectives* (pp. 28–38). Routledge.

Portnoi, L. M., Rust, V. D., & Bagley, S. S. (Eds.). (2010). *Higher Education*. Policy, and the Global Competition Phenomenon. doi:10.1057/9780230106130

Reinalda, B., & Kulesza, E. (2006). *The Bologna Process – Harmonizing Europe's Higher Education: Including the essential original texts*. Verlag Barbara Budrich. https://elibrary.utb.de/doi/book/10.3224/9783847412786

Rosa, M. J., Sarrico, C. S., Tavares, O., & Amaral, A. (Eds.). (2016). *Cross-Border Higher Education and Quality Assurance*., doi:10.1057/978-1-137-59472-3

Schwarz, S., & Westerheijden, D. F. (2004). Accreditation in the Framework of Evaluation Activities: A Comparative Study in the European Higher Education Area. In P. Maassen, S. Schwarz, & D. F. Westerheijden (Eds.), Higher Education Dynamics: Vol. 5. Accreditation and Evaluation in the European Higher Education Area (Vol. 5, pp. 1–41). Springer Netherlands. doi:10.1007/978-1-4020-2797-0_1

Scrivens, E. (1995). *Accreditation: Protecting the professional or the consumer? State of health series*. Open University Press.

Shaluf, I. M., Ahmadun, F., & Mat Said, A. (2003). A review of disaster and crisis. *Disaster Prevention and Management, 12*(1), 24–32. doi:10.1108/09653560310463829

Simon, H. A. (1982). *Models of bounded rationality* (Vol. 3). The MIT Press. doi:10.7551/mitpress/4711.001.0001?locatt=mode:legacy

Steinhardt, I., Schneijderberg, C., Götze, N., Baumann, J., & Krücken, G. (2017). Mapping the quality assurance of teaching and learning in higher education: The emergence of a specialty? *Higher Education*, *74*(2), 221–237. doi:10.1007/s10734-016-0045-5

Stensaker, B. (2011). Accreditation of higher education in Europe – moving towards the US model? *Journal of Education Policy*, *26*(6), 757–769. doi:10.1080/02680939.2010.551785

Sursock, A. (2021). Quality assurance and rankings: some European lessons. In E. Hazelkorn & G. Mihut (Eds.), *Research Handbook on University Rankings* (pp. 185–196). Edward Elgar Publishing. doi:10.4337/9781788974981.00024

The Bologna declaration of June 1999: Joint declaration of the European ministers of education. (1999). http://ehea.info/page-ministerial-declarations-and-communiques

van der Heijden, J., & Kuhlmann, J. (2017). Studying Incremental Institutional Change: A Systematic and Critical Meta-Review of the Literature from 2005 to 2015. *Policy Studies Journal: the Journal of the Policy Studies Organization*, *45*(3), 535–554. doi:10.1111/psj.12191

Chapter 4
Impact of Disasters in Higher Education:
Disruption and Facility Affectations

Jaime Santos-Reyes
https://orcid.org/0000-0002-3758-9862
SEPI-ESIME, ZAC, Instituto Politecnico Nacional, Mexico

Galdino Santos-Reyes
Universidad Tecnologica de la Mixteca, Mexico

Ricardo Tejeida-Padilla
Instituto Politécnico Nacional, Mexico

ABSTRACT

The chapter presents the results of the impact of the two 2017 earthquakes that occurred in Mexico City and the hurricane Otis that hit Acapulco, Guerrero, Mexico. Moreover, the perception of fear of buildings collapsing during the occurrence of the two earthquakes. The case study was centered on university students of the affected cities. The main conclusions were the following: a) Overall, there is a lack of studies or reports on the extent of the impacts caused by disasters in higher educational systems in Mexico. University students are affected by the interruption of their studies and this issue has not been addressed explicitly on how to deal with it. b) The disasters covered in the chapter suggest that university students (and lecturers) exhibited prosocial behavior not only those in the affected areas but also students nationwide. c) Disasters trigger disruption to critical infrastructures such as the water supply systems, energy sector, transportation, communications systems, and d) university students exhibited fear of building collapsing during the 2017 earthquakes.

INTRODUCTION

The 2030 agenda for sustainable development is a plan of action for people, our planet and prosperity (UN-SDGs, 2023). The agenda and the 17 Sustainable Development Goals (SDGs) aim at achieving that aim. The agenda recognizes and reaffirms the urgent need to reduce the risk of disasters. The goals that implicitly or explicitly relate to higher education are the following (UN-SDGs, 2023):

DOI: 10.4018/979-8-3693-1926-0.ch004

Goal 1: No poverty. Target 1.5, "By 2030, build the resilience of the poor and those in vulnerable situations and reduce their exposure and vulnerability to climate-related extreme events and other economic, social and environmental shocks and disasters."

Goal 2: No hunger. Target 2.5, "By 2030, ensure sustainable food production systems and implement resilient agricultural practices that increase productivity and production, that help maintain ecosystems, that strengthen capacity for adaptation to climate change, extreme weather, drought, flooding and other disasters and that progressively improve land and soil quality."

Goal 3: Good health and well-being. Target 3.D, "Strengthen the capacity of all countries, in particular developing countries, for early warning, risk reduction and management of national and global health risks."

Goal 4: Quality of Education. Target 4.3, "By 2030, ensure equal access for all women and men to affordable and quality technical, vocational and tertiary education, including university."

Goal 6: Clean water and sanitation. Target 6.1, "By 2030, achieve universal and equitable access to safe and affordable drinking water for all."

Goal 9: Industry, innovation, and infrastructure: Target 9.1, "Develop quality, reliable, sustainable and resilient infrastructure, including regional and transborder infrastructure, to support economic development and human well-being, with a focus on affordable and equitable access for all."

Goal 11: Sustainable cities and communities: Target 11.5, "By 2030, significantly reduce the number of deaths and the number of people affected and substantially decrease the direct economic losses relative to global gross domestic product caused by disasters, including water-related disasters, with a focus on protecting the poor and people in vulnerable situations."

Goal 13: Climate action. Target 13.3 "Improve education, awareness-raising and human and institutional capacity on climate change mitigation, adaptation, impact reduction and early warning."

The two goals that addressed explicitly the importance of disaster risk reduction in education systems are goals 4 and 13. In line with these goals, several studies have been reported in the literature regarding how universities can contribute, for example, to an Effective Disaster Risk Reduction Education (DRRE) (Chen & Adefila, 2020), and to promote education to disaster risk reduction in general (Seddinghi et al., 2021; Shaw, 2014; Kagawa & Selby, 2014; Adiyoso & Kanegae, 2012; Dufty, 2014; Gokmenoglu et al., 2021, 2023), and voluntarism following disasters (Kawawaki et al., 2023; Rao et al., 2011; Carlton & Mills, 2017; Paciarotti et al., 2018; Wittaker et al., 2015; Rios, 2018; Santos-Reyes & Gouzeva, 2020; Demirbilek & Ozturk, 2023), among others.

However, there are not that many studies addressing explicitly how disasters affect higher education students. It may be argued that students at all levels, are vulnerable to disasters triggered by natural hazards (e.g., earthquakes, hurricanes, among others). Kousky (2016) argued that there are at least three ways children can be affected by disasters: {1} being killed or injured, malnutrition or illness by the lack of food due to the interruption of food supply; {2} psychological effects during and after the emergency disaster (e.g., fear of the building collapsing, among others); and {3} disasters may interrupt students' education when school facilities are destroyed or damaged, or when families are displaced in the aftermath of disasters.

More recently, a study found that disasters affect students' attendance patterns and absenteeism rates, and ultimately dropping out of school. Some of the reasons on why such consequences are given by the study (Hebebci, 2023): economic problems, displaced and sheltered, migration, psychosocial effects, loss of educational materials, security issues, transport problems, and inadequacy of the education system. Hence of vital importance to address the impact of disasters triggered by natural hazard to higher education systems.

However, very few studies have addressed the impact of disasters in higher education in the context of Mexico.

The aim of the study is to present some evidence of how higher education students have been affected by considering the 2017 earthquakes that occurred in Mexico City, and the most recent Hurricane Otis that hit the resort city of Acapulco, Guerrero, Mexico. In short, the chapter presents some results regarding the second and third type of affectation reported by Kousky, i.e., the disruption of university students' education after the 7 & 19 September earthquakes and the 2023 Hurricane Otis. The study will also address university students' perceived fear of the building collapsing during the two earthquakes occurrence (i.e., second type of affectation, see above).

This study may help decision-makers to better address the vulnerability of higher education students when disasters strike worldwide and in Mexico in particular. The aims of the study are the following:

1. To provide some evidence of the affectation of university students' education due to the 2017 earthquakes and the 2023 Hurricane Otis.
2. To investigate university students' perceived fear of building collapsing during the 2017 earthquakes.
3. To shed some light on the prosocial behavior of higher education students in the affected areas.

The chapter is organized as follows: first, a very brief description of the two 2017 earthquakes and the Hurricane Otis is presented. Second, a brief description of the employed materials and methods are described; third, the most relevant results are presented. Fourth, a discussion of the most relevant findings is presented in the section; also, some limitations and future directions are given within the discussion section. Finally, some relevant conclusions are reported.

Earthquakes and Hurricanes: The 07 & 19 September 2017 Earthquakes

Mexico is prone to disasters triggered by earthquakes as demonstrated by two of the most recent earthquakes that occurred in the country affecting Mexico City, Oaxaca, Guerrero, among other states, i.e.: the 07 & 19 September 2017 earthquakes (GEER, 2017, Velázquez-Buccio et al., 2023; Salinas-Jasso et al., 2023; Santos-Reyes & Gouzeva, 2023), Figure 1.

Regarding the 07 September earthquake, it is thought that Mw 8.2 tremor occurred in the Gulf of Tehuantepec, 140 km southwest of Pijijiapan town, Chiapas, at 23:49:18 local time (Velázquez-Buccio et al., 2023). It is believed that 96 people were killed and 110,000 houses were damaged (Godínez-Domínguez et al., 2021).

Twelve days later, a Mw 7.1 earthquake occurred on September 19, 2017, at about 60 km southwest of the state of Puebla, and 120 km southeast of Mexico City, the capital city of the country (GEER, 2017). The consequences of the earthquake in the city were significant in terms of life loss/property/economic in comparison with the earlier one. It is thought that at least 369 people were killed by the earthquake, more than 28 million people, 54 thousand hospitals, 5.7 thousand hospitals were within the exposure of the earthquake (GEER, 2017).

Figure 1. Epicenters of the September 7&19 2017 earthquakes (Rong-Gong & Sanchez, 2017).

Hurricane 'Otis'

On 25 October 2023, the resort city of Acapulco, Guerrero, was hit by Hurricane Otis with a category 5 storm (Graham, 2023) (Figure 2). It is believed that it was one of the strongest ever to land in the Pacific coast of the country. Otis battered the city of Acapulco with winds of 266 km/h causing death and destruction; the resort city's critical infrastructure was also severely affected (Reuters, 2023). At the time of writing, Otis caused the death of 48 people and an estimated economic loss of 16 billion US Dollars (Graham, 2023).

MATERIALS AND METHODS

A questionnaire survey was conducted to a sample size of n = 2400 following the 2017 earthquakes aiming at better understanding of the actions taken by the participants of the study during, and immediately after these events (Santos-Reyes & Gouzeva, 2020, 2023). To address the issue associated with the perception of fear of the building collapsing during the earthquakes, a question was formulated to assess this very important issue, see a later section for details on this.

To analyze the descriptive information, frequency analysis has been conducted. The relationship between the dependent and independent variables have been assessed by employing Pearson Chi-square. All the conducted tests were two-tailed, and the statistically significant results acceptance criteria were yield at $p<.05$.

Regarding Hurricane Otis, on the other hand, data was extracted from several sources, such as online newspapers, government reports, past research studies on higher education and disasters, online databases, among others. The following keywords were used in the literature review process: "Hurricane Otis", "Higher education", "Acapulco and Hurricane Otis", "Consequences of Otis", "University and

Figure 2. Hurricane Otis and the resort city of Acapulco (Harrison, 2023).

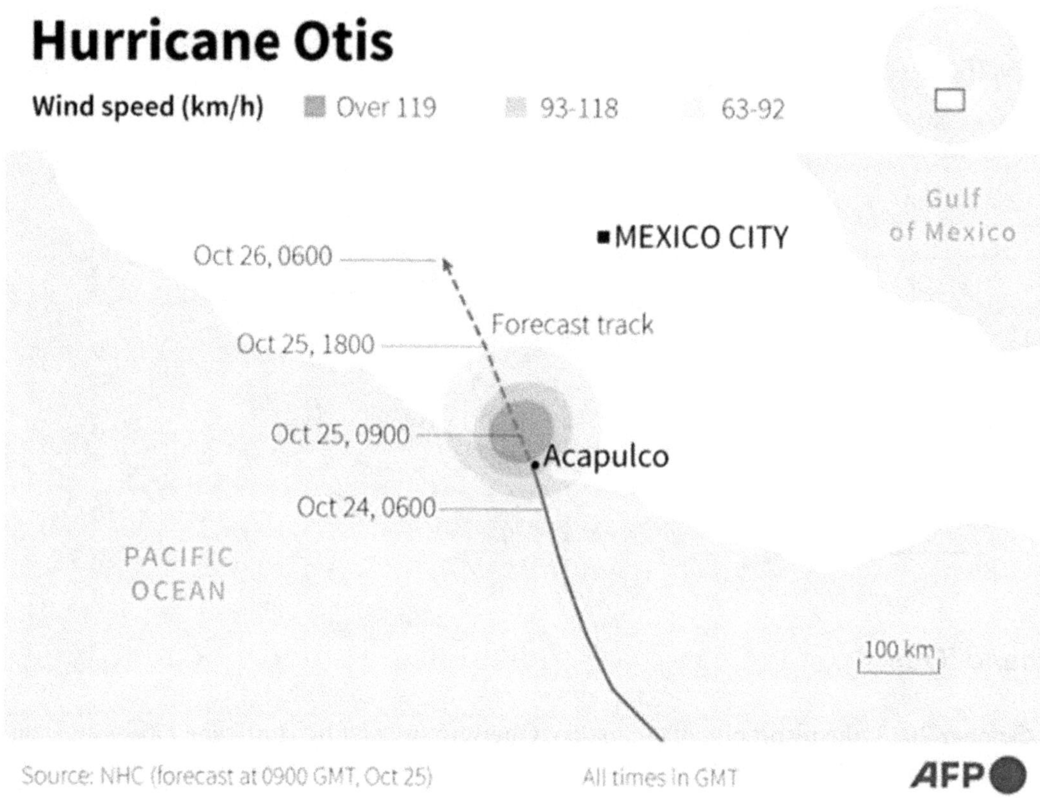

Hurricane Otis", "School affectation and Otis", "Students and Hurricane Otis", "Education affectation and Otis", "Students interview and Otis", among others.

RESULTS

In general, the section presents the most relevant results of the two aims of the present study, i.e.: a) The impact/consequences of the 2017 earthquakes and Hurricane Otis in the context of higher education; and b) the results of an analysis of the degree of fear of the building collapsing experienced by university students during the September earthquakes.

Disasters and Consequences

Tables 1&2 show the universities that aided those affected by the disasters, and the impacts/consequences of these in the context of higher education systems, respectively. It should also be mentioned that several other states were affected, however, only the states of Oaxaca, Guerrero and Mexico City will be considered in the present case study. (See the limitations of the study).

Table 1. Some of the universities that assisted the affected by the disasters considered in the case study

Acronym	Description	City, State
BUAP	Benemérita de la Universidad Autónoma de Puebla [Benemerita Autonomous University of Puebla]	Puebla
ENUFI	Escuela Normal Urbana Federal del Istmo [Federal Urban Normal School of the Isthmus]	Oaxaca
ENEHL	Escuela Normal Experimental de Huajuapan de León [Huajuapan de Leon Experimental Normal School]	Oaxaca
IPN	Instituto Politécnico Nacional [National Polytechnic Institute]	Mexico City
ITI	Instituto Tecnológico del Istmo [Isthmus Technological Institute]	Oaxaca
IESIT	Instituto de Estudios Superiores del Istmo de Tehuantepec [Institute of Higher Studies of the Isthmus of Tehuantepec]	Oaxaca
ITESO	Instituto Tecnológico de Estudios Superiores de Occidente [Technological Institute of Higher Studies of the West]	Guadalajara, Jalisco
UNAM	Universidad Nacional Autónoma de Mexico [National Autonomous University of Mexico]	Mexico City
UABJO	Universidad Autónoma Benito Juárez de Oaxaca [Benito Juarez Autonomous University of Oaxaca]	Oaxaca
UAQ	Universidad Autónoma de Querétaro [Autonomous University of Queretaro]	Querétaro
UTM	Universidad Tecnológica de la Mixteca [Mixteca Technological University]	Oaxaca
UNISTMO	Universidad del Istmo [University of the Isthmus]	Oaxaca
UMAR	Universidad del Mar [Sea University]	Oaxaca
UADY	Universidad Autónoma de Yucatán [Autonomous University of Yucatan]	Yucatán
UAGro	Universidad Autónoma de Guerrero [Autonomous University of Guerrero]	Guerrero
UASLP	Universidad Autónoma de San Luis Potosí [Autonomous University of San Luis Potosi]	San Luis Potosí
UdeC	Universidad de Colima [University of Colima]	Colima
UdeG	Universidad de Guadalajara [University of Guadalajara]	Guadalajara, Jalisco

Fear of the Building Collapsing During the 2017 Earthquakes

The sample size of the study was $n = 2400$ participants of which 43.0% (1027/2400) were university students and 57.0% (1373/2400) were under the category Other (e.g, the general public, workers). When considering the whole sample, the percentage of women were slightly higher than men (55.1% vs. 44.9%). A similar trend was observed for the case of higher education students (i.e, 52.0% (535/1027) vs 48.0% (492/1027)) (Fig. 3). In order to assess the level of fear of the building collapsing during the

Table 2. Some consequences of the 2017 earthquakes and hurricane otis.

Year	Affected State/City (Fig. 1)	Consequences
07 Sep. 2017 Earthquake	Oaxaca	*Regarding University building facilities:* In Oaxaca capital city, the building facility of the Faculty of Fine Arts was severely affected (UABJO, 2017a). After the earthquake, it is believed that 26000 students and around 4000 personnel were affected by the closure of the university (UABJO, 2017b). *Regarding critical infrastructures:* 41 municipalities of the Isthmus of Tehuantepec region were affected by the water supply system, i.e., 800,000 inhabitants were left without drinking water (Carlos-Zavala, 2017).
19 Sep. 2017 Earthquake	Mexico City	*Regarding University building facilities:* In Mexico City, several building facilities were damaged in the Universidad Nacional Autonoma de México (UNAM), e.g., Faculties of Medicine, Dentistry, Humanities, Chemistry, Engineering, the Central library (Valencia-Navarro, 2019). A bridge connecting buildings collapsed at the ITESM (Flat, 2023); six students lost their lives (Santos-Reyes & Gouzeva, 2023). *Regarding critical infrastructures:* Mass transit systems were also affected, e.g., metro lines 1,2,3,4), BRT line 2&4 (Lopez, 2017a). Other modes of public transport were also affected, and mobility disrupted. Water supply shortage affected three large boroughs of the city (Lopez, 2017b), around four million inhabitants were affected (Lopez, 2017c). 1.66 million households were affected by power outage in the city (Expansion, 2017).
	Oaxaca	*Regarding University building facilities:* Escuela Normal Experimental de Huajuapan de León (ENEHL) where a year after the earthquake it was still closed (NVINoticias, 2018). Several university/institute buildings were not damaged but were used as temporary shelters for those affected by the earthquake, e.g.: Escuela Normal Urbana Federal del Istmo (ENUFI), Instituto Tecnológico del Istmo (ITI), Instituto de Estudios Superiores del Istmo de Tehuantepec (IESIT) (Rios, 2018). Three university facilities were affected by the 2017 earthquakes (UTM, UNISTMO, UMAR) (Vélez-Ascencio, 2017) *Regarding critical infrastructures:* 102,000 thousand households were affected by power outage (Expansion, 2017).

Table 2. Cont.

Year	Affected State/City (Fig. 2)	Consequences
25 Oct. 2023 Hurricane "Otis"	Acapulco, Guerrero	*Regarding University building facilities:* In Acapulco city there are 19-23 higher education universities/institutes and all of them were affected (SIC, 2023; Wachauf & Villa-Caña, 2023), e.g., the faculties related to Ecology, Tourism, Sociology, Nursing, Law, Accounting & Administration, Medicine, Psychology, Mathematics, Economics, Nutrition, and several others (Hernández-Campos, 2023). A private university closed indefinitely, and affecting thousands of students (El Sur, 2023; Cancino, 2023). All the collection of 16 municipal libraries of Acapulco city was destroyed (Briseño, 2023). As of 21 November 2023, 106 schools of various educational levels returned to classes (although it is not clear how many universities/institutes) (Yañez, 2023). *Regarding critical infrastructures:* Over 500,000 households lack electricity and after a week 75% of electricity was restored (Yañez, 2023). A month after the disaster, 90% progress is reported in the supply of drinking water through the municipal network (Yañez, 2023). The main roads to/from Acapulco were closed due to falling trees, landslides, and the overflowing of a river next to one of the main highways connecting Acapulco and the capital city of Guerrero State (Yañez, 2023).

Figure 3. Some demographics of the participants of the study

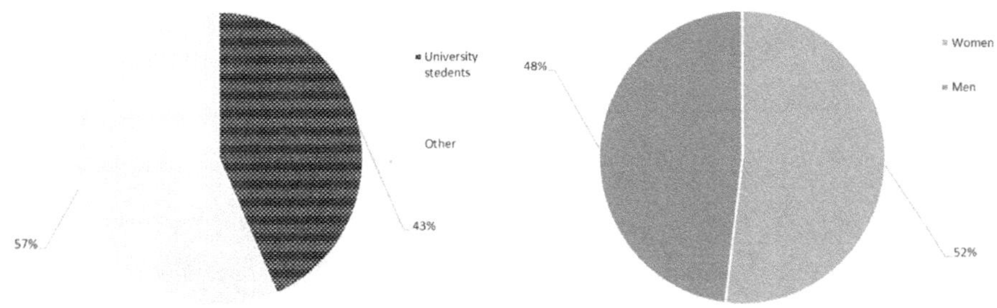

two earthquakes, the university students were asked the following question, "how do you rate your fear of the building collapsing during the earthquake?" (i.e., the question was asked for each of the seismic events). The answers were rated with a Likert-type scale, "Without fear" to "A lot". The results are shown in Figures 4 & 5.

Figure 4 shows that during the 07 September earthquake higher education students experienced "little" fear of the buildings collapsing compared with "a lot" of fear perceived by "Other" participants of the study (χ^2 (4, n = 2,400) = 52.34, $p < .000$, Cramer's $V = .148$). It should be mentioned that the earthquake occurred almost at mid-night (see section Earthquakes & hurricanes). However, the 19 September earthquake occurred while students were still at university facilities, and it was found that "Other" respondents experienced "a lot" of fear of the building collapsing when compared with higher education students during the seismic emergency (χ^2 (4, n = 2,400) = 52.88, $p < .000$, Cramer's $V = .148$; Fig. 5).

Figure 4. Fear of the building collapsing-September 7, 2017, earthquake

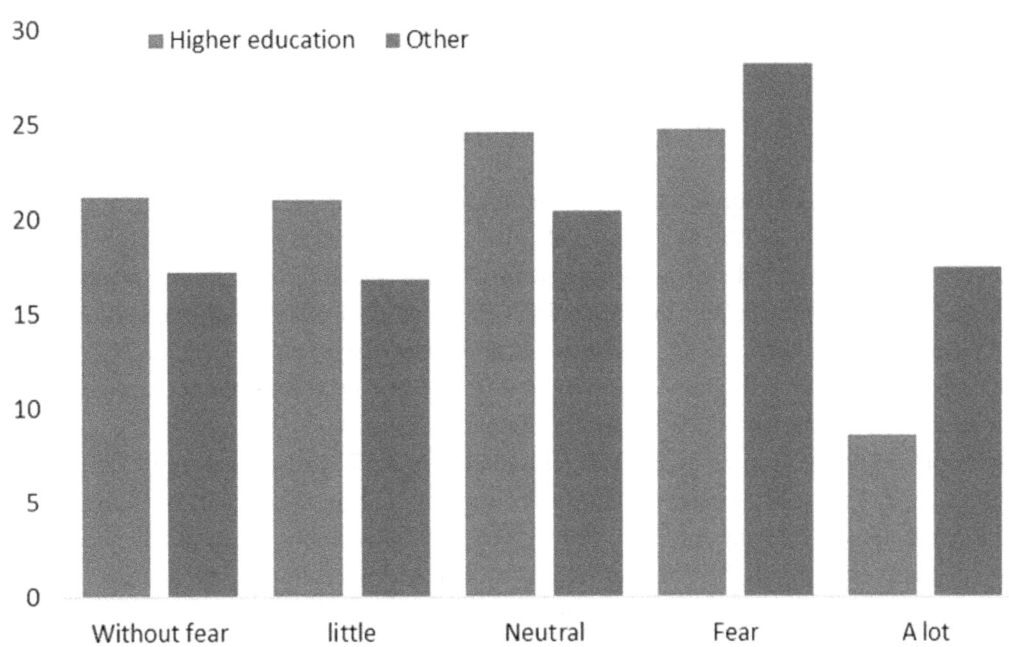

Figure 5. Fear of the building collapsing-September 19, 2017, earthquake

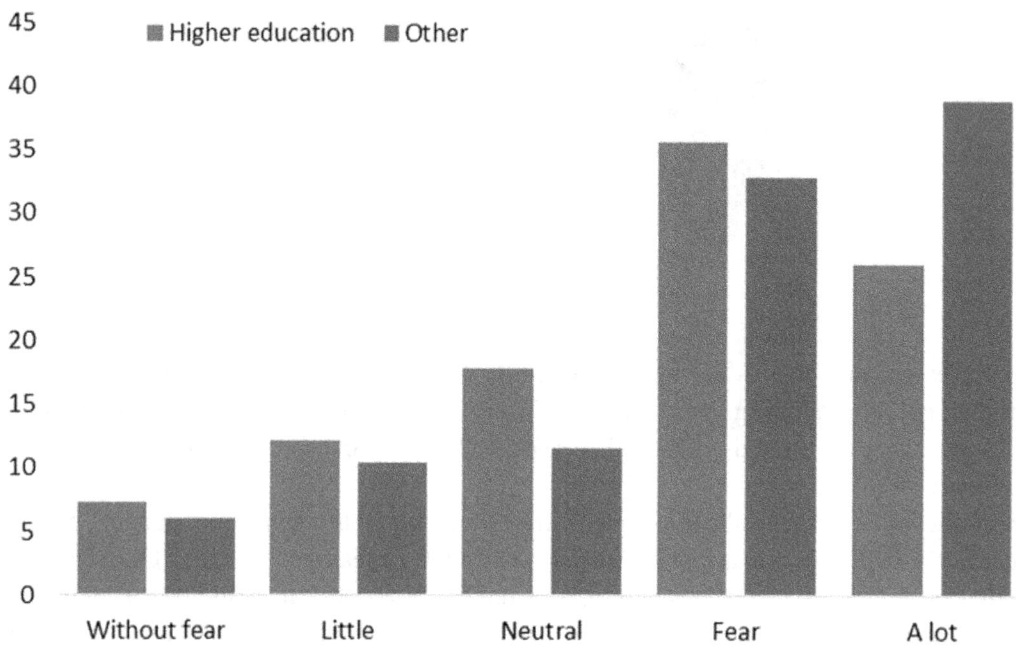

DISCUSSION

Disasters triggered by natural hazards usually have considerable impact in terms of human/property/economic losses (Huang et al., 2024; Jiansheng et al., 2020; CRED/UNISDR, 2020; Shen et al., 2023). The impact of disasters, among others, have led the UN agenda for Sustainable Development Goals (UN-SDGs) to address disaster risk to mitigate the impact of these, e.g., Goals 1-4, 6, 9, 11, & 13 (see the Introduction section).

Disasters, Higher Education, and Cascading Effects

In recent years, a great deal of effort has been put into investigating disaster risk perception, preparedness, and awareness (Hasan et al., 2022; Takayama et al., 2022; Iizuka, 2022), approaches to disaster-resilient universities (Gibbs et al., 2022), disaster risk education (Chen & Adefila, 2020), among others. Moreover, research has also been conducted on the cascading effects on critical infrastructures due to disasters triggered by natural phenomena, such as earthquakes and hurricanes (Damle et al., 2021; Watson et al., 2007), which is also relevant to higher education facilities. However, there are not that many studies on the impact of disasters in higher education in the context of Mexico.

The study presented the impact of the 7 & 19 September 2017 earthquakes and Hurricane Otis in 2023 to higher education buildings and students' education disruption. Moreover, an analysis of the fear of the buildings collapsing during the 2017 September earthquakes have been conducted. Overall, as expected, the results have highlighted that university students are heavily affected by the occurrence of disasters, i.e., these events interrupt their educational process when university facilities are damaged or

Figure 6. Collapse of a bridge connecting buildings (left: Arena Publica, 2017) and a damaged faculty building after Otis (right: Melendez, 2023)

destroyed, families displaced, as demonstrated by the presented case study. These results are consistent with what has been reported by several studies (Hebebci, 2023; Kousky, 2016; Vital, 2018; Arenas-Basurto, 2018; Di Pietro, 2018; Rios, 2018).

For example, a year after the Mw8.2 earthquake, the head of one of the higher education centers in Oaxaca state had this to say,

The school has some difficulties (in using the computer room) as a result of last year's earthquakes, and that this situation needs to be addressed; for example, in the computer classroom the roof needs to be repaired...as is, it represents a risk (to the students). (NVI Noticias, 2018)

Similarly, after Hurricane Otis in Acapulco, the existing 19 to 23 universities were damaged and one of them will close indefinitely starting in July 2024 leaving over one thousand students not knowing what to do with their lives (El Sur, 2023; Cancino, 2023). For example, during an interview with the affected, a student mentioned the following (Cancino, 2023):

We are not being given options to continue with our studies, and no other university wants to accept us. What are we going to do (to recover) the time we spent in this place (university)?

A parent of one of the students mentions the following (Cancino, 2023):

We ask the educational authorities and whoever has jurisdiction in this case to help us. We cannot leave our sons' education like this when they are about to finish their (undergraduate) studies.

The above is consistent with what has been found in a study conducted in Italy following the 2009 L'Aquila earthquake; i.e, the education process disruption reduced the probability that a student would graduate on time (Di Pietro, 2018). More recently, one of the most devastating earthquakes that occurred in Türkiye (and Syria) on February 6th, 2023, had a considerable impact on the higher education system. In the aftermath of the earthquake, it is thought that the Türkiye's Higer Education Council decided to implement a distance education system to all universities, including those not being affected by the seismic event. It was found that the system was inadequate, in particular, deficiencies in "communication, institutional support, and the frequent changes in the system" (Kucukakın et al., 2024).

Two aspects have been highlighted by the beforementioned consequences of the disasters in the context of higher education in the case study, i,e., first, the prosocial behavior showed by students and lecturers from other universities of the country, and second, the phenomena of multi-hazard causing cascading effects.

Prosocial behavior is defined as the expression of voluntary behaviors with the intention to benefit others (Eisenberg et al., 2006; Eisenberg et al., 2009; Carlo, 2006). The findings of the study have highlighted that higher education students affected by the disasters and those elsewhere in the country exhibited prosocial behavior after these events. For example, a group of students of UAGro who were also affected by Hurricane Otis, organized aid brigades for the affected students of other faculties of their university. Moreover, students organized brigades to restore damaged classrooms to make them safe to return to normalcy (Ortega-Morales, 2023). Furthermore, they installed temporary caterings centers to provide, for example, hot beverages and food to supporting brigades such as engineers, architects, veterinarian who assisted pets, among others (Velazquez, 2023).

University students from other states of the country also contributed significantly to collecting goods and taking them to those affected by the hurricane (e.g., those shown in Table 1). For example, UNAM's collection center, delivered 802 tons of humanitarian aid to UAGro for distribution to those in need (UNAM-DGCS, 2023). According to the data from the UAGro, more than 300 tons of humanitarian aid have been collected among external universities "with a deep spirit of solidarity" (Velazquez, 2023).

In a similar vein, after the 7 September 2017 earthquake, three of the major higher education centers in the Isthmus of Tehuantepec, Oaxaca, were used as temporary shelters of hundreds of people. Moreover, university students of the faculty of nursing provided medical support to the injured (Rios, 2018). Similarly, the residents of Mexico City exhibited prosocial behavior after the 19 September earthquake (Santos-Reyes & Gouzeva, 2020).

Prosocial behavior and disasters have also been addressed by scholars; for example, a study has reported that the probability of volunteering decreased with distance from the disaster area, but the probability of monetary donation increased with the distance of the disaster area (Kawawaki, 2023). It has also been found that the degree of prosocial behavior after an earthquake in China increased with "an increasing level of residential devastation but decreased with the passage of time" (Rao et al., 2011). Similar findings have been reported in other studies (Carlton & Mills, 2017; Paciarotti, et al., 2018; Whittaker, et al., 2015). However, it should be highlighted that not all the students (i.e., UAGro) volunteered either in the process of collecting or distributing goods, or in the disaster zone in the present case study. It has also been found that some students cannot volunteering due to family related issues, time, health factors; for some of them, volunteering is seen as boring, time-wasting, among others (Demirbilek & Ozturk, 2024). Undoubtably, prosocial behavior is an area of further research in the context of disasters triggered by natural phenomena.

As already mentioned, disasters such as the ones considered in the present case study trigger disruptions in the supply chain or technological accidents, also known as Natech (Pilone et al., 2021; Maria-Cruz and Suarez-Paba, 2019). For example, the Hurricane Otis did not cause technological accidents, but it caused disruptions in the operations of several critical infrastructures, such as the power supply system, roads, water supply systems, communication systems, health centers, the banking system (Yañez, 2023)

(Table 2). Similarly, the 19 September earthquake caused disruption to key critical infrastructures in the capital city, such as the metro and BRT mass transit systems affecting thousands of people; moreover, millions of inhabitants were left without water (Lopez, 2017a,b,c; Table 2).

Cascading effects of disasters may cause infectious diseases to the affected communities. For example, power cuts may disrupt water treatment facilities, thereby increasing the risk for waterborne disease. Moreover, lack of power supply may also affect health facilities and households. Regarding the former, affectation in the preservation of the vaccine cold chain. Regarding the latter, in 2003, a blackout occurred and left nine million people without power for several hours to two days, causing an increase of diarrheal illness in New York city. The outbreak is believed to have been due to the consumption of meat and seafood after the massive power outage when refrigerators functioning was widely interrupted (Marx, et al., 2006; Watson, et al., 2007).

Fear of the Building Collapsing

According to Coburn, et al. (1992) death tolls in earthquakes are due to three main reasons: structural collapses, non-structural causes, and follow-on disasters. Their study found that 75% of deaths in earthquakes are due to structural collapses. This may be an explanation of why so many deaths are encountered when earthquakes strike in the so-called developing countries; i.e, given that buildings are not designed to withstand earthquakes.

Our results highlighted that higher education students experienced "little" fear of the buildings collapsing compared with "a lot" of fear by "Other" participants during the two earthquakes considered in the present case study. However, there are not that many studies on the perception of the fear of the buildings collapsing during a seismic emergency, but published studies agree on the fear of earthquakes (Cvetkovic et al., 2019; Lovekamp & Tate, 2008; Kung & Cheng, 2012).

Limitations of the Study and Future Work

As with any study, the case study presented in the study has limitations, e.g., the lack of data regarding the consequences of disasters such as the ones covered herein in relation to higher education systems; unlike university students, the UNICEF produces reports on the humanitarian situation of children, for example, six months after the 2017 earthquakes, a report was produced on the impact of these on children (UNICEF, 2018). This raises the question as to whether similar reports should be produced explicitly on the impacts of disasters in higher education systems.

Research is needed to better understand the impact of disasters on higher education in the context of Mexico; for example, the actions taken by higher education students during and after an earthquake with a probabilistic sample. Moreover, conducting studies on prosocial behavior in disasters to better understand, for example, the leading factors that make students to take such actions and its implications. Furthermore, studies on how higher education institutions should built resilience and to minimize the education process disruption in the face of disasters.

CONCLUSION

Disasters triggered by natural hazards usually have considerable impact on higher education systems. Managing educational systems in the aftermath of disasters is challenging and therefore needs to be addressed by key decision-makers. The paper has presented the results of the impact of disasters triggered by the 2017 earthquakes and Hurricane Otis in 2023 to higher education systems in Mexico City, Oaxaca and Acapulco, Guerrero. Moreover, the perception of fear of the building collapsing during the earthquakes of a sample population of university students have also been conducted. The main conclusions were the following:

a) Overall, there is a lack of studies or reports on the extent of the impacts caused by disasters in higher educational systems in Mexico. The findings suggest that university students are affected by the interruption of their studies and this issue has not been addressed explicitly in how to deal with it.
b) The disasters covered in the study suggest that university students (and lecturers) exhibited prosocial behavior not only those in the affected areas but also students nationwide.
c) Disasters trigger disruption to critical infrastructures such as the water supply systems, energy sector, transportation, communications systems. This is more apparent in megacities such as Mexico City.
d) Higher education students exhibited "little" fear of the building collapsing during the 2017 earthquakes.

ACKNOWLEDGMENT

This project was funded under the following grant: SIP-IPN: No-2023, and FOLIO 41/IIIA/2023.

REFERENCES

Adiyoso, W., & Kanegae, H. (2012). The effect of different disaster education programs on tsunami preparedness among schoolchildren in Aceh, Indonesia. *Disaster Mitigation of Cultural Heritage and Historic Cities*, *6*(1), 165–172.

Arena Publica. (2017). *"En caso de sismo, no cruce el puente". Lo que ocurrió en el Tec de Monterrey* ["In case of an earthquake, do not cross the bridge." What happened at Tec of Monterrey]. https://www.arenapublica.com/articulo/2017/09/20/7307/muertos-y-heridos-en-el-tec-de-monterrey-por-terremoto-del-19-de-septiembre-de-2017

Arenas-Basurto, J. G. (2018). Las universidades de México y la cultura de la prevención de los desastres naturales [The universities of Mexico and the culture of prevention of natural disasters]. *Revista de Educación Superior en America Latina, 3, 6-8.* https://rcientificas.uninorte.edu.co/index.php/esal/article/view/10638/10886

Briseño, H. (2023). *"Pérdida total" de las 16 bibliotecas públicas de Acapulco tras 'Otis'* ["Total los" of the 16 public libraries of Acapulco after 'Otis']. https://www.jornada.com.mx/noticia/2023/11/25/estados/perdida-total-de-las-16-bibliotecas-publicas-de-acapulco-tras-otis-1759#:~:text=Acapulco%2C%20Gro.,decenas%20de%20colecciones%20y%20enciclopedias

Cancino, K. (2023). *Universidad de Acapulco Cierra por los daños de "Otis" y deja en la incertidumbre a más de mil alumnos* [University of Acapulco closes due to damage from "Otis" and leaves more tan a thousand students in uncertainty]. https://heraldodemexico.com.mx/nacional/2023/12/22/universidad-de-acapulco-cierra-por-los-danos-de-otis-deja-en-la-incertidumbre-mas-de-mil-alumnos-564799.html

Carlo, G. (2006). Care-based and altruistically based morality. In M. Killen & J. G. Smetana (Eds.), *Handbook of Moral Development* (pp. 551–579). Erlbaum.

Carlos-Zavala, J. (2017). *Sismo dañó Sistema de agua potable en 41 municipios del Istmo de Oaxaca* [Earthquake damaged drinking water system in 41 municipalities of the Isthmus of Oaxaca]. https://oaxaca.eluniversal.com.mx/municipios/29-09-2017/sismo-dano-sistema-de-agua-potable-en-41-municipios-del-istmo-de-oaxaca

Carlton, S., & Mills, C. E. (2017). The Student Volunteer Army: A 'repeat emergent' emergency response organisation. *Disasters, 41*(4), 764–787. doi:10.1111/disa.12225 PMID:28093798

Cheng, Y. F., & Adefila, A. (2020). Enhancing school safety through university engagement in DRR education. *International Journal of Disaster Risk Reduction, 44*, 101386. doi:10.1016/j.ijdrr.2019.101386

Coburn, A. W., Spence, R. J. S., & Pomonis, A. (1992). Factors determining human casualty levels in earthquakes: Mortality prediction in building collapse. *Proc. of Earthquake Engineering, Tenth World Conference*, 5989-5994.

CRE/UNISDR. (2020). *Human cost of disasters-An overview of the last 20 years 2000-2019*. https://reliefweb.int/report/world/human-cost-disasters-overview-last-20-years-2000-2019?gad_source=1&gclid=EAIaIQobChMIlNnznpasgwMVRTqtBh3u-Ae0EAAYASAAEgLlVfD_BwE

Cvetkovic, V. M., Ocal, A., & Ivanov, A. (2019). Young adults' fear of disasters: A case study of residents from Turkey, Serbia and Macedonia. *International Journal of Disaster Risk Reduction, 35*, 101095. doi:10.1016/j.ijdrr.2019.101095

Damle, S., Mani, S. K., & Balamurugan, G. (2021). Natech guide words: A new approach to assess and manage natech risk to ensure business continuity. *Journal of Loss Prevention in the Process Industries, 72*, 104564. doi:10.1016/j.jlp.2021.104564

Demirbilek, O., & Ozturk, S. (2023). Understanding university students WHO do not want to volunteer in disasters: A qualitative study from Turkey. *International Journal of Disaster Risk Reduction, 96*, 103994. doi:10.1016/j.ijdrr.2023.103994

Di Pietro, G. (2018). The academic impact of natural disasters: Evidence from L'Aquila earthquake. *Education Economics, 26*(1), 62–77. doi:10.1080/09645292.2017.1394984

Dufty, N. (2014). Opportunities for disaster resilience learning in the Australian curriculum. *Australian Journal of Emergency Management, 29*(1), 12–16.

Eisenberg, N. (2006). Prosocial behavior. In G. G. Bear & K. M. Minke (Eds.), *Children's Needs III: Development, Prevention, and Intervention* (pp. 313–324). National Association of School Psychologists.

Eisenberg, N., Morris, A. S., McDaniels, B., & Spinrad, T. L. (2009). Moral cognitions and prosocial responding in adolescence. In I. Steinberg & R. Lerner (Eds.), Handbook of Adolescent Psychology (3rd ed., pp. 229–265). Wiley. doi:10.1002/9780470479193.adlpsy001009

El Sur. (2023). *Cerrerá indefinidamente la Universida Americana de Acapulco por los daños de Otis* [The American University of Acapulco will close idenfinetely due to the damage caused by Otis]. https://suracapulco.mx/cerrara-indefinidamente-la-universidad-americana-de-acapulco-por-los-danos-de-otis/#:~:text=El%20rector%20de%20la%20Universidad,un%20riesgo%20para%20la%20comunidad

Expansion. (2017). *Todo sobre el sismo que sacudió el centro del país* [Everything about the earthquake that shook the center of the country]. https://expansion.mx/nacional/2017/09/19/sismo-en-mexico-deja-decenas-de-muertos#uuid00000164-01aa-db07-a166-3fab29f7002d

Flat. (2023). *19S: Edificios dañados por el sismo de 2017 ¿Qué ha ocurrido con ellos?* [19S: Buildings damaged by the 2017 earthquake, what has happened to them?]. https://www.flat.mx/elliving/19-de-septiembre-los-edificios-danados-por-el-sismo-de-2017-que-ha-ocurrido-con-ellos

GEER (Geotechnical Extreme Events Reconnaissance Association). (2017). *Geotechnical engineering reconnaissance of the 19 September 2017 Mw 7.1 Puebla-Mexico City earthquake, version 2.0.* Report No. GEER 055-B. https://learningfromearthquakes.org/2017-09-19-puebla-mexico/images/2017_09_19_Puebla_Mexico/pdfs/UNAM_GEER_Sept2017_v2_Final.pdf

Gibbs, L., Jehangir, H., Kwong, E. J. L., & Little, A. (2022). Universities and multiple disaster scenarios: A transformative framework for disaster resilient universities. *International Journal of Disaster Risk Reduction*, *78*, 103132. doi:10.1016/j.ijdrr.2022.103132

Godínez-Domínguez, E. A., Tena-Colunga, A., Pérez-Rocha, L. E., Archundia-Aranda, H. I., Gómez-Bernal, A., Ruiz-Torres, R. P., & Escamilla-Cruz, J. L. (2021). The September 7, 2017 Tehuantepec, Mexico, earthquake: Damage assessment in masonry structures for housing. *International Journal of Disaster Risk Reduction*, *56*, 102123. doi:10.1016/j.ijdrr.2021.102123

Gokmenoglu, T., Sonmez, E. D., Yavuz, I., & Gok, A. (2021). Turkish Ministry of National Education school-based disaster education program: A preliminary results of the program evaluation. *International Journal of Disaster Risk Reduction*, *52*, 101943. doi:10.1016/j.ijdrr.2020.101943

Gokmenoglu, T., Yavuz, I., & Sensin, C. (2023). Exploring the interplay between curriculum and textbooks in disaster risk reduction education: Insights and implications. *International Journal of Disaster Risk Reduction*, *96*, 103949. doi:10.1016/j.ijdrr.2023.103949

Graham, T. (2023). *A tale of two cities: a month after Hurricane Otis, Acapulco exposes gaps in disaster response.* https://www.theguardian.com/global-development/2023/nov/29/hurricane-otis-mexico-acapulco-disaster-response

Harrison, A. (2023). *Hurricane Otis cuts off Mexico's battered Acapulco.* https://phys.org/news/2023-10-hurricane-otis-weakens-landfall-mexico.html

Hasan, M. K., Uddin, H., & Younos, T. B. (2022). Bangladeshi nursing students' perceived preparedness and readiness for disaster management. *International Journal of Disaster Risk Reduction*, *81*, 103303. doi:10.1016/j.ijdrr.2022.103303

Hebebci, M. T. (2023). *The impact of natural disasters on education.* International Conference on Social Science (IConSoS), Mararis, Turkey.

Hernández-Campos, E. (2023). *Deja huracán Otis daños en 34 planteles de la UAGro* [Hurricane Otis leaves damage to 34 UAGro faculties]. https://www.elsoldeacapulco.com.mx/local/deja-huracan-otis-danos-en-34-planteles-de-la-uagro-10954283.html

Huang, C., Li, N., Zhang, Z., & Liu, Y. (2024). Examining the relationship between meteorological disaster economic impact and regional economic development in China. *International Journal of Disaster Risk Reduction, 100*, 104133. doi:10.1016/j.ijdrr.2023.104133

Iizuka, A. (2022). Disaster perception and preparation among foreign versus local university students in Japan: A comparative study. *Progress in Disaster Science, 15*, 100239. doi:10.1016/j.pdisas.2022.100239

Jiansheng, W., Yuhao, Z., Tian, Y., & Bikai, C. (2020). Evolution of typhoon disasters characteristics and non-structural disaster avoidance measures in the China coastal main functional area. *International Journal of Disaster Risk Reduction, 46*, 101490. doi:10.1016/j.ijdrr.2020.101490

Kagawa, F., & Selby, D. (2014). *Disaster risk reduction in the school curriculum: the present and potental role of development agencies and the implications for the Hyogo Framework for Action 2005-2015 successor.* Unpublished Report commissioned by PLAN International. https://www.preventionweb.net/english/hyogo/gar/2015/en/bgdocs/inputs/Kagawa%20and%20Selby%2C%202014.pdf

Kawawaki, Y. (2023). Giving of time or giving money? An empirical analysis of nationwide prosocial behaviour in times of disaster. *International Journal of Disaster Risk Reduction, 96*, 103888. doi:10.1016/j.ijdrr.2023.103888

Kousky, C. (2016). Impacts of natural disasters on children. *The Future of Children, 26*(1), 73–92. doi:10.1353/foc.2016.0004

Kucukakin, P. M., Tasti, O. Y., Cobanoglu, R., & Gokmenoglu, T. (2024). The post-disaster response of a higher education institution in Turkey: Insights from students and instructors. *International Journal of Disaster Risk Reduction, 102*, 104293. doi:10.1016/j.ijdrr.2024.104293

Kung, Y., & Chen, S. (2012). Perception of earthquake risk in Taiwan: Effects of gender and past earthquake experience. *Risk Analysis, 32*(9), 1535–1546. doi:10.1111/j.1539-6924.2011.01760.x PMID:22300232

Lopez, J. (2017a). *Tras sufrir daños, transporte con suspensiones provisionales en CDMX* [After suffering damage, transportation with povisional suspensions in Mexico City]. https://www.excelsior.com.mx/comunidad/2017/09/24/1190480

Lopez, J. (2017b). *Tláhuac, Xochimilco e Iztapalapa viven escasez de agua tras sismo* [Tláhuac, Xochimilco and Iztapalapa experience water shortages after earthquake]. https://www.excelsior.com.mx/comunidad/2017/09/24/1190460

Lopez, J. (2017c). *Sismo provoca crisis de agua histórica en el Valle de México* [Earthquake causes historic water crisis in the Valley of Mexico]. https://www.excelsior.com.mx/comunidad/2017/10/05/1192854

Lovekamp, W. E., & Tate, M. L. (2008). College student disaster risk, fear and preparedness. *International Journal of Mass Emergencies and Disasters, 26*(2), 70–90. doi:10.1177/028072700802600201

Maria-Cruz, A., & Suarez-Paba, M. C. (2019). Advances in Natech research: An overview. *Progress in Disaster Science, 1*, 100013. doi:10.1016/j.pdisas.2019.100013

Marx, M. A., Rodriguez, C. V., Greenko, J., Das, D., Heffernan, R., Karpati, A. M., Mostashari, F., Balter, S., Layton, M., & Weiss, D. (2006). Diarrheal illness detected through syndromic surveillance after a massive power outage: New York City, August 2003. *American Journal of Public Health, 96*(3), 547–553. doi:10.2105/AJPH.2004.061358 PMID:16380562

Melendez, J. (2023). *No solo los hoteles, escuelas quedaron pulverizadas tras el paso de Otis* [Not only the hotels, but also schools were pulverized after the passage of Otis]. https://www.elgrafico.mx/al-dia/acapulco-huracan-otis-escuelas-sufren-afectadas-quedaron-pulverizadas

NVINoticias. (2018). *Sin reparar, Normal de Huajuapan, Oaxaca* [Without repair, Normal de Huajuapan, Oaxaca]. https://www.nvinoticias.com/general/oaxaca/sin-reparar-normal-de-huajuapan-oaxaca/32465

Ortega-Morales, I. (2023). *La universidad autónoma de Guerrero también tiene afectaciones por Otis* [The Autonomous University of Guerrero is also affected by Otis]. https://ceprovysa.com/?p=68460

Paciarotti, C., Cesaroni, A., & Bevilacqua, M. (2018). The management of spontaneous volunteers: A successful model from a flood emergency in Italy. *International Journal of Disaster Risk Reduction, 31*, 260–274. doi:10.1016/j.ijdrr.2018.05.013

Pilone, E., Casson-Moreno, V., Cozzani, V., & Demichela, M. (2020). Climate change and NaTech events: A step towards local scale awareness and preparedness. *Safety Science, 139*, 105264. doi:10.1016/j.ssci.2021.105264

Rao, L. L., Han, R., Ren, X. P., Bai, X. W., Zheng, R., Liu, H., Wang, Z. J., Li, J. Z., Zhang, K., & Li, S. (2011). Disadvantage and prosocial behavior: The effects of the Wenchuan earthquake. *Evolution and Human Behavior, 32*(1), 63–69. doi:10.1016/j.evolhumbehav.2010.07.002

Reuters. (2023). *Acapulco: nearly 100 dead and missing after Hurricane Otis, officials say.* https://www.theguardian.com/world/2023/oct/30/acapulco-mexico-hurricane-otis-death-toll-damage#:~:text=Acapulco%3A%20nearly%20100%20dead%20and%20missing%20after%20Hurricane%20Otis%2C%20officials%20say,-This%20article%20is&text=The%20number%20of%20people%20dead,risen%20to%20close%20to%20100

Rios, J. U. (2018). Las instituciones de educación superior ante los sismos de Oaxaca y la Ciudad de Mexico [Higher education institutions in te fase of the earthquakes in Oaxaca and Mexico City]. *Revista de Educación Superior en America Latina, 3,* 9-11. https://rcientificas.uninorte.edu.co/index.php/esal/article/view/10638/10886

Rong-Gong, L., & Sanchez, C. (2017). *Mexico City has never catalogued its buildings to know how many are at serious risk of collapse in an earthquake*. https://www.latimes.com/world/mexico-americas/la-fg-mexico-city-earthquake-20171110-htmlstory.html

Salinas-Jasso, J. A., Montalvo-Arrieta, J. C., & Velasco-Tapia, F. (2023). Spatial patterns of shallow landslides induced by the 19 September 2017 Puebla-Morelos earthquake, Mexico. *Bulletin of Engineering Geology and the Environment, 82*(1), 15. doi:10.1007/s10064-022-03030-1

Santos-Reyes, J., & Gouzeva, T. (2020). Mexico City's residents emotional and behavioural reactions to the 19 September 2017 earthquake. *Environmental Research, 186*, 109482. doi:10.1016/j.envres.2020.109482 PMID:32283339

Santos-Reyes, J., & Gouzeva, T. (2023). Adolescents' responses to the 2017 Puebla earthquake in Mexico City. *Journal of Disaster Research, 18*(7), 771–782. doi:10.20965/jdr.2023.p0771

Seddighi, H., Sajjadi, H., Yousefzadeh, S., Lopez, M. L., Vameghi, M., Rafiey, H., & Khankeh, H. R. (2021). Representation of disasters in school textbooks for children with intellectual disabilities in Iran: A qualitative content analysis. *International Journal of Disaster Risk Reduction, 53*, 101987. doi:10.1016/j.ijdrr.2020.101987

Shaw, R. (2014). Overview of Concepts: Education for Sustainable Development and Disaster Risk Reduction. In R. Shaw & Y. Oikawa (Eds.), *Education for Sustainable Development and Disaster Risk Reduction* (pp. 1–13). Springer. doi:10.1007/978-4-431-55090-7_1

Shen, G., Zhou, L., Xue, X., & Zhou, Y. (2023). The risk impacts of global natural and technological disasters. *Socio-Economic Planning Sciences, 88*, 101653. doi:10.1016/j.seps.2023.101653

SIC. (2023). *Universidades en Oaxaca* [Universities in Oaxaca - 77]. https://sic.cultura.gob.mx/lista.php?table=universidad&disciplina=&estado_id=0

Takayama, K., Craig, L., Cooper, A., & Stokar, H. (2022). Exploring deaf students' disaster awareness and preparedness in U.S. higher education settings: Implications for university level DRR policy and programming. *International Journal of Disaster Risk Reduction, 83*, 103409. doi:10.1016/j.ijdrr.2022.103409

UABJO. (2017a). *UABJO reanuda actividades progresivas a partir de este martes 12 de septiembre* [UABJO resumes progressive activities starting this Tuesday, September 12]. https://www.uabjo.mx/uabjo-reanuda-actividades-progresivas-a-partir-de-este-martes-12-de-septiembre

UABJO. (2017b). *Con daños inmuebles de la UABJO por sismos* [With property damage at the UABJO due to earthquakes]. https://imparcialoaxaca.mx/la-capital/59490/con-danos-inmuebles-de-la-uabjo-por-sismos/

UN-SDGs (United Nations-Sustainable Development Goals). (2023). *Sustainable development goals.* https://sdgs.un.org/goals

UNAM-DGCS. (2023). *Envió UNAM más de 800 toneladas de ayuda a damnificados por Otis* [UNAM sent more tan 800 tons of aid to victims of Otis]. https://www.gaceta.unam.mx/envio-la-unam-mas-de-800-toneladas-de-ayuda-a-damnificados-por-otis/

UNDP. (2023). *What are the sustainable development goals?* https://www.undp.org/sustainable-development-goals

UNICEF. (2018). *Mexico earthquakes humanitarian situation report.* https://www.unicef.org/media/82006/file/Mexico-Earthquakes-SitRep-19-March-2018.pdf

Valencia-Navarro, A. (2019). Rehabilitación de inmuebles históricos y artísticos afectados por los sismos [Rehabilotation of historical and artistic properties affected by earthquakes]. *Septiembre 2017 UNAM*, 47-66. doi:10.1016/j.cstp.2019.11.008

Velázquez, I. (2023). *Se unen universitarios para ayudar a Guerrero* [University students join together to help Guerrero]. https://www.reforma.com/se-unen-universitarios-para-ayudar-a-guerrero/ar2702796

Velázquez-Bucio, M. M., Francesca-Ferrario, M., Muccignto, E., Porfido, S., Sridharan, A., Chunga, K., Livio, F., Gopala, S., & María-Michetti, A. (2023). Environmental effects caused by the Mw8.2, September 8, 2017, and Mw 7.4 June 23, 2020, Chiapas-Oaxaca (Mexico) subduction events: Comparison of large intraslab and interface earthquakes. *Quaternary International, 651*, 62–76. doi:10.1016/j.quaint.2021.11.028

Vélez-Ascencio, O. (2017). *Suneo pide 20 mdp para reparar universidades públicas de Oaxaca* [Suneo asks for 20 million pesos to repair public universities in Oaxaca]. https://www.nvinoticias.com/general/oaxaca/suneo-pide-20-mdp-para-reparar-universidades-publicas-de-oaxaca/18756

Vital, L. M. (2018). Looking ahead post-disaster: What Latin American and Caribbean higher education can learn from Haiti. *Revista de Educación Superior en America Latina, 3*, 3-5. https://rcientificas.uninorte.edu.co/index.php/esal/article/view/10638/10886

Wachauf, D., & Villa-Caña, P. (2023). *Resultaron afectados por el huracán "Otis" 127 planteles de bachillerato y de educación superior* [127 highschool and higher education schools were affected by Hurricane "Otis"]. https://www.eluniversal.com.mx/nacion/resultaron-afectados-por-el-huracan-otis-127-planteles-de-bachillerato-y-de-educacion-superior/

Watson, J. T., Gayer, M., & Connolly, M. A. (2007). Epidemics after natural disasters. *Emerging Infectious Diseases, 13*(1), 1–5. doi:10.3201/eid1301.060779 PMID:17370508

Whittaker, J., McLennan, B., & Handmer, J. (2015). A review of informal volunteerism in emergencies and disasters: Definition, opportunities and challenges. *International Journal of Disaster Risk Reduction, 13*, 358–368. doi:10.1016/j.ijdrr.2015.07.010

Yañez, B. (2023). *Día a día: la cronología a un mes del paso del huracán Otis* [Day by day: the chronology one month after Hurricane Otis]. https://politica.expansion.mx/mexico/2023/11/25/que-paso-acapulco-huracan-otis

Chapter 5
The Impact of NGOs in Rebuilding Higher Education in the Recovery Phase

Ahmad Kayaly
WATAN Foundation, Syria

ABSTRACT

This chapter is an extensive investigation into the crucial role that non-governmental organisations (NGOs) play in facilitating the restoration and reconstruction of higher education systems following conflicts, natural disasters, and other crises. Higher education institutions usually experience major hurdles in such challenging situations. This chapter covers the many facets of the influence of NGOs on the reconstruction of higher education, spanning crucial subjects such as accessibility, quality, innovation, and sustainability. The talk thoroughly addresses the practical and ethical concerns, providing insight into the subtle dynamics of the interplay between NGOs and the recovery of higher education. This chapter stresses the relevance of non-governmental organisations (NGOs) in aiding communities that are dealing with crisis-related disturbances to their educational institutions.

INTRODUCTION

In the chaotic aftermath of battles, the devastation of natural catastrophes, and the confusion created by other societal crises, the path to recovery for societies often rests upon the regeneration of their higher education systems. Higher education institutions stand as the citadels of knowledge, innovation, and human progress, serving as the furnace in which the future is fashioned. Yet, when confronted with the immense challenges offered by these tragedies, these organisations find themselves at a crossroads, with the formidable task of restoring their foundations. "The Impact of NGOs in Rebuilding Higher Education in the Recovery Phase" starts on an expanded investigation of the essential role played by Non-Governmental Organizations (NGOs) in supporting the recovery and reconstruction of higher education institutions in the wake of crises. Within these pages, we begin on a journey through the subtle

DOI: 10.4018/979-8-3693-1926-0.ch005

and numerous elements of NGO involvement, putting light on their important contributions to access, quality, innovation, and sustainability in the sphere of post-crisis higher education.

NGOs emerge as important partners in the endeavour to breathe life back into higher education institutions left reeling in the wake of tragedy. Their amazing agility, unrivalled experience, and steadfast devotion position them as key agents of change, bridging the divide between the immediate demands of educational systems and the wellspring of resources and support available from the international community. This partnership concept crosses boundaries and ideologies, enabling a harmonic collaboration between global and local stakeholders, guiding their collective efforts toward the effective recovery of higher education. As we embark on this chapter's intellectual voyage, our compass takes us to study the important dimensions of NGO engagement in higher education recovery. We focus a spotlight on their vital role in extending access and equity, enhancing educational quality, inspiring innovation, and promoting the long-term sustainability of educational efforts. We go deep into their modus operandi, their methods, their achievements, and their tribulations, providing a vivid portrayal of their impact upon the education scene during times of crisis.

Yet, our trip does not only encompass the practical realities of NGO activity. We also navigate the complicated terrain of ethics, morality, and social responsibility that underlie these collaborations. We wrestle with the significant ramifications of the decisions made by NGOs as they attempt to be agents of positive change, oftentimes in circumstances loaded with political, economic, and cultural sensitivity. This chapter seeks to be a complete compass, guiding our readers through the labyrinthine maze of NGO contributions to the resilience and human development of societies besieged by crisis-induced disturbances to their educational institutions. Drawing from a rich tapestry of various case studies, empirical research, and the collective wisdom of individuals in the area, our purpose is to deliver not only knowledge but significant insights. These findings will serve as lighthouses for policymakers, educators, and development practitioners, putting light on the complex and essential junction of NGOs and higher education recovery. In our exploration, we want not only to throw light on the past and present but also to show the route forward towards a more resilient, egalitarian, and enlightened world.

In the wake of disasters, whether they be natural or human-made, the ramifications typically reach far beyond the immediate physical destruction. Societies are left struggling with the enormous issues of restoring not only their infrastructure but also their social fabric. Amidst the instability and uncertainty that follow such revolutions, higher education systems play a key role in determining the future of these communities. Higher education institutions, with their power to nurture knowledge, innovation, and human potential, serve as the linchpin for societal growth. They are the bastions where the seeds of progress are sowed, and the leaders of tomorrow are trained. However, when calamity strikes, these very institutions find themselves exposed, their foundations disturbed, and their capacity to serve their important duties in risk.

Nonetheless, hope comes in the form of Non-Governmental Organizations (NGOs), whose presence and contributions are important in the recovery phase. These organizations, generally driven by a sense of social responsibility and a desire to constructive change, come in to bridge the gaps created by disasters. Their versatility and skill make them important allies in the fight to restore higher education systems.

The impact of NGOs in this context is nothing short of revolutionary. They bring with them not only financial resources but also a wealth of expertise and knowledge. This influx of support allows higher education institutions to not just rebuild but also to innovate and improve. Access to quality education becomes a reality for individuals who may otherwise be left behind in the aftermath of a crisis.

This chapter, in its expanded analysis of NGO contributions to higher education recovery, tries to provide a thorough understanding of these multidimensional dynamics. Drawing upon a plethora of case studies, empirical research, and real-world experiences, it delivers insights that go beyond the surface. These ideas serve as a guiding light for policymakers, educators, and development practitioners as they cope with the delicate convergence between NGOs and higher education recovery. The role of NGOs in rebuilding higher education in the recovery phase after crises cannot be emphasised. These groups are the hidden heroes, working diligently to guarantee that the flame of education continues to burn bright even in the darkest of times. Their efforts extend far beyond the immediate aftermath of disasters, impacting the destiny of societies and paving the road for a more resilient, equitable, and enlightened world.

The Role of NGOs in Higher Education Recovery

In the complicated fabric of post-crisis recovery, the renewal of higher education systems stands as a keystone of optimism and resilience. These institutions, celebrated for their power to spread the seeds of knowledge, creativity, and human development, often bore the weight of conflicts, natural disasters, and other societal crises. When faced with the formidable challenges of rebuilding higher education in the aftermath of such cataclysms, Non-Governmental Organizations (NGOs) emerge as crucial architects of change, weaving the threads of recovery into a sustainable and equitable future.

The first and most urgent function that NGOs assume in higher education recovery is that of first responders and resource mobilizers. In the wake of crises, educational institutions usually find themselves contending with the devastation of their physical infrastructure, the displacement of students and teachers, and the depletion of educational resources. This is where NGOs step in with exceptional agility and the ability to deploy resources fast. These organizations become the lifelines for educational institutions, providing emergency finance for the rebuilding of damaged facilities and the establishment of temporary learning spaces. Their prompt response guarantees that educational institutions can restart operations with minimal interruption, helping students and communities reestablish a feeling of normalcy despite disaster (Tarozzi, 2020). NGOs, driven by a feeling of urgency and a dedication to make a difference, offer a key bridge between the current needs of educational institutions and the resources available from the international community. Their ability to immediately gather finances and assistance is a credit to their agility and dedication to the cause of higher education recovery.

NGOs acknowledge that true resilience in higher education can only be achieved by empowering institutions with the ability to withstand future storms (Abiddin, 2022). Through personalised training programs and coordinated initiatives, they ensure that educational institutions are better prepared to handle not only the aftermath of disasters but also the problems that lie ahead.

Beyond their direct involvement in recovery initiatives, NGOs play a significant role in campaigning and raising awareness regarding the pivotal necessity of higher education in post-crisis circumstances (Abiddin, 2022). These organizations emerge as champions for the rights of affected individuals to access quality education, especially higher education. NGOs routinely work with politicians, governments, and international entities to stress the need of reconstructing educational systems as a cornerstone of broader societal recovery. Through focused advocacy efforts and strategic alliances, NGOs work carefully to generate the required resources and support for higher education recovery measures. They become the amplifiers of the voices of affected communities, arguing for the inclusion of education as a major component of comprehensive post-crisis development initiatives. By raising global awareness of the enduring benefits of investing in higher education, NGOs contribute to gaining vital support and commitment for rebuilding and sustaining these crucial educational institutions. NGOs recognize that education is not a luxury but a vital human right. In the wake of crises, when many parts of life are affected, education provides a bridge to stability, empowerment, and long-term rehabilitation. By fighting for the importance of higher education in the broader context of post-crisis rebuilding, NGOs guarantee that this crucial component receives the attention and funding it deserves.

One of the most revolutionary roles that NGOs play in higher education recovery is that of accelerators for international collaboration. They bridge the distance between global players, such as governments, universities, and philanthropic organizations, and local equivalents, including educational institutions, community leaders, and crisis-affected governments. In doing so, they arrange a harmonious symphony of knowledge, resources, and passion that propels effective and sustainable recovery initiatives (Yousif, 2022). NGOs function as intermediates, crossing cultural, linguistic, and logistical hurdles that may inhibit international cooperation. They allow the interchange of knowledge, the sharing of resources, and the dissemination of best practices, enabling educational institutions to benefit from global experiences and innovations. These joint activities ensure that higher education recovery is not a solitary pursuit but an integral component of bigger international development agendas. In the ever-globalizing world, the issues confronted by higher education institutions in the aftermath of disasters sometimes necessitate international solutions. NGOs acknowledge that no single institution or entity can solve these difficulties in isolation. Therefore, they serve a vital role in bringing together diverse stakeholders from around the world, establishing a feeling of shared responsibility and collaboration.

The complex role of NGOs in higher education recovery is not only supportive but transformative. They are the rapid responders, channeling resources and support for immediate relief. They are the capacity builders, equipping educational institutions with the resilience and agility to succeed in the face of adversity (Schuster, 2021). NGOs are the advocates, highlighting the relevance of higher education on the world stage. Moreover, they are the catalysts for international collaboration, coordinating relationships that drive effective, equitable, and sustainable recovery. As we delve deeper into the various characteristics of NGO involvement, we will investigate their tremendous impact on access, quality, innovation, and sustainability in higher education recovery. The chapters that ensue will shed light on the rich tapestry woven by NGOs in the service of higher education, highlighting their unshakable dedication to constructing a brighter and more resilient future for individuals and communities affected by disasters.

Enhancing Academic Research

One of the fundamental duties that Non-Governmental Organizations (NGOs) have in the realm of higher education is to create and nurture relationships with worldwide academies, research institutes, and universities. This relationship serves as a stimulus for the development and improvement of curricula inside educational institutions in crisis-affected nations. In this tale, we dig into the different elements of this crucial duty, investigating how NGOs bridge the gap between local educational institutions and worldwide centers of intellectual excellence. In the ever-evolving landscape of higher education, the globalization of knowledge and the exchange of ideas have become vital. Today, universities and research institutes around the world are sites of innovation, research, and intellectual growth (Göttlichová, 2020). However, in locations ravaged by crises such as armed wars, natural disasters, or societal upheaval, local educational institutions often find themselves isolated and failing to stay pace with international norms. This is where NGOs step in as facilitators of communication and collaboration. They operate as conduits that bridge the gap between local academic institutions and their worldwide equivalents (Sheikh, 2019). Through collaborations and networks, NGOs offer avenues for local educators and administrators to engage with experts, researchers, and educational leaders from international academies and research institutes.

One of the primary purposes of these partnerships is to facilitate the creation and enhancement of curricula inside higher education institutions. Curricula are the basis of educational programs, detailing what students learn, how they acquire it, and the skills they gain. Therefore, ensuring that curriculum are current, up-to-date, and aligned with global best practices is crucial. NGOs, exploiting their contacts with international academies and research institutions, actively engage in curriculum development efforts. They work together with local universities and faculty members to identify areas where courses may be expanded and modernized. This strategy entails a careful evaluation of existing programs, assessing them against worldwide standards, and identifying gaps or potential for development.

NGOs employ a needs-based approach when approaching curriculum creation. They undertake extensive needs assessments within the local context to understand the specific difficulties and requirements of the educational institutions in crisis-affected regions. These assessments take into account elements such as the availability of resources, the educational background of students, and the specific issues offered by the crisis itself (Reza, 2022). Armed with this information, NGOs engage with worldwide academies and research institutions to design particular solutions. These solutions include a wide range of domains, including:

1. Curriculum Revisions: NGOs encourage the rewriting of existing curricula to reflect contemporary information, new trends, and best practices from worldwide educational institutions. This approach usually comprises modifying course content, changing syllabi, and linking academic programs with corporate demands.
2. Innovative Pedagogy: NGOs support the implementation of innovative educational approaches and pedagogy. They introduce faculty members to cutting-edge approaches, such as active learning, problem-based learning, flipped classrooms, and technology integration. These tactics not only enrich the learning experience but also teach kids to flourish in an ever-changing world.
3. Interdisciplinary Programs: Recognizing the interconnection of knowledge, NGOs advocate the formation of interdisciplinary programs. These programs offer students to explore many disciplines of study, promoting creativity, critical thinking, and a broad grasp of challenging circumstances.
4. Research Opportunities: Collaboration with overseas universities gives chances for research collaborations. NGOs foster cooperative research projects that tackle local challenges and contribute to the global body of knowledge. This not only boosts the research capacities of local universities but also exposes students to research opportunities that might shape their academic journey.
5. Faculty Development: NGOs invest in faculty development programs, ensuring that educators are prepared with the newest information and instructional abilities. This may involve organizing workshops, seminars, and training sessions offered by specialists from international institutions.
6. Quality Assurance: To sustain and increase educational standards, NGOs partner with local institutions to create quality assurance methods. This comprises procedures for curriculum updating, assessment, and accreditation, connecting them with worldwide criteria.

Benefits of International Collaboration

The connection between NGOs and international academies, research centers, and universities delivers considerable benefits for local educational institutions and their students. Let's look into some of these advantages:

1. Global Perspective: By aligning courses with international standards and embracing global best practices, students get a bigger and more global perspective. They are exposed to varied perspectives, cutting-edge research, and a larger knowledge of global concerns and opportunities.
2. Enhanced Quality: Collaboration with international specialists boosts the quality of education. It guarantees that students receive instruction that matches or surpasses international standards, presenting them with the information and skills necessary for success in an increasingly interconnected world.
3. Research Opportunities: Students and faculty members benefit from research cooperation with international universities. They receive access to resources, experience, and finance that can greatly increase their research potential and contribute to academic performance.
4. Career Opportunities: Graduates of universities with internationally-aligned curricula are usually better positioned to pursue careers both locally and abroad. Their education is acknowledged and appreciated on the international stage, offering doors to a wide range of professional alternatives.
5. Innovation and Adaptability: Exposure to innovative teaching methods and transdisciplinary programs enhances creativity, adaptability, and critical thinking among students. These qualities are vital in an ever-changing labour market.
6. Cultural interchange: Collaboration with international partners increases cultural interchange and understanding. Students and faculty members have the opportunity to connect with folks from other backgrounds, fostering a more inclusive and diverse educational environment.

Challenges and Considerations

While the benefits of NGO-led international collaboration in curriculum development are considerable, it is not without its issues and considerations (Vartiainen, 2022). It is vital to understand and handle these issues to guarantee that the combined efforts are effective and sustainable.

1. Cultural Sensitivity: When designing curricula with international influence, it is necessary to create a balance between global standards and local cultural sensitivities. The curriculum should be relevant and respectful to the local context, customs, and values.
2. Resource restrictions: Educational institutions in crisis-affected nations sometimes encounter resource limits. NGOs must collaborate to ensure that the planned curricular enhancements are feasible and sustainable within the local setting.
3. Language Barriers: Language differences might pose hurdles in international collaboration. NGOs should consider language accessibility and give support to overcome language difficulties, especially when cooperating with non-English-speaking organisations.
4. Sustainability: Curriculum development is a continual activity. NGOs must ensure that the enhancements and innovations introduced are sustainable in the long run. This may involve capacity-building initiatives inside local institutions to carry forward the curricular enhancements.

In the ever-evolving landscape of higher education, NGOs play a critical role in developing links between local educational institutions and worldwide academies, research centers, and universities (Şehitoğlu, 2023). Through these ties, NGOs actively engage in the development and upgrading of curricula, ensuring that educational programs stay current, of high quality, and aligned with worldwide best practices. This collaborative approach not only benefits students by providing them with a more comprehensive and globally-oriented education but also empowers local educational institutions to flourish and contribute to the global knowledge economy. It is a monument to the transformative power of cooperation, where NGOs serve as catalysts for change, bridging the gap between local goals and global excellence in higher education.

Extending Access and Equity in Post-Crisis Higher Education

In the turbulent aftermath of crises, including armed conflicts, natural disasters, and societal upheaval, higher education institutions often find themselves in a state of disarray. The disruption to normalcy is palpable, with physical infrastructure damage, displacement of students and faculty, and a scarcity of essential educational resources. However, it is during these moments of adversity that the pivotal role of Non-Governmental Organizations (NGOs) in extending access and equity in post-crisis higher education becomes manifest. It is a harsh reality that in many crisis-affected regions, vulnerable populations and disadvantaged groups are disproportionately affected by the upheaval. These individuals often face formidable barriers to accessing higher education, including financial constraints, geographic isolation, cultural biases, and the absence of tailored support systems. As a result, they are at risk of being left behind in the wake of disaster or conflict, perpetuating cycles of inequality and marginalization.

Recognizing the imperative of addressing this stark inequity, NGOs become catalysts for change, working tirelessly to ensure that access to higher education becomes a beacon of hope for all, regardless of their circumstances. Their commitment extends beyond mere rhetoric; it takes tangible form in a variety of initiatives designed to level the playing field and offer equitable opportunities for education and skill development. One of the primary mechanisms through which NGOs extend access and equity in post-crisis higher education is through the establishment of scholarship programs and financial support systems. These initiatives are designed to alleviate the financial burden that often deters marginalized individuals from pursuing higher education. By offering scholarships, grants, or low-interest loans, NGOs make it possible for disadvantaged students to access quality higher education (Elmagrhi, 2023). These financial support mechanisms not only cover tuition fees but also often include assistance for essential living expenses, textbooks, and other educational resources. This holistic approach ensures that students facing financial hardship can fully engage in their educational journey without the burden of worrying about how to make ends meet. NGOs understand that financial barriers should never be a hindrance to accessing education, especially in the wake of a crisis when economic resources are scarce. Scholarships and financial aid programs provide a lifeline for individuals who aspire to continue their education and contribute to the recovery of their communities.

NGOs are at the forefront of leveraging technology to extend access to higher education in crisis-affected regions. Distance learning and online education platforms offer flexible and accessible pathways to education for individuals who may be geographically isolated or displaced. NGOs collaborate with educational institutions to develop and implement these digital solutions, breaking down geographical barriers and making education available to a wider audience. In regions where physical infrastructure may still be in disrepair, digital education serves as a lifeline, allowing students to continue their studies without the need for physical presence on a campus. Furthermore, these platforms are often designed to be user-friendly and culturally sensitive, ensuring that they are accessible and relevant to the diverse populations they serve. Online education not only extends access but also promotes self-paced learning, allowing students to balance their education with other responsibilities, such as work or family commitments. This flexibility is particularly crucial for individuals in crisis-affected areas who may have limited time or resources to dedicate solely to their studies.

NGOs also establish community-based learning centers in crisis-affected areas, bringing education closer to individuals who may not have access to traditional higher education institutions. These centers serve as hubs for learning, offering a range of educational programs, vocational training, and skill development opportunities. By locating these centers within communities, NGOs reduce transportation barriers and make education more accessible to those who would otherwise face challenges in accessing urban or distant educational institutions. These centers often foster a sense of community engagement and empowerment, promoting education as a tool for individual and collective advancement (Sanadgol, 2021). Community-based learning centers not only facilitate access but also provide a space for individuals to connect, share knowledge, and collaborate on projects that benefit their communities. These centers often become vital community hubs that extend beyond education, serving as platforms for social and economic development.

In their pursuit of extending access and equity, NGOs are not content with merely removing financial or geographical barriers. They actively work to create inclusive and diverse educational environments where individuals from various backgrounds and identities feel welcomed and valued. These organizations often implement policies and programs aimed at increasing diversity among student populations, faculty, and staff. They promote cultural sensitivity and intercultural understanding, recognizing that education should not only be accessible but also respectful of the unique identities and experiences of all learners. Inclusive education goes beyond the removal of physical barriers; it fosters an atmosphere where individuals of different genders, ethnicities, religions, abilities, and backgrounds can engage in meaningful dialogue and collaboration. This diversity enriches the educational experience, preparing students to navigate a globalized world with empathy and cultural competence. NGOs also actively address the specific needs of marginalized and underrepresented groups, such as refugees, internally displaced persons, and individuals with disabilities. They ensure that educational programs and facilities are designed to accommodate the unique challenges these individuals may face. NGOs play a pivotal role in extending access and equity in post-crisis higher education. Through scholarship programs, financial support, distance learning initiatives, community-based learning centers, and a commitment to inclusivity and diversity, these organizations transform education from an exclusive privilege to an inclusive right. By doing so, they ensure that vulnerable populations and disadvantaged groups have equal opportunities for education and skill development during the recovery phase, ultimately contributing to the broader goals of resilience and human development in crisis-affected regions.

NGOs serve as the bridge between adversity and opportunity, recognizing that education is not only a path to personal growth but also a catalyst for societal transformation. Their unwavering dedication to extending access and equity in post-crisis higher education paves the way for a more inclusive, equitable, and prosperous future for all. In this journey of empowerment and enlightenment, NGOs stand as beacons of hope, lighting the way for individuals and communities to rebuild and thrive in the face of adversity.

Enhancing Quality, Innovation, and Sustainability in Higher Education

In the chaotic aftermath of crises, including armed conflicts, natural disasters, and societal upheaval, higher education institutions often find themselves in a condition of chaos. The disturbance to normalcy is obvious, with physical infrastructure damage, dislocation of students and instructors, and a scarcity of vital educational supplies. However, it is during these moments of struggle that the crucial role of Non-Governmental Organizations (NGOs) in increasing access and equity in post-crisis higher education becomes obvious (Adu-Baffoe, 2021). It is a grim reality that in many crisis-hit countries, vulnerable communities and disadvantaged groups are disproportionately affected by the turmoil. These individuals often confront severe challenges to pursuing higher education, including financial constraints, geographic isolation, cultural biases, and the absence of appropriate support networks. As a result, individuals are at risk of being left behind in the wake of tragedy or conflict, continuing cycles of inequity and marginalization.

Recognizing the importance of addressing this stark injustice, NGOs become catalysts for change, working relentlessly to guarantee that access to higher education becomes a beacon of hope for all, regardless of their circumstances. Their dedication extends beyond simple rhetoric; it takes real form in a range of initiatives meant to level the playing field and create equitable opportunities for education and skill development (Darabi, 2016). One of the key strategies through which NGOs enhance access and equity in post-crisis higher education is through the introduction of scholarship programs and financial support networks. These efforts are aimed to ease the financial burden that often deters vulnerable populations from seeking higher education. By granting scholarships, grants, or low-interest loans, NGOs make it possible for disadvantaged students to access quality higher education. These financial support systems not only cover tuition rates but also often offer assistance for critical living expenses, textbooks, and other educational materials. This holistic approach ensures that students facing financial difficulties can actively engage in their educational path without the weight of worrying about how to make ends meet. NGOs realise that financial obstacles should never be a hindrance to obtaining education, especially in the wake of a crisis when economic resources are low. Scholarships and financial aid programs provide a lifeline for individuals who wish to continue their education and contribute to the healing of their communities.

NGOs are in the forefront of leveraging technology to extend access to higher education in crisis-affected countries. Distance learning and online education platforms offer flexible and accessible avenues to education for those who may be physically isolated or displaced. NGOs cooperate with educational institutions to develop and execute these digital solutions, breaking down geographical barriers and making education available to a wider audience. In locations where physical infrastructure may still be in decay, digital education acts as a lifeline, allowing students to continue their studies without the necessity for physical presence on a campus (Iliško, 2021). Furthermore, these platforms are generally built to be user-friendly and culturally sensitive, ensuring that they are accessible and relevant to the different populations they serve. Online education not only widens access but also supports self-paced learning, allowing students to match their education with other responsibilities, such as work or family commitments. This flexibility is particularly vital for those in crisis-affected countries who may have little time or resources to concentrate completely to their studies.

NGOs also create community-based learning centers in crisis-affected areas, bringing education closer to individuals who may not have access to formal higher education institutions. These centers serve as hubs for learning, offering a range of educational programs, vocational training, and skill development possibilities. By establishing these centers within communities, NGOs eliminate transportation obstacles and make education more accessible to those who might otherwise encounter difficulty in attending urban or remote educational institutions. These centers frequently generate a sense of communal engagement and empowerment, advocating education as a tool for individual and social growth. Community-based learning centers not only facilitate access but also provide a platform for individuals to network, share information, and cooperate on projects that benefit their communities. These centers typically become significant community hubs that transcend beyond education, acting as platforms for social and economic growth.

In their aim of broadening access and equity, NGOs are not content with only eradicating financial or geographical barriers. They actively try to establish inclusive and varied educational environments where persons from various origins and identities feel accepted and respected. These organizations generally employ policies and initiatives targeted at boosting diversity among student populations, professors, and staff. They foster cultural sensitivity and intercultural understanding, emphasising that education should not only be accessible but also respectful of the varied identities and experiences of all learners. Inclusive education goes beyond the removal of physical obstacles; it develops an atmosphere where persons of different genders, races, religions, abilities, and backgrounds can engage in meaningful discourse and collaboration. This diversity improves the educational experience, preparing students to navigate a worldwide world with empathy and cultural competence. NGOs also aggressively address the special needs of underprivileged and underrepresented groups, such as refugees, internally displaced persons, and individuals with disabilities. They guarantee that educational programs and facilities are tailored to suit the unique issues these persons may experience. NGOs have a vital role in improving access and equity in post-crisis higher education. Through scholarship programs, financial support, distant learning initiatives, community-based learning centers, and a commitment to tolerance and diversity, these organizations change education from an exclusive privilege to an inclusive right. By doing so, they ensure that vulnerable populations and disadvantaged groups have equitable opportunity for education and skill development during the recovery phase, ultimately contributing to the greater aims of resilience and human development in crisis-affected regions.

NGOs act as the connection between adversity and opportunity, realising that education is not just a road to personal progress but also a catalyst for society transformation. Their unrelenting drive to improving access and equity in post-crisis higher education sets the path for a more inclusive, egalitarian, and prosperous future for all. In this path of empowerment and enlightenment, NGOs stand as beacons of hope, illuminating the way for individuals and communities to rebuild and prosper in the face of tragedy.

Practical and Ethical Considerations in NGO-Higher Education Partnerships

The recovery of higher education institutions in the aftermath of crises extends beyond the basic restoration of infrastructure and access. It also involves the imperative of increasing the quality of education, fostering innovation in teaching methods and curriculum creation, and promoting long-term sustainability. Non-Governmental Organizations (NGOs) have emerged as essential catalysts in this transformative process, actively striving to guarantee that post-crisis higher education is not only a return to the status quo but an opportunity for growth and resilience. One of the major goals of NGOs in post-crisis higher education recovery is to enhance the quality of education given by institutions. They acknowledge that a focus on quality is crucial for not only restoring but also improving higher education systems. NGOs cooperate with educational institutions to examine and enhance their curricula, ensuring that they remain relevant to the evolving needs of students and communities. This approach frequently entails revising course content, adding contemporary challenges, and connecting academic programs with industry and societal demands. By doing so, NGOs equip graduates with the skills and information needed to contribute effectively to the rehabilitation and development of their communities. In the wake of crises, the educational landscape can change swiftly, with fresh difficulties and opportunities that must be handled (Okine, 2021). NGOs play a critical role in helping institutions adapt their curricula to these altering realities, ensuring that students are well-prepared to meet the particular difficulties offered by post-crisis environments. Innovation in teaching methods and pedagogy is another hallmark of NGO involvement in higher education recovery. These organizations advocate the use of current and successful teaching practices, which may include active learning, problem-based learning, technological integration, and experiential education. By supporting pedagogical innovation, NGOs not only enhance the learning experience but also educate instructors with the tools to engage students more effectively, fostering critical thinking, creativity, and flexibility (Brophy, 2020). In the context of post-crisis recovery, creative teaching methods can be extremely beneficial. They allow instructors to adjust their methods to the individual demands and obstacles that kids may experience after suffering trauma or displacement. Innovative pedagogy fosters an inclusive and supportive learning environment, where kids can thrive and build resilience.

Recognizing that the quality of education is integrally related to the capacities of educators, NGOs invest in faculty development projects. These programs give educators with possibilities for professional improvement, such as training in innovative teaching approaches, research skills, and curriculum design. Empowered faculty members are more equipped to give high-quality education and assist students' academic and personal development. Additionally, these educators become ambassadors of innovation, encouraging positive change inside their institutions and communities. NGOs frequently develop a network of support and collaboration among educators, encouraging the exchange of best practices and fostering a culture of continual improvement. Faculty development efforts not only assist individual educators but also contribute to the overall betterment of higher education institutions.

Innovation is a driving element in post-crisis higher education recovery. NGOs actively foster innovation not only in teaching but also in research and entrepreneurship. They encourage students and staff to engage in research projects that address local concerns and help to the creation of new solutions. Moreover, NGOs often encourage the formation of entrepreneurial and innovation centers within educational institutions. These centers serve as hubs for innovation and cooperation, establishing an environment where students may translate their ideas into real solutions that benefit their communities. Innovation in higher education extends beyond the classroom. It comprises research that addresses critical concerns, such as catastrophe recovery, environmental sustainability, and social equality (Wiśniewska, 2022). By funding research programmes, NGOs allow higher education institutions to become engines of innovation and positive change in their regions.

Sustainability is a fundamental element in the activities of NGOs active in higher education rehabilitation. These organizations know that the programmes they support must be sustainable in the long run to have a lasting impact. NGOs actively partner with local communities and governments to build frameworks that promote the sustainability of higher education projects. They participate in talks with stakeholders to understand the unique requirements and difficulties of the region and involve them in the decision-making process. By creating collaborations with local institutions, NGOs ensure that recovery efforts coincide with the broader development goals of the community and that educational activities are interwoven into the fabric of local society. This collaborative strategy ensures that higher education institutions become active contributors to the sustainable development of their communities. By working directly with local stakeholders, NGOs assist institutions identify and address urgent issues, establishing a sense of ownership and responsibility among all parties involved.

NGOs also contribute in the capacity building of higher education institutions to ensure their long-term viability. This may involve enhancing governance frameworks, financial management, and strategic planning. Efficient and effective governance is vital for the viability of educational institutions. NGOs provide information and support to help institutions create transparent and accountable governance systems that can withstand the difficulties posed by crises and transitions. Financial management is another key part of institutional sustainability. NGOs support institutions in implementing effective financial procedures, including budgeting, fundraising, and resource distribution. These policies ensure that institutions have the financial stability essential to weather economic changes and uncertainties. Strategic planning is the cornerstone of institutional sustainability. NGOs engage with educational institutions to build strategic plans that match with their mission, vision, and long-term goals. These plans provide a blueprint for institutions to negotiate the challenging terrain of post-crisis recovery and continuous development. NGOs have a crucial role in boosting the quality, fostering innovation, and promoting sustainability in post-crisis higher education. By focusing on curricular enhancement, pedagogical innovation, faculty development, and the promotion of innovation and entrepreneurship, they transform education into a dynamic and adaptive force for recovery and progress. Moreover, via partnership with local communities and governments and capacity building for institutional sustainability, NGOs ensure that the beneficial changes brought about by higher education recovery projects survive, boosting resilience and progress in crisis-affected regions.

The role of NGOs in higher education recovery extends far beyond the immediate aftermath of crises. It creates the framework for sustainable educational ecosystems that allow individuals and communities to thrive in the face of adversity. NGOs serve as catalysts for positive change, guiding educational institutions on a path of expansion, innovation, and long-term resilience. forth this transformative voyage, they embody the spirit of optimism and progress, ushering forth a brighter future for everybody. As we dive deeper into the multidimensional role of NGOs in higher education recovery, it becomes obvious that their contributions are significant and far-reaching. Their focus to boosting quality, fostering innovation, and supporting sustainability demonstrates a dedication to the enduring well-being of persons and society affected by disasters. In the chapters that follow, we will study individual case studies, empirical facts, and the collective wisdom of those involved in this crucial endeavour. Through these insights, we hope not only to inform but to inspire policymakers, educators, and development practitioners, directing them toward a more resilient, innovative, and sustainable future for higher education in post-crisis situations.

Practical and Ethical Considerations in NGO-Higher Education Partnerships

The junction of Non-Governmental Organizations (NGOs) and higher education recovery is a dynamic and complicated environment. While the contributions of NGOs to post-crisis higher education are evident, it is necessary to acknowledge and address the practical and ethical factors that shape these collaborations (Herawatia, 2020). In this extended examination, we go deep into the rich geography of practical obstacles and ethical issues that define the link between NGOs and higher education recovery.

Practical Challenges

1. Resource Allocation and Coordination:

Resource distribution and coordination are a significant practical barrier in NGO-higher education relationships. NGOs generally operate with limited financial resources, and properly allocating these resources to guarantee fair access and impact can be a challenging undertaking. Ensuring that funds are directed to the regions of highest need while avoiding duplication of efforts involves precise planning and coordination. Coordinating efforts among numerous NGOs and stakeholders can sometimes be problematic. The extensive web of groups engaging in post-crisis rehabilitation might lead to potential overlaps or gaps in support. Effective communication and collaboration among these institutions are crucial to optimize the utilisation of available resources and prevent inefficiencies.

2. Cultural Sensitivity and Local Context:

Navigating the cultural and contextual peculiarities of the regions in which NGOs work is a key practical problem. Each community affected by a crisis has its own distinct set of cultural norms, customs, and values that must be respected. The imposition of foreign models or solutions might lead to resistance or ineffectiveness, undermining the fundamental goals of rehabilitation and restoration. Achieving cultural sensitivity and customising programs to the local context is a continuous and evolving effort. NGOs must participate in continual communication with community people, educators, and students to ensure that their activities are culturally appropriate and respectful of local identities.

3. Sustainability:

Ensuring the long-term sustainability of educational programmes post-disaster is a real challenge that deserves careful attention. Many NGOs give short-term money or support, but developing a sustainable framework often needs engagement with local governments and communities, which may be complex and time-consuming. Sustainability extends beyond the financial aspect and covers the establishment of systems and processes that can continue beyond the initial intervention. Building local capacity, establishing leadership within educational institutions, and ensuring that programs can operate independently are critical components of sustained recovery initiatives (Dube, 2019).

4. Logistics and Infrastructure:

In regions affected by crises, logistical issues can greatly delay the delivery of educational services. Transportation may be disrupted, communication networks may be unreliable, and access to reliable infrastructure might be a considerable impediment. NGOs must struggle with these practical challenges to guarantee that their initiatives reach individuals in need. Overcoming logistical obstacles typically demands innovative problem-solving and adaptability. NGOs may need to explore alternative delivery methods, such as distant learning, mobile classrooms, or community-based learning centers, to ensure that education reaches children even in the most challenging environments.

Ethical Considerations

1. Cultural Sensitivity and Respect:

Ethical considerations are inextricably linked with the cultural sensitivity and respect that must support any NGO operations in higher education recovery. Approaching their work with humility and respect for local cultures, beliefs, and values is not simply a practical need but an ethical duty. The imposition of foreign ideals or practices can be ethically problematic and lead to unforeseen repercussions, including cultural degradation or community resistance. Cultural sensitivity involves continual involvement with local communities, a dedication to learning their perspectives, and a readiness to alter programs to line with cultural norms. It is crucial to engage local educators and community leaders as cultural liaisons to bridge the gap between external NGOs and the communities they serve.

2. Power Dynamics and Dependency:

NGO-higher education relationships might occasionally unwittingly generate power asymmetries. NGOs, generally well-resourced and influential, can hold major sway in decision-making processes. This tendency can develop to reliance among educational institutions and communities, creating ethical problems concerning agency and self-determination. Ethical partnerships attempt to balance power dynamics by ensuring that decision-making procedures are inclusive and transparent. Empowering local stakeholders, including students, staff, and community residents, to participate actively in crafting recovery activities is crucial. NGOs must aggressively seek the feedback of those affected by their work and build systems for shared decision-making to alleviate power asymmetries.

3. Transparency and Accountability:

Maintaining transparency and accountability is a core ethical obligation for NGOs participating in higher education rehabilitation. Transparency requires being open and honest about objectives, financing sources, and decision-making processes. Accountability guarantees that the actions of NGOs align with the best interests of the impacted communities. To sustain these ethical norms, NGOs should establish clear and transparent communication procedures. This includes freely providing information about financing sources, budgets, and expenditures, as well as being frank about obstacles and constraints (Channy, 2019). Accountability methods, such as frequent reporting and independent audits, can further strengthen ethical behaviours and guarantee that resources are used effectively and responsibly.

4. Beneficiary Participation:

Ethical partnerships promote beneficiary engagement as a strategy of ensuring that programs are relevant, effective, and respectful of the agency of those affected. In higher education recovery, including students, faculty, and communities in decision-making processes can strengthen the relevance and sustainability of programs. Failure to involve beneficiaries in decision-making might generate ethical problems around paternalism and exclusion. Ethical NGOs actively seek the participation and feedback of the individuals and communities they serve, recognising their views and experience in influencing the healing process.

5. Crisis Profiteering:

Ethical problems also revolve on the possibility for crisis profiteering, where NGOs exploit crises for fundraising or branding goals, placing their organizational interests over the well-being of affected populations. Maintaining ethical standards needs a genuine commitment to the well-being of afflicted populations rather than profiteering on their pain. Ethical NGOs are devoted to ensuring that their efforts are inspired by a real desire to assist and support those in need. They refrain from sensationalizing problems for fundraising purposes and focus on offering genuine, impactful help that matches with their mission and principles.

In negotiating the practical and ethical aspects of NGO-higher education relationships, stakeholders must engage in open dialogues, set clear standards and frameworks, and promote the well-being and agency of affected populations. Ethical partnerships are built on trust, mutual respect, and a shared commitment to the long-term resilience and growth of higher education systems in crisis-affected regions. The complexity of post-crisis recovery demands that practical and ethical issues are addressed in combination. By appreciating the delicate interplay between these difficulties and considerations, NGOs and educational institutions can work together to negotiate the terrain of higher education recovery with integrity, empathy, and a persistent commitment to the well-being of those they serve. In this spirit, they can contribute to the repair and renewal of higher education systems, ensuring that they emerge from crises stronger, more resilient, and better suited to play their crucial role in society.

WATAN Foundation's Role in Higher Education

WATAN Foundation, with its unrelenting commitment to education, has emerged as a beacon of hope for Syrian and Turkish students wanting to continue their studies and rebuild their lives amidst suffering. Through its Education program, WATAN has been essential in giving scholarships, allowing access to quality education, and extending support during times of crisis, particularly highlighted by its response to the 2023 Turkey-Syria earthquakes. The Education program at WATAN is built in the belief that education is not simply about obtaining knowledge; it is about cultivating character, values, and trust (WATAN Foundation, 2024). WATAN administers and supports schools in Syria and Turkey, striving to contribute to children' holistic development, finding unique skills, and fostering them. The program mentors students from high school through university, giving numerous supportive activities and workshops that foster cohesion and address the issues faced by Syrian and Turkish students in integrating into the Turkish academic environment.

The objective of WATAN's Education program is to provide access to inclusive and excellent education for children and teens affected by crises in protective and safe surroundings. It focuses on personality formation, information gain, and empowerment at all levels. The program emphasises the psychological, social, cognitive, and professional needs of students, intending to build future entrepreneurs, academics, and leaders.

Areas of Expertise

WATAN's Education program spans a wide spectrum of interventions:

1. Early Childhood Development: WATAN emphasises the crucial relevance of early childhood development. The program delivers comprehensive Early Childhood Development (ECD) services to caregivers and children in secure nurturing care environments. These services are meant to address the developmental needs of children aged 0-8, with a specific focus on boosting school readiness for 3-5-year-olds. This early foundation sets the basis for lifetime learning.
2. Formal Education: Operating schools that offer formal education for primary, intermediate, and secondary stages is a major component of WATAN's work. The program not only helps school running expenses but also offers educational materials, supplies, and works on rebuilding school infrastructure. This guarantees that kids have access to quality formal education in a safe and conducive atmosphere.
3. Informal Education: Recognizing the multiplicity of educational demands, WATAN's curriculum extends to informal education. This comprises remedial programmes, Accelerated Learning Programs (ALP), catch-up classes, literacy and numeracy classes, and Self-Learning Programs (SLP). Alongside these initiatives, the program ensures the provision of curricular textbooks and supporting materials, catering to students who may benefit from alternate educational pathways.
4. Higher Education: WATAN's devotion to education extends to higher education. The charity assists Syrian refugee adolescents and high school graduates with scholarships and language training, enabling them to pursue higher education at the university level. This not only boosts their educational journey but also raises their chances of successful integration into the Syrian and Turkish job markets, affording them stronger prospects for the future.
5. Capacities and Life Skills Building: WATAN understands that the quality of education is inherently connected to the capabilities of educational workers. To achieve a high standard of education, the program engages in faculty development programmes. These programs provide educators chances for professional improvement, including training in innovative teaching approaches, research skills, and curriculum design. By training faculty members with the essential tools and expertise, WATAN helps to the creation of a vibrant and informed teaching community, ultimately boosting the quality of education given by institutions.

Expanding on WATAN Foundation's Role in Higher Education

WATAN Foundation's involvement in higher education and its broader educational initiatives is a testament to its unwavering commitment to providing hope, support, and opportunities for individuals affected by crises, particularly Syrian and Turkish students facing adversities in the aftermath of the 2023 Turkey-Syria earthquakes. The foundation's multidimensional approach comprises a wide range of educational initiatives aimed at supporting holistic development, inclusion, and empowerment. At the heart of WATAN's Education program is the conviction that education extends beyond the acquisition of knowledge; it is about cultivating character, values, and trust. This attitude supports WATAN's approach to education, where the focus extends beyond standard classroom learning to embrace the broader development of pupils. By administering and supporting schools in Syria and Turkey, WATAN aspires not only to convey education but also to identify and foster the unique abilities of pupils. Through mentoring and different supporting activities and workshops, the program builds cohesion and addresses the issues faced by Syrian and Turkish students as they integrate into the Turkish university environment. The aim of WATAN's Education program is built in the principles of access, inclusivity, and excellent education. It attempts to offer children and youth affected by crises with access to education in protected and safe environments. The curriculum is designed to cater to numerous areas of a student's growth, including personality building, information accumulation, and empowerment at all levels. By recognising the psychological, social, cognitive, and professional needs of students, WATAN strives to mold future entrepreneurs, researchers, and leaders, ultimately contributing to the rebuilding of Syrian society. WATAN Foundation's passion to education is reflected in its varied approach to support students and communities affected by disasters. By providing scholarships, quality education services, and psychosocial support, WATAN encourages individuals to overcome barriers and thrive in an inclusive and loving learning environment. The foundation's unwavering commitment to education is a tribute to its belief in the transformational power of knowledge and its purpose to rebuild communities and create a better future for all.

CONCLUSION

In the complicated landscape of post-crisis recovery, the chapters of human history are replete with stories of resilience, reconstruction, and renewal. Among these storylines, the rebirth of higher education emerges as a beacon of hope—a testament to the resilient spirit of individuals, communities, and nations. In this chapter, we have conducted a sweeping journey, documenting the enormous role of Non-Governmental Organizations (NGOs) in the rekindling of higher education throughout the recovery phase. As we now reflect upon the end of this investigation, it becomes increasingly obvious that NGOs occupy a vital role in the orchestration of the recovery symphony. They stand as the connection between the shattered remnants of educational institutions and the aspirations of innumerable individuals attempting to reconstruct their lives via knowledge and skill gain. This position is not only pragmatic but profoundly symbolic, embodying the unity of purpose that connects humanity in times of suffering. Yet, beyond the practical dimensions of their involvement, we have also explored the ethical and moral basis of NGO-higher education partnerships. In a world distinguished by deep political, economic, and cultural sensitivities, the choices made by NGOs bear tremendous implications. Their quest of education as a catalyst for change necessarily engages with intricate webs of power, privilege, and responsibility. This chapter has investigated these ethical issues, asking us to ponder the difficult balance between kindness and paternalism, autonomy and intervention. As we close this chapter, it is crucial to remember that the tale of NGOs in higher education recovery is not static; it is an evolving narrative, a continuing symphony. The challenges and catastrophes that define our world are dynamic, demanding adaptive answers and inventive tactics. In this dynamic world, NGOs must continue to play a key role, modifying their techniques, deepening their commitment, and polishing their ethical compasses. Our trip through the nexus between NGOs and higher education recovery offers more than a retrospective perspective; it acts as a compass for the future. It is a monument to the endurance of the human spirit and the power of education to illuminate even the darkest of roads. It presents lessons that policymakers, educators, and development practitioners can draw upon as they navigate the tumultuous waters of post-crisis recovery.

In closing, we return to the essential concept that connects this chapter together—the enormous impact of NGOs on the recovery of higher education. It is a monument to the human capacity for collaborative action and humane partnership. It reminds us that even in the face of the most overwhelming circumstances, the desire for knowledge, the pursuit of greatness, and the yearning for a greater future survive as unshakable beacons of hope. NGOs serve as the catalysts, the enablers, and the champions of this quest, and their impact reverberates far beyond the bounds of the recovery phase, illuminating the route to a more resilient, equitable, and enlightened world for centuries to come.

REFERENCES

Abiddin, N. Z., Ibrahim, I., & Abdul Aziz, S. A. (2022). Non-governmental organisations (NGOs) and their part towards sustainable community development. *Sustainability (Basel), 14*(8), 4386. doi:10.3390/su14084386

Abiddin, N. Z., Ibrahim, I., & Aziz, S. A. A. (2022). The role of non-governmental organisations (NGOs) for community development: Way forward. *Journal of Management Information & Decision Sciences, 25*(2).

Adu-Baffoe, E., & Bonney, S. (2021). The Role of Non-Governmental Organizations in Basic Education Delivery in Ghana: Implications for Theory, Policy, and Practice. *International Education Studies, 14*(4), 35–47. doi:10.5539/ies.v14n4p35

Brophy, M. (2020). The Role of NGOs in Supporting Education in Africa. *Journal of International and Comparative Education (JICE),* 45-56.

Channy, D. Y., & Ogunniran, M. O. (2019). A comparative analysis of the role of international organizations in financing higher education system: A case study of Nigeria and Cambodia. *US-China Education Review, 9*(1), 1–17. doi:10.17265/2161-623X/2019.01.001

Darabi, S., Azizi, N., Salimi, J., & Shirabegi, N. (2019). Elements and Dimensions of Sustainability in Iran's Higher Education System, and Comparison of Public and Non-Governmental Academic Approaches. *Iranian Journal of Educational Sociology, 2*(3), 12-26.

Dube, T., Zikhali, W., & Dube, S. P. (2019). An analysis of the application of servant-leadership model in the educational development programmes of selected non-governmental organizations in Zimbabwe. *International Journal of Innovative Business Strategies, 5*(2), 332–340. doi:10.20533/ijibs.2046.3626.2019.0044

Elmagrhi, M. H., & Ntim, C. G. (2023, April). Non-financial reporting in non-profit organisations: The case of risk and governance disclosures in UK higher education institutions. *Accounting Forum, 47*(2), 223–248. doi:10.1080/01559982.2022.2148854

Göttlichová, M. (2020). Higher Education and Innovation in the Non-profit Sector. In Strategic Innovative Marketing and Tourism: 8th ICSIMAT, Northern Aegean, Greece, 2019 (pp. 331-340). Springer International Publishing. doi:10.1007/978-3-030-36126-6_36

Iliško, D. (2021). Towards sustainability as a frame of mind in higher education: Thinking about sustainability rhizomatically. Handbook on teaching and learning for sustainable development, 366-375.

Okine, S. (2021). *Non-Governmental Organisations' Impact on Educational Policy in Rural Ghana* [Doctoral dissertation]. Walden University.

Reza, F. (2022). The role of NGOS in promoting education: Successes and challenges. *International Journal of Education, Learning and Development, 10*(1), 24–43. doi:10.37745/ijeld/vol10.no1.pp24-43

Sanadgol, A., Doshmangir, L., Majdzadeh, R., & Gordeev, V. S. (2021). Engagement of non-governmental organisations in moving towards universal health coverage: A scoping review. *Globalization and Health, 17*(1), 1–19. doi:10.1186/s12992-021-00778-1 PMID:34784948

Schuster, J., & Kolleck, N. (2021). Between capacity development and contestation: A systematic review of the involvement of inter-and non-governmental actors in inclusive education. *International Journal of Inclusive Education*, 1–24. doi:10.1080/13603116.2021.1994033

Şehitoğlu, R., Tekin, B., & Güner, O. (2023). *Education Diplomacy in Higher Education Institutions: An Evaluation of International Student Centers as Think Tanks*. Academic Press.

Sheikh, S. S. S., Melati, W. P., Ester, R., & Puteh, F. (2019). The roles of academic NGOs in empowering the higher learning system in Malaysia. *Seuntai Kata*, 7.

Tarozzi, M. (2020). Role of NGOs in global citizenship education. The Bloomsbury handbook of global education and learning, 133-148. doi:10.5040/9781350108769.0020

Vartiainen, H., Vuojärvi, H., Saramäki, K., Eriksson, M., Ratinen, I., Torssonen, P., Vanninen, P., & Pöllänen, S. (2022). Cross-boundary collaboration and knowledge creation in an online higher education course. *British Journal of Educational Technology, 53*(5), 1304–1320. doi:10.1111/bjet.13186

WATAN Foundation- Education sector. (n.d.). https://www.watan.foundation/expertise/education-program

Wiśniewska, S., & Wiśniewski, K. (2022). *Socially responsible activities addressed to internal stakeholders of non-governmental organisations*. Academic Press.

Yousif, N. B. A., Grondys, K., Gad, S., & Elsayed, W. (2020). Knowledge management in non-governmental organizations (NGOs). *Administratie si Management Public, 1*(35), 90–108. doi:10.24818/amp/2020.35-06

KEY TERMS AND DEFINITIONS

Academic Research: A methodical examination and analysis carried out to verify information and draw fresh insights, crucial for fostering innovation and progressing knowledge in higher education establishments.

Accessibility in Education: The extent to which education is accessible to all persons, including those from marginalized or disadvantaged groups, ensuring equitable opportunities for learning and progress.

Education Quality: Refers to the high standard of education, encompassing the efficiency of teaching, learning settings, and curricula, with the goal of achieving the best learning results for students.

Equity in Education: The concept of ensuring justice by offering tailored resources and opportunities to students according to their individual requirements, with the goal of attaining equal results for everyone.

Ethical Considerations: The moral principles and values that guide the actions and choices of NGOs and higher education institutions, emphasizing integrity, transparency, and the well-being of impacted populations.

Higher Education Recovery: The restoration, reconstruction, and enhancement of higher education institutions and systems that have been affected by conflicts, natural catastrophes, or other crises.

Innovation in Education: It involves implementing novel concepts, techniques, technology, or products to improve the efficiency and engagement of learning and teaching practices.

Non-Governmental Organizations (NGOs): Independent, non-profit entities that operate at different levels (local, national, or international) to offer aid, support, and specialized knowledge in areas such as education, free from government influence.

Practical Considerations: The logistical, operational, and financial factors in the relationship between NGOs and higher education institutions throughout the recovery phase.

Sustainability in Higher Education: It refers to the capacity of educational systems to function in a manner that fulfills present requirements without jeopardizing the capability of future generations to fulfill their own demands, encompassing environmental, economic, and social sustainability.

Chapter 6
Sustaining a Globally Diverse Community Through a Time of Crisis: Challenges and Consequences

Elizabeth Stroble
https://orcid.org/0000-0002-9064-2407
Webster University, USA

ABSTRACT

This chapter describes how Wester University, a global university system, navigated the dimensions of crisis arising from recent pandemic conditions. Placing the university's strategies for sustainability in a broader context, the author considers the impact to students and institutions of threats to health and safety, economic disruption, and needed reckoning with issues of isolation, racism, and discrimination. The author reflects on the pandemic's effects on the accelerating threats to higher education's academic and business models, drawing insights from research and experience for rebuilding after crisis.

INTRODUCTION

With its main campus in St. Louis, Missouri, USA, Webster University is an institution with a distinctive history and mission. Founded in 1915 to address gender inequities in access to education, Webster has expanded to function as a global university system with students studying at residential campus locations in North America, Europe, Central Asia, and Asia as well as in a robust online learning environment. A commitment to meeting the unmet needs of the globally diverse students who enroll and graduate from Webster in St. Louis and around the world has resulted in an alumni community of over 200,000 individuals.

As a global university system, Webster coped with the recent COVID-19 pandemic and learned what is likely always true: crises do not occur in isolation, and the consequences are not always immediately realized nor apparent. As the university addressed the ongoing physical health and safety of campus

DOI: 10.4018/979-8-3693-1926-0.ch006

communities around the globe, it was also necessary to address the disruptions in students' economic circumstances and the necessary reckoning with the devastating and disparate impacts of isolation, racism, and discrimination. For many students, the pandemic was simply one of several threats to their own and their families' physical and mental well-being. The challenges and consequences of crises, in their many dimensions, continue for higher education systems that seek to strengthen and sustain community. Strategies for a successful future demand multidimensional approaches that address intersecting threats and challenges, tapping the institution's core strengths and enduring values.

Prior to the pandemic, the landscape for American higher education was already shifting. The impacts of changing demographics of college-going populations, disruptive socioeconomic conditions, geopolitical disruptions and tensions, skeptical public perception, an increasingly competitive landscape, and politically charged policy and regulatory environments challenged the viability of long-standing traditions and functions. Pandemic conditions accelerated the threats to the academic and business models, and developments in transformative technologies and generative artificial intelligence create dynamic conditions for unknown futures. Sustaining a global university system through layers of crisis, notwithstanding needed actions for rebuilding, was and is a work in progress.

The Crises Unfold: Pandemic Conditions

In January 2020, Webster University's Student Health Service shared the first informational message about COVID-19 with the Webster community. Webster's sites in China were the first to alert the university to the impact of the coronavirus. The university's global presence provided an early awareness of the impending health threats as the virus spread globally from east to west, arriving in St. Louis, Missouri, USA, site of the main campus by March 2020.

As the coronavirus took hold in the heart of the Midwestern United States, the threat required Webster leadership to "to balance the felt urgency for action with the need to gather accurate information and provide space for deliberation and engagement. The starting point in setting priorities lay in engaging and consulting with key stakeholders in order to create a common vision for health, well-being, security, and safety of all constituencies" (Stroble et al., p. 190). Leadership adopted broad criteria with these areas of focus: "(a) saving human lives; (b) efficiency and effectiveness; and (c) equity and fairness. These individual criteria were ultimately aimed toward achieving and maintaining sustainability and vibrancy of the University" (Stroble et al., p. 191).

The mechanisms leadership put in place relied upon participation by faculty and staff, with leadership by administrators with specific knowledge and responsibility for the areas of work. In March, the university established the COVID-19 task force. By May, three working groups had been formed. The three working groups—safety, security, and communications; transition to campus-based operations; and academic continuity and support—met almost daily in Spring 2020. Each group had access to public health officials locally and globally; the university's faculty expertise in healthcare and institutional partnerships with healthcare agencies and organizations proved particularly valuable.

As the task force and the working groups navigated a particularly dynamic public health environment, they honed the ability to plan and respond from a platform of information with attentiveness to the specific contexts and directives of diverse governmental entities. Little unanimity could be found in public discourse about the best ways to assure safety and still serve students. The campus community's opinions reflected a range of stances, requiring the individuals leading the COVID-19 response to listen and lead. Circumstances changed from day to day and week to week in all the locations where Webster's

Figure 1. Webster University COVID-19 timeline highlights

Spring 2020

Spring athletics cancelled

Closed Student housing on main campus

Cancelled Study Abroad

Began modified operations

Planned and held virtual Commencement

St. Louis County and City Public Health issued "stay at home" orders

Developed and provided guidelines and resources for community

Established confidential protocol for reporting infections

April 2020: CARES task force established— made federal relief funding available to students

Launched crowdfunding in May 2020: Gorloks Now! And Front-Line Responders funds

Modified operations continued through June 15, 2020; required face coverings

Fall 2020

Phased in flexible in-person transition for fall 2020 semester

Enacted daily health screenings, signing of attestations by employees

Installed hand-sanitizers

Created signage to encourage social distancing, hand washing, mask-wearing behaviors

Changed academic calendar to accelerate the fall semester

Required students required to sign pledge, agreeing to follow safety measures

Extended limitations on gatherings of people on main campus indefinitely

November 2020—Resumed remote instruction because of pandemic developments

Spring 2021

Provided HEERF applications for students to enable them to receive federal relief funding

Faculty and students volunteered at vaccination clinics

April 2021—Held public vaccination clinics on campus

May 2021—Lifted numerous safety protocols and implemented "honor code" for vaccination status

Held 2021 Commencement virtually

Fall 2021

July 2021—Resumed safety measures and required vaccinations by fall, required masks again

September 2021—Provided Federal CARES Act HEERF III application live for students

Spring 2022

March 2022—Made masks optional in US classrooms'

students lived. The international campuses and campuses on military bases and operations responded to their local conditions and government notices in coordination with the main campus task force and work groups.

Over the two years that Webster experienced various COVID-19 disruptions, a number of decisions were made and communicated to accomplish the priorities established by Webster's leaders. A timeline of these measures, with a focus on those impacting the main campus, reveals the force with which the pandemic disrupted the university's usual operations, providing both challenge and opportunity during the crisis and afterwards.

The actions taken addressed conditions of physical safety and more, indicating the strength of the university's sense of community and the longstanding commitment to "meeting unmet needs." The timeline illustrated in Figure 1 provides a snapshot of selected plans implemented on the main campus to meet those evolving needs.

The Crises Evolve: More Than a Pandemic

Because the pandemic's effects extended beyond threats to the physical health and safety of Webster communities around the globe, effective responses required multi-layered approaches. The associated effects of a global pandemic and the steps taken to arrest it had a direct impact on individuals and communities economically, socially, and psychologically. The disparate effects on the most vulnerable populations significantly enlarged the complex needs to be met. The layers of crisis demanded actions beyond public health solutions. In this environment, the university sought to maintain and build a sense of community, to meet needs through new approaches to philanthropy, address needed reckoning with issues of race and gender, and make global connections for a sustainable future. Balancing these needs with the steps required to assure safety meant that technology tools—traditional and innovative—would be key to sustaining the university.

Technology for Instruction

Webster's leadership committed to continuing to provide instruction and tapped its longstanding strength in online education. Webster's development of fully online programs in the 1990s had advanced the institution's historic goal of meeting an unmet need. The asynchronous online offerings in Webster's master's degree programs provided important access for working adults who could not access night or weekend courses, especially for women in the workforce who were juggling work and family commitments while still pursuing better employment opportunities. Access enabled service members who had enrolled in Webster programs at their bases or installations to continue degree progress when deployed. Webster's online programming was and continues to be unique to the community, having been built from the ground up by faculty and faculty/staff support/instructional design teams and customized to meet the specific needs of Webster students. Numerous third-party sources have recognized the quality of the university's online programs.

As a result of these in-house capabilities, Webster pivoted quickly to all classes being fully online at the start of the pandemic in 2020. Even departments with little or no history of providing online instruction were equipped with course shells and support that provided them with the tools to continue. Faculty experimented with synchronous platforms like Webex and then Zoom. Webster's approach was pragmatic—get everyone online and operating in as short a time as possible. The Faculty Development Center and the Online Learning Center and several staff members from Webster's extended campuses assisted in providing open office hours to support faculty. Between March 13 and June 16, 2020, 1760 hours of support services were provided. Participants were present in 1399 of them, an 80% utilization rate.

Faculty and staff with greater online experience convened online coffee hours to provide quick tutorials on Zoom tools. In addition to Webster's traditional online mode, greater flexibility grew, including WebNet+ which enabled students to dial-in to real-time classes occurring at set times and hybrid approaches that met face-to-face once per week coupled with online learning. Webster's WebNet+ offerings grew in Fall 2020 to 328 courses across 51 sites. The university's online programs now employ both synchronous and asynchronous methods for both graduate and undergraduate degree programs and certificates.

Those supporting this monumental undertaking to meet students' instructional needs observed that the effort was very humanizing. Egos were set aside as was natural reluctance to adopt new technologies rapidly in full view of students. The spirit of "we are all in this together" and a shared faculty-staff commitment to supporting one another was buoyed by the strength of Webster's faculty who commit to life-long learning as part of their identity. Leaders of the faculty development center opined, "What makes Webster faculty special? They figured out how to apply their pedagogical principles in a virtual setting, developing tools to enhance relationships." They observed that the pandemic pivot was "a stretching moment to create a community of learners—a preview of higher ed in the future—particularly hybrid models" (E. Palmore & L. Dister, personal communication, September 1, 2020).

In 2020 as universities navigated pandemic challenges in real time, *The Economist* Intelligence Unit issued a report sponsored by Microsoft Higher Education that investigated the "impact of the new higher education model spurred by covid-19 on teaching and learning experiences, engagement, performance, and value" (Bray, 2020b, p. 2). Accompanying the report, the unit also published a companion "faculty playbook" which included a case study about Webster's adaptions from a foundation of natural capabilities and innovations (Bray, 2020a).

The researchers' interviews and survey research focused on higher education faculty and students in the United States, United Kingdom, Australia, and Germany, inquiring into the extent that the pandemic catalyzed and accelerated remote and hybrid learning models and with what impact on engaging student learning. They concluded that a "key silver lining" in their findings was the "shattering of higher education's stereotype of being slow to respond to change and wedded to tradition. If anything, the sector's response to the pandemic has underscored its ability to pivot rapidly and embrace a new pedagogical model" (Bray, 2020b, p. 18).

Technology: Meeting Unmet Needs

The recognition of the needs of students and community, beyond continuity of instruction, spurred greater learning, leadership, and innovation on the part of the university community. It became important to identify the ways that the university could be helpful to students and staff—through emergency aid fundraising, through partnerships with donors and funding agencies, and through forums that provided information and resources. Webster's mission calls for ensuring an education designed to transform students for global citizenship and individual excellence. Pandemic conditions provided a real-life laboratory for these transformative experiences. In turn, the community not only gained experience in mobilizing quickly as the main campus closed and classes were fully online in a week's time but also built the capacity to attend to multiple realities across campus locations and within one another's lives.

Among the silver linings in Webster's online education adaptability was the recognition that digital tools could be used to engage the community to address layers of pandemic need. Among these was the need to overcome the isolation experienced by members of the community when lockdowns, remote and modified operations, online instruction, and restrictions on events and gatherings removed both formal and informal interactions.

A strong sense of community was supported as the head of our Global Marketing and Communications unit launched a podcast that featured stories from across the global system—how the pandemic was affecting individuals at each campus and how groups and individuals were showing leadership to help everyone remain safe through these challenging conditions. The consistent message was that members of the community would keep each other safe, relying on individuals' concerns for each other and the community, using preventative and self-reported measures rather than required universal testing.

As the university experimented with virtual platforms like Zoom to build community, members of the community were invited to share their art, dance, theatre, and music as a starting place in a forum called #WebsterAtHome. Together, people participated in performances, attended academic presentations, engaged in conferences across the Webster network, provided counseling and career advice, and shared photos of pets, as a few examples. Visual artists showcased their home workspaces, theatre students produced cabarets, and dancers prepared video portfolios. Faculty, staff, and students stayed connected through technology and enlarged their communities by including those who could not be physically present. Technology bridged the boundaries of time and space, and, ironically, the global community grew closer as they were physically isolated. Zooming became a proxy for physical proximity. For many students, the screen became the new dorm room.

Support services were enlarged as the Academic Resource Center expanded its services to function virtually. After receiving donor support, the renamed and renovated Reeg Academic Resource Center also expanded in-person capacity, enabling it to be more effective in St. Louis and across the network. In 2020, Webster's summer bridge program—Transition Academic Program—continued in a virtual format. Webster students volunteered to serve learners in the region, tutoring virtual K-12 students in literacy and mathematics. The innovations Webster adopted in self-reporting, social distancing, and masking resulted in very high levels of safety and subsequent high levels of adherence to vaccination and masking requirements. Athletic competition continued when the campus reopened in Fall 2020, resulting in the commissioner's cup award for athletic competitiveness and sportsmanship. Students showed they could be counted on to do what is in the interests of being present and keeping everyone safe.

The needs of the main campus community were many, including emergency support for students and staff. The pandemic economic disruptions resulted in job losses and loss of needed income for students and family members as well as loss of services in the community. Emergency expenses for health services and childcare rose as families juggled working from home amid cycles of school openings and closures, as well as supporting their children's online learning. Those who were working on the front lines in emergency services and health care were challenged to continue degree programs because of the rapid changes and demands of their working conditions. International students who could not return to their homes required support for summer living expenses.

Members of the community rose to the challenge these needs presented. During the Summer of 2020, Webster's Advancement team initiated their very first use of a digital crowdsourced fundraising drive. They asked friends, alumni, and members of the university community to help fund just-in-time needs with just-in-time dollars. The results were stunning—in addition to meeting the emergency needs of students and employees the drive energized many individuals who made a gift for the very first time. Digital fundraising perfectly matched a solution with a need. Lead donors initiated the efforts to establish targeted funds for specific needs, and Webster invested institutional dollars to create a time-limited matching program designed to incentivize unrestricted pledges and gifts and continue an evergreen fund. In 2020-2021 $315,985 was committed from 844 gifts, benefitting hundreds of individuals. The matching program yielded $327,909. The ability of these online crowd-sourced means of seeking philanthropic support proved so successful that their use expanded as pandemic conditions waned. These mechanisms are uniquely suited to supporting short-term needs for student organizations and athletic teams. The digital platform perfectly marries a solution with a need and has the added benefit of building a culture of philanthropy by engaging first-time donors at all levels while engaging those who benefit from donor support in the effort. By these actions, the advancement team contributes to building a virtuous cycle of giving. Webster's digital fundraising now supports an annual giving day as well as the evergreen fund that persists in funding students' emergency needs.

Societal Impacts

Beginning in Spring 2020, Webster's Global Marketing and Communications (GMC) unit accelerated and enlarged their communications to the members of the community through almost-daily e-mail messages, as well as a dedicated website that aggregated information from the Webster global campuses. Because accurate and timely information was needed to guide individuals' actions, these communications were a critical component of Webster's proactive approach. The dedicated website functioned as a clearinghouse for needed information and resources, including the health self-assessments and required public reporting of infection counts. It is important to note that throughout the two years of pandemic conditions, Webster's infection rate was remarkably and significantly low (Nay, 2020).

While the number and type of COVID-19 communications grew, the GMC team mobilized to communicate about the issues of racial equity and justice as a time of racial reckoning gripped the United States and the world. The May 2020 killing of George Floyd by police in Minneapolis, Minnesota, later judged to be murder, prompted protests in the United States and globally, reflecting a pent-up frustration and anger expressed in outrage over the killings of unarmed members of the African American community. The 2016 killing of Michael Brown and its aftermath in nearby Ferguson, Missouri, were close in memory of St. Louisans and the Webster main campus community. The need for dialogue with and among the community was clear and consistent with Webster's core value of diversity, equity, and inclusion and historic commitment to access and success for a diverse community of students.

After the 2016 events in Ferguson, Webster organized its first annual diversity, equity, and inclusion conference—a one-day event featuring expert speakers representing varied communities and viewpoints. From that initial convening for critical conversations about race, gender, religion, and ethnicity, the annual conference has grown in its size and impact. In 2020, the conference had just convened with four days of panels and speakers, welcoming 1600 participants. As the pandemic conditions present in 2021 created continued risk for large gatherings, Webster organizers shifted the conference to an online conference platform, enlarging the pool of topics, presenters and participants. To this day, a hybrid format continues because of its advantages for participation and impact.

As awareness grew that the pandemic effects were more severe for racial and ethnic minority groups, with increased risk of becoming ill and dying from COVID-19, attention focused on inequities in access to healthcare that contributed to disparate outcomes for these populations. The systemic barriers and structures of racism were increasingly apparent in these pandemic conditions, and the Webster community committed to showing concern for students and employees who were isolated from the natural bonds of community. Expressing that concern took many forms, as illustrated in Figure 2.

In a partnership to provide direct support for the resources and activities designed to support BIPOC students and colleagues, the Office of Diversity, Equity, and Inclusion and Advancement launched a new giving society, the Jackson & Thomas Society. Named for two African American women who graduated from Webster in 1950, the society welcomes gifts to support the activities of the DEI office as they meet needs of the community. A special appeal built a strong foundation of support from members of the Webster community, alumni, partners, and friends. The fund, which was created in the time of pandemic crisis necessity, continues as the community's needs evolve.

Collaborations between the DEI Office and Global Marketing and Communications took advantage of new capabilities in videoconferencing to offer a new series called "Webster Speaks." In weekly sessions, the Chief Diversity Officer moderated live discussions with national experts, professionals in the community, faculty, staff, and students to explore issues of race, equity, and inclusion. The series was broadcast for more than 20 episodes, drawing hundreds of viewers. Again, technology tools proved critically important in strengthening bonds of community, overcoming feelings of isolation and marginalization, and providing needed information and resources.

Making Global Connections

The new digital tools used for new purposes were instrumental in the university's ability to make global connections, preserving instructional offerings and building a competitive advantage in international

Figure 2. Webster's support for BIPOC (Black, Indigenous, People of Color) students and employees (Stroble,et al., 2021)

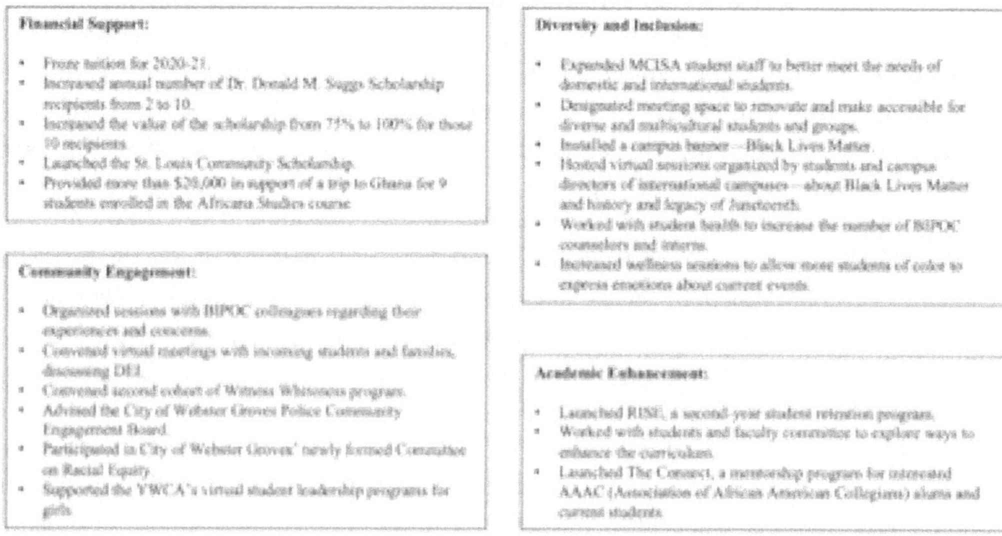

locations when competing institutions had little to no capacity for online education. These tools also helped to form stronger bonds with international partners with global missions, especially those whose work focused on pandemic effects on vulnerable populations. Such was the case with the expanded partnership with the African Renaissance and Diaspora Network (ARDN), an NGO based in the United States and with a global orientation.

Having first signed an MOU with ARDN in 2017, Webster took a more active role in 2020 and 2021 to advocate for the achievement of the Sustainable Development Goals with a focus on #5, Gender Equality. That activity included the Chancellor's co-chairing ARDN's higher education initiatives and helping to launch ARDN's international Red Card campaign at the United Nations in February 2020, just prior to the March closing of the main campus. The campaign was designed to raise awareness about domestic violence and gather personal pledges from people around the world to help raise awareness of discrimination and violence against women. As raising a red card in the global games of soccer/football signals a serious infraction of the rules of the game and the need to desist, ARDN's campaign asks individuals to sign a pledge and raise a physical red card to call for the end of all forms of violence and discrimination against girls and women.

Because of Webster's mission focus on global citizenship, the teamwork with ARDN included participation in four different online summits featuring intergenerational speakers across the African Diaspora. These summits were designed to build knowledge and engagement to achieve the UN Sustainable Development Goals. The virtual summits reached worldwide audiences as speakers shared their lived experiences and highlighted the direct and indirect harm the coronavirus created for girls and women around the world. The summits provided needed information about the challenges of access to healthcare and mental health support and the economic and personal costs women and girls were experiencing. The theme of racial injustice was strong across the diaspora communities and speakers.

In September 2021, Webster launched its own Red Card campaign, consisting of rallies, educational forums, invited speakers, flash mobs, athletic events, library exhibits, and partnerships with community groups and local city government. The campaign, led by a committee of faculty and staff, included the main campus as well as campuses across the United States and internationally. Scheduled events and social media provided resources to prevent and address gender-based harassment and discrimination while popularizing the Sustainable Development Goals. Hundreds of engagements resulted from the campaign, which culminated in the awarding of an honorary degree and Commencement address by ARDN CEO Dr. Djibril Diallo when in-person ceremonies resumed in 2022.

From Sustaining to Rebuilding a Global University System: Silver Linings, Lessons Learned, Signs of Resilience

As the pandemic's pervasive impact was felt worldwide, both the immediate and longer-term impacts on higher education became topics of interest and inquiry. Was it wise to return to on-campus conditions? Should testing be required or voluntary? How about masks? Vaccines? Could online learning provide a quality education? How should institutions engage learners? How much emergency relief should be provided to institutions and to students? How should growing mental health challenges be met? What role should higher education play in combatting systemic racism as the Black Lives Matter movement gained momentum? Could internships and study abroad be provided virtually?

Of course, the speculations about the future of higher education were also present. It had become commonplace to describe U.S. higher education as being at an "inflection point" prior to the pandemic, and pandemic conditions accelerated and often exacerbated the forces that were already taking form to threaten the academic and business models of traditional institutions.

As a result of pandemic conditions, Webster experienced direct and indirect impacts on required costs. These included needed investments in disinfection, plexiglass, software licenses, equipment, maintenance of physical spaces, and enhanced and new technologies. Similarly, Webster experienced opportunity costs as it was necessary to adjust physical spaces, first emptying and limiting students' housing. The university's metropolitan and extended campuses, which primarily served Black and underserved populations, experienced a 20% drop in enrollment that was directly correlated with the loss of employment in these communities. To preserve service members' safety, military bases and installations closed access to civilians; enrollment plummeted.

Loss of revenue was a common experience as enrollment decreased and costs mounted. Webster provided refunds for room and board for the remainder of Spring 2020, when residence halls were closed. When the residence halls re-opened, safety considerations limited rooms to single-occupancy. International students experienced delays in gaining student visas even when international travel became possible. Comfort with returning to in-person activities ranged from enthusiasm to reluctance.

This environment required stamina, commitment, and creativity to balance the requirements of safety and the needs of students to persist and complete their degrees. While there are many lessons learned, the importance of teamwork, communications, and strong partnerships cannot be overstated. The increased capacity for teamwork, communications, partnerships, and transformative uses of technology were elements of the resilience that preserved safety and community during the years of pandemic conditions.

It is early to consider the consequences of pandemic conditions and associated crises for rebuilding higher education systems, and drawing conclusions about cause-effect relationships is always fraught. Yet, we can see the signs of ongoing consequences that seem to extend beyond a single institution or system. We can observe the social, economic, and political contexts reflected in our higher education systems and their impacts on our opportunities and limitations in advancing our missions.

Among those contexts are changing work conditions and preferences among faculty and staff. Similarly, institutions have seen changes in students' preferences for program delivery options and programs themselves. While travel conditions for global mobility have eased, concerns for safety have taken on new importance. Economic disruption has affected the financial health and sustainability of higher education. Affordability has been threatened in multiple ways. For institutions that have focused on expanding equity, the impact of judicial and legislative actions is felt in essential functions: admissions, financial aid, employment, and industry partnerships. The challenges of disinformation demand greater attention to media literacy; gaining accurate information tests public dialogue, which is increasingly polarizing and confrontational.

What merits thoughtful reflection is the cost and consequence to community itself—locally and globally. Has the isolation that kept us physically secure at the same time eroded our trust and confidence in each other? Evidence mounts in the United States and Europe that pandemic conditions have affected many students' mental health and sense of well-being (Hu et al., 2022; Kiltz et al., 2023). How have these effects targeted particular groups of students and equitable opportunities for their participation and success (Farnell et al., 2021)? Have the relief measures taken by governments to maintain our viability during crisis times now left institutions without a path forward during trying economic circumstances? And are the growing numbers of leadership transitions a byproduct of the toll taken by the pandemic on leaders and institutions?

How will the building blocks of resilience tapped during the pandemic help higher education institutions rebuild their own strength of community in the days and years ahead? Explicit attention to sustaining a globally diverse community is essential to the path forward for Webster and may be instructive for higher education leaders helping their systems adapt, recover, and prosper.

"We Did Not Stand Still"

These words became a rallying cry for the dynamics of sustaining a global university system during the pandemic years. They signified the resilience that typified Webster for more than one hundred years, evolving from a small Midwestern Catholic women's college to an independent comprehensive university educating more than 13,000 women and men on three continents. They reflected the characteristic spirit of innovation at Webster. A bias for action and a commitment to mission were embodied in "We did not stand still."

As institutions managed what was required to maintain operations through rapidly changing circumstances, the circumstances also called for putting in place the seeds of future opportunity. The calls to seize higher education's defining moments were ubiquitous; terms like "pivot," "unprecedented," and "the new normal" became cliché. Not all were sanguine about the status and outlook for universities and their students. The research underway pointed to areas of promise as well as peril for the sector (Farnell et al., 2021).

Surveys completed in Fall 2020 (Barker et al., 2020) indicated Americans were divided on whether colleges made the right decision to bring students back to campus. Skepticism raised questions regarding students' safety and, at the same time, the comparability of online classes. College graduates were most likely to question their educational value. Concern about the general direction of higher education tended to fall along political party lines. And the impact of changing conditions caused a "decline in college enrollment [that]was more pronounced among Black and Asian young adults, the foreign born and those living in the Northeast" (p.3). Fewer men aged 18-24 were enrolled, while women's enrollment was unchanged. Economically vulnerable students delayed graduation because of the pandemic (Dua et al., 2020, pp. 1-2).

In 2024 it is easier to see that "online learning is here to stay" (Ubell, 2021), and where it was less favored in the United States and Europe prior to the pandemic, investments by institutions and governments have increased the capacity to provide online learning and to do it well (Guidi et al., 2023). A McKinsey report proclaimed,

Remote and online learning are here to stay. The need is to determine what combination of remote and in-person learning delivers the highest educational quality and equity. As institutions refine this hybrid model, they have a once-in-a-generation chance to reconfigure their use of physical and virtual space. (Dua et al., 2020, para. 22)

Learners of many types have gained comfort, confidence, and competence with online tools and virtual environments. While a 2021 Accenture study of 6500+ prospective and current postsecondary students aged 16-65+ found a mix of students who preferred in-person to online instruction, the learners overwhelmingly preferred digital solutions for enrolling, registering, paying bills, and gaining IT support as examples (Fisher et al., 2021). According to Gaetz (2023),

Today, students hold these administrative functions to a higher digital standard, as 96% say a high-quality digital experience is important to their satisfaction with their campus. This priority reflects a broad set of shifts impacting today's student body—they are digitally savvy, demographically and geographically diverse, and approach higher education with new needs. (para.1)

Student needs and expectations have evolved. Many believe "universities should improve their non-academic and administrative support, including mental health and financial counseling. They expect more virtual services, from IT and financial services to counseling. Since the pandemic, they also expect remote and hybrid class models" (Gaetz, 2023, para. 10). These evolving needs present both challenge and opportunity. Rebuilding university systems will prove most successful as institutions are able to make needed investments in innovation and new growth (Bar Am et al., 2020). Innovations will be required for adaptations in programs and services that respond to a range of students' shifting and growing needs. Drawing on examples from the fundamental changes in restaurants and dining in Fall 2020, the Education Advisory Board suggested, "In the drive to differentiate, institutions often hope for a first-mover advantage—and there are many opportunities for this in the pandemic" (Adams, 2020, p. 5). Standing still is not an option.

While innovative uses of technology were instrumental to providing educational programs to students across the network, the benefits opened by other uses of technology to advance Webster's mission and core values were immeasurable. Due to the engagement with ARDN, our commitment to equity issues took on new dimensions with greater knowledge of varying conditions by communities here and abroad (Stroble, 2021, 2022). The intentional sharing of pandemic experience stories, the virtual DEI conference, and webinars like Webster Speaks expanded our connections with members of the global Webster community and beyond. Even in the two years of pandemic conditions, faculty, staff, and students stayed connected academically and operationally with topics of global concern such as refugee support and human trafficking. Through these activities the university was building a platform for global mobility when study abroad and international recruitment would return.

Webster's history of taking education to the student had taken on new dimensions in 2019 as a new campus was launched for several hundred students in Tashkent, Uzbekistan, in partnership with the Republic of Uzbekistan. Four years later, more than 4000 students are enrolled there, and a second campus will soon be built for Webster students. In Fall 2022, Webster welcomed more new students than in recent memory, including more international students. And in Fall 2023, the university returned to pre-pandemic levels including growth in enrollment of international students in the United States and in Central Asia (Stroble, 2022). Applications for study abroad steadily increase as the university pilots new models for providing these experiences for Webster students and a significant new number of affiliate partners. It is important for internationalization to innovate and take on new forms, including the use of hybrid models to expand access (Guidi et al., 2023).

Instrumental to these results were the developments to strengthen Webster's academic offerings in high-demand fields, including the sciences, healthcare, data analytics, cybersecurity, and emerging communications fields as examples. Webster created a new honors college and restructured academic programs to align better with workforce needs. In place of Webster's large and expansive College of Arts and Sciences, there are now two more focused colleges: The College of Science and Health and The College of Humanities and Social Sciences. These new units will provide important coherence for faculty teaching and scholarship and increase the visibility for growth in health professions and science programs. The accompanying growth in STEM programs is instrumental to attracting more international students and the developing partnerships to address healthcare needs in Central Asia. Continued investments in technology are apparent in the renovations made in the School of Communications building, now equipped with cutting edge state-of-the art equipment and media support.

As university systems take advantage of tailwinds that propel innovation and growth, of course, headwinds persist. Economic challenges continue for many students and families as well as funding agencies, raising issues of affordability and the sustainability of the business model for higher education regardless of the base funding support—governments and/or student tuition. Concerns about the burden of student debt and how to provide debt relief highlight affordability questions. As federal relief funds for institutions and students ended, institutions experienced growing demands on institutional funding at a time that revenues were reduced. Enrollments declined and some traditional markets are no longer viable. It is commonplace to refer to an impending "demographic cliff" for traditional age enrollments. At the same time, skepticism has grown about the value of degrees, and employers may be less likely to subsidize enrollments by their workforce. It is not surprising that media attention to mergers and closings persist (Selingo, 2022).

An end-of-year report by WGU Labs (2023) shares their perspective on the potential impact of AI and other emerging trends on the "7 wicked problems" of higher education. Those seven problems include:

- Postsecondary access
- Cost of education
- Student belonging
- Tech-enabled learning
- Diverse learning paths
- Learning-to-work transition
- The future of learning.

Developments in AI and the growing uses of ChatGPT have, without question, changed higher education in 2023 (McMurtrie, 2023). The year 2024 is likely a pivotal year for transformative changes not imagined during the health threat, isolation, economic and societal disruptions, and changing global dynamics of the pandemic years. The "new normal" is a work in progress.

Each institution's rebuilding story is unique to that institution—its mission and core values and how the institution builds a future from the learnings of pandemic crises. Higher education systems that adapt, that value and capture innovative, global, and diverse approaches to meet the individual and professional needs of students are best positioned to survive and thrive. It is an advantage—and in Webster's case, an imperative—to strengthen virtual capacity, embrace a distributed sense of place, and welcome engagement with diverse people, communities, and markets.

ACKNOWLEDGMENT

With gratitude I acknowledge the skills and talents of Webster University graduate student and research assistant, Chris Groh. Absent his assistance, this project would not have been completed. I acknowledge the contributions of co-authors cited in this chapter as well as the organizations with whom I have partnered during my tenure as Chancellor. Finally, this chapter contains the narrative and insights gained from being a member of the resilient community of Webster University. With gratitude for their exceptional perseverance, I share what I have observed.

REFERENCES

Adams, M. (2020). *What higher ed can learn from restaurants.* EAB. https://eab.com/insights/resource/strategy/higher-ed-lessons-restaurants-competition/

Bar Am, J., Furstenthal, L., Jorge, F., & Roth, E. (2020, June 17). *Innovation in a crisis: Why it is more critical than ever.* McKinsey & Company. https://www.mckinsey.com/capabilities/strategy-and-corporate-finance/our-insights/innovation-in-a-crisis-why-it-is-more-critical-than-ever

Bray, M. (2020a). *Flattening the multimodal learning curve: A faculty playbook.* The Economist Intelligence Unit. https://impact.economist.com/perspectives/sites/default/files/eiu163_-_msft_multimodal_learning_playbook_v4_1.pdf

Bray, M. (2020b). *Bridging the digital divide to engage students in higher education.* The Economist Intelligence Unit. https://impact.economist.com/perspectives/sites/default/files/eiu_microsoft_higher_education_report.pdf

Dua, A., Law, J., Rounsaville, T., & Viswanath, N. (2020, October 26). *Reimagining higher education in the United States.* McKinsey & Company. https://www.mckinsey.com/industries/education/our-insights/reimagining-higher-education-in-the-united-states#/

Farnell, T., Skledar Matijević, A., & Šćukanec Schmidt, N. (2021). *The impact of COVID-19 on higher education.* European Commission. https://data.europa.eu/doi/10.2766/069216

Fisher, S., Brodie, J., & Pollman, P. (2021). *Serving all students: A survey of US learner mindsets from age 16 to 65+.* Accenture. https://www.accenture.com/content/dam/accenture/final/a-com-migration/r3-3/pdf/pdf-165/accenture-serving-all-students.pdf

Gaetz, R. (2023, December 14). *Serving the digital student.* LinkedIn. https://www.linkedin.com/pulse/serving-digital-student-ryan-gaetz-vee4e

Guidi, E., Jensen, T., & Marinoni, G. (2023). *Shaping teaching & learning and internationalization beyond the pandemic.* International Association of Universities. https://www.iau-aiu.net/IMG/pdf/final__shaping_teaching_learning_and_internationalization_beyond_the_pandemic_2_1_.pdf

Hu, K., Godfrey, K., Ren, Q., Wang, S., Yang, X., & Li, Q. (2022). The impact of the COVID-19 pandemic on college students in the USA: Two years later. *Psychiatry Research, 315*, 114685. Advance online publication. doi:10.1016/j.psychres.2022.114685 PMID:35872401

Kiltz, L., Trippenzee, M., Fleer, J., Fokkens-Bruinsma, M., & Jansen, E. P. W. A. (2023). Student well-being in times of COVID-19 in the Netherlands: Basic psychological need satisfaction and frustration within the academic learning environment. *European Journal of Psychology of Education.* Advance online publication. doi:10.1007/s10212-023-00680-x

Labs, W. G. U. (2023, December 15). *From wicked problems to equitable solutions: WGU labs end of the year impact report.* https://www.wgulabs.org/posts/from-wicked-problems-to-equitable-solutions-end-of-year-report#learning-to-work-transition

McMurtrie, B. (2023, March 6). *ChatGPT is already upending campus practices: Colleges are rushing to respond.* The Chronicle of Higher Education. https://www.chronicle.com/article/chatgpt-is-already-upending-campus-practices-colleges-are-rushing-to-respond

Nay, B. (2020, October 9). *St. Louis area schools, universities continue to meet COVID-19 challenges.* Ladue News. https://www.laduenews.com/business/st-louis-area-schools-universities-continue-to-meet-covid-19-challenges/article_83e5d380-a5b8-5896-a5a2-2e2c5c2be6d2.html

Parker, K., Barroso, A., & Fry, R. (2020, October 26). *Americans are divided on whether colleges that brought students back to campus made the right decision.* Pew Research Center. https://www.pewresearch.org/short-reads/2020/10/26/americans-are-divided-on-whether-colleges-that-brought-students-back-to-campus-made-the-right-decision/

Selingo, J. J. (2022, September 28). *How the pandemic changed higher education.* FutureEd. https://www.future-ed.org/how-the-pandemic-changed-higher-education/

Smith, J. (2020). *Wake 'em up! Super-charge your student engagement.* The Part-Time Press.

Stroble, E. J. (2021, July 29). *Innovation & technology during pandemic & post-pandemic* [Conference presentation]. International Association of University Presidents Triennial 2021 Virtual Conference: Innovation and Inclusion, Key Priorities for Higher Education in a Post Pandemic World, CETYS University, Mexicali, Mexico.

Stroble, E. J. (2022, September 29). *Key leadership challenges for post pandemic universities: Strategy and design of better, different, and resilient institutions* [Conference presentation]. 2022 International Association of University Presidents Regional Meeting: Recovery and Transformation: New Challenges for Higher Education Leaders, Autonomous University of Bucaramanga, Bucaramanga, Colombia.

Stroble, E. J., Schuster, J. Z., & Flewellen, V. C. (2021). Rooted in resilience: A global university responds to twin pandemics. *Journal of Higher Education Management, 36*(1), 189–197. https://issuu.com/aaua10/docs/twin_pandemics

Ubell, R. (2022). *Staying online: How to navigate digital higher education.* Routledge.

KEY TERMS AND DEFINITIONS

Artificial Intelligence (AI): AI refers to the study and practice devoted to a combination of computer science and robust data sets. By combining these two fields, experts and study the ability and application of machines for capabilities ranging from reasoning and problem-solving, representing knowledge, communicating, and learning.

Asynchronous Learning: By contrast, asynchronous means not happening at the same time. The methods used for this kind of instruction typically occur through web-based methods, and students can access these classes with more flexibility and no set class times.

ChatGPT: ChatGPT is a proprietary tool from the company OpenAI that allows uses to generate text based on prompts. ChatGPT's artificial intelligence is generative which means it is not creating data but gathering it from many sources. By collecting data, it refines the tool's ability to respond to a wide range of queries and research questions and support tasks posed by users. The responses to prompts are not connected to the data it gathers, which can affect the validity of those responses.

Demographic Cliff: This term is used to describe the impending decline in the number of traditional age college students in the United States, with varying impacts in regions of the country. This decrease in this population can be predicted by studying birth rates and residential data, looking at the time prospective students would have been born.

Synchronous Learning: Synchronous means happening at the same time. This phrase refers to online methods of instruction that occur simultaneously between instructor and class members, typically through use of video.

U.N. Sustainable Development Goals: In 2015, the member states of the United Nations adopted the 2030 Agenda for Sustainable Development which comprises seventeen goals. The goals of the plan, and their specific targets, address the threats to the world's sustainable future. They range from ending poverty to assuring quality education. Included are goals related to climate change, health, economic growth, and justice.

WebNet+: This term refers to a specific synchronous learning method used by Webster University to make live in-person classes accessible to students connecting to the class in real-time by using the internet from their home or workplace.

Zoom: Zoom is a proprietary platform that supports synchronous audio and video connections for the purposes of meetings and instruction. Zoom sessions may also be recorded to support asynchronous learning.

Chapter 7
Exist or Exit:
Resilience During and Post COVID-19 Pandemic Crisis in Higher Tertiary Institutions

Mubango Hazel
University of Venda, South Africa

Hlanganipai Ngirande
https://orcid.org/0000-0003-2604-2156
University of Venda, South Africa

Khathutshelo Khashane
University of Venda, South Africa

ABSTRACT

The COVID-19 pandemic affected the entire world causing every aspect of the higher education ecosystem to undergo a tipping point. The total mandatory lockdown resulted in all academic activities from learning to administrative activities coming to a standstill as institutions did not know how to go around it. Employees' job security, indefinite suspension of lectures, disruption of the university calendar, employee work-life balance, and business continuity were all operational challenges faced by universities during the pandemic. This chapter aims to establish the resilience strategies implemented by higher education institutions in South Africa and Zimbabwe as they strived to continue delivering services to stakeholders during and after the COVID-19 pandemic crisis. Digital transformation as a primary resilience strategy enabled HEIs to implement online learning and work from home for academic and non-academic staff for university business continuity.

DOI: 10.4018/979-8-3693-1926-0.ch007

INTRODUCTION

The Covid-19 pandemic had a global impact, leading to a significant transformation in many aspects of the higher education ecosystem. Higher education institutions serve as the central source of knowledge for the progress of any country, and their sustainability and survival were jeopardised by the COVID-19 pandemic. The absolutely mandated lockdown culminated in all academic activity from learning to office operations coming to a standstill since institutions were unable to figure out how to go around it. Tertiary institutions encountered various operational issues during the pandemic, including employees' job security, indefinite suspension of lectures, interruption of the university calendar, employee work-life balance, and business continuity. Due to the crisis, Higher Education Institutions (HEIs) were compelled to transition to remote mode, necessitating rapid adaptation by lecturers, students, and administrative personnel to the new reality (Bartolic *et al.,* 2022; Biwer *et al.,* 2021). The digitalisation of procedures at higher education institutions facilitated the continuation of teaching and learning, as well as administrative responsibilities, through online learning, blended learning, and remote work, respectively.

BACKGROUND

Out of the total of 1.91 billion learners worldwide, 1.6 billion experienced adverse effects due to the closure of schools, colleges, and universities. This accounts for approximately 91% of all enrolled learners globally, (Gyamerah, 2020; UNESCO, 2020). According to UNESCO, the closure of higher education institutions (HEIs) and schools in 185 countries had an impact on 9.8 million African students enrolled in HEIs (Tamrat, & Teferra, 2020). COVID-19 has been a severe challenge to the higher education industry, leading to financial losses and uncertainty for many institutions, employees, and students as economies strive to recover. The worldwide instability caused heightened uncertainty in all sectors of the market economy, including the labour market, affecting students as well (Baker et al., 2020). The economic slowdown induced by the pandemic led to a reduction or complete loss of income for students who lost their part-time and full-time jobs, internships, or job offers due to business closures (Aucejo et al., 2020). National lockdowns led to a significant decrease in the income and financial security of numerous students and their families, rendering them unable to cover tuition costs. On the other hand, due to the tuition fee deficits, several higher education institutions lowered salaries or terminated employees, resulting in a decrease in the number of permanent employees. Several universities faced challenges in offering needed financial assistance to underprivileged and struggling students because of the decrease in tuition revenue and other funding sources (Dennis, 2020; Singh et al., 2022).

HIGHER EDUCATION BEFORE THE PANDEMIC

Education systems before the pandemic were traditional from the administrative processes to lectures. Lectures were done face to face where every student was expected to attend physically and hand in assignments as hard printed copies. The student had the priviledge of having physical interactions with the lecturer during and even after the lesson. On the administrative side, the application process of student enrolment was done physically with potential students visiting the university campuses to make the application. Payment of school fees, writing of exams, and even exam processing were mostly done in the physical space. The face-to-face lectures ensured that all students benefited from learning.

Covid-19 and the Higher Education Institutions Workplace

Higher education institutions worldwide found themselves in a novel situation following the implementation of Covid-19 boundaries. Universities globally promptly suspended or terminated all on-campus activities, such as face-to-face teaching, laboratory research, exams, sports, leisure activities, and conferences, due to concerns about the rapid transmission of the virus (Grange, 2020). The shutdown of educational institutions during the COVID-19 pandemic impeded global educational growth and learning, as well as the functioning of education systems and other social activities performed by educational institutions (UNESCO, 2021). Several universities responded by cancelling face-to-face lectures, and graduation ceremonies, and restricting postgraduate students' access to lab facilities (Manyuchi *et al.,* 2021).

As the pandemic in the United States passed the two-year mark since the initial wave of school shutdowns, academic normalcy remained unattainable for many students, lecturers, and parents. Alongside the rise in COVID-19 cases in late 2021, universities have seen significant deficits in staff, elevated levels of absenteeism and quarantines, and periodic closures. Moreover, students and educators faced difficulties related to mental health issues, increased levels of violence and misconduct, and worries about disrupted educational processes (Kuhfeld et al., 2022). The UK Office for National Statistics (ONS) research revealed that 29% of students expressed discontent with their student experience during lockdown. Additionally, 65% of participants expressed concerns regarding their accommodation and a decline in their general life satisfaction (Zimmermann et al., 2020). Contrarily, most students in the Republic of North Macedonia were content with the university's implementation of online platforms during the COVID-19 outbreak and the assessment process for teaching and learning, (Stankovska, 2022). China's government implemented the policy "Suspending Classes Without Stopping Learning and Suspending Classes Without Stopping Teaching" to combat the pandemic by temporarily stopping online teaching in schools and transitioning to online education (Ministry of Education of the People's Republic of China, 2020).

The African region exhibited the most significant proportion of HEIs that implemented campus closures as a result of the COVID-19 pandemic. Regarding the transition of HEIs to online teaching and learning, Africa was the sole region where teaching was temporarily halted or terminated before transitioning to online learning methods. Nevertheless, about 29% of African HEIs managed to swiftly adapt to online teaching and learning, as well as virtual methods for carrying out administrative tasks (Toquero, 2020). According to a recent African survey, around 39% of privately owned educational institutions and 18% of state institutions successfully shifted to online learning right away (Waithima, 2021). In Zambia, to assist remote learning during the school shutdown, online 'E-learning' and 'Smart Revision' platforms were created (UNESCO, 2022).

On May 1, 2020, the President of South Africa declared a gradual relaxation of the lockdown measures to alert level 4. This permitted only final-year students to receive face-to-face lecture instruction. Some HEIs decided to permit students to resume in-person classes, while other universities decided against conducting face-to-face sessions for the rest of the 2020 academic year (Ngqakamba, 2020; van Schalkwyk, 2022). Consequently, students were compelled to engage in online and blended learning as it emerged as the sole feasible remedy during the COVID-19 pandemic. In addition, office staff were required to telecommute and alternate their return to the workplace to mitigate the transmission of the virus.

The pandemic has widely been recognised as a disruptive factor in the smooth functioning of operations at HEIs. The higher education community experienced the adverse effects of the pandemic, including disruptions to teaching and learning activities, as well as infections and the loss of human capital due to fatalities. The cancellation or postponement of on-campus events at universities included face-to-face lectures, workshops, conferences, sports, graduation ceremonies, and other important activities. The primary objective was to prevent students and employees from acquiring the highly transmissible illness (Mudenda, *et al.,* 2020; Sahu, 2020).

Organisational Resilience

Organisational resilience is a growing topic in contemporary business practices following the recent economic recession and pandemics. It refers to an organisation's ability to effectively handle unexpected changes and adapt to previously unknown dangers in the business environment (Biggs *et al.,* 2012). Business resilience can be demonstrated by an organisation's ability to both patiently wait for a return to normalcy and adapt and develop in the face of disruptions, (Dahles & Susilowati 2015).

Organisational resilience refers to the capacity of an organisation to efficiently adapt, respond, and ultimately implement transformative measures to modernise in the face of unexpected disruptions that could potentially jeopardise its survival (Lengnick-Hall *et al.,* 2011). This concept emphasises greatly on "bouncing back" to the initial equilibrium and considerably makes use of coping strategies (Heeks & Ospina, 2019). Resilience in higher education institutions (HEIs) entails more than just coping with and resolving the impact of shocks. It involves surpassing mere recovery, acquiring new abilities, and capitalising on new opportunities (Lengnick-Hall *et al.,* 2011; Wildavsky, 1988). Additionally, it necessitates learning from these disruptive events (Giustiniano *et al.,* 2018).

There is a shift from a focus on providing relief to adopting a proactive strategy (Sakurai, & Chughtai, 2020). The effectiveness of proactive resilience during the COVID-19 pandemic primarily relies on solutions driven by technology (Djalante *et al.,* 2020; Ting *et al.,* 2020). Universities had to devise technologically-driven resilience methods to navigate and overcome the unprecedented challenges posed by the Covid-19 pandemic, to restore their regular operations both during and after the pandemic.

THE COVID-19 EXPERIENCE OF HIGHER EDUCATION INSTITUTIONS IN SOUTH AFRICA

The education system, like other crucial sectors, experienced significant repercussions, and South Africa is no exception. The closing of higher educational institutions in South Africa in March 2020 was a deliberate government measure aimed at curbing the spread of COVID-19, in line with global policies (Nwosu, 2021). Due to the intermittent closures of schools in South Africa, almost one-third of the academic year was missed in 2020. As a result, the nation's 13 million students fell behind in curriculum coverage and faced the potential of developing significant skill shortages that may persist beyond 2030, (Hoadley, 2020). The closure of educational institutions disrupted academic activities, resulting in the cancellation and delay of international and local conferences, workshops, and sports tournaments (Yaya, Out & Labonté 2020). Furthermore, practical-oriented courses were significantly impacted, leading to the postponement and cancellation of examinations, particularly internal assessments. Occasionally, it also led to a modification of assessment techniques.

Also, the South African universities for example the University of Cape Town, the University of Pretoria, the University of Venda, the University of South Africa, the University of Kwa Zulu Natal, and the Nelson Mandela Metropolitan University were all affected by mandatory closures of their academic institutions. The institutions were required to adhere to the lockdown mode, and all aspects of their campuses and operations-including conferences, graduation ceremonies, research workshops, seminars, and tutorials-were restructured. All staff members were told to work remotely by employing e-learning resources, while every student was told to go back to their homes (Mapanga, 2020).

Due to the significant decline in student enrollment, certain institutions that heavily depend on the admission of both local and international students experienced a sense of job insecurity among their staff. This was because there was a risk that the university might downsize, meaning that they would reduce the number of employees (Charoensukmongkol & Phungsoonthorn, 2020). Moreover, the financial solvency of institutions, especially those relying on such funding, has been emphasised as a result of the imminent sharp decline in tuition payments from international and local students (Plass, 2020). Universities were required to serve as dependable entities in combating a crisis of such magnitude by formulating measures to ensure the survival of institutions and the efficiency of workers.

The COVID-19 Experience of Higher Education Institutions in Zimbabwe

The first confirmed case of COVID-19 in Zimbabwe was reported on March 21, 2020 (Dzobo, Chitungo & Dzinamarira, 2020), coinciding with the beginning of the first semester of the 2020 school year in March. Consequently, on March 30, 2020, an official declaration was made mandating a nationwide lockdown of 21 days, requiring all individuals to remain in their homes (Mackworth-Young *et al.,* 2020; Mbunge *et al.,* 2020). Zimbabwean institutions were temporarily shut down in March 2020 to mitigate the transmission of COVID-19 and safeguard the student body, (Rwodzi *et al.,* 2021). In March 2020, the University of Zimbabwe implemented the first closure among major universities in the country due to the outbreak of the coronavirus. The Ministry of Higher and Tertiary Education, Science, Innovation, and Technology Development (MHTESITD) instructed all other tertiary institutions to cease any activities that are impacting the universities' admissions process, including application selection, meetings, and new student orientation., mandated the temporary suspension of end-of-semester examinations by all universities, (Zimbabwe Council for Higher Education (ZIMCHE), 2020). Furthermore, the university academic calendar is impacted by pauses in students' learning activities, disruptions of internal evaluations, and the cancellation of planned assessments and examinations (Yaya *et al.,* 2020).

COVID-19 had a major impact on teaching and learning in higher education in Zimbabwe, including delayed syllabus coverage, inadequate resources for standardised learning, low computer technology literacy among educators and students, and further costs for e-learning tools, WIFI, and internet access (Magaisa et al., 2023). Zimbabwean universities such as the University of Zimbabwe, Zimbabwe Open University, Ezekiel Guti University, Women's University in Africa, Midlands State University, Great Zimbabwe University, Chinhoyi University of Technology, and National University of Science and Technology among others were not spared either from these pandemic effects on their institutions. At most universities, students had only completed two weeks of in-person lectures when the pandemic lockdown commenced whilst some students participating in industrial attachment were unable to find placements or had their contracts terminated early due to company closures (Mahere, 2021). During the lockdown time, when their premises were locked, the universities were required to devise methods to maintain their educational services (Tsekea & Chigwada, 2020). In September 2020, universities gradually reopened, with priority given to final-year students. Higher education had to evolve in the use of ICT and worldwide networking due to the impact of COVID-19. This included online learning platforms like Google Classroom and student mobile libraries (Magaisa et al., 2023).

RESILIENT STRATEGIES FOR ABSORBING COVID-19 SHOCKS IN HEIS

UNESCO devised mitigatory methods for educational institutions in nations that have implemented lockdowns as a containment strategy. These encompass the implementation of appropriate technology (Rose, 2020). To bounce back from the brink of the epidemic, higher education institutions expedited the modernization of their operations. Before the pandemic, the utilisation of ICT in university systems and procedures was not regarded with seriousness. However, the epidemic drastically changed this perspective, as the need for business continuity became increasingly crucial. The use of digitalisation as a primary resilient strategy facilitated the integration of online teaching and learning, blended learning, and remote work for both academic and non-academic personnel.

Digital Transformation a Panacea to Higher Education Institutions' Resilience

The COVID-19 pandemic expedited the process of digital transformation (DT), leading to the widespread use of information and communication technology (ICT) by most institutions after the pandemic. Digital transformation (DT) refers to the utilization of digital technologies to significantly enhance the performance and reach of an organization (Westerman *et al.,* 2014). The implementation of digital transformation in higher education institutions has been examined from social, organisational, and technological viewpoints (Castro Benavides *et al.,* 2020). Within the given framework, noticeable transformations have taken place in many university environments. DT procedures have influenced various dimensions within a HEI, including teaching, infrastructure, curriculum, administration, research, business process, human resource, extension, DT governance, information, and marketing views (Castro Benavides *et al.,* 2020).

Digital transformation refers to the process of implementing organisational changes that enable people, processes, and the entire business model to effectively utilize technology as a means of creating value for both consumers and collaborators (Rodrigues, 2017; Schwab, 2016). Higher education institutions utilised technology to manage a broader array of jobs and obligations, including teaching and learning, research, and other support services that ensure the university's operation (Arunasalam, 2016).

Consequently, the process of digitalisation plays a crucial role in HEIs, as it can entice students and enhance their academic experiences through the use of digital resources, such as course materials and training methods (Gurung & Rutledge, 2014). It provides job flexibility and enhances employees' working experience. Implementing digitalisation in HEIs facilitates the easy identification and monitoring of obstacles in work processes, hence reducing the risk associated with university operations. DT is considered the incorporation of digital technologies into educational activities, encompassing teaching, learning, and organisational aspects (Thoring, *et al.,* 2018).

The rapid and continuous progress of information and communication technology has had a profound influence on academic discourse, specifically by altering the daily activities of teaching, research, and scholarship (Peimani & Kamalipour, 2021). The utilisation of digital educational resources is seen as facilitating novel responsibilities for educators and learners, fostering adaptable and stimulating methods of acquiring knowledge, and promoting autonomy and collaboration (Fleaca, 2011). The primary objectives of digital educational resources in higher education are to enhance human resources, facilitate teaching and innovation, streamline administration, improve access and market openness, optimise the building process, contribute to society, and support research (Sandhu, 2018; Thoring *et al.,* 2018; Wang *et al.,* 2023).

For at least forty years, digital technology has been widely present in educational platforms in South Africa, although its implementation has been inconsistent. Over the past two decades, its use has grown rapidly (Fataar & Norodien-Fataar, 2021). The digital technology landscape in Zimbabwe has a distinct characteristic since not all institutions possess a robust ICT infrastructure. This can be attributed to the ongoing economic crisis, (Maunganidze *et al.* 2021). Consequently, HEIs in Zimbabwe and South Africa effectively utilised the advancements and possibilities offered by digital technology to maintain their services to students and ensure the uninterrupted operation of university enterprises. South African and Zimbabwean HEIs modernized their processes by implementing a work-from-home (WFH) flexible working arrangement, which was adopted by staff. The modernization in the two countries enabled the use of online teaching and learning, as well as blended learning methodologies.

THE ONLINE LEARNING

The COVID-19 pandemic has led to significant changes in higher education pedagogy, with many higher education institutions worldwide adopting mixed, online, and remote teaching as a primary reaction (Kachintorn, 2020; Somabut & Tuamsuk, 2021). HEIs implemented significant technological adaptations in response to the pandemic, such as adopting distance learning or online teaching methods (Rashid & Yadav, 2020), to compensate for the time during the pandemic. Despite initial discomfort, reluctance, and detachment, many HEIs around the world adopted online learning as a 'new normal' (Izumi, Sukhwani, Surjan, & Shaw, 2020). Africa had positive outcomes from e-learning, including the acquisition of technical skills and the development of proactive coping techniques (Cynthia *et al.,* 2022). HEIs implement online learning platforms, such as MOODLE, as a substitute for in-person instruction (Mhandu *et al.,* 2021). Online learning offers a significant benefit by enabling learners to engage in high-quality learning when geographical separation and time constraints hinder or prevent in-person learning (Turner, 2020).

Academics in South Africa were granted permission to recommence classes starting in June 2020. Following the announcement of the pandemic in Africa, South African Higher Education Institutions (HEIs) implemented measures about their operating procedures, such as the delivery of in-person classes and utilisation of on-site support services. Universities implemented a comprehensive strategy known as the "no student left behind" approach, with the overarching objective of preserving lives during the COVID-19 epidemic. Each university has reorganised the academic calendar to offer equitable and unbiased opportunities for all students to complete the academic year, even if it extends into the early months of 2021, (2020-04-20-no-student-will-be-leftbehind-by-e-learning-universities-sa).

Online Learning and Learning Management Systems

Learning Management Systems (LMS) serve multiple functions, such as disseminating learning resources, delivering timely notifications, providing access to online learning modules, and facilitating conversations and comments through features like discussion forums and chat rooms (Mshayisa & Ivala, 2022). Learning Management Systems, such as Moodle, WebCT, Blackboard, Sakai, and KEWL, are widely used by educational institutions in South Africa and Zimbabwe (Cynthia *et al.,* 2022). As an illustration, the University of the Western Cape (UWC), which was one of the first institutions in South Africa to adopt e-learning, implemented Computer-based Education (CBE) in 1980. The University of Pretoria implemented WebCT to enhance and facilitate student learning, (University of Pretoria, 2013). The University of South Africa (UNISA) employed Sakai as a blended learning tool on the myUNISA platform to provide students with a more equitable learning experience (Goosen, 2012). The Nelson Mandela Metropolitan University (NMMU) and the University of Venda (UNIVEN) have utilised the Moodle platform.

Meanwhile, in Zimbabwe, universities such as Zimbabwe Open University (ZOU), Women's University in Africa (WUA), and Great Zimbabwe University (GZU) implemented off-campus Learning Management Systems (LMS). These systems included My Vista, My Hope, and Google Classroom, respectively. These online teaching and learning models were integrated with the Email platform for instructional purposes. Nevertheless, the majority of students who had difficulties with internet connectivity and power opted to utilise alternative platforms such as WhatsApp. In addition, they had a preference for working during the night to minimize disruptions and take advantage of affordable overnight mobile data. To ensure universal participation in online education, several institutions distributed data to each student, as reported by (Cynthia *et al.,* 2022).

South Africa's universities were able to readily embrace technology-mediated teaching and learning approaches due to their relatively abundant resources, which surpass those of institutions in other African countries like Zimbabwe, (Fataar & Norodien-Fataar, 2021). For example, the University of Johannesburg, which actively engages in current discussions regarding the Fourth Industrial Revolution (4IR), swiftly and effortlessly transitioned to online learning. Furthermore, the University of South Africa (UNISA), which is recognised as one of the largest e-learning institutions in Africa (Mashile & Matone, 2012), effectively utilised its established online platform to maintain pedagogical continuity (Chibuwe & Munoriyarwa, 2023). Nevertheless, in Zimbabwe, some universities such as the University of Zimbabwe and Midlands State University have adopted numerous tactics to enhance the modernisation of their Learning Management Systems (LMS). These efforts include implementing a 'bring your own device' policy (BYOD) and enforcing a no laptop, no admissions policy (Mishra et al., 2020; www.uz.ac.zw).

In Zimbabwe and South Africa, WhatsApp played a significant role as a platform for the efficient transmission of instructions and lecturer-student engagement due to its accessibility and adaptability. It symbolised the growing significance of being a platform for educational instruction, a position it did not have before the pandemic. Consequently, it has evolved into a structured environment that is seamlessly included in the educational activities and acquisition of knowledge. The WhatsApp platform was not only easily accessible but also demonstrated clear use for every student (Basitere, & Mapatagane, 2018; Chibuwe, & Munoriyarwa, 2023; Benson & Morgan 2018; Rwodzi *et al.,* 2021). In addition to being adaptable and cost-effective. (Rudhumbu, et al., 2021). The adoption of the WhatsApp platform attributed to the lecturers' heightened empathy towards the students, as they lacked substantial organisational support, aside from data, a cell line, and lessons on Google Classroom management, to efficiently conduct online teaching.

Online Learning and Mobile Network Operators' Partnership

To ensure that learners have access to the Internet at a reasonable cost, universities in South Africa and Zimbabwe collaborated with the Internet service providers and their governments, specifically through their ministries. They have addressed this crucial issue by negotiating for zero-rated access to certain educational and informational websites. This initiative aimed to promote equal and inclusive access to teaching and learning. These collaborations provided pupils with modernised PCs, data packages, and smartphones (Moyo & Munoriyarwa, 2021; The Zimbabwe Herald, 2020; Sonn, *et al.,* 2021).

Certain universities even went to the extent of providing equipment, such as tablets or smartphones, directly to their students' homes to facilitate their continued online participation. To ensure uninterrupted learning, certain institutions provided their students with data bundles to mitigate any connectivity issues or significant barriers, (Gaidi, 2020; Beans, Maireva, & Muza, 2020). Universities like Kwa-Zulu Natal in South Africa and Women's University in Africa in Zimbabwe provided data bundles to the students and employees to overcome hurdles that could hinder the start of online teaching, learning, and working (Tamrat & Teferra, 2020). The method facilitated the seamless online collaboration for lecturers, students, and staff by eliminating any limitations or disadvantages. However, the data that was provided to the lecturers was inadequate whereas the data for online learning was too steep for the students, (Chibuwe & Munoriyarwa, 2023).

In the same vein, the Zimbabwean Ministry of Higher and Tertiary Education, Science, Innovation, and Technology Development offered financial support and provided universities with ICT equipment and bandwidth to help them manage the negative effects of the Covid-19 pandemic (ZIMCHE, 2020). However, the provision of zero-rated access to the internet by Mobile Network Operators (MNOs) was limited to university students and exclusively applied to university websites. If a student had to utilise Zoom, WhatsApp, Google, or Google Classroom, they would have to bear extra costs to access these services.

To minimise the need for physical movement between offices, higher education institutions have implemented electronic communication methods such as email, WhatsApp, telephone, and video-conferencing tools for non-academic staff. These platforms were used to distribute information and conduct meetings. Virtual working necessitated the enhancement of technological infrastructure in colleges to provide compatibility with platforms such as Zoom, video conferencing, and Google Meet, among others. Emphasis was placed on investing in digital technologies, ensuring that funding was promptly allocated to technological investments that would have otherwise required a lengthy fundraising process (Beech & Anseel, 2020). All staff members received training on the efficient utilisation of online management systems, learning management systems, and the utilisation of open educational resources. The investments were undertaken to guarantee the continuity of academic activities in case of more student demonstrations that may potentially lead to a compulsory closure of universities (Koopman & Koopman, 2020).

DIGITALISATION AND WORK-FROM-HOME

Due to the emergence of the epidemic, remote work became imperative for a significant number of individuals (Shao *et al.,* 2021; Zhang *et al.,* 2021). Indeed, to guarantee the uninterrupted provision of services, governments and enterprises global institutions swiftly implemented methodologies and frameworks to facilitate remote work for individuals, (Bonacini *et al.,* 2021). Telecommuting, often known as remote work or work-from-home (WFH), is a flexible employment arrangement that allows employees to do their job duties outside of the traditional office setting (Glossary, 2021).

Consequently, there has been a significant surge in the utilisation of computers and the internet during the lockdown, even though several individuals were caught unprepared for the obligatory shift to remote labor (Zhang, *et al.,* 2021). The swift progress in technology has facilitated the transition to WFH, as novel and enhanced digital platforms for online and remote services have been created to maintain operations and engagements (Renu, 2021). Both academic and non-academic staff, as well as students, actively pursued internet-enabled employment and educational opportunities. In the post-COVID-19 era, academics need to possess advanced technical skills and be well-prepared to achieve success (Cabero-Almenara *et al.,* 2021). The process of digitalisation has enhanced the preparation and delivery of lectures, as well as the overall teaching and learning experience while working from home. The modernisation of information technology (IT) operations has had a profound impact on various aspects of education, including institutional procedures, teaching and learning, communication, and the interaction between lecturers and students, intending to enhance these areas. Effective remote work requires institutional management support, encompassing the optimal utilisation of infrastructure, services, and laws (Aczel, *et al.,* 2021).

Both lecturers and administrative staff in South Africa adopted the practice of WFH. The pandemic forced academics and employees to work remotely, a situation for which universities and academics were ill-equipped (Jansen, 2020). Consequently, stakeholders in the South African education sector, particularly university administrators, academic staff, and students, made significant strides in embracing WFH and online learning options, indicating a shift towards virtual work (Mhlanga & Moloi, 2020).

In Zimbabwe, several activities such as blended learning, online lectures, student recruitment, application processing, staff meetings, and student inquiries were all conducted remotely using technology from home. For instance, the University of Zimbabwe effectively carried out its academic operations by attaining the highest number of students enrolled using digital technology while working remotely, (www.uz.ac.zw). All individuals employed in academic and non-academic roles who were working from home expressed satisfaction with the freedom provided by this new way of operating. The implementation of WFH practices resulted in enhanced service provision within higher education institutions, (Guti & Sibanda 2023).

Nevertheless, several members of the academics observed that WFH leads to mental weariness, which negatively impacts the effectiveness of teaching and learning, and hampers academic productivity levels (Ugwuanyi, Okeke & Shawe, 2021; Van Niekerk & Van Gent, 2021). The majority of remote workers were not provided with sufficient training to effectively utilise the new remote work system. In addition to the convenience of working from home, WFH has had a considerable impact on the balance between work and family life by blurring the boundaries between work and personal obligations (Perrigino & Raveendhran, 2020). Particularly when one has worked extended hours and has utilised their living space, (Sols, 2016).

CONCLUDING THOUGHTS FOR CONSIDERATION

The impact of the pandemic on HEIs in South Africa and Zimbabwe was identical, resulting in the closure of universities and the disruption of the academic calendar in both countries. The techniques employed to alleviate the impacts of the pandemic were mostly similar, while their execution varied due to the distinct economic resources and capabilities of each country. The development of online learning and remote work was considered a solution for developing resilient strategies in higher education institutions in both nations during the pandemic, as they embraced digitalisation (Rosak-Szyrocka *et al.,* 2022). Although universities strongly advocated for the usage of Google Classroom and Moodle as the designated LMS platform, WhatsApp was found to be the most widely utilised (Rudhumbu *et al.,* 2021). The process of digitalisation has revealed various vulnerabilities in education systems in the two countries, such as inadequate infrastructure and resources to facilitate immediate teaching and learning, unequal student access to laptops and online resources, the challenge of teleworking for administrative tasks, lack of ICT skills and the inability to conduct practical courses online, (Makwasha et al., 2023; Maphosa, Dube, & Jita, 2020; Maphosa, Sisasenkosi, & Dhlamini, 2020; Supiano, 2020).

Higher education institutions must prioritise the utilisation of digital technologies to enhance their operational procedures, streamline educational service delivery, and reduce complexity in operation (Bond *et al,* 2018; Gama, 2018; Thoring *et al.,* 2018; Tay & Low 2016; Zulfikar, 2018). Zimbabwe must place education as a top priority and allocate greater resources towards enhancing its ICT infrastructure within the education system to ensure its ability to withstand future crises. DT transformation has been integrated into university processes and systems and is now part of its ecosystem. Digital transformation guarantees the continued existence of universities in the face of crises however, those universities that do not fully embrace DT become vulnerable and end up exiting the higher education arena.

REFERENCES

Aczel, B., Kovacs, M., van der Lippe, T., & Szaszi, B. (2021). Researchers working from home: Benefits and challenges. *PLoS One*, *16*(3), e0249127. doi:10.1371/journal.pone.0249127 PMID:33765047

Arunasalam, N. (2016). *Technology-enhanced learning in transnational higher education*. Academic Press.

Atamba, C., Abazie, G. A., Nweze, C. N., Bonney, B., Mohamed, H. H., & Onwubuya, G. C. (2022). A Systematic Review of African University Student's Learning in the Era of COVID-19 Pandemic: Lessons from China. *Creative Education*, *13*(11), 3581–3592. doi:10.4236/ce.2022.1311228

Aucejo, E. M., French, J., Ugalde Araya, M. P., & Zafar, B. (2020). The impact of COVID-19 on student experiences and expectations: Evidence from a survey. *Journal of Public Economics*, *191*, 104271. doi:10.1016/j.jpubeco.2020.104271 PMID:32873994

Baker, S., Bloom, N., & Davis, S. (2020). *COVID-induced economic uncertainty*. National Bureau of Economic Research., doi:10.3386/w26983

Bartolic, S., Matzat, U., Tai, J., Burgess, J. L., Boud, D., Craig, H., & Guppy, N. (2022). Student vulnerabilities and confidence in learning in the context of the COVID-19 pandemic. *Studies in Higher Education*, *47*(12), 2460–2472. doi:10.1080/03075079.2022.2081679

Basitere, M., & Mapatagane, N. (2018, June). Effects of a Social Media Network Site on Student's Engagement and Collaboration: A Case Study of WhatsApp at a University of Technology. In *ECSM 2018 5th European conference on social media.*

Beans, H., Maireva, C., & Muza, C. (2020). Zimbabwe higher education institutions' preparedness in responding to COVID-19-induced disruptions to education. *Journal of New Vision in Educational Research*, *1*(2), 267–281.

Beech, P. (2020). These new gadgets were designed to fight COVID-19. In *World Economic Forum* (Vol. 5). Academic Press.

Benavides, L. M. C., Tamayo Arias, J. A., Arango Serna, M. D., Branch Bedoya, J. W., & Burgos, D. (2020). Digital transformation in higher education institutions: A systematic literature review. *Sensors (Basel)*, *20*(11), 1–22. doi:10.3390/s20113291 PMID:32526998

Benson, V., & Morgan, S. (2018). Measuring the social impact: How social media affects higher education institutions. In *Social Media Marketing: Breakthroughs in Research and Practice* (pp. 1167–1184). IGI Global., doi:10.4018/978-1-5225-5637-4.ch055

Biggs, R., Schlüter, M., Biggs, D., Bohensky, E. L., BurnSilver, S., Cundill, G., Dakos, V., Daw, T. M., Evans, L. S., Kotschy, K., Leitch, A. M., Meek, C., Quinlan, A., Raudsepp-Hearne, C., Robards, M. D., Schoon, M. L., Schultz, L., & West, P. C. (2012). Toward principles for enhancing the resilience of ecosystem services. *Annual Review of Environment and Resources*, *37*(1), 421–448. doi:10.1146/annurev-environ-051211-123836

Biwer, F., Wiradhany W., oude Egbrink, M., Hospers, H., Wasenitz, S., Jansen, W., & de Bruin, A. (2021). Changes and Adaptations: How University Students Self-Regulate Their Online Learning During the COVID-19. *Pandemic. Frontiers in Psychology, 12*, 1-12. doi:10.3389/fpsyg.2021.642593

Bonacini, L., Gallo, G., & Scicchitano, S. (2021). Working from home and income inequality: Risks of a 'new normal' with COVID-19. *Journal of Population Economics*, *34*(1), 303–360. doi:10.1007/s00148-020-00800-7 PMID:32952308

Bond, M., Marín, V. I., Dolch, C., Bedenlier, S., & Zawacki-Richter, O. (2018). Digital transformation in German higher education: Student and teacher perceptions and usage of digital media. *International Journal of Educational Technology in Higher Education*, *15*(1), 48. Advance online publication. doi:10.1186/s41239-018-0130-1

Cabero-Almenara, J., Barroso-Osuna, J., Gutiérrez-Castillo, J. J., & Palacios-Rodríguez, A. (2021). The Teaching Digital Competence of Health Sciences Teachers. A Study at Andalusian Universities (Spain*). International Journal of Environmental Research and Public Health*, *18*(5), 2552. doi:10.3390/ijerph18052552 PMID:33806483

Carvalho, S., Rossiter, J., Angrist, N., Hares, S., & Silverman, R. (2020). *Planning for school reopening and recovery after COVID-19.* Center for Global Development.

Castro Benavides, L. M., Tamayo Arias, J. A., Arango Serna, M. D., & Branch Bedoya, J., W., & Burgos, D. (2020). Digital transformation in higher education institutions: A systematic literature review. *Sensors (Switzerland)*, *20*(11), 1–23. PMID:32526998

Charoensukmongkol, P., & Phungsoonthorn, T. (2021). The effectiveness of supervisor support in lessening perceived uncertainties and emotional exhaustion of university employees during the COVID-19 crisis: The constraining role of organisational intransigence. *The Journal of General Psychology*, *148*(4), 431–450. doi:10.1080/00221309.2020.1795613 PMID:32691689

Chibuwe, A., & Munoriyarwa, A. (2023). *Emerging methods and challenges associated with teaching and learning media studies during the COVID-19 pandemic induced lockdowns in Zimbabwe and South Africa*. doi:10.1177/21582440231167113

Chida, P. (2020). Government shuts down varsities and colleges. *NewsDay*. https://www.newsday.co.zw

Chikoko, V. (2020). This technology we all wash with: The efficacy of leveraging WhatsApp in delivering a master's module. In *Technology-based Teaching and Learning in Higher Education during the Time of COVID-19*. CSSALL Publishers (Pty.) Ltd. http://alternation.ukzn.ac.za/Files/books/series-02/00-Font-Matter.pdf

Cloete, N., & Moja, T. (2005). Transformation tensions in higher education: Equity, efficiency, and development. *Social Research*, *72*(3), 693–722. doi:10.1353/sor.2005.0033

Cynthia, A., Anulika, A. G., Nneoma, N. C., Belinda, B., Habuba, M. H., & Chinemerem, O. G. (2022). A Systematic Review of African University Student's Learning in the Era of COVID-19 Pandemic: Lessons from China. *Creative Education*, *13*(11), 3581–3592. doi:10.4236/ce.2022.1311228

Dahles, H., & Susilowati, T. P. (2015). Business resilience in times of growth and crisis. *Annals of Tourism Research*, *51*, 34–50. doi:10.1016/j.annals.2015.01.002

Daniels, R. J., & Shadmi, K. (2020). *Universities' vital role in the pandemic response, universities respond to the pandemic with sound science and advice. We can do more: "A Vital Mission" in the COVID-19 Special Edition issue of Hopkins Bloomberg Public Health Magazine*. https://dev.magazine.jhsph.edu/2020/universities-vitalrole-pandemic-response

Dennis, M. J. (2020). The impact of COVID-19 on the world economy and higher education. *Enrollment Management Report*, *24*(9), 3. Advance online publication. doi:10.1002/emt.30720

Di Pietro, G., Biagi, F., Costa, P., Karpinski, Z., & Mazza, J. (2020). *The likely impact of COVID-19 on education: Reflections based on the existing literature and recent international datasets*. Publications Office of the European Union.

Djalante, R., Shaw, R., & DeWit, A. (2020). Building resilience against biological hazards and pandemics: COVID-19 and its implications for the Sendai Framework. In *Progress in Disaster Science 6* (Vol. 100080). Elsevier Ltd. doi:10.1016/j.pdisas.2020.100080

Dzinamarira, T., Mukwenha, S., Eghtessadi, R., Cuadros, D. F., Mhlanga, G., & Musuka, G. (2020). COVID-19 response in Zimbabwe: A Call for Urgent Scale-up of Testing to meet National Capacity. *Clinical Infectious Diseases*, *23*(10), 134–143. PMID:32865204

Dzobo, M., Chitungo, I., & Dzinamarira, T. (2020). COVID-19: A perspective for lifting lockdown in Zimbabwe. *The Pan African Medical Journal*, *35*(13, Supp 2), 1–10. doi:10.11604/pamj.supp.2020.35.2.23059 PMID:32528624

Eiffel Corp. (2007). *Use of Learning Management Systems by South African Universities*. Retrieved from www.eiffel-corp.co.za/newsletters/eiffel_newsletter07.htm

Englund, C., Olofsson, A. D., & Price, L. (2017). Teaching with Technology in Higher Education: Understanding Conceptual Change and Development in Practice. *Higher Education Research & Development, 36*(1), 73–87. doi:10.1080/07294360.2016.1171300

Faria, J. A., & Nóvoa, H. (2020). Digital transformation at the University of Porto. *Proceedings of the International Conference on Exploring Services Science*.

Farley, A., Jain, A., & Thomson, D. (2011). Blended learning in finance: Comparing student perceptions of lectures, tutorials and online learning environments across different year levels. *Economic Papers, 30*(1), 99–108. doi:10.1111/j.1759-3441.2010.00094.x

Fataar, A., & Norodien-Fataar, N. (2021). Towards an e-learning ecologies approach to pedagogy in a post-COVID world. *Journal of Education, 84*(84), 155–168. doi:10.17159/2520-9868/i84a08

Fleaca, E. (2011). Embedding digital teaching and learning practices in the modernisation of higher education institutions. *Proceedings of the SGEM2017 International Multidisciplinary Scientific Geo Conference: SGEM*. 10.5593/sgem2017/54/S22.006

Gaidi, F. (2020). *The Impact of COVID-19 on Universities in Africa*. ACCORD. https://www.accord.org.za/analysis/the-impact-of-covid-19-on-universities-in-africa/

Gama, J. A. P. (2018). Intelligent educational dual architecture for university digital transformation. *Proceedings of the 2018 IEEE Frontiers in Education Conference (FIE)*, 1–9. 10.1109/FIE.2018.8658844

Giustiniano, L., Clegg, S. R., Cunha, M. P., & Rego, A. (2018). *Introduction to Theories of Organisational Resilience*. Edward Elgar Publishing. doi:10.4337/9781786437044

Grange, L. L. (2020). Could the COVID-19 pandemic accelerate the uberfication of the University. *South African Journal of Higher Education, 34*(4), 1–10. doi:10.20853/34-4-4071

Gurung, B., & Rutledge, D. (2014). Digital learners and the overlapping of their personal and educational digital engagement. *Computers & Education, 77*, 91–100. doi:10.1016/j.compedu.2014.04.012

Guti, O., & Sibanda, S. (2023). Prospects and Challenges of the WFH practices: A case of Zimbabwe Open University. *International Journal of Research Publications., 117*(1), 4–17. doi:10.47119/IJRP1001171120234419

Gyamerah, K. (2020). The impacts of COVID-19 on basic education: How can Ghana respond, cope, and plan for recovery. *School of Education*. https://schoolofeducation.blogs.bristol.ac.uk/2020/03/31/the-impacts-of-covid-19-on-basiceducation-how-can-ghana-respond-cope-and-plan-for-recovery/

Heeks, R., & Ospina, A. V. (2019). Conceptualising the link between information systems and resilience: A developing country field study. *Information Systems Journal, 29*(1), 70–96. doi:10.1111/isj.12177

Hoadley, U. (2020). *Schools in the time of COVID-19: Impacts of the pandemic on curriculum*. Resep Non-Economic Working Paper, Stellenbosch University.

Izumi, T., Sukhwani, V., Surjan, A., & Shaw, R. (2020). Managing and responding to pandemics in higher educational institutions: Initial learning from covid-19. *International Journal of Disaster Resilience in the Built Environment, 12*(1), 51–66. doi:10.1108/IJDRBE-06-2020-0054

Jansen, J. (2020). Data or bread? A policy analysis of student experiences of learning under lockdown. *Southern African Review of Education with Education with Production, 26*(1), 167–181.

Kachintorn, U. (2020). *University Transformation in Response to the Changes in Disruption Era*. Mahidol University.

Koopman, O., & Koopman, K. (2020). The Rise of the University without Classrooms after COVID-19. *CSSALL, 1*, 153–168. doi:10.29086/978-0-9869936-1-9/2020/AASBS01

Kuhfeld, M., Sokand, J., Lewis, K., & Morton, E. (2022). *The pandemic has had devastating impacts on learning. What will it take to help students catch up?* https://www.brookings.edu/articles/the-pandemic-has-had-devastating-impacts-on-learning-what-will-it-take-to-help-students-catch-up/

Lengnick-Hall, C. A., Beck, T. E., & Lengnick-Hall, M. L. (2011). Developing a capacity for organizational resilience through strategic human resource management. *Human Resource Management Review, 21*(3), 243–255. doi:10.1016/j.hrmr.2010.07.001

Lues, R., Padayachee, A., & Jager, H. (2020). *Universities of technology in the post-COVID-19 landscape*. https://www.universityworldnews.com/post.php?story=20200713153430109

Mackworth-YoungC. R. S.ChingonoR.MavodzaC.McHughG.TemboM.Dziva ChikwariC. (2020). 'Here, we cannot practice what is preached': early qualitative learning from community perspectives on Zimbabwe's response to COVID-19. Bull World Health Organ. *E-pub: 20, 1-19* doi:10.2471/BLT.20.260224

Magaisa, E., Moyo, T., Wilson, E. C., Shava, G., & Makokoro, P. P. (2023). Campus and classroom-based higher education: The effects of covid-19 on teaching and learning in Zimbabwe. *International Journal of Research and Innovation in Social Science. International Journal of Research and Innovation in Social Science, 7*(4), 1259–1266. doi:10.47772/IJRISS.2023.7504

Makwasha, L., Takavarasha, S., & Mubango, H. (2023). Blended learning in the wake of ICT infrastructure deficiencies: The case of a Zimbabwean University. *African Conference on Information Systems and Technology*, 27. https://digitalcommons.kennesaw.edu/acist/2023/presentations/27

Manyuchi, M. M., Mutusva, T. N., & Sukdeo, N. (2021). *An assessment of the impact of COVID-19 on Zimbabwean Universities Students*. In *International Conference on Industrial Engineering and Operations Management, Rome*, Italy.

Mapanga, C. (2020). *Level 4 Lockdown: Students won't return to campus, risk is too great - Blade Nzimande.* https://www.news24.com/news24/SouthAfrica/News/level-4-lockdownstudents-wont-return-to-campus-

Maphosa, V., Dube, B., & Jita, T. (2020). The UTAUT evaluation of WhatsApp as a tool for lecture delivery during the covid-19 lockdown at a Zimbabwean University. *International Journal of Higher Education, 9*(5), 84–93. doi:10.5430/ijhe.v9n5p84

Maphosa, V., Sisasenkosi, H., & Dhlamini, N. (2020). An evaluation of the acceptance of Moodle by the faculty at a rural university in Zimbabwe during the COVID-19 period. Academic Press.

Mashile, E. O., & Matoane, M. C. (2012, October). E-Learning Development: the case of UNISA. In *E-Learn: World Conference on E-Learning in Corporate, Government, Healthcare, and Higher Education* (pp. 1152-1157). Association for the Advancement of Computing in Education (AACE).

Maunganidze, F., Bonnin, D., & Rugguna, S. (2021). Economic crisis and professions: Chartered accountants in Zimbabwe. *SAGE Open*, *11*(1). doi:10.1177/2158244021994802

MbungeE.FashotoS. F.AkinnuwesiB.GurajenaC.MetfulaA. (2020). Challenges of Social Distancing and Self-Isolation during COVID-19 Pandemic in Africa: A Critical Review. *Social Science Research Network*, 1-13.

McInnes, R., Aitchison, C., & Sloot, B. (2020). Building online degrees quickly: Academic experiences and institutional benefits. *Journal of University Teaching & Learning Practice*, *17*(5), 2. doi:10.53761/1.17.5.2

Mhandu, J., Mahiya, I. T., & Muzvidziwa, E. (2021). The Exclusionary Character of Remote Teaching and Learning during the COVID-19 Pandemic. An Exploration of the Challenges Faced by Rural-Based University of KwaZulu Natal Students. *Cogent Social Sciences*, *7*(1), 1947568. Advance online publication. doi:10.1080/23311886.2021.1947568

Mhlanga, D., & Moloi, T. (2020). COVID-19 and the digital transformation of education: What are we learning on 4IR in South Africa? *Education Sciences*, *10*(180), 1–11. doi:10.3390/educsci10070180

Ministry of Education of the People's Republic of China. (2020). *Guidance on the Organization and Management of Online Teaching in Colleges and Universities during the Epidemic Prevention and Control Period.* http://www.moe.gov.cn/srcsite/A08/s7056/202002/t20200205_418138.html

Mishra, L., Gupta, T., & Shree, A. (2020). Online teaching-learning in higher education during lockdown period of the COVID-19 pandemic. *International Journal of Educational Research Open*, *1*, 100012. doi:10.1016/j.ijedro.2020.100012 PMID:35059663

Mittal, P. (2020). Impact of COVID-19 on higher education in India. In *Regional/National Perspectives on the Impact of COVID-19 on Higher Education* (pp. 18–21). International Association of Universities.

Moyo, D., & Munoriyarwa, A. (2021). 'Data must fall': Mobile data pricing, regulatory paralysis and citizen action in South Africa. *Information Communication and Society*, *24*(3), 365–380. doi:10.1080/1369118X.2020.1864003

Mpungose, C. B. (2020). Is Moodle or WhatsApp the preferred e-learning platform at a South African university? First-year students' experiences. *Education and Information Technologies*, *25*(2), 927–941. doi:10.1007/s10639-019-10005-5

Mshayisa, V. V., & Ivala, E. N. (2022). No student left behind: Students' experiences of a self-paced online learning orientation in undergraduate studies during COVID-19 pandemic. *Education Sciences*, *12*(6), 386. doi:10.3390/educsci12060386

Mudenda, S., Zulu, A., Phiri, M. N., Ngazimbi, M., Mufwambi, W., Kasanga, M., & Banda, M. (2020). Impact of coronavirus disease 2019 (COVID-19) on college and university students: A global health and education problem. *Aquademia*, *4*(2).

Mukeredzi, T. (2020, March 23). Under pressure, the government closes universities and schools. *University World News*. https://universityworldnews.com/post

Nangia, M., & Mohsin, F. (2020). Revisiting talent management practices in a pandemic driven VUCA environment - a qualitative investigation in the Indian IT industry. *J. Crit. Rev.*, *7*(7), 937–942.

Ngqakamba, S. (2020). *Universities gear up to welcome more students on campus—But online classes still the new normal for many*. News24. https://www.news24.com/news24/southafrica/news/universitiesgear-up-to-welcome-more-students-on-campus-but-online-classes-still-the-new-normal-for-many-20200901

Nwosu, C. O. (2021). Childcare and Depression During the Coronavirus Pandemic in South Africa: A Gendered Analysis. *PLoS One*, *16*(8), e0255183. doi:10.1371/journal.pone.0255183 PMID:34358235

Peimani, N., & Kamalipour, H. (2021). Online Education and the COVID-19 Outbreak: A Case Study of Online Teaching during Lockdown. *Education Sciences*, *11*(2), 72. doi:10.3390/educsci11020072

Perrigino, M. B., & Raveendhran, R. (2020). Managing remote workers during quarantine: Insights from organizational research on boundary management. *Behavioral Science & Policy*, *6*(2), 87–94. doi:10.1177/237946152000600211

Plass, P. (2020). *Strategies of higher education admissions professionals to adapt to the recent drop in foreign student enrollment at us universities*. Academic Press.

Rashid, S., & Yadav, S. (2020). Impact of Covid-19 Pandemic on Higher Education and Research. *Indian Journal of Human Development.*, *14*(7). doi:10.1177/0973703020946700

Renu, N. (2021). Technological advancement in the era of COVID-19. *SAGE Open Medicine*, *9*, 20503121211000912. doi:10.1177/20503121211000912 PMID:33786181

Rodrigues, L. S. (2017). Challenges of digital transformation in higher education institutions: A brief discussion. In *Proceedings of the 30th international business information management association conference, IBIMA 2017 Vision 2020: Sustainable economic development, innovation management, and global growth.* IBIMA. https://www.scopus.com/inward/record.uri?eid=2-s2.0-85048618825&partnerID=40&md5=65525232d18dbc0ae37a733eb45b100d

Rosak-Szyrocka, J., Żywiołek, J., Zaborski, A., Chowdhury, S., & Hu, Y. C. (2022). Digitalisation of higher education around the Globe during covid-19. *IEEE Access : Practical Innovations, Open Solutions*, *10*, 59782–59791. doi:10.1109/ACCESS.2022.3178711

Rose, P. (2020). *Rethinking Education in the Time of COVID-19: What can we contribute as researchers.* https://www.ukfiet.org/2020/rethinking-education-in-thetime-of-covid-19-what-can-we-contribute-as-researchers/

Rudhumbu, N., Parawira, W., Bhukuvhani, C., Nezandoyi, J., Majoni, C., Chikosha, F., Zvokuomba, K., & Chingwanangwana, B. (2021). Insight into online teaching behaviour of lecturers in Zimbabwean universities during the Covid-19 era and beyond: Issues and challenges. *The International Journal of Information and Learning Technology.*, *38*(5), 518–539. doi:10.1108/IJILT-07-2021-0104

Rwodzi, A., & Sibanda, T. (2021). The "New Normal" Learning and Teaching Modes in Educational Institutions in the COVID (ized) Era: An Appraisal of Zimbabwe and Botswana. *Patrolling Epistemic Borders in a World of Borderless Pandemics: Epistemological Policemen in COVID-19 Afflicted 21st Century*, 183.

Sahu, P. (2020). Closure of universities due to Coronavirus Disease 2019 (COVID-19): Impact on education and mental health of students and academic staff. *Cureus*, *12*(4), 1–6. doi:10.7759/cureus.7541 PMID:32377489

Sakurai, M., & Chughtai, H. (2020). Resilience against crises: COVID-19 and lessons from natural disasters. *European Journal of Information Systems*, *29*(5), 585–594. doi:10.1080/0960085X.2020.1814171

Sandhu, G. (2018). The role of academic libraries in the digital transformation of the universities. *Proceedings of the 2018 5th International Symposium on Emerging Trends and Technologies in Libraries and Information Services (ETTLIS)*, 292–296, 10.1109/ETTLIS.2018.8485258

Schwab, K. (2016). *The fourth industrial revolution*. Currency.

Shao, Y., Fang, Y., Wang, M., Chang, C. H. D., & Wang, L. (2021). Making daily decisions to work from home or to work in the office: The impacts of daily work-and COVID-related stressors on next-day work location. *The Journal of Applied Psychology*, *106*(6), 825–838. doi:10.1037/apl0000929 PMID:34138589

Shipman, K., Burrell, D. N., & Huff Mac Pherson, A. (2023). An organisational analysis of how managers must understand the mental health impact of teleworking during COVID-19 on employees. *The International Journal of Organizational Analysis*, *31*(4), 1081–1104. doi:10.1108/IJOA-03-2021-2685

Sibanda, F., & Muyambo, T. (2020). Programme Delivery and Assessment in the Context of the COVID-19 Crisis at One Higher Education Institution in Zimbabwe. In *Technology-based Teaching and Learning in Higher Education during the Time of COVID-19*. CSSALL Publishers (Pty.) Ltd. http://alternation.ukzn.ac.za/Files/books/series-02/00-Font-Matter.pdf

Solís, M. S. (2016). Telework: Conditions that have a positive and negative impact on the work-family conflict. *Academia (Caracas)*, *29*(4), 435–449. doi:10.1108/ARLA-10-2015-0289

Somabut, A., & Tuamsuk, K. (2021). Online Teaching and Learning Ecology in Thai Higher Education during the COVID-19 Pandemic. In L. Waller & S. Waller (Eds.), *Higher Education - New Approaches to Accreditation, Digitalisation, and Globalisation in the Age of Covid* (pp. 1–14). IntechOpen Ltd.

Sonn, I. K., Du Plessis, M., Jansen Van Vuuren, C. D., Marais, J., Wagener, E., & Roman, N. V. (2021). Achievements and Challenges for Higher Education during the COVID-19 Pandemic: A Rapid Review of Media in Africa. *International Journal of Environmental Research and Public Health*, *18*(24), 12888. doi:10.3390/ijerph182412888 PMID:34948499

Stankovska, G., Memedi, I., & Grncarovska, S. P. (2022). *Impact of COVID-19 on higher education: challenges and opportunities*. Bulgarian Comparative Education Society.

Supiano, B. (2020). Pandemic Pedagogy and the Limits of Compassion. *The Chronicle of Higher Education*.

Tamrat, W., & Teferra, D. (2020). COVID-19 Threat to Higher Education. *Industry and Higher Education*, (102), 28–30.

Tay, H. L., & Low, S. W. K. (2017). Digitalisation of learning resources in a HEI–a lean management perspective. *International Journal of Productivity and Performance Management, 66*(5), 680–694. doi:10.1108/IJPPM-09-2016-0193

The Zimbabwe Herald. (2020). *Zimbabwe government pushes for online education.* Harare: Zimpapers. https://www.herald.co.zw/govt-pushes-for-online-education/

Thoring, A., Rudolph, D., Vogl, R., (2018). The digital transformation of teaching in higher education from an academic's point of view: An explorative study. *Proceedings of the Applications of Evolutionary Computation, 10924*, 294–309.

Ting, D. S. W., Carin, L., Dzau, V., & Wong, T. Y. (2020). Digital technology and COVID-19. *Nature Medicine, 26*(4), 459–461. doi:10.1038/s41591-020-0824-5 PMID:32284618

Toquero, C. M. (2020). Challenges and opportunities for higher education amid the COVID-19 pandemic: The Philippine context. *Pedagog. Res., 5*(4), 1–5. doi:10.29333/pr/7947

Tsekea, S., & Chigwada, J. P. (2020). Covid-19: Strategies for positioning the university library in support of e-learning. *Digital Library Perspectives, 37*(1), 54–64. doi:10.1108/DLP-06-2020-0058

Turner, C. (2020). *A looming financial meltdown for America's schools.* https://www.wlrn.org/2020-05-26/a-looming-financial-meltdown-for-americas-schools

Ugwuanyi, C. S., Okeke, C. I. O., & Shawe, T. G. J. (2021). South African Academics' Perception of the Impact of Work from Home (WFH) on Effective Teaching and Learning in Universities. *Libr. Philos. Pract.* 5701. https://digitalcommons.unl.edu/libphilprac/5701/

UN. (2020). *Business as Unusual: How COVID-19 Could Change the Future of Work.* https://news.un.org/en/story/2020/05/1064802

UNESCO. (2020). *COVID-19 Impact on Education.* https://en.unesco.org/covid19/educationresponse

UNESCO. (2021). *A Snapshot of Educational Challenges and Opportunities for Recovery in Africa.* UNESCO.

UNESCO. (2022). *COVID-19 in Sub-Saharan Africa: monitoring impacts on learning outcomes Zambia report.* UNESCO.

University of Pretoria. (2013). *Web-supported Learning.* http://web.up.ac.za/default.asp?ipkCategoryID=9587&subid=9587&ipklookid=1

Unwin, T., Kleessen, B., Hollow, D., Oloo, L. M., Alwala, J., Mutimucuio, I., & Muianga, X. (2009). *Digital Learning Management Systems in Africa: rhetoric and reality Digital Learning Management Systems in Africa: rhetoric and reality.* http://www.gg.rhul.ac.uk/ict4d/LMS.pdf

Van Niekerk, R. L., & Van Gent, M. M. (2021). Mental health and well-being of university staff during the coronavirus disease 2019 levels 4 and 5 lockdowns in an Eastern Cape university. South Africa. *South African Journal of Psychiatry, 27*, a1589. doi:10.4102/sajpsychiatry.v27i0.1589

van Schalkwyk, F. (2021). Reflections on the public university sector and the covid-19 pandemic in South Africa. *Studies in Higher Education, 46*(1), 44–58. doi:10.1080/03075079.2020.1859682

Waithima, A. (2021). *The impact of covid-19 on higher education, its business implications and proposed recovery strategy.* Academic Press.

Wang, K., Li, B., Tian, T., Zakuan, N., & Rani, P. (2023). Evaluate the drivers for digital transformation in higher education institutions in the era of industry 4.0 based on decision-making method. *Journal of Innovation & Knowledge, 8*(3), 100364. doi:10.1016/j.jik.2023.100364

Westerman, G., Bonnet, D., & McAfee, A. (2014). *Leading digital: Turning technology into business transformation*. Harvard Business Press.

Wildavsky, A. B. (1988). *Searching for Safety*. Transaction Publishers.

Yaya, S., Otu, A., & Labonté, R. (2020). Globalisation in the time of COVID-19: Repositioning Africa to meet the immediate and remote challenges. *Globalization and Health, 16*(1), 1–7. doi:10.1186/s12992-020-00581-4 PMID:32580728

Zhang, C., Yu, M. C., & Marin, S. (2021). Exploring public sentiment on enforced remote work during COVID-19. *The Journal of Applied Psychology, 106*(6), 797–810. doi:10.1037/apl0000933 PMID:34138587

Zimbabwe Council for Higher Education. (2020). *Summary of Universities' Responses on Survey on Covid 19 Preparedness Strategies and Challenges in Higher Educations in Zimbabwe*. Author.

ZimmermannM.BledsoeC.PapaA. (2020). *The Impact of the COVID-19 pandemic on college student mental health: a longitudinal examination of risk and protective factor*s. doi:10.31234/osf.io/2y7hu

Zulfikar, M. W., Bin Hashim, A. I., Umri, H. U. B. A., & Dahlan, A. R. A. (2018). A business case for digital transformation of a Malaysian-based university. *Proceedings of the 2018 International Conference on Information and Communication Technology for the Muslim World*, 106–109.

KEY TERMS AND DEFINITIONS

Digital Transformation: The process of changing from a traditional business model to a digital one by applying digital technologies to change the operating method, leadership, working process, and company culture to significantly enhance organisation performance.

Digitalisation: The use of digital technologies to change a business model to achieve efficiency and provide value-producing opportunities.

Higher Education Institution: An institution of post-secondary school learning like a university that is active in teaching and research through the conduct of regular classes or distance education systems leading to the award of a degree or diploma.

Online Learning: Using ICT tools to teach students content and skills and to support their learning both within and outside of the conventional classroom environment.

Resilience: The capacity of an organisation to withstand and adaptable or to recover quickly from difficulties caused by environmental factors.

Strategies: High-level plans to achieve one or more goals under conditions of uncertainty.

Work From Home: A concept where the employee can do his or her job from home using ICT technologies.

Chapter 8
A Case Study on the Role of Distance Education in Disaster and Crisis Management

Ebru Ozge Aksu
Anadolu University, Turkey

Hakan Kilinc
https://orcid.org/0000-0002-4301-1370
Anadolu University, Turkey

ABSTRACT

The purpose of the research is to examine the perceptions of educators in the disaster area regarding the positive and negative aspects of distance education and their views on how the process is experienced in the field of education, in the Kahramanmaras-centered earthquake that occurred in 11 provinces of Turkey in February 2023. After determining the basic research questions for the purpose of the research, a case study design was adopted with a qualitative approach in order to obtain the expert opinions and suggestions of educators providing distance education in the disaster area during the data collection phase. The educators were interviewed using Google Forms, and they were asked to answer the semi-structured open-ended questions directed to them. The answers given by the educators were subjected to content analysis. Some of the themes obtained in this direction can be listed as follows: dynamic and updated education process, raising awareness of the people living in the region, a potential to benefit from lifelong learning, and access to true and reliable information resources.

INTRODUCTION

The 7.5 and 7.8 magnitude Kahramanmaraş-centered earthquakes occurred on the East Anatolian Fault Line in Turkey on February 6, 2023. The devastating earthquakes caused loss of many lives and damages, mainly across 11 cities of Turkey and Syria, and the earthquakes were felt in a wide area including Lebanon, Cyprus, Iraq, Israel, Jordan, Iran, and Egypt. Aftershocks of the earthquakes continued for

DOI: 10.4018/979-8-3693-1926-0.ch008

months. The 6.4 and 5.8 magnitude Hatay-centered earthquakes occurred on February 20, 2023, and then, the 5.6 magnitude Adana-centered earthquake occurred on July 25, 2023.

The 2023 earthquakes occurred on the East Anatolian Fault (EAF), which is a major tectonic structure in the eastern Mediterranean, separating the Arabian and Anatolian tectonic plates along a boundary that spans nearly 600 km. Together with the North Anatolian Fault (NAF), the EAF enables the Anatolian plate (most of mainland Türkiye) to escape westward at about 20 mm yr–1 from the collision between the Arabian and Eurasian plates in eastern Türkiye. This long-term motion of Anatolia is facilitated by repeated earthquakes along the North and East Anatolian Faults.

We are living in a highly complex-dynamic system that is in a vast complex-dynamic system, the Universe, as part of a complex-dynamic system at micro and macro scales. Here, the term "dynamic" refers to difference and repetition (Deleuze, 1994), i.e., becoming instead of being in a constant and unchanged state, and "complex" does not necessarily mean complicated (Hagen-Zanker, 2008), but refers to a non-linear system concept that a reductionist approach falls short to understand the system in its entirety, not only the elements or processes studied, but also the relationships, and a holistic approach is needed (Khorram-Manesh, 2016; Turner & Baker, 2019).

Unfortunately, one of the main areas affected by the earthquake was education. Due to the magnitude and effect of the earthquake across the country, the Higher Education Council announced that it was decided that it would be appropriate to complete the 2022-2023 academic year spring semester through distance education in the whole country. Distance education determines using the internet to access learning material, communicate with the instructor and other students in the learning process, and improve the experience (Ally, 2004). Today, distance education takes place mainly as web-based learning and mobile learning. Distance education is one of the most practical ways to provide equal education opportunities when suitable place, time, and conditions are not possible (Uzun et al., 2020). In the post-earthquake period in Türkiye, distance education was implemented, as in the pandemic period, during the holidays so that students can continue their education (Telli & Altun, 2021). However, a disaster environment such as an earthquake has many factors hindering education. Due to many factors, such as infrastructure problems, the psychological and socioeconomic status of students, technical, institutional, and political actions should be taken together for the distance education process to succeed (Pregowska, et al., 2021; Tanik Onal & Onal, 2023).

Since the reflection of the results of the earthquake affects the whole society, the possible problems that the students, who are not directly affected by the earthquake, may encounter, as well as the students who are victims of the earthquake, should be taken into consideration during the distance education process. For example, in the post-earthquake period, psychological conditions such as the fear of a repeat of the earthquake and the fear of losing loved ones can be seen in all students (Yildirim, 2023). For this reason, it is necessary to consider the psychological state of all students in the distance education process. Based on this point, this study aims to reveal what roles distance education should play in times of earthquakes and disasters.

THE PROBLEM

The Office of Press and Public Relations of the Council of Higher Education of the Republic of Turkey announced the Council's decision regarding 2022-2023 spring term on February 11, 2023. The Council stated in their decision that 2022-2023 spring term would be completed by means of distance education.

This decision resulted with reappearance of heated discussions on distance education in Turkey. We have recently witnessed a considerably long period of global compulsory distance education during Covid-19 pandemics. Therefore, these discussions can be understood and justified in terms of sociocultural deprivation of students from campuses, on one hand, but on the other hand, benefits of distance education in times of disasters and crises are obvious and clear for uninterrupted educational attendance of students living in the disaster and crisis area.

Space and Time Dimension of the Events

We agree with Badiou in terms of time comprehension in the 21st century. In our times of modernity, in the western world, we started to perceive time as a linear flow that moves fast into the future, which differs from cyclical time perception of our ancestors, who were living in the old times, within agricultural communities, before modernity (Badiou, 2007). This can be attributed to industrial revolution, the evolution of commodity markets as well as science and technology, which have dramatically been influencing nature, our natural environment, and our lifeworld that is mostly determined by free market economy through the speed of production, consumption, communication, and logistics, etc. This time perception also shapes our sociocultural experiences and our lifeworld.

When we consider the space and time dimension of February 2023 earthquakes, we may say that the event(s) happened in the southeastern Turkey with great impacts mainly across eleven cities of Turkey, and Syria. Turkey is one of the most important seismic active countries in the word. Turkey is between three major tectonic plates and compressed towards the north by Arabian and African plates. This region is the most active area of the Mediterranean region in terms of earthquake activity (Kalafat, Zülfikar & Okuyan-Akcan, 2021). According to the experts (in example: Korkmaz, 2006; TMMOB Jeoloji Mühendisleri Odası, 2021; Palutoğlu & Şaşmaz A., 2017), an earthquake was expected in the area with a high probability for decades, but the exact time of the event was vague and unknown. Notwithstanding uncertainty of the time of the earthquake(s), the experts constantly warned the public and competent authorities, and even a couple of earthquake simulations were carried out by AFAD (2023) before February 6, 2023. So, this was not a question of if, but when, and we may propose that the event was expected, at least partially, i.e., spatially.

We know that time is of the essence to take protective action during disasters and the first seconds of an earthquake matter to warn people to take timely action. Earthquake Early Warning (EEW) system is an integrated system of hazard monitoring, forecasting, disaster risk assessment, communication and preparedness activities, systems and processes that enables individuals, communities, governments, businesses, and others to take timely action to reduce disaster risk in advance of hazardous events. There was no public earthquake early warning system in place in Turkey (for the region) at the time of the events. Google's Android Earthquake Alert system also failed (bbc.com). According to the news, the 7.8 magnitude Kahramanmaras-centered earthquake, the first that initiated a sequence of many earthquakes in the disaster area, happened at 4:17 AM local time when most people were still in their beds and sleeping. Unfortunately, many people didn't have time to wake up and/ or take protective actions when the event happened. This might have increased the number of loss of lives and injuries. The second unfortunate reason which increased the number of loss of lives and injuries was that many buildings collapsed or heavily damaged. The third unfortunate reason was that military's earthquake response was relatively slow when compared to other earthquake military responses in the past.

Since the earthquake with devastating impacts on 11 cities of Turkey coincided with the opening of 2022-2023 spring term, the decision of the Council of Higher Education of the Republic of Turkey is legally justified in accordance with the national and international laws and legislations.

Human Nature and Socioecological Systems

We agree with Vogel (1988) in this context. In example, there is a new semi-synthetic fabric that is produced from woodchops and called Lyocell, branded as Tencel, in fashion industry, which is an industry that has devastating impacts on the environment. This fabric is manufactured by applying water and chemicals in a closed loop, which means that nearly all chemicals and water reused during the entire production process of the fabric. The fabric can be mixed with other materials such as cotton, wool, silk etc., which changes the texture of the fabric. The wide-ranging texture options of the fabric increase the demand. This fabric is said to be environmentally friendly and sustainable when compared to the other fabrics available on the market. When we consider the production process only, the brand's argument can be somewhat justified and agreed, but when it comes to consumption, any environmentally friendly and sustainably manufactured commodity may rapidly turn into a monster of excessive consumption that paradoxically consume and destroy nature with multiple adverse effects. So, we may conclude that consumption dimension of any product of any industry is at least as important as the production process. We must be aware of the fact that energy and resources of nature are not unlimited and must be used for each and every product we consume regardless of how environmentally friendly and sustainable a product is. This applies to any current environmentalist movements, including, but not limited to Veganism.

The decline in the consumption of petroleum and other fossil fuels has led to a reduction in greenhouse gas emissions and global warming slightly decelerated during the novel Covid-19 pandemic (Rad et.al., 2021). The Covid-19 pandemic is an irrefutable evidence of anthropogenic climate change (Tian, An, Chen & Tian, 2020; Bauwens et al., 2020; Usman, Husnain & Riaz, 2021; Aloi et al., 2020.), even if we keep on denying the fact that we influence nature as it influences us and that there is a mutual and interactive relationship between nature and humankind. Although a disaster like an earthquake is a natural disaster, i.e., not an anthropogenic disaster, understanding the concept of mutual human-nature relationship and that we are living as a part of a socioecological system provides a deeper comprehension and consciousness on human nature, human experience, and natural or anthropogenic events and disasters.

Multi-Hazard Approach and Distance Education in Disaster and Crises Management

Multi-hazard approach, which is adopted in disaster and crises management for establishing disaster-resilient communities, is a rational approach for assessment of secondary hazard and risks that may lead to disasters (Disaster and Emergency Management Presidency (AFAD) Dictionary) and includes prevention and mitigation of damages and loss of lives in case of secondary disasters by analyzing multi-hazard risks. The experiences that revealed the inefficiency of the response plans that are prepared for a specific hazard paved the way for multi-hazard risk analysis and risk reduction before disasters. 2015-2030 Sendai Disaster Risk Reduction Framework adopted in 2015 in United Nations Disaster Risk Reduction Conference that was held in Japan/Sendai. Sendai Disaster Risk Reduction Framework aims to improve disaster resilience of communities through prevention and reduction of disaster risks (Sendai FDRR 2015-2030). Integration of distance education into disaster and crises management can be implemented for prevention and reduction of multi-hazard disaster risks.

Case Study on the Role of Distance Education in Disaster and Crises Management

In our Case Study on the Role of Distance Education in Disaster and Crises Management. The aim of the study was to examine the perception of educators on negative and positive aspects of distance education in the disaster area and how the process was experienced in the field of education in the context of disaster and crises management after Turkey/Kahramanmaraş-centered earthquake hit the entire region with an impact mainly across eleven cities of Turkey in February 2023.

The study focused on and limited to the following context:

- contribution(s) of distance education to disaster and crisis management;
- contribution(s) of distance education to students and educators for dealing with the adverse effects and relieving the symptoms of the disaster;
- contribution(s) of distance education to provide equal opportunities in education;
- problems encountered in the process of distance education; and
- opinion(s) and advice(s) of experts involved in distance education in the disaster area.

METHOD

Data Collection and Tools

After main research questions were identified in line with the aim of the study, case study method, one of the qualitative research designs, was adopted for seeking opinion and advice of experts involved in distance education in the disaster area. Educators were interviewed through "Google Forms" and asked to answer semi-structured open-ended research questions in the data collection process. We asked the participants to answer the following semi-structured open-ended questions:

Q1- How do you interpret contribution(s) of distance education to disaster and crisis management?

Q2- How distance education affects students and educators who are living in the disaster area for dealing with the adverse effects and relieving the symptoms of the disaster?

Q3- What are the key points regarding distance education in the disaster area?

Participants

The participants of the study were five educators, who were involved in distance education, in the earthquake region. Demographical information of the participants and interview time stamp is presented in Table 1.

Table 1. Demographical information of the participants and interview time stamp

Participants	Gender	Title	Age	Service (Years)	Time Stamp
Participant-1	Male	Dr. Academician	39	12	4/27/2023 17:33:11
Participant-2	Male	Associate Professor	35	10	4/27/2023 19:15:13
Participant-3	Male	Academician	40	5	5/3/2023 11:36:26
Participant-4	Female	Academician	50	29	5/3/2023 12:12:41
Participant-5	Male	Research Assistant	34	5	5/3/2023 17:47:01

Data Analysis

We analyzed the responses of the participants by using Nvivo-14 qualitative data analysis software. After reading and coding the responses, we generated the themes. We also focused on the discourse and emotions that are conveyed by the discourse of the participants, not only the content of their opinions and thoughts, in theme generation process. In this process, we concluded that the responses of the participants could distinctly be divided into two main themes as positive and negative. Direct references were assigned to the themes after coding. We compared the themes while we were reporting the findings. Then we visualized the data through concept map and word cloud features of Nvivo-14 software.

Limitations and Strengths

The interviews were made with the five participants through Google Forms in extraordinary conditions of the state of emergency. We did not contact with the participants or interfere with their responses verbally or in writing after we provided questions through Google Forms.

We did not include the political context in the study, intentionally, and limited the context of the study with the role of distance education in disaster and crises management. This limitation and the fact that we could not conduct face-to-face interviews with the participants in the state of emergency constitute the limitations of our study. We considered that the political context would be another research context, *per se*, and instead, we reviewed and evaluated national and international legal documents and legislation. This limitation of the study to the role of distance education in disaster and crises management enabled us to focus on the students and educators in the disaster area. The fact that the participants are experts in the field of education and involved in distance education in the disaster area has a strengthening effect on this study. We hope to contribute to further studies on the role of distance education in disaster and crises management.

Ethics

This study is supported by the Ethics committee document received by Anadolu University.

FINDINGS

The findings of our study, which was conducted to examine the perception of educators on negative and positive aspects of distance education in the disaster area and how the process was experienced in the field of education in the context of disaster and crises management, were presented and interpreted in the section of findings and interpretation of our study. The conceptual framework of our study was outlined by research questions that were determined after document and social media search to examine the news, opinions on social media, national and international laws, and legislation. The findings of our study were supported by direct references from the participants.

How Do You Interpret Contribution(s) of Distance Education to Disaster and Crisis Management?

We saw that the responses of the educators to our first question could be reviewed in two main themes, i.e., positive opinion and negative opinion. There are 21 subthemes and themes under the positive opinion in total. The responses were coded in two main subthemes: "advantages in terms of education dimension" and "advantages in terms of communication dimension".

Advantages in terms of education dimension can be listed as in the following table:

There were 3 themes under "negative opinion" subtheme as in the following: "No contribution", "face-to-face education should continue", and "different precautions should be taken".

Table 2. Contribution(s) of distance education to disaster and crisis management (positive opinion(s))

Advantages in Terms of Education Dimension (Frequency)	Percentage (%)	Advantages in Terms of Communication Dimension (Frequency)	Percentage (%)
Dynamic and updated education process (3)	23.076	Facilitating education through information and communication technologies (4)	57.143
Raising awareness of the people living in the region (2)	15,384	Access to true and reliable information resources	
A potential to benefit from lifelong learning (3)	23.076	Information exchange with the competent organizations	
Space and time independency in education (5)	38.464	Contribution to education of the working class with unsynchronous contents	
Equal and flexible education opportunity for everyone		Establishing contact(s) with the field expert(s)	
Continuous education in a comfortable environment		Mitigating negative impacts (3)	42.857
Not to be obliged to continue educational activities in damaged buildings		Opportunity for informing and providing psychological support	
Providing uninterrupted education		Space and time independency in education	
Increasing attendance (of students) in education		Establishing communication with senior experts	
Providing minimum continual attendance in education and learning			

Table 3. Contribution(s) of distance education to disaster and crisis management (negative opinion)

Theme	Frequency	Percentage (%)
No contribution	1	33.33
Face-to-face education should continue	1	33.33
Different precautions should be taken	1	33.33

How Distance Education Affects Students and Educators Who Are Living in the Disaster Area for Dealing With the Adverse Effects and Relieving the Symptoms of the Disaster?

We saw that the responses of the educators to our second question could again be reviewed in two main themes, i.e., positive opinion and negative opinion. There were 6 main subthemes and 7 themes under the positive opinion, i.e., 13 subthemes and themes in total. The main subthemes can be listed as in the following table.

There were 3 themes under "negative" subtheme" as in the following: "obstacle encountered for continual education", "basic needs that have not been met", and "being exempt from education". The theme "basic needs that have not been provided" was coded under "obstacle encountered for continual education".

Table 4. Effect(s) of distance education on psychology of students and educators (positive opinion(s))

Theme	Frequency	Percentage (%)
Reaching more students simultaneously	3	21.428
Sharing communication channels for resolving the problems	2	14.285
A guiding role on what to do	4	28.574
Psychological support		
Psychologically comforting students		
An opportunity to receive psychological support from experts		
(Students) thinking that they were not forgotten		
Comprehending that life should go on		
Becoming socialized	3	21.428
Students not being isolated from society		
Continuity of social life		
Preventing (students) to feel lonely		
A solution to the problem of not having enough number of experts	2	14.285
Experts reaching more students simultaneously		

Table 5. Effect(s) of distance education on psychology of students and educators (negative opinion)

Theme	Frequency	Percentage (%)
Obstacle encountered for continual education	2	28.571
Basic needs that have not been met	3	42.8580
Being exempt from education	2	28.571

Table 6. Key Points regarding distance education process (positive opinion(s))

Theme	Frequency	Percentage (%)
Mitigating traumas through communication with the peers	2	13.333
Reaching more students simultaneously	2	13.333
Easy unsynchronous use of digital contents	3	20.002
Right to education	4	26.666
(Students) not being deprived of education		
Advantages in terms of cost(s) and management	4	26.666
By providing low-cost education, mitigating economic burden through education		
Giving time to current administrations while conducting recovery actions		

What Are the Key Points Regarding Distance Education in the Disaster Area?

We saw that the responses of the educators to our second question could again be reviewed in three main themes, i.e., positive opinion, negative opinion, advice(s).

There were 26 themes, of which 8 were under theme "positive", 3 were under theme "negative", and 15 were under theme advice(s). Themes under "positive" can be listed as in the following table.

There were 2 themes under "negative" subtheme, and they can be listed as in the following: "no key points", "(students) should return to school after one term".

There were 7 themes under "advice(s)" subtheme, and they can be listed as in the following table.

CONCLUSION, DISCUSSION, AND SUGGESTIONS

Pain and desperation drastically increased both in the region and throughout Turkey when disaster response, search and rescue activities delayed after Turkey/Kahramanmaraş-centered earthquake occurred

Table 7. Key points regarding distance education process (negative opinion)

Theme	Frequency	Percentage (%)
(Students) should return to school after one term	4	80
No key points	1	20

Table 8. Key points regarding distance education process (advices)

Theme	Frequency	Percentage (%)
Establishing technological infrastructure	5	16.129
Computer access		
Internet access		
Internet speed		
Supporting technological infrastructure with physical means		
Providing training to the entire society	4	12.903
Open-access education resources		
Publishing and broadcasting educational resources on social media		
Benefiting from lifelong educational opportunities	5	16.129
Trainings on amateur walkie-talkies	3	9.677
Getting beyond classical curriculum	3	9.677
Being informed about psychological approaches	2	6.453
Flexibility in lesson hours	5	16.129
Ensuring that share their experiences and feelings are shared	4	12.903
Learning support through emotional awareness		

on the East Anatolian Fault Line in Turkey on February 6, 2023. The decision of the Council of Higher Education of the Republic of Turkey regarding compulsory distance education raised many criticisms and debates in the public. Especially, the decision on removing students from their dormitories resulted with a great reaction in public opinion.

As Koç (2023) and Yıldırım (2023) stated, the concerns and criticisms that came to the forefront in the news and the posts on Twitter (now called as: "X") after the decision of compulsory distance education can be summarized as in the following: face-to-face education rights of students, limited or no means of students to access distance education, deprivation of education due to deficient infrastructure, digital divide problem, hastily removing students from dormitories, students deprived of face-to-face education during Covid-19 lockdown, adverse effects of compulsory distance education on social lives of university students, face-to-face education demands of the students, aggravating traumas due to distance education, the decision might have been understood is disaster response was not delayed, and students were not removed from their dormitories, etc.

According to the findings of our study, four educators out five participants shared their opinions and advice on contribution(s) of distance education to disaster and crisis management. The educator, who stated that there is no contribution of distance education to disaster and crisis management and other actions should be taken, provided a very short and vague response. This was interesting since all educators experienced compulsory distance education during Covid- 19 pandemic crisis throughout Turkey and the participant is a senior educator in terms of working years in his profession. Therefore, we may conclude that he experienced distance education for a considerable time. Notwithstanding the fact that there are different opinions, i.e., both positive and negative, on compulsory distance education during Covid-19 pandemic lockdown in the literature, this was interesting. However, we could not conduct a face-to-face interview with the educator and kindly ask additional questions to clarify his opinion whereas we consider that negative opinions are as significant as positive opinions for inferring conclusions.

Four educators out five participants shared their opinions and advice on the effect(s) of distance education on psychology of students and educators. According to the educator who shared his negative point of view, the fact that basic needs were not met was an obstacle encountered for continual education. It is understood that distance education loses its meaning under certain conditions, especially where and when basic needs are not met. In this regard, Ingrassia et al., (2014) and Kurt and Gulbahce (2019) emphasize that earthquakes and other disasters have many negative effects on students.

When this response with another response of the educator is taken into consideration, a similar conclusion can be made. The educator indicated that the students or educators in the disaster area wished to be with their families and this was important for support. The response of the educator could be related with one of the contributions of distance education. It is understood that the educator does not consider distance education in relation with space and time independency in education. As far as it is understood, the educator used this expression in the context of educational exemption of students and educators, i.e., interruption of education. While four participants, who provided positive opinion, highlighted continuity of education by distance education, one participant, who provided negative opinion, highlighted interruption of education for a certain time. As in Tanik Onal and Onal (2023), we also consider that there were many problems in the disaster area and the trauma had adverse impacts on both educators and students in the disaster region, and the educator might be away from his family and could not leave the disaster area. Since a face-to-face interview could not be conducted, additional questions could not be asked for the educator to share his opinion in detail. The same concept can be interpreted from different points of views in various ways based on our experience of the events we encounter.

Two educators out of five participants highlighted positive aspects of distance education in their responses for the key points of distance education process. The educator, who provided negative opinion, stated that there are no key points, and they should return to education. It was understood from the expression of the educator that he means that they should return face-to-face education. This indicates the perception of the educator regarding distance education. The perception of the educator matches with the negative public opinions on distance education. When all responses of the educator are considered as a whole, it is understood that the educator had an adverse experience of distance education. Considering the flexible structure offered by distance education (Tarrayo, Paz & Gepila, 2023), it is possible to state that the distance education process benefits both educators and students.

It is understood from the findings of the study and the literature that there are many advantages and contributions of distance education to disaster and crisis management, but there are also various problems when face-to-face education is interrupted such as being deprived of campus, digital divide, etc. Temporary housing and basic needs should priorly be provided and met, and distance education can only be meaningful, if these are provided and met.

In disasters such as earthquakes, floods, etc., when electric lines and telecommunication systems are damaged, technological infrastructure and internet access and having technological means are primary requirements for distance education. It is highly important that these systems should quickly be operational in the disaster area. Students need technological means to attend distance education such as computers, laptops, smartphones. So, fighting against digital divide for equality of education opportunity becomes prominent.

It is important for central and local governments to act against digital divide in terms of disadvantaged groups in order to provide equal opportunity to all segments of the society. The fight against this issue may provide an advantage for integration of distance education into disaster and crisis management as well as equality of opportunity in education. Activities and implementations on this issue may increase the capacity and resilience of individuals in terms of disaster and crisis management and may provide informed and active participation of every segment of society in disaster and crisis management.

The inclusion of topics on climate change, environmental issues, and disasters in the curriculum through an environment and climate change education program prepared by the General Directorate of the Ministry of National Education of the Republic of Turkey in 2022 is a significant step in this field, and providing environment and climate change courses to the society can be beneficial as a contribution to disaster resilient communities. In addition, awareness can be raised on natural and anthropogenic disasters and environmental issues in every segment of the society through distance education programs to be provided via communication channels, such as television and/or internet, as a part of disaster preparedness activities. Socioecological systems approach can be applied in educational programs.

For the future, with the help of artificial intelligence, places that are vulnerable to disasters can be determined by region and distance education measures can be taken in advance accordingly. The infrastructure issue can be given priority in line with the results to be obtained from this.

Author's Note

The current study utilized the dataset of the first author's term project.

REFERENCES

AFAD. (2023). *Disaster and Emergency Management Presidency Dictionary*. https://www.afad.gov.tr/aciklamali-afet-terimleri-sozlugu

Ally, M. (2004). Foundations of educational theory for online learning. *Theory and Practice of Online Learning*, *2*(1), 15–44.

Aloi, A., Alonso, B., Benavente, J., Cordera, R., Echániz, E., González, F., Ladisa, C., Lezama-Romanelli, R., López-Parra, Á., Mazzei, V., Perrucci, L., Prieto-Quintana, D., Rodríguez, A., & Sañudo, R. (2020). Effects of the COVID-19 lockdown on urban mobility: Empirical evidence from the city of Santander (Spain). *Sustainability (Basel)*, *12*(9), 3870. doi:10.3390/su12093870

Badiou, A. (2007). *The Century*. Polity Press.

Bauwens, M., Compernolle, S., Stavrakou, T., Müller, J. F., Van Gent, J., Eskes, H., Levelt, P. F., van der A, R., Veefkind, J. P., Vlietinck, J., Yu, H., & Zehner, C. (2020). Impact of coronavirus outbreak on NO2 pollution assessed using TROPOMI and OMI observations. *Geophysical Research Letters*, *47*(11). Advance online publication. doi:10.1029/2020GL087978 PMID:32836515

Creswell, M. J. (1986). Why Objects Exist but Events Occur. *Studia Logica*, *45*(4), 371–375. doi:10.1007/BF00370270

Deleuze, G. (1994). *Difference and Repetition*. Columbia University Press.

Hager-Zanker, A. (2008). *Measuring the performance of geosimulation models by map comparison* [Doctoral Thesis]. Maastricht University. doi:10.26481/dis.20081210ah

Ingrassia, P. L., Foletti, M., Djalali, A., Scarone, P., Ragazzoni, L., Della Corte, F., ... Fisher, P. (2014). Education and training initiatives for crisis management in the European Union: A web-based analysis of available programs. *Prehospital and Disaster Medicine*, *29*(2), 115–126. doi:10.1017/S1049023X14000235 PMID:24642198

Jeoloji Mühendisleri Odası, T. M. M. O. B. (2021). Fay üzerinde yaşayan illerimiz: Hatay raporu-6. *Jeodergi*, *21*(1), 1–10.

Kalafat, D., Zülfikar, A. C., & Okuyan-Akcan, S. (2021). Seismicity of Turkey and real-time seismology Applications in determining earthquake hazard. *Academic Platform Journal of Natural Hazards and Disaster Management*, *2*(2), 96–111. doi:10.52114/apjhad.1039670

Khorram-Manesh, A., Lupesco, O., Friedl, T., Arnim, G., Kaptan, K., Djalali, A. R., Foletti, M., Ingrasia, P. L., Ashkenazi, M., Arculeo, C., Fischer, P., Hreckovski, B., Komadina, R., Voigt, S., Carlström, E., & James, J. (2016). Education in disaster management: What do we offer and what do we need? Proposing a new global program. *Disaster Medicine and Public Health Preparedness*, *10*(6), 854–873. doi:10.1017/dmp.2016.88 PMID:27435533

Koç, N. (2023). The views of associate degree students from child development on distance education after the earthquake. *Journal of Educational Technology and Online Learning*, *6*(3), 756–770. doi:10.31681/jetol.1326200

Korkmaz, H. (2006). Antakya'da zemin özellikleri ve deprem etkisi arasındaki ilişki. *Coğrafi Bilimler Dergisi*, *4*(2), 49–66.

Kurt, E., & Gulbahce, A. (2019). Van depremini yaşayan öğrencilerin travma sonrası stres bozukluğu düzeylerinin incelenmesi. *Atatürk Üniversitesi Sosyal Bilimler Enstitüsü Dergisi*, *2*(1), 957–972.

Palutoğlu, M., & Şaşmaz, A. (2017). 29 Kasım 1795 Kahramanmaraş depremi, güney Türkiye. *Maden Tetkik ve Arama Dergisi, 155*(2), 191-206.

Pregowska, A., Masztalerz, K., Garlińska, M., & Osial, M. (2021). A worldwide journey through distance education—From the post office to virtual, augmented and mixed realities, and education during the COVID-19 pandemic. *Education Sciences*, *11*(3), 118. doi:10.3390/educsci11030118

Rad, A. K., Zarei, M., Pourghasemi, H. R., & Tiefenbacher, J. P. (2021). The Covid-19 crisis and its consequences for global warming and climate change. *Computers in Earth and Environmental Sciences*, *2022*(1), 377–385. doi:10.1016/B978-0-323-89861-4.00006-3

Tanık Önal, N., & Önal, N. (2023). Online learning and teaching experiences of preservice science teachers during Covid-19 pandemic. *Journal of Educational Technology and Online Learning*, *6*(2), 295–314. doi:10.31681/jetol.1078486

Tarrayo, V. N., Paz, R. M. O., & Gepila, E. C. Jr. (2023). The shift to flexible learning amidst the pandemic: The case of English language teachers in a Philippine state university. *Innovation in Language Learning and Teaching*, *17*(1), 130–143. doi:10.1080/17501229.2021.1944163

Telli, Y. G., & Altun, D. (2023). Türkiye'de deprem sonrası çevrimiçi öğrenmenin vazgeçilmezliği, *Üniversite Araştırmaları Dergisi, 6*(2), 125-136.

Tian, X., An, C., Chen, Z., & Tian, Z. (2021). Assessing the impact of COVID-19 pandemic on urban transportation and air quality in Canada. *The Science of the Total Environment, 765*(1), 144–270. doi:10.1016/j.scitotenv.2020.144270 PMID:33401062

Turner, J. R., & Baker, R. M. (2019). Complexity Theory: An Overview with Potential Applications for the Social Sciences. *Systems, 7*(4), 1–22. doi:10.3390/systems7010004

Usman, M., Husnain, M., Riaz, A., Riaz, A., & Ali, Y. (2021). Climate change during the COVID-19 outbreak: Scoping future perspectives. *Environmental Science and Pollution Research International, 28*(35), 49302–49313. doi:10.1007/s11356-021-14088-x PMID:33934308

Uzun, G. Ö., Es, A. Ç., & Evram, G. (2020). Examination of attitudes of university students in distance education according to some variables. *Near East University Online Journal of Education, 3*(2), 104–115.

Vogel, S. (1988). Marx and alienation from nature. *Social Theory and Practice, 14*(3), 367–387. doi:10.5840/soctheorpract198814317

Yildirim, S. (2023). 6 Şubat kahramanmaraş depreminin psikososyal etkisi ve depremzedelere yönelik sürdürülebilir müdahalenin önemi üzerine bir gözlem araştirmasi. *Anasay, 24*(1), 133–153. doi:10.33404/anasay.1286368

KEY TERMS AND DEFINITIONS

Case Study: An intensive analysis of an individual unit (such as a person or community) stressing developmental factors in relation to environment.

Content Analysis: Analysis of the manifest and latent content of a body of communicated material (such as a book or film) through a classification, tabulation, and evaluation of its key symbols and themes in order to ascertain its meaning and probable effect.

Crisis: The turning point for better or worse in an acute disease or fever.

Crisis Management: An organization's process and strategy-based approach for identifying and responding to a critical event.

Disaster: A sudden calamitous event bringing great damage, loss, or destruction.

Distance Education: A method of study where teachers and students do not meet in a classroom but use the Internet, email, mail, etc., to have classes.

Earthquake: A shaking or trembling of the earth that is volcanic or tectonic in origin.

Chapter 9
Learning From Chaos:
The Complexity of Students' Needs in Online Learning

Vicki A. Hosek
Illinois State University, USA

Jay C. Percell
Illinois State University, USA

ABSTRACT

Online course experiences of undergraduate students continue to grow due to both a pre- and mid-COVID-19 pandemic increase in online courses offered and taken at universities. In this study, students identified the benefits and challenges of an online course format and their online learning needs for achievement of learning objectives. While the majority of the students in this study had previous online course experiences, they were not confident in their ability to reach learning objectives in that course format. Students identified a lack of specific digital and life skills as well as instructor issues with communication and organization of online course content as their primary concerns. Suggested improvement of practices based on findings include: a focus on assessment and development of the digital literacy and life skills of students, clear communication of course requirements while ensuring students' realistic understanding of course commitments, and professional development opportunities for instructors that are closely reflective of students' online learning needs.

INTRODUCTION

Before the COVID19 pandemic hit the United States of America, universities across the country saw a steady increase in enrollment of college students in distance education courses. In 2018, 34.5% (6.9 million) of U.S. college students were enrolled in a distance education course (U.S. Department of Education, 2022) which reflects a 6% increase from 2016. Pre-pandemic, American universities were already increasing online course offerings for several reasons. A 2015 survey of university administrators and instructors showed that 63% of chief academic leaders believed that online learning was "critical to their

DOI: 10.4018/979-8-3693-1926-0.ch009

long-term strategy" (Allen & Seaman, 2016, p.5). Importantly, 71.4% of those leaders believed that the learning outcomes in online education are "the same or superior to those in face-to-face instruction" (Allen & Seaman, 2016, p.5). Also noteworthy, 60% of higher education faculty reported that they "accept the value and legitimacy of online learning" (Allen & Seaman, 2016, p.6).

While administrators and instructors saw the value of online teaching and learning, we found there was a noted absence in the research of the students' perspectives about their online learning needs to ensure their academic objectives could be met. The pandemic forced undergraduate instructors and students to move fully online for much of the 2019 academic year, the entirety of the 2020 academic year, and the fall of 2021. This unique situation prompted us to survey students about their experiences and perspectives regarding their online learning needs. Our findings provide an understanding of why students choose and/or do not choose to take online courses and what university administrators and instructors can do to support students' needs who either choose or have to take an online course. One student in our study explained, "I didn't sign up for this." While this study was conducted during the pandemic when students had no choice but to take fully online courses, its implications are not limited to times of crisis. Our findings highlight the value of seeking students' perspectives which can inform university administrators and instructors about how to recognize and meet students' needs to ensure impactful and effective online course design and delivery.

OVERVIEW OF THE LITERATURE

In an effort to understand the needs of undergraduate students participating in fully online courses, we conducted an overview of the literature. In the previous research, we found a quantitative emphasis on: the students' interests, the amount of student interactions and participation throughout a course, and students' achievement based primarily on assessment scores. We also noted a less prevalent focus on course structure and issues of access. Below, we present that research. Then, we identify the gaps in the literature which directed our study.

Students' Interests

Researchers most often utilized quantitative measures such as Likert scales, learning analytics, trace data, and data mining of learning management systems (LMS) to identify what students sought and valued in online courses. According to The Ruffalo Noel-Levitz National Online Learners Priorities Report (2017) which utilized a Likert scale survey, students' interest in online courses centered on "convenience, flexible programming, and the ability to fit education into their work schedule" (p.17). Overall, 74% of the approximately 128,000 students surveyed stated that they were satisfied with their online courses and depth of learning. They also found that when choosing the online format, students tended to value flexible scheduling over classroom interaction and discussion. In addition, the survey showed that the greatest challenges students faced were locating excellent quality online instruction, instructor provision of clearly defined student assignments in the syllabus, and faculty responsiveness to student needs (Noel-Levitz, 2017; Martin and Bolliger, 2018).

Interactions and Assessments

In many studies, a notably prevalent focus was on student-instructor interactions and how those interactions impacted student attendance & retention levels in an online course (Cochran et al., 2014; Kinlaw et al., 2012; Martin & Bolliger, 2018; Joksimović et al., 2015). Those studies showed that the greater the number of interactions with instructors, the more positive the students' perspectives about and satisfaction with online courses. In addition, researchers noted that students responded more positively about an online course when interactions with instructors were immediate (Chakraborty & Nafukho, 2015; Ladyshewsky, 2013; Martin et al., 2019) and continuous throughout the course (Martin & Bolliger, 2018; Moallem, 2015). Other studies of students' beliefs about online learning success also showed positive correlations between number and quality of interactions, amount of instructor feedback and engagement, and students' perceived learning (Han & Johnson, 2012; Kupczynski et al., 2010; Lammers & Gillaspy, 2013). llgas and Gulbahar (2015) pointed to an additional positive correlation between providing multiple opportunities for evaluation and student satisfaction. In that same study, they found technical issues and accessibility to be significant barriers for students in online courses.

Another quantitative focus in previous research was on final assessment and/or grade received as an indicator of student needs being met in an online course (Chen, Lambert, & Guidry, 2010; Ladyshewsky, 2013). Researchers correlated higher grades with higher student satisfaction in the course. Similar to earlier cited studies, they also pointed to student-instructor engagement as critical to students' success in online classrooms. Primarily, the levels of engagement were determined by learning analytics where the number of times students participated within an LMS was recorded and analyzed.

COURSE STRUCTURE

Course structure was a focus we found in previous research concerning students' needs. Several researchers found that students valued online courses where collaborative activities were incorporated (Kemp & Grieve, 2014; Martin & Bolinger, 2018; Yang, Yeh, & Wong, 2010) while others showed that students viewed group work to be a negative component of online courses resulting in lower participation and lower course satisfaction (Erdmann et al., 2017; Lowenthal & Dunlap, 2018). We also identified a strong focus in the literature concerning the impact of content and course structure on students' perceptions about taking an online course versus a face-to-face (F2F) course. Studies showed students most highly valued instructional design where the structure of the course and the roles of the students and instructors were made clear. This specifically included instructor expectations about number and depth of interactions required by students and details about timing and types of assessments students would need to complete (Joksimović, Gašević, Kovanović, Riecke, Hatala, 2015; Kupczynski et al., 2010).

Issues of Access

An additional focus in the research concerned issues of access in online learning environments. Researchers explained that students needed not only physical access to computer equipment and the internet, but also access to opportunities to learn how to meaningfully and critically use digital platforms (Hosek, 2022;

Jung & Lee, 2020; Ortagus, 2017; Roswell et al., 2017; Van Dijk, 2006). This speaks directly to the need for the development of students' digital literacy to help ensure equitable access to online learning opportunities (Hosek, 2022; Buchholz, 2020). In fact, research showed that students tended to overestimate their personal digital literacy (Bhatt & MacKenzie, 2019; List, 2019; Oncul, 2020; Ortagus, 2017; Ng, 2012), revealing either a missing awareness of the skills needed to be successful in online courses or a lack of opportunity to develop those skills. Oncul (2020) created an online test to quantitatively assess the digital literacy of students in order to provide guidance for instructors by giving them a gauge as to students' digital needs.

GAP IN RESEARCH AND VALUE OF OUR STUDY

In the literature reviewed, we found that very few studies provided an opportunity for students to self-identify and explain their online course needs and the kinds of support they believed were necessary to achieve their learning objectives. The majority of studies we reviewed used quantitative means to determine results including learning analytics and limited answer surveys (Likert scales). While such analyses can provide informative foundational information, they were largely absent of first-person voice and experiences.

We conducted our study during the COVID 19 global pandemic which provided us with the unique opportunity to learn about how students felt about the online course format in real time. We sought students' perspectives about the benefits, challenges, support (or lack there of) and needs they experienced after all of their courses were forced online. Importantly, this study provides insights leading to an expanded understanding of the online classroom format and the ways it does and does not deliver according to students' first-hand experiences.

Theoretical Framework

The theoretical framework for our study is grounded on our understanding of complexity theory and chaos theory. In combination, these theories informed our approach to the data, analysis, and discussion about the implications for teaching practices, online course design, and future research. Our framework is unique to this study. The COVID 19 pandemic proved to be a massive disruption to education worldwide making the application of complexity and chaos theories to this phenomenon appropriate.

Complexity Theory

First, we draw on complexity theory to gain an understanding of the dynamics at work at a large, public Midwestern university, Lorenso State University (LSU), during the time of our study. We consider LSU to be a complex adaptive system. According to Waldrop (1992) "'Complex adaptive systems' scan and sense the external environment and then make internal adjustments and developments in order to meet the demands of the changing external environment" (p.294). Morrison (2008) defined such changes as the "ability of a system to evolve from within" (p.18).

Based on expert researchers' work in the field (Cochran-Smith, 2014; Houchin & McLean, 2005; Mason, 2008; Morrison, 2008; Radford, 2008; Waks, 2011; Waldrop, 1992), we identified the following key components of a complex system:

- an open system that is "nested in and intersect(s) with other systems" (p.6) but is boundaried by initial conditions and parameters (Cochran-Smith et al., 2014),
- continual occurrence of nonlinear, variant, short term and redundant interactions, and
- dynamic interactions that provide feedback "loops" (p.6) and yield learning, change and growth (Cochran-Smith et al., 2014).

The university system is complex because the individuals operating and navigating within the system are diverse and complex. Their actions, responses, and decisions are non-linear and dynamic so there can be no prescribed formula with how to predict them. Waks (2011) explained that "Systems are complex to the extent that there are strong interactions among the parts, and interactions are likely to be strong to the extent that they involve many and diverse agents following diverse local rules or strategies" (p.67). We submit that in the case of our study, those "diverse agents" are the university administrators, instructors, and students which operate individually and also work and interact together to form a dynamic university system.

Chaos Theory

Next, because of the timing of our study (during the COVID 19 pandemic), we draw on our understanding of chaos theory as a crisis resulting from the dramatic shift in course format and delivery. One contention of chaos theory is that complex systems, whether macro in the sense of a university or micro in the sense of individuals working within a larger system, can benefit from the chaos that caused the need for change (Hunter & Benson, 1997; Morrison, 2008;). Morrison (2008) explained: "The closer one moves towards the edge of chaos, the more creative, open-minded, imaginative, diverse and rich are the behaviors, ideas and practices of individuals and systems, and the greater the connectivity, networking and information sharing between participants" (p.19). In this way, chaos theory has the potential to "contribute to a better understanding of some of the phenomena we observe in the natural world" (Hunter & Benson, 1997, p.88). As a result of COVID 19, the university's face-to-face course system stability was impacted which "can prompt phase transitions at the edge of chaos" (Waks, 2011, p.68) pushing instructors and students to quickly react and respond. The result of chaos within a complex system is unpredictable, but the challenge of that instability can lead to interactions that spur the growth of qualities or characteristics within individuals that help them adapt and survive (Houchin & McClean, 2005; Hunter & Benson, 1997; Waks, 2011). Consequently, we employed complexity theory in light of chaos theory.

Additional Considerations

In the case of our study, we recognize that the depth of interactions between instructors and students can be impacted by a multitude of factors that are not always in the control of the individuals. This adds to the complexity of communication and participation in online classrooms. Because of this, we drew on McLaren's (2017) understanding of critical theory in educational practice as we considered the individual participants and each one's diverse needs in online courses. McLaren (2017) explained that an essential component of critical theory is recognizing that the social spaces that students experience in school need to be critically examined to ensure that the needs of all students are considered when making decisions about content and methodology. He explained that critical pedagogy focuses on teaching and learning that is "relevant, critical, and transformative" (McLaren, 2017, p.80). Learning becomes dialectical when student voices, cultures, and experiences are recognized, prioritized, and considered essential. As such, we chose an open-ended survey format that provided a confidential space for students' first person perspectives and worked to ensure their voices were the priority in our coding, discussion of findings, and suggested implications.

We also recognize that utilizing complexity theory to ground this study places us in a dichotomous position whereby the cause of the disruption, or the chaos (the COVID pandemic), created a unique environmental change to the university course delivery system that may never be replicated. Previous researchers who have applied complexity theory to educational studies point to the lack of "causal explanations with implications for practice" (Cochran-Smith et al., 2014, p.2) as a weakness for its use. We contend that examination of the resulting perspectives and lived experiences of the students provides pivotal insight about their needs that can be beneficially used to shape future online teaching and learning environments. As such, our study points to the value of learning about the needs of students in an effort to foster growth, evolution, and survival of both individuals within the teaching and learning community as well as the university community as a whole. In addition, previous researchers point to the value of incorporating criticality when utilizing a research framework grounded on complexity theory to ensure equitable representation of all voices (Cochran-Smith et al., 2014; Morrison, 2008; Reed & Harvey, 1992). We believe our use of a critical theoretical lens when analyzing our data helps ensure the consideration and implications of the diverse perspectives and challenges that face equally diverse individuals.

Based on our understanding and use of complexity and chaos theories to ground this study, we submit that its purpose is not to develop a standardized approach to dealing with unpredictable situations in higher education where chaos ensues. Rather, the goal is to understand the diverse skills, needs, and perspectives of students and connect those needs with future teaching and course design suggestions that could lead to both personal and academic growth. Using complexity theory and the employment of a critical lens to enlighten and guide us, we identify the diverse opportunities for emergent growth that occurred, or did not occur, as a result of the unexpected move to fully online courses and make recommendations for future practice.

METHODOLOGY

This article focuses on one part of a larger qualitative study. Qualitative methods were used for several reasons. Previous studies focused on quantitative methods to identify students' needs in online classrooms rather than first-hand student voice. Qualitative methods help "expand rather than confine understanding" (Lutrell, 2010, p.37). Considering this in light of our theoretical framework, we believe that students' needs and values are complex and diverse which prompted us to choose data collection and analysis methods that support students' voices, perspectives and beliefs about online courses and their academic and personal needs regarding the online course format.

In this study, we sought to answer the following questions:

1. What do students identify as the benefits and challenges of the online course format?
2. What kinds of support do students believe they need to achieve their learning objectives in an online course?
3. How can universities and instructors support students while taking an online course?

Participants

Lorenso State University (LSU) is a large, public, state university in the rural Midwest of the United States. At the time of this study, 89% of the students were considered full-time with 78% of the students receiving financial aid. According to the university website, LSU has a high retention rate of 79% from academic year to year. While economically diverse, this university is not considered to be ethnically diverse as 71% of students identified as white, 11% as Hispanic, 9% as black or African American, and the remaining as "other."

Data Collection and Analysis

To begin, we created an open-ended questionnaire that was distributed via email through a university-wide listserv to all undergraduate students at LSU (see Appendix A for survey questions). Each participant was randomly assigned a number which served in place of their name to ensure confidentiality. One hundred seventy-two students participated in this study.

Each author independently coded the questionnaire responses using NVivo coding with analytic memos followed by theming (see Saldana, 2016). NVivo coding limited the liberty we could take with the data thus increasing reliability by staying closely aligned with the participants' experiences (Saldana, 2016). The memos served as guides for our discussions and interpretations of the data (member-checking) leading to qualitative themes based on the connections and interconnections of the coded data (Lutrell, 2010; Savin-Bader & Howell-Major, 2013). Then, we numerically converted our themes by identifying the number of participant responses within each theme and divided that by the overall number of participants who responded to each respective question resulting in percentages. We present that data in Tables 1, 2, and 3.

Limitations

Purposeful convenience sampling was used for this study. As such, our study is not generalizable. Because the study took place during a time of worldwide crisis, it is unlikely to be replicable. At the same time, we believe that our theoretical perspective frames this study in such a way as to make the findings highly relevant for guiding future online teaching practices that closely attend to the needs of online students.

Throughout this study, both authors were teaching courses from which students were recruited to participate. While we believe teaching during this time provided us with a depth of first hand experience and knowledge about students' needs, we recognize that, while unintentional, those experiences could impact data reduction and analysis (Lutrell, 2010).

Findings

Below we present the results of our questionnaire. First, we provide information students gave about their previous online classroom environment experiences. Next, we present students' beliefs about the benefits and challenges of the online course format. Then, we provide the specific needs and support students identified as essential to successfully accomplishing their learning objections in an online course format.

Previous Online Classroom Environment Experiences

Participants were asked to provide the degree of experience with online classroom environments and explanations about those experiences. According to our survey, the majority of students had previous experience with online components of courses. Primarily, their experiences focused on using the university LMS to access and turn in online assignments, access course materials, and take exams and/or quizzes. Approximately half of the students surveyed chose to take an online course prior to this study. They provided numerous reasons for choosing the online format. Thirty-nine percent of students needed the flexibility the format offered for work, to complete at their own pace, and/or to lighten their course load in later semesters. Thirty-one percent stated that the online format was the only option offered for the course they needed. Ten percent of students believed taking the course online saved them resources such as not having to commute. An additional 10% chose the format because it was recommended by other students. Seven percent wanted the condensed online version offered in summer sessions. Finally, 3% of students believed the online format was less overwhelming.

The students who did not take an online course previously were asked why they chose not to. Responses showed that 33% of students believed they could not learn without the instructor physically present. They stated that they needed to be able to ask questions and receive an immediate response. Another 26% of students stated their inability to focus was a factor in not choosing the online course format. They attributed this to a lack of ability to make and keep a schedule, a lack of accountability, and procrastination. Fourteen percent of students believed they learned better in person and 14% of students stated "no option" to take an online format as a reason. Nine percent of students said they wanted in-person connections with fellow classmates and the instructors. Finally, 4% of students stated "the content doesn't translate to online."

Benefits of the Online Course Format

Students were asked what they believed the benefits of having online courses were. We identified the following major themes in their responses: time, flexibility, and comfort. Over one-third of students felt that online courses afforded them the benefit of more time for other things such as jobs and family. A similar percentage of students felt that flexibility, specifically in relation to location of taking the courses and their personal schedules, was of great benefit. Another benefit cited by 18% of students was comfort. This included feeling less academic and social pressures, feeling safer outside of the physical classroom, and being able to protect their personal health. Nine percent of students believed there were no benefits to taking online courses. Also noteworthy, 6% of students noted greater access to resources and materials such as recorded lectures and notes as a benefit of online courses. See Table 1 for details.

CHALLENGES OF THE ONLINE COURSE FORMAT

Students were asked what they saw as the challenges of taking an online course. The following themes were identified: self-discipline/focus, motivation, instructor issues, access and content adaptability. Approximately 37% of students stated that they lacked the self-discipline needed for online courses. They cited their tendencies to forget, "not do enough," and/or procrastinate as personal challenges that could

Table 1. Students' self-identified benefits of taking 100% online courses

Benefit	Percent of Students	Indicators	Indicator Percent of Total
Time	34%	More time for job/family Self-pacing More Free Time	15% 14% 5%
Flexibility	28%	Schedule Choose location No specifics provided	15% 9% 4%
Comfort	18%	Less academic/social pressure Physical safety and health	10% 8%
No benefit	9%	No specifics provided	9%
Greater access	6%	More resources such as materials and lectures/notes	6%
Learn life skills	5%	Adapting Self-discipline	3% 2%

impede their success. In addition, they worried about their ability to focus on the course objectives as well as holding themselves accountable to complete all components of the courses.

Thirty-five percent of students identified issues with instructors as a major challenge. In fact, in a separate survey question when asked if they felt their instructors possessed the digital skills necessary to effectively teach online, only 41% answered yes while 29% believed that some did and some did not and 23% believed they did not. Instructor issues that students identified included lack of effective online teaching methods, lack of clarity about deadlines and details of assignments, and assumptions that students had the time and skills they needed to be successful in the online format.

Twenty-five percent of students believed that access was also a challenge. Specifically, they explained that getting help from instructors and peers was more difficult online. They cited that the access was impersonal resulting in a shallower depth of interactions and socialization. In addition, they stated it could be "difficult to find a location" conducive to taking class and completing coursework outside of the physical classroom. In addition, technology access such as a reliable internet connection posed a challenge for many. Finally, 3% of students did not believe certain courses or aspects within those courses could translate to impactful learning when moved online. Specifically, they explained that groups were more difficult and a lack of "hands-on" and "real" experiences posed a great challenge. See Table 2 for additional details.

In light of the benefits and challenges that students' identified, they were asked if they believed their learning objectives could be met in an online course format. Fifty percent of students surveyed did not believe their own learning objectives could be met while twenty-nine percent believed they could be met and 19% were unsure.

Students' Self-Identified Skills and Support Needed for Success

Students were asked what skills they needed to be successful in a fully online course. We identified the following themes from their answers: self-discipline, organization and time management, determination, and technology skills. Approximately 30% of students surveyed identified self-discipline and focus as essential to their online learning success. Twenty-nine percent felt that organization and time management skills were important. Twenty-five percent believed that motivation and determination were key

Table 2. Students' self-identified challenges of taking 100% online courses

Challenges	Percent of Students	Indicators	Indicator Percent of Total
Self-Discipline and focus	37%	Low Motivation Procrastination Other Time management	14% 10% 8% 5%
Instructor issues	35%	Online teaching methods Adaptability No hands-on experience	23% 6% 6%
Issues of access	25%	Interactions and socialization Technology issues Inconvenient location	13% 10% 2%
F2F courses cannot be replicated	3%	Inauthentic Limited collaboration	2% 1%

to success. This included answers such as such as being "a hard worker," "taking charge of your own learning," and being able to "teach yourself." The remaining percentage of answers included being "tech savvy," having strong communication skills, and being adaptable with their learning styles.

Students were asked what kinds of support they believed they needed from their instructors to be successful in online courses. The themes we identified include: communication with instructors, resources, empathy, and flexibility. Approximately 45% of students surveyed identified communication and access as key support that they needed from their instructors to be successful. This included consistent and timely feedback, constant "checking in" on students' progress, and extensive details about the course structure as well as instructions about all course components. Thirty-eight percent of students stated that a multitude of resources were essential to meet their learning needs. Such support included tutorials, recorded lectures, notes, PowerPoints, detailed schedules and "to do" lists. Eight percent of students listed understanding and leniency with due date and grading as support they valued. Also important to approximately 7% percent of students was flexibility from instructors such as allowing for self-pacing through materials and time extensions on exams and assignments. See Table 3 for details.

DISCUSSION

We found students' beliefs about being successful during this time in this new course system were contingent upon the university and instructors meeting the needs they identified in our questionnaire. Importantly, students stated that many of the skills deemed as necessary for their successful achievement of course learning were underdeveloped and, in some cases, absent. This connection was made through cross-analysis of the students' answers to multiple questions in our survey. That analysis helped us identify the relationships across the students' answers in order to deepen our understanding of students' perspectives and experiences. Below, we discuss those connections and parallels that provided valuable information about students' needs in the online course format leading to our identification of the implications of this study.

Table 3. Students' self-identified support needed in fully online courses

Support	Percent of Students	Indicator	Percent of Indicator
Communication	45%	Consistent, constant and timely feedback Clear course structure including extensive details	29% 16%
Resources	38%	Extra (no details provided) Tutorials Recorded Lectures Access to instructor PowerPoints	13% 13% 5% 6% 1%
Empathy	10%	Understanding Leniency	7% 3%
Flexibility	7%	Self-pace Time extensions	5% 2%

Experience Without Confidence

All students in our study stated they had previous experience with online components of courses. In fact, 75% of the students surveyed took a fully online course prior to this study. Students stated that the primary type of online learning experience they had, regardless of course format, was superficial engagement on the university LMS and reading PowerPoints. This supports previous studies of student and instructor use of online classroom platforms which found a lack of critical and meaningful engagement in those online classroom environments (Hosek, 2018; Cabero-Almenara et al., 2021; Dzikite et al., 2017; Ruggiero & Mong, 2015). Importantly, 50% of the students did not believe their learning objectives could be met in an online classroom. This is significant particularly because so many students had previous firsthand online course experiences from which to draw that conclusion. The data showed that students were keenly aware of the skills needed for successful online learning and the skills that they did and did not have. This finding provides an important student perspective about learning outcomes in online courses that in combination with the earlier cited Allen and Seaman (2016) study about university administrators' perspectives could lead to valuable guidance about online course instructional design (see Implications section below).

THE BENEFITS AND CHALLENGES OF THE ONLINE COURSE FORMAT

When asked why students took an online course, the greatest number of students stated they needed the flexibility to be able to work (outside of school) and/or complete coursework in their own time. In fact, students considered the flexibility that an online format offers as one of the greatest benefits of taking an online course (See Table 1). Interestingly, students did not identify flexibility as a necessity for success in an online course. Importantly, when compared to the challenges of the online format that students identified, it appeared that the flexible nature of an online course created obstacles preventing them from capitalizing on the potential benefits of a flexible course format.

Focus and Self-Discipline

In our survey, the greatest number of students identified their inability to focus and a lack of self-discipline as the biggest challenge in online course (See Table 2). Specifically, they cited "forgetting," procrastination, lack of personal accountability, time management, and self-motivation to complete the work as issues. This presented a juxtaposition between the greatest challenge and the greatest benefit (flexibility) that students identified creating a possible roadblock for achieving learning objectives.

Instructor Issues

Many students also attributed their lack of confidence in the online format to issues with instructors. Specifically, they stated that instructors didn't provide clarity and details about course schedules, deadlines, and assignments. Without that guidance and direction, the students felt they needed to rely more heavily on their own management of the course material, organization, determination, and scheduling of their time; all skills that a majority of students believed they lacked. In addition, students worried that they would have to teach themselves the course material. In fact, when asked what prevented them in the past from taking an online course, the largest number of students stated that they "can't and don't want to teach themselves" and they "have to ask an instructor questions in person to learn."

Support Students Need

Students identified several kinds of support they needed to be successful. Importantly, instructors appeared to be the main resource for that support. The following needs occurred the most in the students' answers: tutorials, extra materials, notes, and recorded lectures (38%); timely and constant communication (45%); clear course structure including schedules, to-do lists, and detailed instructions (15%); flexibility (9%); and empathy (7%). See Table 3 for details. When considered in light of students' self-identified lack of discipline and focus, their answers provide us with a stronger understanding of how to guide university administrators and instructors. It was clear from the responses that students were aware of their weaknesses and the ways those weaknesses could impact their ability to achieve learning objectives. Below, we provide direction about how to address the challenges and meet the needs our study identified.

IMPLICATIONS

We recognize that the students and instructors in our study did not originally choose to teach and learn in fully online environments, and in that sense, they did not "sign up" for what they experienced. However, much can be learned from that unique situation. Below, we present the implications of this study.

Assessment and Development of Digital Literacy and Life Skills

One major implication we draw from this study is that students recognized they lacked specific skills that would benefit them in an online course. In particular, students could not capitalize on one of the major affordances of an online course format: flexibility. Students recognized they struggled with digital and life skills which made optimizing flexibility not only an issue, but an obstacle in an online course. It was clear students believed their self-management skills were underdeveloped or absent. Learning ways to hold themselves accountable, manage materials and responsibilities, develop a schedule, and identify what motivates them in online platforms were essential skills needed for success in online courses. In fact, vanLaar et al. (2017) identified information management and self-direction as key 21st century digital skills both within and beyond the classroom.

It is critical to understand the diverse skills, needs, and perspectives of students and to connect those needs with future teaching and course design suggestions leading to students' personal and academic growth. Providing time and attention on the front end of an online course to assess and develop those skills could help students resolve this challenge and take advantage of the benefit of flexibility. Development of digital skills such as online time management, information management, and self-motivation in online learning environments could bridge the gap that vanLaar et al. (2017) identified between higher education graduate skills and those needed in the 21st century workplace. Assessing students' digital literacy early in a course could help instructors shape the curriculum and materials to not only support students, but also provide opportunities for them to increase their digital literacy, a workplace skill that is deemed invaluable (Alexander et al., 2019; vanLaar et al., 2017).

As noted earlier in this article, universities are continuing to increase their online course offerings. University administrators should parallel that increase with an increased awareness of the skills students identified as necessary for successful accomplishment of learning objectives and work to ensure that instructors are prepared to meet the needs of students in their online classrooms.

REALISTIC UNDERSTANDING OF COURSE COMMITMENTS

Another important implication we draw from this study is the need for all stakeholders to understand the type and level of commitment needed to be successful in an online course. This understanding extends to university administrators, instructors and students alike. Students repeatedly stated that the amount of work and time required for one online course was often overwhelming. While some of this issue could potentially be resolved by instructors taking time early in a course to assess and develop skills that help alleviate learning curves for online skills, there are other ways to address this student concern. Instructors must have a strong and realistic understanding of the time they are requiring of students. Equally important is that instructors clearly convey the level and depth of that required commitment to their students. Students repeatedly stressed the need for clear and continual communication about course expectations. Providing greater details for all of the course components could help students understand the amount of time and commitment needed to achieve course objectives. In addition, instructors themselves must have a clear and honest view of their own digital literacy and potential issues of technology access their students may have. It behooves instructors not to conflate their own personal online experience and access with that of their students (Gonzales, 2016).

This understanding must extend to university administrators as well. Administrators should be attuned to the level of an instructors' experience with technology integration and prior experience in delivery of an online course. That level of experience can greatly impact an instructor's ability to accurately assess a student's digital literacy as well as the level of time commitment necessary for a student in the online course.

Professional Development Reflective of Students' Needs

Students in this study pointed to several instructor issues as challenges for their learning in online courses. Specifically, they noted that previous online instructors they encountered struggled with: adapting course materials, assignments, and teaching methods to an online format; providing clarity and details; and diversifying content mediums. In fact, in previous research, instructors expressed a lack of confidence in their technology integration skills which they believed impacted their effectiveness of using online platforms meaningfully (Crompton et al., 2016). In this study, students noticed that lack of confidence and stated their frustration with instructors who primarily focused on substitutionary and superficial technology integration.

Professional development for instructors who are teaching online courses is pivotal to the success of all stakeholders. Having a strong understanding of instructors' personal technology skills could point to professional development (PD) needs that are essential to ensure both the instructors' teaching and the students' learning objectives can be realistically met in an online course format. Initially, this includes assessing instructors' digital literacy which can point to areas where instructors fall short with technology integration skills. In turn, such assessment can guide university administrators about the kinds of PD needed (Aitchison et al., 2019). This is essential for successful instructional course design that focuses on engaging in meaningful and critical technology integration with students leading to the development of key 21st century workforce skills (Alexander et al., 2019; Olszewski & Crompton, 2020).

CONCLUSION

Throughout this study, we utilized a theoretical framework that was grounded on complexity and chaos theories. Complexity theory places great importance on understanding both the micro and macro elements of a system and how their interactions and relationships with each other impact the ability of the system to grow and adapt, thus becoming more complex in a way that benefits all entities within that system. In this study, we focused on the diverse perspectives of the students (micro) interacting within the university system (macro) as all adapted to the major shift in the course delivery system. In a complex system, a chaotic event can trigger adaptive responses needed for survival. In the case of our study, that chaotic event was the COVID 19 pandemic which triggered the move of LSU to fully online courses.

Out of that chaos, our study provided a means to identify ways that can and should shape how instructors and students approach online teaching and learning moving forward, and how university administrators can improve course outcomes for students, instructors, and the university as a whole. Using the findings from this study, universities can make internal adjustments, adaptations, and changes that "are just as powerful as the external environment" (Morrison, 2008, p.19; Waldrop, 1992). Specifically, incorporating assessment and development of students' digital literacy and life skills, developing a realistic understanding of course commitments, and provision of professional development for instructors that is reflective of students' needs can lead to more effective and impactful learning in online courses moving forward.

REFERENCES

Aitchison, C., Harper, R., Mirriahi, N., & Guerin, C. (2019). Tensions for educational developers in the digital university: Developing the person, developing the product. *Higher Education Research & Development, 39*(2), 171–184. doi:10.1080/07294360.2019.1663155

Alexander, B., Ashford-Rowe, K., Barajas-Murph, N., Dobbin, G., Knott, J., McCormack, M., Pomerantz, J., Seilhamer, R., & Weber, N. (2019). *Horizon report 2019 higher education edition.* EDUCAUSE. https://library.educause.edu/resources/2019/4/2019-horizon-report

Allen, I. E., & Seaman, J. (2016). *Online report card: Tracking online education in the United States.* Babson Survey Research Group. https://onlinelearningconsortium.org/read/online-report-card-tracking-online-education-united-states-2015/

Bhatt, I., & MacKenzie, A. (2019). Just google it! Digital literacy and the epistemology of ignorance. *Teaching in Higher Education, 24*(3), 302–317. doi:10.1080/13562517.2018.1547276

Buchholz, B. A., DeHart, J., & Moorman, G. (2020). Digital Citizenship During a Global Pandemic: Moving Beyond Digital Literacy. *Journal of adolescent & adult literacy: a journal from the International Reading Association, 64*(1), 11–17. doi:10.1002/jaal.1076

Buchholz, B. A., DeHart, J., & Moorman, G. (2020). Digital citizenship during a global pandemic: Moving beyond digital literacy. *Journal of Adolescent & Adult Literacy, 64*(1), 11–17. doi:10.1002/jaal.1076 PMID:32834710

Cabero-Almenara, J., Guillén-Gámez, F. D., Ruiz-Palmero, J., & Palacios-Rodríguez, A. (2021). Digital competence of higher education professors according to DigCompEdu. statistical research methods with ANOVA between fields of knowledge in different age ranges. *Education and Information Technologies, 26*(4), 4691–4708. doi:10.1007/s10639-021-10476-5 PMID:33758572

Chakraborty, M., & Nafukho, F. M. (2015). Strategies for virtual learning environments: Focusing on teaching presence and teaching immediacy. *Journal of Online Learning Research and Practice, 4*(1), 8–37.

Chen, P. S. D., Lambert, A. D., & Guidry, K. R. (2010). Engaging online learners: The impact of web-based learning technology on college student engagement. *Computers & Education, 54*(4), 1222–1232. https://www.learntechlib.org/p/67431/. doi:10.1016/j.compedu.2009.11.008

Cochran-Smith, M., Ell, F., Ludlow, L., Grudnoff, L., & Aitken, G. (2014). The challenge and promise of complexity theory for teacher education research. *Teachers College Record, 116*(4), 1–38. doi:10.1177/016146811411600407

Crompton, H., Olszewski, B., & Bielefeldt, T. (2016). The mobile learning training needs of educators in technology enabled environments. *Professional Development in Education, 42*(3), 482–501. doi:10.1080/19415257.2014.1001033

Dzikite, C., Nsubuga, Y., & Nkonki, V. (2017). Lecturers' competencies in information and communication technology (ICT) for effective implementation of ICT-integrated teaching and learning in textiles and clothing degree programmes. *International Journal of Educational Sciences, 17*(1-3), 61–68. doi:10.1080/09751122.2017.1305756

Erdmann, J., Rummel, N., Christmann, N., Elson, M., Hecking, T., Herrmann, T., Ulrich Hoppe, H., Krämer, N. C., Kyewski, E., & Wichmann, A. (2017). Challenges in implementing small group collaboration in large online courses. *International Society of the Learning Sciences,* 625-628. https://repository.isls.org/bitstream/1/301/1/94.pdf

Gonzales, A. L. (2016). The contemporary US digital divide: From initial access to technology maintenance. *Information Communication and Society, 19*(2), 234–248. doi:10.1080/1369118X.2015.1050438

Han, H., & Johnson, S. D. (2012). Relationship between students' emotional intelligence, social bond, and interactions in online learning. *Journal of Educational Technology & Society, 15*(1), 78–89.

Hosek, V. A. (2018). *Locating the critical component in technological, pedagogical, and content knowledge (TPACK): An examination of how graduate students recruit TPCAK and critical digital literacy into classroom practices* [Doctoral Dissertation, Illinois State University]. ProQuest.

Hosek, V.A. (n.d.). *It's personal: Examining and meeting the changing needs of students in online learning environments.* Pathways to Research in Education, EBSCO.

Hosek, V. A., & Percell, J. C. (2022). Reaching Critical Depths: Developing Teacher Candidates' Critical Pedagogy Online. In E. Mikulec & T. Ramalho (Eds.), *Best Practices in Teaching Critical Pedagogy Online*. DIO Press, Inc.

Houchin, K., & MacLean, D. (2005). Complexity theory and strategic change: An empirically informed critique. *British Journal of Management, 16*(2), 149–166. doi:10.1111/j.1467-8551.2005.00427.x

Hunter, W. J., & Benson, G. D. (1997). Arrows in time: The misapplication of chaos theory to education. *Journal of Curriculum Studies, 29*(1), 87–100. doi:10.1080/002202797184215

Ilgaz, H., & Gülbahar, Y. (2015). A snapshot of online learners: e-Readiness, e-Satisfaction and expectations. *International Review of Research in Open and Distance Learning, 16*(2), 171–187. doi:10.19173/irrodl.v16i2.2117

Joksimović, S., Gašević, D., Kovanović, V., Riecke, B. E., & Hatala, M. (2015). Social presence in online discussions as a process predictor of academic performance. *Journal of Computer Assisted Learning, 31*(6), 638–654. doi:10.1111/jcal.12107

Jung, I., & Lee, J. (2020). A cross-cultural approach to the adoption of open educational resources in higher education. *British Journal of Educational Technology, 51*(1), 263–280. doi:10.1111/bjet.12820

Kemp, N., & Grieve, R. (2014). Face-to-face or face-to-screen? Undergraduates' opinions and test performance in classroom vs. online learning. *Frontiers in Psychology, 5*. Advance online publication. doi:10.3389/fpsyg.2014.01278 PMID:25429276

Kinlaw, C. R., Dunlap, L. L., & D'Angelo, J. A. (2012). Relations between faculty use of online academic resources and student class attendance. *Computers & Education, 59*(2), 167–172. https://www.learntechlib.org/p/66746/. doi:10.1016/j.compedu.2011.12.028

Kupczynski, L., Ice, P., Wiesenmayer, R., & McCluskey, F. (2010). Student perceptions of the relationship between indicators of teaching presence and success in online courses. *Journal of Interactive Online Learning, 9*(1), 23–43. https://www.learntechlib.org/p/109406/

Ladyshewsky, R. (2013). Instructor presence in online courses and student satisfaction. *International Journal for the Scholarship of Teaching and Learning*, *7*(1), 1–23. doi:10.20429/ijsotl.2013.070113

Lammers, W. J., & Gillaspy, J. A. (2013). Brief Measure of Student-Instructor Rapport Predicts Student Success in Online Courses. *International Journal for the Scholarship of Teaching and Learning*, *7*(2), 1–13. doi:10.20429/ijsotl.2013.070216

List, A. (2019). Defining digital literacy development: An examination of pre-service teachers' beliefs. *Computers & Education*, *138*, 146–158. doi:10.1016/j.compedu.2019.03.009

Lowenthal, P., & Dunlap, J. (2018). Investigating students' perceptions of instructional strategies to establish social presence. *Distance Education*, *39*(3), 281–298. doi:10.1080/01587919.2018.1476844

Luttrell, W. (Ed.). (2010). *Qualitative educational research: Readings in reflexive methodology and transformative practice*. Routledge.

Martin, F., & Bolliger, D. U. (2018). Engagement matters: Student perceptions on the importance of engagement strategies in the online learning environment. *Online Learning : the Official Journal of the Online Learning Consortium*, *22*(1), 205–222. doi:10.24059/olj.v22i1.1092

Martin, F., Ritshaupt, A., Kumar, S., & Budhrani, K. (2019). Award-winning faculty online practices: Course design, assessment and evaluation, and facilitation. *Higher Education*, *42*, 34–43. doi:10.24059/olj.v23i1.1329

Mason, M. (2008). Complexity theory and the philosophy of education. *Educational Philosophy and Theory*, *40*(1), 4–18. doi:10.1111/j.1469-5812.2007.00412.x

McLaren, P. (2017). Critical pedagogy: A look at the major concepts. In A. Darder, R. D. Torres, & M. P. Baltodano (Eds.), *The critical pedagogy reader* (3rd ed., pp. 56–78). Routledge.

Moallem, M. (2015). The impact of synchronous and asynchronous communication tools on learner self-regulation, social presence, immediacy, intimacy and satisfaction in collaborative online learning. *The Online Journal of Distance Education and e-Learning : TOJDEL*, *3*(3), 55–77. https://www.tojdel.net/journals/tojdel/articles/v03i03/v03i03-08.pdf

Morrison, K. (2008). Educational philosophy and the challenge of complexity theory. *Educational Philosophy and Theory*, *40*(1), 19–34. doi:10.1111/j.1469-5812.2007.00394.x

National Center for Education Statistics. (2022). Undergraduate Enrollment. In *Condition of Education*. U.S. Department of Education, Institute of Education Sciences. Retrieved May 31, 2022, from https://nces.ed.gov/programs/coe/indicator/cha

Ng, W. (2012). Can we teach digital natives digital literacy? *Computers & Education*, *59*(3), 1065–1078. doi:10.1016/j.compedu.2012.04.016

Olszewski, B., & Crompton, H. (2020, June). Educational technology conditions to support the development of digital age skills. *Computers & Education*, *150*(C), 103849. Advance online publication. doi:10.1016/j.compedu.2020.103849

Öncül, G. (2021). Defining the need: Digital literacy skills for first-year university students. *Journal of Applied Research in Higher Education, 13*(4), 925–943. doi:10.1108/JARHE-06-2020-0179

Ortagus, J. C. (2017). From the periphery to prominence: An examination of the changing profile of online students in American higher education. *The Internet and Higher Education, 32*, 47–57. doi:10.1016/j.iheduc.2016.09.002

Radford, M. (2008). Complexity and truth in educational research. *Educational Philosophy and Theory, 40*(1), 144–157. doi:10.1111/j.1469-5812.2007.00396.x

Reed, M., & Harvey, D. L. (1992). The new science and the old: Complexity and realism in the social sciences. *Journal for the Theory of Social Behaviour, 22*(4), 353–380. doi:10.1111/j.1468-5914.1992.tb00224.x

Rowsell, J., Morrell, E., & Alvermann, D. E. (2017). Confronting the digital divide: Debunking brave new world discourses. *The Reading Teacher, 71*(2), 157–165. doi:10.1002/trtr.1603

Ruffalo Noel Levitz. (2017). *National student satisfaction and priorities report*. Ruffalo Noel Levitz. Retrieved May 31, 2022, https://learn.ruffalonl.com/rs/395-EOG-977/images/2017_National_Student_Satisfaction_Report_1.0

Ruggiero, D., & Mong, C. J. (2015). The teacher technology integration experience: Practice and reflection in the classroom. *Journal of Information Technology Education, 14*, 162–178. doi:10.28945/2227

Saldaña, J. (2016). The coding manual for qualitative researchers. *Sage (Atlanta, Ga.).*

Savin-Baden, M., & Howell-Major, C. (2013). *Qualitative research: The essential guide to theory and practice. In Qualitative research: The essential guide to theory and practice*. Routledge.

Van Dijk, J. A. (2006). Digital divide research, achievements and shortcomings. *Poetics, 34*(4-5), 221–235. doi:10.1016/j.poetic.2006.05.004

Van Laar, E., Van Deursen, A. J., Van Dijk, J. A., & De Haan, J. (2017). The relation between 21st-century skills and digital skills: A systematic literature review. *Computers in Human Behavior, 72*, 577–588. doi:10.1016/j.chb.2017.03.010

Waks, L. J. (2011). Teacher education programs as complex organizations. *Complicity: An International Journal of Complexity and Education, 8*(1), 65-69. https://journals.library.ualberta.ca/complicity/index.php/complicity/article/view/10026

Waldrop, M. (1992). *Complexity: The Emerging Science at the Edge of Order and Chaos*. Simon & Schuster Editions.

Yang, Y. F., Yeh, H. C., & Wong, W. K. (2010). The influence of social interaction on meaning construction in a virtual community. *British Journal of Educational Technology, 41*(2), 287–306. doi:10.1111/j.1467-8535.2009.00934.x

APPENDIX A

Student Survey Questions

1. Prior to going entirely online, how did you use technology in your courses during the current semester? (Please include examples...smart board, online discussion, etc)
2. Do you feel you/your professor possess the skills necessary to effectively teach online? If so, why? If not, why not?
3. What do you think are the benefits of having online courses?
4. What do you think are the challenges of online courses?
5. What skills do you believe are needed to be successful as an online student?
6. What kind of support do you believe you need from your instructors/professors to be successful when taking an online course?
7. Do you think your learning objectives (goals) for your courses can be met in an online course format? If so, why? If not, why not?

Chapter 10
Supporting Academics to "Revive, Thrive, and Drive" During Times of Challenge and Uncertainty

Wendy M. Goff
https://orcid.org/0000-0002-0665-2793
Swinburne University of Technology, Australia

Robert New
https://orcid.org/0009-0000-4537-0265
Swinburne University of Technology, Australia

ABSTRACT

The COVID-19 pandemic has resulted in change across the world. The higher education academic workforce was largely ignored during the COVID-19 pandemic, and there has been little to no acknowledgement of the increased workload, extensive lockdowns, pivots to online learning, and the 'business as usual' expectations that were thrust upon academics during this time. In this chapter, the authors use Aristotelian ethics, the concept of human flourishing, and the United Nations Sustainable Development Goals to provide insight into how human resource in the higher education sector can be supported to recover and thrive amidst ongoing unprecedented global challenges and uncertainties. Drawing on VanderWeele's domains and pathways for human flourishing, the authors demonstrate how universities and colleges around the world can support their academic workforce to revive, thrive, and drive sustainability, productivity, economic growth, and recovery during times of challenge and uncertainty. The chapter concludes with recommendations for future directions and research.

INTRODUCTION

The impact of the COVID-19 pandemic is still being experienced throughout the world. At the time of writing this chapter there have been 772, 386, 069 confirmed COVID-19 cases and a total of 6, 987, 222 deaths despite 13, 595, 721, 080 vaccine doses being administered worldwide (World Health Organisa-

DOI: 10.4018/979-8-3693-1926-0.ch010

tion, 2020). Alongside the substantial health challenges, the COVID-19 pandemic has also resulted in some significant and long-lasting economic and social impacts that are yet to be truly understood or realised. In fact, according to the International Bank for Reconstruction and Development/The World Bank "many of the consequences of the pandemic may not fully emerge for years (or even decades)" (Schady et al., 2023, p.1).

Human capital, or "the wealth embodied in people's health, knowledge, skills, and emotional and mental well-being." (Schady et al., 2023 p.16) has already suffered an enormous impact because of the COVID-19 pandemic, with sharp decreases in human capital at critical stages of lifespan experienced worldwide. This is concerning for countries across the globe as human capital is associated with both economic and non-economic wealth and prosperity. Schady and colleagues suggest that the erosion of human capital due to the pandemic occurred at three critical stages of the life cycle – *early childhood (0-5 years); school-aged children (6-14 years);* and *early adulthood (15-24 years)* (Schady et al., 2023). They further suggest that these are the three key groups that need to be aggressively targeted by government policy worldwide to support and aid global recovery.

Whilst aggressively targeting identified impacted groups through directing government policy is an important notion, this becomes extremely complex when the development of human capital is dependent on the human capital of others. All three critical stages identified by Schady et al. (2013) as being impacted by the pandemic fall into this category. Therefore, it is important that any policy targeting these groups also consider this dependence and the source of human capital. A good example of such dependence is in the Higher Education (HE) context where the identified group of young adults are not only dependent on the human capital of academics (*the source*) to help them to accumulate their own human capital, but they are also dependent on academics (*the source*) to support them to make the transition from *accumulating* to *using* human capital (Schady et al., 2023).

In this chapter we focus on recovery for the early adult critical stage of life development. In doing so, we draw on Aristotelian ethics, the concept of human flourishing, and the United Nations Sustainable Development Goals to provide insight into how human resources in the higher education sector can be supported to recover and thrive amidst ongoing unprecedented global challenges and uncertainties. Drawing on VanderWeele's (2017) domains and pathways for human flourishing we demonstrate how universities and colleges around the world can support their academic workforce to revive, thrive and drive sustainability, productivity, economic growth, and recovery and in turn support, shape and nurture the human capital of early adults. We conclude the chapter with some recommendations for future research directions.

ARISTOTELIAN ETHICS AND HUMAN FLOURISHING

Aristotelian Virtue Ethics is agent-focused and forefronts person (*how should I be?*) rather than action (*what should I do?*) (Fisher & Dimmock, 2017). According to Aristotle, humans are natural agentic beings who live through and engage in purposive and meaningful action. Purposeful and meaningful action as defined by Aristotelian theory is action that is a rational goal orientated pursuit of human excellence or the human good (Fisher & Dimmock, 2017). In other words, "the good is not simply a matter of psychological experience but is located in our actions in the world and particularly in the quality of those actions" (Flowers, 2012, p.11).

For Aristotle, *eudaimonia* is the highest good that people strive towards in their journey of becoming. *Eudaimonia* is the 'ultimate result' in a "pursuit of virtue, excellence, and the best within us" (Huta & Waterman, 2014; pp. 1426). It is not merely a path to a goal; it is the goal itself. According to Aristotle, *eudaimonia* is both the realisation and the actualisation of human happiness, fulfilment, and potential, and can be described as living a good and virtuous life (Aristotle, ca. 350 B.C.E./1925). In Aristotelian Virtue Ethics both moral and intellectual virtues underpin *eudaimonia* which can also be described as human flourishing.

Moral virtues, according to Aristotle, are shaped through experience and developed, refined, and changed according to time and context (Aristotle, ca. 350 B.C.E./1925). For Aristotle, moral virtue evolves through repeated experiences and habitual behaviour that condition character and build personal tendencies and dispositions to act, react and respond to both context and phenomenon in specific ways (Aristotle, ca. 350 B.C.E./1925). Intellectual virtue is closely related to moral virtue but is underpinned by the notion of 'an informed deliberateness' or what Aristotle explains as a 'deliberate excellence' (Solopova, 2016). In other words, it is underpinned by knowledge and truth; the "purpose of the moral act is to achieve the good, while the purpose of thought [the intellectual] is to achieve the true" (Solopova, 2016, p. 523). Through this mixture of reasoning and action, fluency in the practice of virtue is achieved (Sprod, 2001). It is this which leads to flourishing.

For Aristotle, human flourishing is enacted through a human's virtuous action; their pursuit of the good and the truth (Buckingham, 2021). Flourishing requires both internal and external conditions including social interaction within community. According to Aristotle, flourishing is the highest good of human undertaking and the goal or purpose of all human action and activity (Aristotle, ca. 350 B.C.E./1925).

In the following section, the United Nations Sustainable Development Goals (SDG's) are introduced. The SDG's are aligned with the Aristotelian concept of human flourishing in two ways; the focus of the goals to build the external conditions for flourishing and the combined human efforts of establishing and working towards achieving the goals by the global community. The SDG's are initially explained and progress towards achieving the goals is then outlined.

THE UNITED NATIONS SUSTAINABLE DEVELOPMENT GOALS

The United Nations Sustainable Development Goals were borne from a united global agenda and commitment to end poverty, promote peace, preserve the rights and dignity of all human beings, and protect the planet by the year 2030 (United Nations General Assembly (UN), 2015). Underpinning the '2030 Agenda' are a set of 17 sustainable development goals, housed in targets, and indicators that represent the global call to action and provide a blueprint for countries to follow (UN DESA, 2023). The SDG's, together with the accompanying targets and indicators, were developed through an extensive consultation with global civil society and various stakeholders worldwide, with a specific emphasis on prioritising the voices and insights of the most disadvantaged and marginalised individuals from around the globe (UN, 2015). According to the United Nations (2023), the "defining principle of the 2030 Agenda for Sustainable Development is a shared promise by every country to work together to secure the rights and well-being of everyone on a healthy, thriving planet" (UN, 2023, p.4).

Each year since the conception of the SDG's the UN prepares a report outlining progress and highlighting areas that require attention to achieve the goals. Unfortunately, the 2023 Sustainable Development Goals Report paints a bleak picture. The lasting impacts of the COVID-19 pandemic, along with ongoing war and conflict, and an intensifying climate crisis, have had significant impact across the globe (UN, 2023). This impact has, in some ways, shifted focus away from the combined human efforts of the global community, and as a result, building the external conditions for human flourishing globally has been significantly disturbed. According to the UN (2023), "an unprecedented effort" along with "a renewed sense of common purpose across the international community" is now necessary to address the progress lost towards achieving the goals by 2030 (UN, 2023, p. 2).

THE IMPORTANCE OF HIGHER EDUCATION

Although 'ensuring inclusive and equitable quality education and promoting lifelong learning opportunities for all' is an explicit goal (goal 4) of the SDGs (UN, 2015), education is an important feature across the SDGs and underpins many of the 17 goals. Higher Education in particular, has an important role to play in supporting the achievement of all SDG's particularly around the notion of preparing and provisioning for enactment. It also has a significant role to play in relation to innovation and human preparedness, particularly relative to working towards and achieving the SDGs that are focused on the planet.

Global post-pandemic recovery will also be heavily reliant on Higher Education, particularly in relation to future global prosperity, economic growth, and development, as well as in fostering and building human capital in a changed global world (Jensen-Pennington, 2021). However, during the COVID-19 pandemic Higher Education Institutions (HEI's) were not immune to repercussions. Like other parts of society, the impact on HEI's was abrupt, and was received without contingency or time for preparation. In many countries it also resulted in a loss of public financing and resulted in forced restructure and job loss across the sector (United Nations Educational, Scientific and Cultural Organization (UNESCO), 2020).

During the pandemic, for many HEIs there was also an unforeseen shift in the mode of teaching with a rapid move from classroom-based pedagogies to online-remote learning. Many HEI's across the globe adopted a 'business as usual' mantra and were reliant on the academic workforce to make these changes and continue to deliver coursework for students with minimal disruption. There were also various challenges for academic research that emerged during this time. These challenges included limitations to fieldwork, restrictions on face-to-face contact, restrictions to specific contexts, the closure of laboratories and in some instances, the closure of universities or entire faculties or schools (Mukhopadhyay, 2023). As such, the disturbance caused by covid disrupted the practice of virtues through introducing new situations for which a virtue needed to be established, and a reconsideration of the prior status quo. In short, the practices which led to flourishing in humans and HEIs were no longer as functional as they were previously. This meant adaptation had to occur, and HEIs needed to be at the forefront of such a process, given their status and role in society.

Historically, HEIs have been viewed as instruments for social change and the academic workforce as the expert drivers of such change, particularly in the aftermath of major social disruptions (Schrum, 2019). From an Aristotelian perspective this makes sense, particularly in relation to the rebuilding of human capital and supporting human flourishing. Without significant human understanding and contribution to future global prosperity, economic growth, innovation, and development, recovery from social disruption is not only challenging but in reality it may be impossible. HEIs and the academic workforce have a crucial role to play in this human preparation of understanding and contribution, however, this may pose difficult given pre-pandemic developments in HEIs across the globe.

Over the past decade a major development in the HE landscape has been the infiltration of the sector by industry and the private sector workforce. This infiltration has occurred through monetary incentive and funding and has, in most instances, been driven by a lack of Government investment into the sector. The result of these developments has been the morphing of some HEIs from places of education, personal growth, and learning, to extended corporate "training grounds for the private workforce" (Krishnamoorthy & Kearing, 2021, p. 262). Post-pandemic recovery offers a critical opportunity to re-evaluate these developments and to both recognise and operationalise HEIs as tools for nurturing human flourishing so that the next generation are prepared to understand and solve global challenge.

A significant component of recognising and operationalising HEIs as drivers of social change is the academic workforce. The COVID-19 pandemic and its ensuing aftermath has posed some significant challenges for HEIs around the world, particularly in relation to nurturing the flourishing of the academic workforce. The 'business as usual' mantra coupled with the lack of support and acknowledgement of academic effort throughout the pandemic has resulted in a tired and in some instances, burnt out academic workforce that are quietly rejecting the 'hustle culture' of the academy and have turned their efforts towards self-preservation and survival (Forrester, 2023). If HEIs do not act to rectify this damage and support academics to flourish, there may be damaging and irreversible consequences for future global prosperity, economic growth, and development, particularly in relation to supporting young people to make the transition from *accumulating* to *using* human capital.

DOMAINS AND PATHWAYS FOR HUMAN FLOURISHING

Human flourishing is embedded in human action but can be viewed through a temporal lens (lifespan), or as an action (to flourish) (Wolbert et al., 2015). According to Duraiappah and colleagues:

Human flourishing is both the optimal continuing development of human beings' potentials and living well as a human being. It means being engaged in relationships and activities that are meaningful, that is, aligned with both an individual's own values and humanistic values, in a way that is satisfying to them. Flourishing is conditional on the contribution of individuals and requires an enabling environment. (de Ruyter et al., 2022, p. 74)

In relation to supporting the human flourishing of the academic workforce, the focus is on the operationalise of action. It involves creating a secure base and an enabling environment that provisions for activity that is both subjectively and objectively meaningful, is valuable to the individual and to others, and facilitates and nurtures successful academic engagement through trust and autonomy. According to White this can also be described as "success in worthwhile activities and relationships which [academics] have freely engaged in and which they pursue wholeheartedly" (2007, p. 25). For the academic workforce to flourish HEIs need to strategically repair the damage caused during the COVID-19 pandemic and invest in its human capital. VanderWeele's (2017) pathways and domains for supporting human flourishing provide guidance on how this might be achieved in practice.

In relation to human flourishing VanderWeele identifies five broad domains of human life that an individual would be doing or being well in: "(*i*) happiness and life satisfaction; (*ii*) health, both mental and physical; (*iii*) meaning and purpose; (*iv*) character and virtue; and (*v*) close social relationships"

(VanderWeele, 2017, p. 8149). He suggests that a sixth domain, *financial and material stability,* underpins the other five domains and represents flourishing over life course.

Alongside the six domains, VanderWeele suggests there are four prominent pathways to human flourishing. These four pathways are identified as family, work, education, and religious community and are key determinants of the six domains of flourishing (VanderWeele, 2017). According to VanderWeele, the pathways are not intended as an all-inclusive list with respect to the key determinants of flourishing but are four common pathways and "if efforts were made to support, improve, and promote participation in these pathways, the consequences for human flourishing would be substantial" (VanderWeele, 2017, p. 8151). In the remainder of this chapter, we draw on VanderWeele's six domains and 'work' pathway to present a model of support for academic flourishing.

A MODEL OF SUPPORT FOR ACADEMIC FLOURISHING

As previously discussed in this chapter, the academic workforce has a crucial role to play in post-pandemic recovery efforts globally and HEIs have a vital role to play in developing human capital and progress globally towards achieving the UNs SDGs. However, to do this effectively HEIs need to move away from the model of corporate training and back towards recognising the academy as a driver of recovery and an instrument for social change. A key component of this recognition is acknowledging the impact that the COVID-19 pandemic has had on the academic workforce and formulating a way to support academics to flourish. The Model of Support for Academic Flourishing is offered as a framework from which HEIs can commence this process.

The proposed Model of Support for Academic Flourishing positions academics as the centre of support. This is an important component of the model that has been designed to ensure that focus is centred primarily on the academic workforce. The work pathway as identified by VanderWeele (2017) envelopes the model to further support this explicit focus on academics and the academic workforce.

VanderWeele's six broad domains of human life represent points of support. They provide the context to prioritise the lives of the academic workforce in their entirety (Aristotle, 2009). They also act as a point of reference for HEIs to ensure that adequate support that is focused on human flourishing is in place.

Three specific contexts are highlighted as the main settings for resource and support delivery: *services that are embedded in the university; services that are provided by partners but are offered at the university;* and *off-site partners who are referred to by the university.* These three contexts are again explicitly aligned to the six broad domains of life and provide HEIs with a way to evaluate whether the HEI is adequately supporting academic flourishing.

Alongside the three specific contexts there are four key cultural drivers embedded in the model: *an enabling environment, autonomy, secure-base leadership,* and *trust.* These cultural drivers are explicitly designed to provide academics with the necessary environment, time, freedom, and support required to develop and pursue their academic endeavours. The four key cultural drivers are also vehicles in which HEIs can target internal funding and research support initiatives. In the following section we further elaborate on these four key components of the model.

An Enabling Environment

In outlining SDG 4 *Ensure inclusive and equitable quality education and promote lifelong learning opportunities for all*, UNESCO provides a useful definition of an enabling environment. According to UNESCO, in an educational context an enabling environment is met if four of the following criteria is met: "(a) has institutional arrangements supporting breadth of skills, (b) has budget supporting breadth of skills, (c) has policies promoting positive school climate, and (d) has policies for schools to engage with communities." (UNESCO, 2015). Whilst this definition is explicitly aligned with schooling, in the Model of Support for Academic Flourishing it has been adapted for the HE context. This adaptation is outlined as follows: an enabling environment is met if four of the following criteria is met: The HEI (a) has institutional arrangements supporting the breadth of skills of academic, (b) has budget supporting breadth of skills, (c) has policies promoting positive university climate, and (d) has policies for the university to engage with relevant research, teaching and industry communities (Adapted from UNESCO, 2015).

Autonomy

In relation to the notion of autonomy in the academic context Noorda suggests that:

Academic autonomy is more than freedom from outside interference, and more than the freedom to do things yourselves. It has and should have serious implications for university structures and cultures, strategic plans, and operational practices. (Noorda, 2013, p. 9)

In relation to the Model of Support for Academic Flourishing autonomy is rooted in the social good. It is embedded in the wider purpose of HEIs and aligned with Schrum's notion that HEIs are instruments for social change and that the academic workforce are the expert drivers of such change (Schrum, 2019). If a HEI has aligned its sole purpose with training for the private workforce, within the Model of Support for Academic Flourishing it cannot be viewed as supporting the autonomy of academics. This is not to say that a HEI cannot support workforce preparation, but it cannot be the primary focus of the HEI. To support autonomy within the Model of Support for Academic Flourishing, HEIs must have a primary focus on the social good.

Secure Base Leadership

Another key cultural driver of the Model of Support for Academic Flourishing is a Secure Base Leadership. This type of leadership is directly aligned with Bowlby's (1988) conception of the human need for a Secure Base. Bowlby emphasised that humans need to have confidence that their needs will be responded to when required and that this responsiveness is both consistent and secure (Bowlby, 1988). Although Bowlby's conception referred explicitly to the attachment between infant and parent, Bowlby acknowledged that for humans to flourish and reach their potential, in each pathway of their life they must have a Secure Base in which to leave and return. In the HEI this Secure Base includes the wider HEI as well as the 'relational other' that supports academics in their day-to-day work. Kohlrieser and colleagues (2012) describe Secure Base Leadership as:

The way a leader builds trust and influences others by providing a sense of protection, safety and caring and by providing a source of inspiration that together produce energy for exploration, risk taking and seeking challenge. (Kohlrieser et al., 2012)

This secure-base from the leader facilitates the development of a mental framework which leads to a 'broaden-and-build' cycle through the attachment security (Davidovitz et al., 2007 p.633). Such a situation is a key component for the successful reviving, and future thriving, of academics in HEI.

Trust

Although the notion of trust might be considered as central components of the cultural drivers previously outlined, the Model of Support for Academic Flourishing positions trust as an independent driver that requires an explicit focus. According to Aristotelean theory trust is a virtue that supports humans to flourish or to achieve *eudaimonia* (D'Olimpio, 2016). Trust, when understood as a virtue is practiced via rational, responsive, and compassionate habitual action that is embedded in good will.

The virtuous moral agent is not only trustworthy but is motivated by goodwill to 'trust'. In other words, trust "is a relational attitude, and its manifestation is contextual" (D'Olimpio, 2016 p.195). In the HEI, building such relational attitude amongst and with the academic workforce requires an acknowledgement of past wrongs and treatments, as well as an explicit focus on healing. The 'business as usual mantra' will not suffice if the academic workforce are to lead recovery and social change.

Enclosing the Model of Support for Academic Flourishing is a recognition of context. Drawing on VanderWeele's (2017) work pathway to human flourishing the HEI is acknowledged as the pathway of action, that is, all activity targeting support towards and for academic flourishing, is the responsibility of the wider institution. Figure 1 provides a visual representation of the Model of Support for Academic Flourishing.

APPLICATION OF THE MODEL OF SUPPORT FOR ACADEMIC FLOURISHING

A starting point for HEIs wanting to employ the Model of Support for Academic Flourishing is the work pathway. Both establishing and ensuring that the purpose of the university is well communicated and understood allows people to make informed decisions in relation to work pathway 'fit'. It also articulates the direction of the HEI and positions and re-positions the workforce in relation to 'choice of focus'. If strategically mapped out alongside government commitment to the SDG's it might also highlight different avenues for funding and support HEIs to take on a more rigorous advocacy in this space.

A key component of the work pathway is also articulating the vision for the academic workforce as well as for students and other workforce roles within the HEI context. Again, this might be embedded in the universities strategic planning and would be forward thinking in relation to operationalising the vision. This would be shared within the immediate and wider community and might also be open to community consultation prior to adoption.

The four cultural drivers of the model should align with and drive the identified and communicated purpose. They provide a structure to evaluate the existing cultural drivers, as well as a way for the HEI to target resource to support change. This evaluation could be achieved through an internal or external audit that provides opportunity for academic input and voice. This could also be done in conjunction or

Figure 1. The model of support for academic flourishing

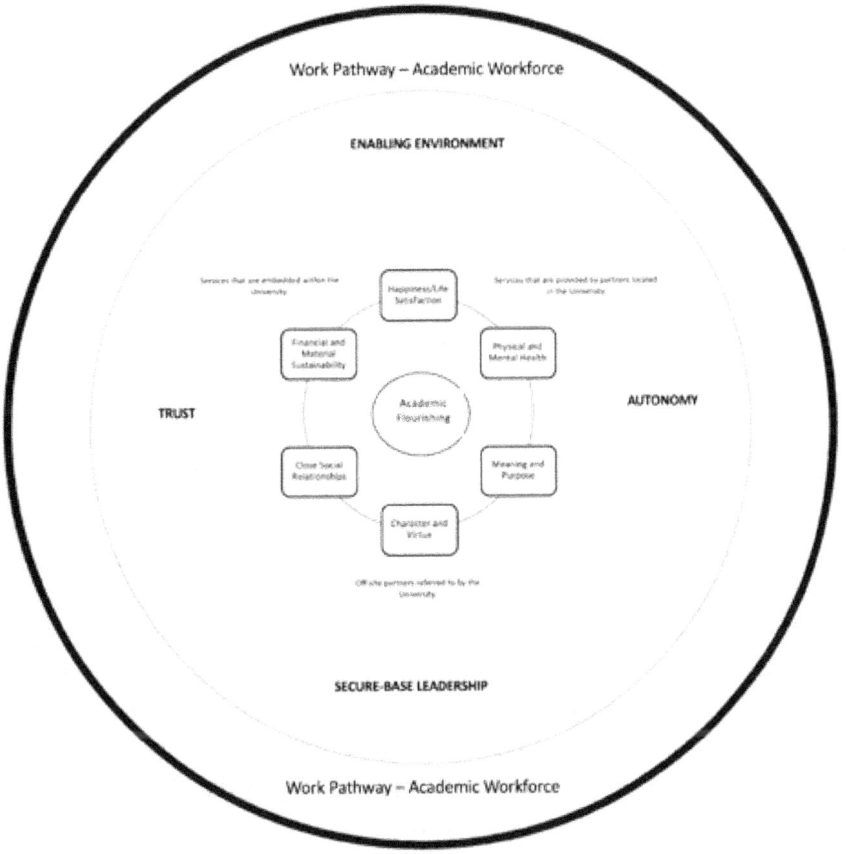

in consultation with the academic workforce and draw on a process of re or co-designing each of the four cultural drivers. This could be a particularly powerful process in establishing a Secure Base Leadership across the HEI if leaders across the university were tasked with this role.

The three targeted areas provide HEIs with a contextual understanding of how and where to provision support and resource. Once established these support structures can be built on to provide a wraparound support system that changes and adapts to needs over time. They can also be enhanced through the adoption of different partners and services and extended to nurture the human flourishing of students or the wider community. These targeted areas are embedded in the cultural drivers and should be closely aligned with supporting the six domains of human flourishing. The three targeted areas provide HEIs with a roadmap for starting this support, but this roadmap should be a living map that adapts and changes to meet changing needs over time.

The six domains of human flourishing provide the framework for leadership to work with individual and teams of academics. The framework is aligned with the contextual resources and three targeted areas, but it also provides a way for leadership to pinpoint where support to flourish might be necessary. This framework could form the structure of workplace discussion, performance review process, or be used as guide from which to structure one-to-one consultation. From a Secure Base Leadership perspective, it could also assist in structuring and growing the relationship between academic and HEI leadership and provide a basis for building trust.

The six domains of human flourishing might also be drawn on by HEIs to establish where, when and how to provide support to academics during times of challenge or heightened need. It provides a structure to assist leadership to identify the domains of academic life that might need some additional support and positions them well to provide such support.

CONCLUSION

In this chapter we have advocated that the academic workforce are critical components of post-pandemic global recovery but have also acknowledged that to-date the needs of the academic workforce have been largely ignored. We have also recognised that to support young adults (students) and societies to change, adapt and flourish during times of uncertainty and challenge that academics too must flourish. We have suggested that HEIs have a responsibility to their academic workforce to know their purpose and to also acknowledge the unsustainable expectations that were placed on the academic during the pandemic.

Drawing on Aristotelian ethics, the concept of human flourishing, the United Nations Sustainable Development Goals and VanderWeele's (2017) domains and pathways for human flourishing we have presented model that can be drawn on by HEIs to support their academic workforce to revive, thrive and drive sustainability, productivity, economic growth, and recovery during times of challenge and uncertainty. To guide and assist HEIs to better understand how to support the academic workforce in this changed and damaged landscape we have presented The Model of Support for Academic Flourishing. The Model provides a structure and several embedded frameworks that can been drawn on by HEIs to nurture the flourishing of their academic workforce.

The Model of Support for Academic Flourishing is extensive and whilst we contend that this may be a limitation for the implementation of the model in its entirety, HEI's can isolate and draw on different components of the model in different ways. For example, academic flourishing at the grassroots level might be targeted first. This would involve leadership within the university working closely with individual academics and teams to ensure that the six domains of human flourishing were addressed and well supported.

Another limitation may lie in the purpose of the HEI. The Model of Support for Academic Flourishing might not align with the purpose of the HEI, particularly if that purpose is to train people to join the corporate workforce. The Model of Support for Academic Flourishing is a model for social change and recovery and recognises the importance of academics as the drivers of this process. More specifically it recognises the role of the academic workforce in building the social capital and capacity of young people and supporting their transition from accumulating to using their capital. It might be difficult for HEIs to draw on the model effectively if leading social change and recovery are not the HEIs key purpose.

HEIs have an important role to play in supporting government to meet their commitments to the SDGs, however, they are not positioned well to play this role if the academic workforce is not flourishing. Recognising the urgency and criticality of nurturing academic flourishing not only supports government to get back on track with their commitments to the SDGs but it also highlights how and where government might invest in the HEIs context for maximum impact.

Academics need to be supported to revive after the past two years of expectations and 'business as usual' mantra, but they also need to be supported to thrive so they can drive recovery and the social change necessary for a strong and flourishing society. The Model of Support for Academic Flourishing that has been outlined in this chapter is a tool to support this flourishing.

REFERENCES

Aristotle. (1925). *Nichomachean ethics: Book II* (W.D. Ross, Trans.). The Internet Classics Archive. http://classics.mit.edu/Aristotle/nicomachaen.2.ii.html (Original work published 350 B.C.E.)

Aristotle. (2009). *The Nicomachean Ethics (D. Ross, Trans., L. Brown, Ed.)*. Oxford University Press.

Bowlby, J. (1988). *A secure base: clinical applications of attachment theory*. Routledge.

D'Olimpio, L. (2018). Trust as a virtue in education. *Educational Philosophy and Theory*, *2*(50), 193–202. doi:10.1080/00131857.2016.1194737

Davidovitz, R., Mikulincer, M., Shaver, P. R., Izsak, R., & Popper, M. (2007). Leaders as attachment figures: Leaders' attachment orientations predict leadership-related mental representations and followers' performance and mental health. *Journal of Personality and Social Psychology*, *93*(4), 632–650. doi:10.1037/0022-3514.93.4.632 PMID:17892336

De Ruyter, D., Oades, L. G., Waghid, Y., Ehrenfeld, J., Gilead, T., & Singh, N. C. (2022). Reimagining Education: The International Science and Evidence based Assessment. New Delhi: UNESCO MGIEP. doi:10.56383/XBXZ7711

Fisher, A., & Dimmock, M. (2017). *Ethics for A-Level* (1st ed.). Open Book Publishers. https://www.jstor.org/stable/j.ctt1wc7r6j

Forrester, N. (2023). Fed up and burnt out: 'quiet quitting' hits academia *Nature*, *615*, 751-753. doi:10.1038/d41586-023-00633-w

Jensen-Pennington, H. (2021). Envisioning Higher Education for the Future. In H. van't Land, A. Corcoran, & D. C. Iancu (Eds.), *The Promise of Higher Education*. Springer. doi:10.1007/978-3-030-67245-4_58

Kohlrieser, G., Coombe, D., & Goldsworthy, S. (2012). *Care to Dare: Unleashing Astonishing Potential Through Secure Base Leadership*. Jossey-Bass.

Krishnamoorthy, R., & Keating, K. (2021). Education Crisis, Workforce Preparedness, and COVID-19: Reflections and Recommendations. *American Journal of Economics and Sociology*, *80*(1), 253–274. doi:10.1111/ajes.12376 PMID:34230670

Lee, M. T. (2019). Promoting Human Flourishing Beyond Foundational Concerns. *Humanistic Management Journal*, *4*(2), 235–237. doi:10.1007/s41463-019-00065-7

Mukhopadhyay, U. (2023). Impact of COVID-19 pandemic on academic performance and work–life balance of women academicians. *Asian Journal of Social Science*, *51*(1), 62–70. doi:10.1016/j.ajss.2022.07.003 PMID:35937579

Noorda, S. (2013). Academic Autonomy as a Lifelong Learning Process for Universities. *Leadership and Governance in Higher Education,* (4), 2013. https://lead-project.org/sites/default/files/2017-04/Academic%20Autonomy%20for%20Universities.pdf

Schrum, E. (2019). *The Instrumental University*. Cornell University Press.

Solopova, M. (2016). Aristotle on the Intellectual Virtues: On the Meaning of the Notions of Consideration (γνώμη) and Consideration for Others (συγγνώμη) in *Nicomachean Ethics*. *Russian Studies in Philosophy*, *54*(6), 519–534. doi:10.1080/10611967.2016.1293439

Sprod, T. (2001). *Philosophical Discussion in Moral Education: The Community of Ethical Inquiry* (1st ed.). Routledge. doi:10.4324/9780203470749

UNESCO. (2015). *Enabling Environments*. https://uis.unesco.org/en/glossary-term/enabling-environment

UNESCO. (2020). *Education in a post-COVID world: Nine ideas for public action*. https://en.unesco.org/sites/default/files/education_in_a_post-covid_world-nine_ideas_for_public_action.pdf

United Nations General Assembly. (2015). *Transforming Our World: The 2030 Agenda for Sustainable Development.* Resolution Adopted by the General Assembly on 25 September 2015, 42809, 1-13. https://www.refworld.org/docid/57b6e3e44.html

VanderWeele, T. J. (2017). On the promotion of human flourishing. *Proceedings of the National Academy of Sciences of the United States of America*, *31*(31), 8148–8156. doi:10.1073/pnas.1702996114 PMID:28705870

White, J. (2007). What schools are for and why. Philosophy of Education Society of Great Britain IMPACT pamphlet, 14, 1-50. doi:10.1111/j.2048-416X.2007.tb00116.x

Wolbert, L., de Ruyter, J., & Schinkel, A. (2015). Formal Criteria for the Concept of Human Flourishing: The First Step in Defending Flourishing as an Ideal Aim of Education. *Ethics and Education*, *10*(1), 118–129. doi:10.1080/17449642.2014.998032

Chapter 11
Educational Leadership Through the Lens of Crisis:
Quest for Paradigm Shift From Classical to Transformative Leadership Approaches

George Ngwacho Areba
https://orcid.org/0000-0001-5661-3090
Kisii University, Kenya

ABSTRACT

Globally, the 21st century has been characterized by the emergence of several crises including the most recent COVID-19 pandemic. This has altered various systems and structures socially, economically, and politically. It is apparent that classical leadership approaches of response to these crises aren't a panacea anymore to navigate through these challenges hence need to explore novel, and blended transformative strategies to act as a buffer against the ever-evolving new crises. It is during the crisis that leadership becomes conspicuous and needed. It's not possible to discern when or if a crisis will occur, but education leadership could influence how the university responds to such upheavals. Drawing from a scholarly wealthy stance of leadership approaches, this discourse aims to equip higher education leadership with blended transformative strategies on how to respond to crises competently, develop practices that lessen the probability of such occurrences, and stimulate self-confidence in leadership for lasting strategies.

INTRODUCTION

Globally, higher institutions of learning have been defined as hubs for knowledge, skills, learning, teaching, research, innovation, community development or service, and maturation. The attainment of this multilateral mission is anchored on the identification, articulation, and exhibition of defined shared ideals. Such shared ideals encompass ethics and morality, creativity and innovation, community service, ethical governance, partnerships, diversity, academic autonomy equity and inclusion, societal responsibility, and stewardship of institutional and donor resources among other shared ideals (Sutin & Jacob,

DOI: 10.4018/979-8-3693-1926-0.ch011

2016). Notwithstanding this noble role of higher institutions of learning in society, in the recent past, they have been caught in an ill-fated yet ever-increasing environment of crises thus the same universities have been turned into places of disorder, uncertainty, dejection, and dread. From the most recent COVID-19 pandemic to issues of climate change and student unrest these unfortunate circumstances 'crises' present both opportunities and challenges and demand for transformative leadership to explore novel, innovative, and transformative strategies to act as a buffer against the ever-evolving new crises.

Seeger and Ulmer (2001) remark that crises are high-impact happening and circumstances that frequently undermine the institution's core operations and mission. Catastrophes naturally lead to interruption and have the probability of threatening any institution's fundamental mandate, vision, and mission. Heath and Millar (2004) postulate that a crisis is classically described as an unexpected but foreseeable occurrence that has definite or possible effects on shareholders' welfare as well as the repute of the institution affected by the catastrophe. Coombs (2015) stretches the concept further by defining a crisis as the discernment of an erratic occurrence that jeopardizes significant prospects of shareholders and might utterly affect an institution's productivity and engender undesirable results. As per this description, the discernment of disaster by others is as vital emblematically as the catastrophe itself. Additionally, Pauchant and Mitroff's (1992) effort to differentiate between crisis and incident creates the parameters of this specific study, where disasters are perceived to be interruptive to the whole institution, as opposed to a localized minor event. Catastrophes offer the opportunity for reflective moments for the leader and the followers in the organization.

In the wake of the 21st century, tertiary and higher institutions of learning have been steering through a continuing and protracted myriad of political instability, high rates of inflation, across-the-board racial turbulence, and the current effect of global warming, COVID-19, and other new emerging challenges. Every of these challenges in one way or another destabilizes the normal operations of tertiary and higher institutions of learning and hence needs attention from those entrusted with management and leadership obligations in the respective institutions.

Several of these ill-fated happenings may be categorized as catastrophes, which Ulmer et al. (2018) describe as explicit, unanticipated, and out-of-routine occurrences or sequences of happenings that cause extreme heights of confusion and concurrently subject an institution to both threats and prospects to its high set priority in terms of objective. Notwithstanding the 'unparalleled 'crash of catastrophes emerging in these hard economic times, higher institutions of learning are not utterly new to these calamities. Several past occurrences have defined the manner in which researchers and academicians perceive and comprehend calamities in higher institutions of learning and this has influenced the approach tertiary and higher institutions of learning engage in preparation for ill-fated circumstances. Moreover, encounters with emergencies at the institutional, national, and international levels have made universities more responsive to the effects of catastrophes.

Calamities are challenging circumstances that have the capacity to propel personnel, leadership, and institutions to better levels of performance if managed meticulously. Crisis management and leadership in universities remain a matter of great concern and of magnified attention from both academicians and researchers given that leadership ought to plan prior and efficiently mitigate the ever-emerging and widespread crises (Brown, Brown, & Billings, 2015; Gigliotti & Fortunato, in press; Menghini, 2014; Muffet-Willett, 2010; Fortunato, 2008; Bataille & Cordova, 2014; Booker, 2014; Coombs, 2015; Varma, 2011).

As forementioned, the management of disasters points to leadership in tertiary and higher institutions of learning. Moments of crises bring forth unprecedented and unique challenges to universities' vice-chancellors. Fisher (1984) proposes that the vice-chancellors in higher institutions of learning are required to be knowledgeable in several spheres, with the capacity to manage extensively the various tasks of their respective institutions. This cadre of leaders plays a significant role in navigating the institution and its personnel through turbulent times of disasters (Smircich & Morgan, 1982) throughout the period of calamities, leaders' communication and behavior are regularly on sharp focus by their followers and related stakeholders who wait for managerial directions on the way out of such crises (Menghini, 2014).

Owing to the inevitable nature of higher institutions of learning crises and the significant part tertiary and higher institutions vice-chancellors or presidents play during these times of institutional disorder and uncertainties, it is worthwhile to empower them with requisite competencies to enhance their situational awareness to guarantee apt responsibility prior, during, and after disasters. Additionally, it is prudent for them to know how to manage both policy and operational concerns during and after such circumstances and know the type of skills, values, and competencies that are crucial for a leader working in crisis in higher institutions of learning. The end focus is to enable the leader to manage the transition from instantaneous response to enduring recovery or normalcy. This is the proposed gist of this discourse for a transformative leader amidst crisis in higher institutions of learning.

LEADERSHIP PARADIGM SHIFTS AND CRISIS MANAGEMENT

With changing times, leadership perspectives have evolved, depicting ever-changing contexts, dynamics, circumstances, institutional needs, and societal changes among other prompters. Ensuing is a brief glimpse of the evolutionary leadership strides made over time amidst unprecedented emerging crises.

Classical Leadership Perspectives and Crisis Mitigation

The clamor for blending transformative leadership perspectives and classical leadership approaches to fill leadership gaps needed to mitigate against the ever-new and unique emerging crises draws reminisces to initial traditional and classical perspectives concerning leadership and their capacity to empower leaders especially higher education leadership to navigate through the 21st-century crises efficaciously (Kumar, et al, 2014). Such is the 'Great Man' theory of leadership advanced by the scholar known as Carlyle in around the 1800s presently referred to as trait perspective to leadership (Spector, 2016). In his conceptualization of leadership, he argues that every moment of history is characterized by unique leaders of the time such as Obama, Reagan, Carter, Mandela, Kwame Nkrumah, Nyerere, and Jomo Kenyatta among other great leaders of the time. Hence the capability to lead is inherent (you are either born a leader or not) Carlyle held that leadership cannot be developed or nurtured. His perspective on leadership was informed by the argument that:

Specific individual qualities are responsible for certain forms of conduct and behavior designs are constant across diverse circumstances as aforementioned leaders are 'born' possessing already leadership qualities implying that leadership is in a way predetermined in some or likely in others whereas unlikely, if not impossible for others (Kumar, Adhish & Deoki, 2014).

Given the changing trends and dynamisms globally psychologists have keenly studied and argued against the trait perspective to leadership. From the 1940s to 1970s, mental specialist Melvin Ralph argued that management is the outcome of the collaboration between the leader or manager and the prevailing social situation or circumstance and not just merely the outcome of a predetermined isolated of traits. This sits well with the call of who is this leader in crises. Is it the one born to tackle crises or prepared potentially to handle the crisis? (Oakleaf, 2016).

As dynamics in leadership and management keep on changing as demonstrated by the digital transformations engrossed in aspects such as social media, e-learning, and e-commerce among others, there is a need to acquire suitable, relevant, and related traits to thrive as a leader and manager not necessary born technologically savvy but natured along. Within this understanding, the capacity to inspire is debatably more vital than the capacity to lead or born a leader.

As aforementioned, the context and the circumstances similarly play a significant role in ascertaining whether one is a leader or not. In many instances, it is the engagement between the circumstances and traits that distinguish a leader from a non-leader. Carlyle's concepts motivated initial and subsequent studies on leadership including post-modern transformative approaches to leadership and management.

It is noteworthy to mark that the 'Great Man' dimension to leadership is the basis for subsequent other classical perspectives on leadership emerging from the Industrial Revolution by proponents of the likes of Tylor, Fayol, and Max Weber and focused more on productivity, efficiency, and outcomes of staff as well as of the institution wholistically. They generally didn't pay attention to the humanity aspects of the staff or their interactive enduring qualities or disparities and the impact on work-related performance aspects or efficacy. The classical approaches to management and leadership draw as aforementioned chiefly from three proponents namely: Taylor, Weber, and Fayol whose viewpoints are keen on the effectiveness of staff and on enhancing an institution's output through what can be measured or data-based performance.

The classical viewpoint is frequently faulted for overlooking human needs and desires in the work environment and fundamentally doesn't take into account errors owing to unique circumstances brought by crises like the just concluded COVID-19 and other related calamities. This classical approach has permeated through modern work environments which highly utilize the system of measurement to establish how efficiently a process is running notwithstanding the human factor and prevailing conditions of the time thus its major weakness. Characteristically, Taylor as one of the proponents in this viewpoint was engrossed in the effectiveness and output of personnel as opposed to work environment and human-related concerns. Through a scientific approach to management and bringing efficiency in production regardless of the prevailing human and contextual concerns, Taylor presumed that employee productivity could be entirely enhanced forgetting the environmental and human factors in productivity. He utilized the scientific approach of quantifying tasks to come up with procedures for personnel training and management which equally were devoid of context and human-related components.

Adding to scientific tenets of management and leadership devoid of context and human-related concerns is the proponent by the name of Max Weber who originated the bureaucratic approaches to leadership and management, which heavily has a keen eye on the thematic aspects of rules and processes in addition to expertise in the overall management of the institution's affairs and operations. Weber's model similarly emphasizes on effectiveness and clarity of roles in an institution just like Taylor ignores the salient components of humanity in the workplace and the changing contexts and hence not appropriate for managing emerging crises in work environments.

Fayol is similarly a follower of classical management perspectives equally anchoring his tenets on the effectiveness of personnel, however as a departure from his preceding proponents he looks at it from a managerial angle by paying attention to the improvement of management in terms of its efficacy rather than enhancing every personnel's efficacy. Fayol as an outcome of this perspective came up with four functions of management namely: planning, organizing, leading, and controlling.

Human Relations Approach to Leadership and Crisis Management

The behavioral approach to leadership takes a much more diverse slant from the classical approach. It started in the 1920s with proponents like Follett, Mayo, and Abraham Maslow. These models changed focus from qualities (traits) to behavioral concerns. Scholars delved into staff and administrators' modes of interaction. The emphasis was on visible activities, communication techniques, and the process of decision-making. In a nutshell, the perspective underscores the significance of mental and social components in the work environment. Staff are considered social beings with exceptional desires and behavioral inclinations as opposed to viewing them simply as money-making objects. The main tenets encompass Staff Incentives: Acknowledging that efficiency and incentives can be augmented via constructive social engagements; Constructive Social Bonds: Recognizing the staff as an exceptional entity and nurturing constructive relations within the institution. This perspective stresses the worth of staff welfare, communication, and relational dynamics (Jones & Pund, 2010).

The Hawthorne research done in the 1920s marked a momentous contribution to the behavioral approach of leadership and management. Spearheaded by Mayo, the research was carried out at the Western Electric Company in the Hawthorne factory Illinois. As opposed to the classical scientific approach which perceives staff as an extension of machines, Elton stresses the social context in which staff work. This is a better approach for a leader operating in crisis as he/she is likely to choose the suitable approach to mitigate the prevailing crisis. He contended that productivity is influenced not only by a worker's capability and expertise but also by the work environment. The Hawthorne research established that staff were more robustly driven by psychosocial aspects compared to monetary inducements.

In his quest to understand human behavior and its relationship to management and leadership, Abraham Maslow a well-known psychologist, introduced the concept of the Hierarchy of Needs, which enumerates 5 crucial types of human wants often given a pyramidal approach hierarchically namely: Physiological, safety, love, esteem and self-actualization, which demonstrated that staffs were inspired through several bottom-level to top-level needs (Maslow, 1970). This theoretical approach has been utilized by leaders, especially during crises in work environments to better comprehend 'soft' aspects of staff motivation like team participation and target setting to navigate through crises (Jones & Pund, 2010).

Further theoretical approaches to behavior science and management encompass McGregor's Theories X and Y (Sorensen & Yaeger, 2018). These are two contrasting perspectives of staff management developed in the 1950s and enhanced in the 1960s. These perspectives are all about how administrators view and engage with their staff which largely centres on the views administrators hold about their staff and how staff respond to those perceptions. In Theory X, the leaders presume that staff are innately lazy and, thus, dislike their job and if possible, could evade duty, have no ambition, are individual goal oriented, and work basically to earn income to sustain themselves. The leader in this scenario is likely to utilize rewards and sanctions as inspirational components to realize institutional objectives. The administrator in theory Y is more inclined to laissez-faire and accords staff more autonomy in performing their duties and tasks. It presumes that staff are self-driven and can take up duties and tasks in the absence of persistent supervision. Workers here are viewed as intrinsically motivated beyond monetary rewards and generally tend to do better when allowed to act independently and responsibly (Sorensen & Yaeger, 2018). Leaders in this viewpoint emphasize more on recognition of staff efforts, job satisfaction, and individual staff development. McGregor's concept of institutional leadership is a demonstration of how behavior-management perspectives delve more into the humanistic facets of management and inspire administrators to take time to comprehend how psychological aspects can enhance or deter staff productivity especially when faced with a crisis.

Largely, the behavioral science management approaches are more engrossed with staff welfare and support leadership and management techniques that perceive the staff as self-driven and who desire to engage in tasks and responsibilities given productively. This school of thought in leadership thus supports management perspectives that are less engrossed in micromanaging and are more engrossed in nurturing relations with staff to assist them in attaining their set targets as efficiently and competently as possible even in a crisis. The welfare component of the staff in this perspective is a strong quality of management during crisis moments if utilized diligently.

Following in close succession humanistic approach to management are Situational or Contingency perspectives to management (Vroom, Yetton & Jago, 1973). These models portent that leadership success is contingent on the situation or circumstance prevailing at that moment. Diverse circumstances call for diverse leadership approaches. For example, a leader confronted by a crisis needs a different approach as opposed to a routine approach to issues. The contingency model was originated by Fiedler, in this model he postulates that leaders may not change their approach to leadership thus he believes in getting the right leader for the right situation. For a Leader to be effective it depends on the situation hence believes in choosing the right leader for a particular circumstance or adapting the circumstance to suit the leader's style of management. Going by this assertion it then implies that an institution facing a crisis may determine if the leader is capable of managing the crisis or require to change leadership to whoever can handle the crisis at hand. By determining situational variables and the leadership approach, one can envisage leadership efficacy or failure in diverse situations.

On the other hand, the situational leadership model also referred to as Hersey and Blanchard's situational leadership theory, accentuates that diverse circumstances need a change of leadership styles. The proponents of this perspective argue that successful leaders adjust their approach depending on the workforce level of maturity and the particular situation prevailing at that moment. This approach in leadership is recommended in dealing with crises whenever they arise in the work environment.

The following leadership behaviors characterize this perspective: The leader directs staff with explicit instructions to follow; The leader explains decisions arrived and asks the followers to buy in the ideas for the best interest of the institution. Participating- The leader collaborates with the staff to undertake the required tasks; Delegating- the leader empowers the workforce to contribute to organizational decisions. In a nutshell, the situational approach acknowledges the significance of leadership, the workforce, and the circumstances, permitting leadership to adapt their approaches to fulfill specific situational needs. These are the requisite qualities needed of a leader handling a crisis. Conversely as aforementioned while the contingency dimension emphasizes choosing a suitable leader for a particular circumstance, situational leadership stresses a leader adapting and tailoring leadership approaches to suit the situation and staff's needs.

Modern Leadership Perspectives and Crisis Management

The classical leadership movement ushered in the humanistic approach to leadership and the same has been given impetus through the evolution of post-modern perspectives or approaches to management. The transactional leadership perspective is regarded as a modern approach to management. This approach is distinguished by its clear emphasis on clarity and distinctive tasks, rules, and rewards (Bass, 1997). The main aspects of this approach include contingent rewards where leadership sets clear achievable targets and incentives or punishment depending on performance; Management by Exception (Active): Leadership monitors aggressively nonconformities from set standards and intervenes whenever required; Management by Exception (Passive): Leadership intervenes only when challenges surmount the ability of the followers (Bass, 1990). Evidently, transactional leadership succeeds in crisis crisis-free work environment where routine tasks and observance of established policies are given priority. Critics of the shortcomings of transactional perspectives have advocated for a transformative approach to leadership and management for modern managers confronted with unique challenges that demand exceptional leadership qualities to mitigate. Transformational leadership originated with proponents like Downton in 1973 who originally coined the terminology transformative. It was later advanced by Burns in 1978 and later on, the concept was enhanced by Bass (Bass and Avolio, 1994).

Transformational leaders are visionary and strive to motivate their workforce to accomplish their targets above their supposedly untapped potential. There are four main components of transformative leader qualities also referred to as the '4 I, s' namely: Individual consideration- Where leadership takes time to comprehend the unique needs of staff and relevant stakeholders and provide individual required support; Inspirational motivation- The leader communicates the envisaged vision, inspires staff towards achieving the conceptualized vision and enhances resilience among staff to achieve the vision; Idealized influence-where the leader role models through exhibiting commitment, ethics, and integrity and Intellectual stimulation where leadership endears to inspire creativity, critical thinking, and problem-solving among staff members (Bass,2006).

During crises, the transformative approach to leadership and management has proved to be thriving predominantly as compared to other approaches. It is made possible through its big-picture perspective approach where the transformational leader can focus beyond immediate foreseeable challenges. In this dimension, they take a grip on the wider context and comprehend the scope of the impending or prevailing crisis. Rather than opting to give direct solutions, they strive to get input from those who have some understanding of the situation. They encourage contributions from various perspectives and support collaboration in mitigating any crisis.

The transformative leader creates adaptive and flexible plans. They harness shared skills, strategies and knowledge then adjust leadership responses to the crisis accordingly as desirable (Wango, 2009). Transparency and authenticity take the lead in a transformative environment. Transformational leadership communicates openly, promoting trust even amidst crisis. The clamor by education stakeholders and scholars to blend various leadership approaches as the best approach to crisis management has metamorphosized into what is perceived as transformative leadership.

Blending Leadership Approaches for Crisis Management

During turbulent times often referred to as crisis moments, efficacious leadership is usually of paramount importance and blended approaches to management which can enable educational managers to mitigate crises effectively are crucial (Mitroff, 2004; Ruben and Gigliotti, 2021). Successful leaders in modern times quite often find it prudent to blend elements from various leadership approaches as they weigh options and adapt the approaches that work best according to the dynamic demands of modern institutions and the scope or magnitude of the crisis. For instance, while classical leadership approaches stress on control and stability of an institution, transformative leadership emphasizes empowering and inspiring the staff. In this regard, the human relations approach to management plays a significant role in supporting staff's emotional and mental welfare during a crisis. In turbulent times, channels of communication which enhanced with a lot of clarity are essential. Human relations leadership ascertains that correct information reaches all staff punctually. Human Relations leadership is always ready to implement crisis-related policies, like health protocols, and other safety measures which were crucial during the outbreak of pandemics such as COVID-19.

Irrespective of what type of crisis is being handled by leadership, efficacious crisis management needs the coordination and cooperation of several stakeholders within the institution, including support from human relations ICT, finance, and legal among others. It should be observed that during crises, a human-centric approach that regards relations and staff welfare is vital for operational leadership and crisis management. Crisis management demands a response informed by specific behavior. In this context, leadership must make swift decisions, boldly adapt the same, deliver steadfastly on their mandate, and engage intensely with the workforce. Concerning trait approaches and crisis management, leaders endowed with certain traits like resilience and confidence are likely to manage crises better (Spector, 2016 & Oakleaf,2016). Nevertheless, it should be noted that traits alone are inadequate in crises hence the need for a blending approach to management.

Situational or Contingency perspectives to management similarly are key to transformative leadership for managing crises if blended ingeniously. As aforementioned, these models by Fiedler portent that leadership success is contingent on the situation or circumstance prevailing at that moment thus diverse circumstances call for diverse leadership approaches. For example, a leader confronted by a crisis needs a different approach as opposed to a routine approach to issues. Crisis circumstances call for flexibility on the part of the leader. Successful leaders change their mode of leadership depending on or contingent on the crisis at that moment. This elevates them to being transformative in the manner in which they approach issues. Transformational leaders navigate through crises successfully because they can blend various leadership approaches, visualize the bigger picture, communicate the vision, engage stakeholders collaboratively, inspire creativity, and come up with adaptive plans for the staff and their learners if operating within an educational context (Wango, 2009).

The Hawthorne approaches to leadership and management advanced by proponents like Elton and others as argued prior, find space in transformative leadership, especially in crisis moments (David & Cortes,2019) Given its tenets based on psychosocial aspects, whenever job-related challenges arise, it offers ready-needed support to manage and recuperate from such challenges.

In a nutshell, the Hawthorne Studies have transformed management approaches by stressing the socio-psychological components of human behavior in institutions, and crisis management has remained a critical skill for any leader currently. Elton's strong focus on the social context in which staff work is presumed as a better approach for a leader operating in crisis as he/she is likely to choose the suitable approach to mitigate the prevailing crisis. He contends that productivity is influenced not only by a worker's capability and expertise but also by the work environment.

McGregor's concept of institutional leadership under theory X and Y is a demonstration of how behavior-management perspectives delve more into the humanistic facets of management and inspires administrators to take time to comprehend how psychological aspects can enhance or deter staff productivity especially when faced with crisis. The transformative leader in higher education blending this perspective is called upon to understand the workforce and permeate leadership with humanity where necessary.

Maslow concept of the Hierarchy of Needs, which observes that staff are inspired by several bottom-level to top-level needs similarly finds space for higher education transformative leaders (Maslow, 1970). This theoretical approach has been utilized by leaders, especially during crises in work environments to analyze the areas in which staff need to be motivated or what is referred to as the 'soft' aspects of staff motivation like team participation and target setting to navigate through crises. It should be observed that during turbulent times, it's not all about rigid rules and stringent measures but about inspiring resilience, adaptability, and collaboration.

The transactional leadership approach as observed earlier is distinguished by its clear emphasis on clarity and distinctive tasks, rules, and rewards thus transactional leadership succeeds in a crisis-free work environment where routine tasks and observance of established policies are given priority. Where there is stability, the transformational leader can engage transactionally with the workforce. However, it should be noted that this approach is subject to a lack of creativity and problem-solving skills much needed in crisis management.

Transformative Approach to Crisis Management

As already reaffirmed in circumstances characterized by crises, transformative leadership emanating from the blending of various leadership approaches thrives exceptionally well because it can blend various leadership approaches, visualize the bigger picture, communicate the vision, engage stakeholders collaboratively, inspire creativity, and come up with adaptive plans by consolidating collective knowledge and skills from the faculty and students (Mitroff, 2004; Bass and Avolio,1994). It is prudent for leadership in higher education to know how to manage both policy and operational concerns during and after such circumstances and know the type of skills, values, and competencies that are crucial for a leader working in crisis in higher institutions of learning. The end focus is to enable the leaders to manage the transition from instantaneous response to enduring recovery or normalcy.

Requisite Competencies for Transformative Leadership Crisis Management

In their research on matters regarding higher education, Booker (2014), Ruben and Gigliotti,(2021) postulate some leadership competencies that are most vital for each stage of the crisis. Such competencies and strategies most commonly associated with transformative leadership during crises include:

- Communication Competencies: Clear and transparent or open communication is crucial during crises. Transformative leadership ought to communicate empathetically and transparently during a crisis. The open communication should be with all stakeholders. Clarity in messaging, providing concise, accurate information, and addressing concerns punctually are critical in handling any crisis. Providing regular updates and responding to concerns about crises builds trust among stakeholders (Gigliotti, 2019; Fairhurst, 2007; Fairhurst &Putnam, 2004; Sellnow & Seeger, 2020).
- Compassion and Empathy: As a leader the capacity to comprehend the human effect of the crisis is vital. During a time of crisis, empathy matters. Crisis circumstances induce stress, fear, and anxiety. Leadership must sympathize with the challenges facing their workforce. It has been observed that compassionate leadership enhances resilience and trust among the workforce and all stakeholders (DuBrin, 2013).
- Collaboration and Networking: Transformative leaders should have the capacity to enhance collaborations across the workforce and relevant stakeholders; link with other educational leaders, share practices, challenges, and solutions; and collaborate with relevant local, national, and international organizations. Leaders who engage in all-inclusive problem-solving

strategies with tall stakeholders are more likely to model collaborative behavior. Cooperative efforts foster resilience during crises thus leading to innovative solutions and improved desired outcomes (Sutin & Jacob, 2016).

- Flexibility and Adaptability: As aforementioned, the all mark of transformative leadership is the ability of the leader to blend various leadership approaches, visualize the bigger picture, communicate the vision, engage stakeholders collaboratively, inspire creativity, and come up with adaptive and flexible plans by consolidating collective knowledge and skills from the faculty and students. Transformative leadership doesn't get trapped in a rut. Transformative leaders are able to think outside the box and adapt fast to varying circumstances (Bass and Avolio, 1994).
- Psychosocial Support Skills: Transformative leaders handling crises need to engage in trauma-informed techniques. They need to take cognizance of the emotional effect of disasters on faculty and students and implement trauma-informed strategies to foster their well-being (Gortner and Pennebaker, 2003).
- Resilience: Transformative leaders must have the capacity to model resilience and inspire the same in faculty members and relevant stakeholders. Progressive and persistent community-building strategies by leadership foster a sense of community within an educational institution. Robust relations among the faculty members, students, and other associated stakeholders contribute to resilience which is a component of crisis management (Nichols, 2014).
- Ambidexterity strategies and skills: Transformative leaders are encouraged to balance between exploration (creativity, innovation,) and exploitation (stability, efficiency) skills and strategies during crises. Ambidextrous leadership qualities foster crisis management approaches and organizational resilience (Gigliotti, 2019).
- Capacity to stimulate the workforce intellectually: Transformative leaders faced with crisis need to be endowed with the ability to inspire creativity, critical thinking, and problem-solving skills among the workforce and other related stakeholders and cultivate a culture of constant learning (Craig, 2018).
- Ability to learn from Past Case Studies and Innovative Practices: Transformative leaders have the capacity to evaluate the present situation in relation to past case studies from diverse catastrophes (conflict-driven, natural, or health-related) and pick out effective approaches and adopt suitable workable solutions for current crisis (Stern, 2009).
- The Grip of Policies and Funding Dynamics: Transformative leaders operating in a crisis need to have a thorough understanding of how government and relevant bodies 'policies impact crisis response and recovery in the education sector. He or she needs to collaborate with relevant policy designers to advocate for required support. Policy goes along with financial resilience. Leadership should pursue sound policy funding practices, strategies, grants, and subsidies among other financial options to enhance the institution's financial resilience during crises (Winn, 2017).
- Competencies in Instructional Adaptation Strategies: Many times, pandemics like COVID-19 have underscored the significance of remote learning in the educational sector. There is a need for leadership to be invigorated with knowledge and tools for virtual teaching and learning in addition to exploring operative virtual platforms, and blended learning models, and providing access to such devices for utilization during pandemics or crises. Leadership should invest and plan professional development programs for the faculty, and train educators to circumnavigate virtual platforms efficiently. These efforts will cushion faculty and students against disparities in access to

education occasioned by the digital divide in a technological environment when it becomes necessary owing to crises like COVID-19 (Wallace et al., 2020; Hodges et al., 2020).

- Emotional stability: Leaders and followers can be destabilized and be overwhelmed emotionally especially during crises. At such times, there is a need for leaders who have taken the time to cultivate the values of dependability, and tolerance and are able to regulate their emotions and can circumvent overreactions (Spector, 2019).
- Decisiveness: Crises by nature demand firm decisions from leadership. A transformative leader is capable of making decisions is confident with the decisions made and does not lament over the same (Boin et al., 2017).
- Interpersonal Skills: Exceptional interpersonal skills are crucial for guiding efficiently with or without any crisis. Transformative leaders are conversant on how to interrelate well with the workforce and other leaders within the institution (Fortunato, 2008; Leeper & Leeper, 2006; Len-Rios, 2010).
- Task competence: Transformative leaders are skilled and competent in their line of duty. The workforce looks up to leadership for modeling and mentorship whenever they are faced with challenges on how to go about their tasks and duties during moments of crisis (Nichols,2014).
- A stance of agility: crises need the ability to adapt swiftly and efficiently to changing situations contrary to the sluggish moving and premeditated dimensions of the so-called old-style higher education setting. One of the key objectives for transformative leaders is to nurture a culture of readiness that permits agile and rapid—yet also rational and well-thought, value-based, and stakeholder-centered decision-making when faced with uncertainty or complex situations (Eichinger and Lombardo,2004).
- Capacity to manage the 'new norm' or post-crisis: After the crisis, Transformative leaders ought to guide their followers through the difficult transition which is usually characterized by traumatic experiences due to crises to a new normal. This encompasses adjusting roles, processes, and prospects (Sellnow and Seeger,2020).

Selected Transformative Leadership Success Case Studies

Job Steve has always been cited as a role model and an epitome of a transformative leader success story owing to his inspiring and charismatic qualities (Bass and Riggio, 2006; Bryant, 2003). Meta-analytic evidence enlists many scholarly publications depicting him as charismatic, visionary, brilliant, and iconic. The New York Times contended that he spearheaded a cultural change in the manner music, movies, and mobile technologies were expected in the digital era (Isaacson, 2011). Jobs of his own volition was intrinsically motivated to transform the globe, as declared in his mantra that he desired to put a ding in the world (Isaacson, 2011& Jacobs,2013).

Scholars too credited transformative approaches to Jobs. Vroom, a renowned professor asserted in 2010 that job is the best model for transformative leadership stands out prominently and has prompted many leaders to engage in tasks that they might never have thought to do initially ('A design, a dream', 2010). After Jobs's demise, Directors of a Board related to Apple company released a statement declaring that the universe is an immensely good place due to Jobs's effort in transformative approach to leadership described as 'leadership beyond expectations' (Bass, 1985).

Former American President Obama stated that Job was amongst the supreme American innovators who could think inversely, courageous enough to hope that he could change the universe, and adequately gifted to do it. Further Barack says Job transformed people's lives, redefined the whole industry, and accomplished one of the scarcest wants in the history of humankind: he generally transformed the way every one of us perceives duty and work (Gardner & Thornhill, 2011).

A case study by Northouse (2018) on transformative leadership specifically sought to explore how chief executives of various vast growing companies are transformative in the manner and approach to their customers and employees. Purposely the study was conducted in order to establish whether and how the selected Chief Executive Officers exhibited transformative leadership in the challenging and complex work atmosphere. The aforementioned case study involved ten CEOs of vast growing companies and eighteen employees. Customer satisfaction data survey was scrutinized to include customer perceptions.

The significant main outcome of the research was that many of the CEOs (82%) and employees indicated that the most recurring transformative leadership tasks that the CEOs undertook were personalized consideration, which entails listening keenly to personal needs, coaching or counselor, unique treatment and being caring; idealized influence, that involves high sense of integrity and trust building (Bass & Avolio, 1997). Entirely all Chief Executive Officers (100%) learned to be transformative informally through learning from earlier practices. Many of the employees (64%) reported that their CEOs enabled their transformative skills, whereas 50% indicated that the absence of organizational processes and collaboration was hampering their transformative initiatives.

A major recommendation originating from this case study is that Chief executives and those aspiring to become CEOs of any organization or institution should engage in idealized influence and personalized consideration initiatives if they are yearning to be transformational in nature and succeed.

Based on other premier pieces of evidence, a transformative approach to leadership has been established to have had a profound influence on productive output in many institutions, in which performance ranks high. Undeniably, outcomes like additional effort, organizational citizenship behaviors, worker or task involvement, trust in the supervisors, enhanced leader-member communication (Hoch et al., 2018), empowering psychologically and identity of team members with the team leader (Koh, Lee & Joshi, 2019), team members incentives (Judge & Piccolo, 2004), and several others offer paybacks to both the institution and team members where transformative leadership is practiced.

Montano et al. (2017) lately established constructive associations between transformative leadership and general employee well-being (such as positive feelings and life fulfillment), and team members' mental effectiveness. Moreover, transformative leadership has been connected to reduced employee adverse psychological health status which encompasses fatigue, stress, health issues, and maladaptive sentimental signs like depression, anger, and anxiety (Montano et al., 2017). Transformative leadership behaviors are similarly connected to lesser levels of employee stress and exhaustion (Harms et al, 2017). In a nutshell, the existing meta-analytic evidence indicates that transformative leadership is connected to improved employee well-being.

Lee et al. (2020) established that transformative leadership is linked to both invention and creativity. When transformative leadership was divided into four segments, every behavior was discreetly connected to creativity (meta-analytic associations between .20 to .22), and each of these behaviors accounted for nearly comparable proportions of the explicated incongruity in creativity. To validate these findings on transformative leadership and creativity, Koh, Lee, and Joshi's (2019) meta-analysis studied transformative leadership and a number of outcomes linked to creativity. The proponents correspondingly reported a robust productive correlation with creativity ($\rho = .33$) similarly with creative distinctiveness ($\rho = .37$), self-efficacy ($\rho = .34$), and inherent drive for creativity ($\rho = .42$). Moreover, transformative leadership was also established to have a productive connection with invention climate ($\rho = .62$).

A meta-analysis steered by Judge and Piccolo (2004) revealed that transformative leadership enhanced employees' job fulfillment and their supervisors. Equally, Hoch's et al. (2018) meta-analysis established a sturdy and productive connection between transformative leadership and employees' job fulfillment including their team leaders. In regards to commitment, Hoch et al. (2018) discovered that transformative leadership was absolutely connected to both sentimental commitment ($\rho = .42$) and organizational commitment ($\rho = .43$) and was responsible for 18% of the discrepancy in both sentimental and organizational commitment among workers. Additionally, Jackson, Meyer, and Wang's (2013) meta-analysis studied the connection between transformative leadership and emotional, continuance, and normative commitment. As in Hoch's et al. (2018) study, transformative leadership exhibited a productive connection with sentimental commitment ($\rho = .45$) similarly to normative commitment ($\rho = .34$).

Judge and Piccolo (2004), in their meta-analysis, established that transformative leadership behaviors were responsible for 8.4% of the discrepancy in leader efficiency correspondingly being meaningly connected with organizational performance (Judge and Piccolo, 2004). These case studies validate transformative leadership's important role in leadership output and provide a rationale for unceasing efforts by scholars to explore more on this type of leadership.

Transformative Leadership Implementation Challenges and Mitigations

The transformative style of leadership faces both challenges and opportunities in the ever-altered educational environment which transformational leadership ought to take centre stage in mitigating. Such include the following:

Ineffective Communication

Insufficient communication may ignite chaos, resistance, rumors, and confusion. Consequently, if utilized well, communication is a critical element of transformative leadership. It is imperative to set up a consistent and clear communication strategy to mitigate against this hurdle. Team leaders ought to utilize several means available to share the change. Avail frequent updates about the progression of the changes and resolve any fears or queries from the team members (Burnes, 2011).

Deficiency of Resources

Transformative leadership is synonymous with change which often calls for extra resources like personnel, finances, and technological expertise. Nevertheless, institutions might encounter hiccups in regard to scarce resources, which hinder the execution of the intended changes. To juggle through this hurdle, there is a need to conduct a wide-ranging evaluation of the resources required for the intended change and secure the much-needed resources in order of priority. If need be, can explore alternate choices, like partnerships with other organizations (Hubbart, Jason,2023).

Absence of Clarity in Established Change Objectives

For a Transformative leader, it is easier to manage and evaluate the achievement of any change process when having clear objectives and you have aligned team members to a mutual goal. To navigate this hurdle, it is imperative to plainly outline the objectives of the change and share the same with all stakeholders. Additionally, it is prudent to frequently assess and update the objectives as the need arises (Errida, & Lotfi,2021).

Inadequate Team Leader Support

The transformative leadership approach needs robust leadership backing from leadership to team members to succeed. Devoid of this kind of backing, team members might not be inspired to own the changes proposed or intended. To mitigate this challenge, it is vital to get assistance from the prevailing leadership. The team leaders should perceptibly show their passion for the change and aggressively talk about its necessity to their team members.

Change Resistance

A prime challenge in Transformative Leadership is the tendency to resist change by team members. Perceptibly staff resist change, as they perceive it as disruptive to their normal routine of doing things and are often troubled by the uncertainty that comes with it. To mitigate this scenario, it is crucial for the team leader to talk about the changes to be introduced, the rationale, and the paybacks to be accrued from the intended changes. The involvement of the team members in the change process is important by considering their views and resolving their worries. Moreover, there is a need for the team leader to capacity-build the team members to enable them to acclimatize to the intended changes (Galli, 2018).

Absence of Experienced Change Agents

The absence of role models to spearhead change may derail any meaningful effective implementation of Transformative leadership. In any work environment, there ought to be a team of knowledgeable agents to guide, motivate and support the organization's team members during the change process (Newman, 2000).

Fatigue Associated Changes

When staff are overawed by several changes occurring concurrently within a limited period of time, this may result in depression, fatigue, resistance, and reduced efficiency. To address this hurdle, it is vital to prudently strategize the sequencing and timing of changes. Team leaders to desist from executing numerous changes at ago and give adequate time for team members to acclimatize to established change. Moreover, the leader is required to diligently assess and manage the effect of changes introduced on staff and give the required support and resources to assist them handle the situation (Moran, & Brightman,2001).

Deficiency of Team Members Involvement

Team members' involvement is key to productive change management in a transformative leadership environment. Involved team members have a higher likelihood to adapt to change and contribute productively to its execution. To navigate through this hurdle, engage all team members in the intended change components from the start. Get their views, resolve their anxieties, and offer them room to give their contribution concerning the intended changes. Finally, appreciate and motivate them in their accomplishments and offer continuous training and any needed support (Burnes & Jackson,2011).

Unclear Roles and Responsibilities

Transformative leadership to be successful, calls for clarity of roles and responsibilities so as to ensure that everybody is responsible for their actions and moves. Nevertheless, an institution may be faced with the challenges of conflicting roles and responsibilities, which may result in chaos and incompetence. To mitigate this hurdle, it is imperative to plainly outline the responsibilities and roles of every staff engaged in the change process. Team leader to set organizational expectations and have staff be accountable for their individual performances. Additionally offer support to team members through training to assist them in comprehending their responsibilities and roles (Chowdhury and Shil,2022; Hussey,2000).

Future Research Directions

There is insubstantial up-to-date literature on growing global crisis leadership skills gaps in higher education learning institutions. An elaborate comparison between the continuous utilization of classical management and leadership approaches vis-a-vis transformative strategies to respond to these crises needs to be re-examined in detail for productive long-term solutions to ever-evolving new crises in higher education settings. It is important in the future to engage in more scientific studies focused on the perceptions of leadership in crisis and the perceptions of the led that is academic staff, non-academic staff, learners, and alumni as opposed to the present approach anchored on leadership alone. As Pacanowsky and O'Donnell-Trujillo (1983) observe, both executor and 'observer must be seen as jointly contributing to delivering the organization's mandate. Similar research in other educational levels such as primary and secondary ought to be carried out so as to validate these outcomes.

CONCLUSION

The high-stakes circumstances of institutional interruption offer a chance for leadership to perfect and relook into the ideals and values that are most reliable and resonate with the vision and mission of higher institutions of learning during crises. For this effort to be fruitful, it will be extremely vital for stakeholders, leadership in education, and researchers to explore novel strategies, blend various existing leadership approaches, be innovative, and utilize transformative strategies to cushion leaders and managers against the ever-evolving new crises. As aforementioned, to retool education leadership with transformative blended innovative strategies on how to respond to upheavals in education competently, develop conventions and practices that lessen the probability of such occurrences in the future, and stimulate self-confidence in leadership for lasting strategies, safety, and prosperity for higher institutions of learning is crucial.

REFERENCES

Bass, B. M. (1990). From transactional to transformational leadership: Learning to share the vision. *Organizational Dynamics*, *18*(3), 1990. doi:10.1016/0090-2616(90)90061-S

Bass, B. M. (1997). Does the transactional/transformational leadership paradigm transcend organizational and national boundaries? *The American Psychologist*, *52*(2), 130–139. doi:10.1037/0003-066X.52.2.130

Bass, B. M., & Avolio, B. J. (1994). *Improving Organizational Effectiveness through Transformational Leadership*. Sage Publications.

Bass, B. M., & Riggio, R. E. (2006). *Transformational leadership*. Erlbaum. doi:10.4324/9781410617095

Bataille, G. M., & Cordova, D. I. (2014). *Managing the unthinkable: Crisis preparation and response for campus leaders. Sterling*. Stylus.

Booker, L. (2014). Crisis management: Changing times for colleges. *Journal of College Admission, 222*, 16–23.

Bordia, P., Hunt, E., Paulsen, N., Tourish, D., & DiFonzo, N. (2004). Uncertainty during organizational change: Is it all about control? *European Journal of Work and Organizational Psychology, 13*(3), 345–365. doi:10.1080/13594320444000128

Brown, N. A., Brown, K. A., & Billings, A. C. (2015). 'May no act of ours bring shame': Fan- enacted crisis communication surrounding the Penn State sex abuse scandal. *Communication & Sport, 3*(3), 288–311. doi:10.1177/2167479513514387

Bryant, S. E. (2003). The Role of Transformational and Transactional Leadership in Creating, Sharing and Exploiting Organizational Knowledge. *Journal of Leadership & Organizational Studies, 9*(4), 32–44. doi:10.1177/107179190300900403

Burnes, B. (2011). Introduction: Why does change fail, and what can we do about it? *Journal of Change Management, 11*(4), 445–450. doi:10.1080/14697017.2011.630507

Burnes, B., & Jackson, P. (2011). Success and failure in organizational change: An exploration of the role of values. *Journal of Change Management, 11*(2), 133–162. doi:10.1080/14697017.2010.524655

Chowdhury, A. & Shil, N.C. (2022). Understanding change management in organizational context: revisiting literature. *Management and Entrepreneurship: Trends of Development, 1*(19), 28-43.

Coombs, W. T. (2015). *Ongoing crisis communication: Planning, managing, and responding* (4th ed.). Sage.

David, S. B., & Cortes, A. H. (2019). Principles of Management. OpenStax.

DeRue, D. S., Nahrgang, J. D., Wellman, N. E. D., & Humphrey, S. E. (2011). Trait and Behavioral Theories of Leadership: An Integration and Meta-Analytic Test of Their Relative Validity. *Personnel Psychology, 64*(1), 7–52. doi:10.1111/j.1744-6570.2010.01201.x

Eichinger, R. W., & Lombardo, M. M. (2004). Learning agility as a prime indicator of potential. *People and Strategy, 27*(4), 12.

Errida, A., & Lotfi, B. (2021). The determinants of organizational change management success: Literature review and case study. *International Journal of Engineering Business Management, 13*, 1–15. doi:10.1177/18479790211016273

Fairhurst, G. T. (2007). *Discursive leadership: In conversation with leadership psychology*. Sage Publications, Inc. doi:10.4135/9781452231051

Fairhurst, G. T., & Putnam, L. (2004, February 1). Organizations as Discursive Constructions. *Communication Theory*, *14*(1), 5–26. doi:10.1111/j.1468-2885.2004.tb00301.x

Fisher, J. L. (1984). *Power of the president*. Macmillan.

Fortunato, J. A. (2008). Restoring a reputation: The Duke University lacrosse scandal. *Public Relations Review*, *34*(2), 116–123. doi:10.1016/j.pubrev.2008.03.006

Galli, B. J. (2018). Change management models: A comparative analysis and concerns. *IEEE Engineering Management Review*, *46*(3), 124–132. doi:10.1109/EMR.2018.2866860

Gigliotti, R. A. (2019). *Crisis leadership in higher education: Theory and practice*. Rutgers University Press.

Gigliotti, R. A., & Fortunato, J. A. (2014 in press). Crisis leadership: Upholding institutional values. In B. D. Ruben, R. De Lisi, & R. A. Gigliotti (Eds.), *A guide for leaders in higher education: Core concepts, competencies, and tools. Sterling*. Stylus.

Gortner, E., & Pennebaker, J. W. (2003). The archival anatomy of a disaster: Media coverage and community-wide health effects of the Texas A&M bonfire tragedy. *Journal of Social and Clinical Psychology*, *22*(5), 580–603. doi:10.1521/jscp.22.5.580.22923

Harms, P. D., Credé, M., Tynan, M., Leon, M., & Jeung, W. (2017). Leadership and Stress: A Meta-Analytic Review. *The Leadership Quarterly*, *28*(1), 178–194. doi:10.1016/j.leaqua.2016.10.006

Heath, R. L., & Millar, D. P. (2004). A rhetorical approach to crisis communication: Management, communication processes, and strategic responses. In D. P. Millar & R. L. Heath (Eds.), *Responding to crisis: A rhetorical approach to crisis communication* (pp. 1–17). Erlbaum.

Hoch, J. E., Bommer, W. H., Dulebohn, J. H., & Wu, D. (2018). Do Ethical, Authentic, and Servant Leadership Explain Variance Above and Beyond Transformational Leadership? A Meta-Analysis. *Journal of Management*, *44*(2), 501–529. doi:10.1177/0149206316665461

Hodges, C., Moore, S., Lockee, B., Trust, T., & Bond, A. (2020, March 27). *The difference between emergency remote teaching and online learning*. https://er.educause.edu/articles/2020/3/the-difference-between-emergency-remote-teaching-and-online-learning

Hubbart, J. A. (2023). Organizational Change: The Challenge of Change Aversion. *Administrative Sciences*, *13*(7), 162. doi:10.3390/admsci13070162

Hussey, D. E. (2000). *How to Manage Organizational Change* (Vol. 28). Kogan Page Publishers.

Isaacson, W. (2011). *Steve Jobs*. Simon & Schuster.

Jacobs, G., van Witteloostuijn, A., & Christe-Zeyse, J. (2013). A theoretical framework of organizational change. *Journal of Organizational Change Management*, *26*(5), 772–792. doi:10.1108/JOCM-09-2012-0137

Jones, G., & Pund, L. (2010). *Leadership and Management in the Early Years*. CPI Antny Rowe.

Judge, T. A., & Piccolo, R. F. (2004). Transformational and Transactional Leadership: A Meta- Analytic Test of Their Relative Validity. *The Journal of Applied Psychology, 89*(5), 755–768. doi:10.1037/0021-9010.89.5.755 PMID:15506858

Koh, D., Lee, K., & Joshi, K. (2019). Transformational Leadership and Creativity: A Meta- analytic Review and Identification of an Integrated Model. *Journal of Organizational Behavior, 40*(6), 625–650. doi:10.1002/job.2355

Kumar, S., Adhish, V. S., & Deoki, N. (2014). Making sense of theories of leadership for capacity building. *Indian Journal of Community Medicine, 39*(2), 82. doi:10.4103/0970-0218.132721 PMID:24963223

Lee, A., Legood, A., Hughes, D., Tian, A. W., Newman, A., & Knight, C. (2020). Leadership, Creativity and Innovation: A Meta-Analytic Review. *European Journal of Work and Organizational Psychology, 29*(1), 1–35. doi:10.1080/1359432X.2019.1661837

Len-Rios. (2010). *Image repair strategies, local news portrayals and crisis stage: A case study of Duke University's lacrosse team crisis. International Journal of Strategic Communication.*

Menghini, R. J. (2014). *Presidential responses to crises at public university campuses: What leaders do and how others perceive their actions* [Unpublished doctoral dissertation]. University of Pennsylvania, Philadelphia, PA.

Mitroff, I. I. (2004). *Crisis leadership: Planning for the unthinkable.* Wiley.

Montano, D., Reeske, A., Franke, F., & Hüffmeier, J. (2017). Leadership, Followers' Mental Health and Job Performance in Organizations: A Comprehensive Meta-Analysis from an Occupational Health Perspective. *Journal of Organizational Behavior, 38*(3), 327–350. doi:10.1002/job.2124

Moran, J. W., & Brightman, B. K. (2001). Leading organizational change. *Career Development International, 6*(2), 111–118.

Muffet-Willett, S. L. (2010). *Waiting for a crisis: Case studies of crisis leaders in higher education* [Unpublished doctoral dissertation]. University of Akron.

Newman, K. L. (2000). Organizational transformation during institutional upheaval. *Academy of Management Review, 25*(3), 602–619. doi:10.2307/259313

Nichols, A. L., & Cottrell, C. A. (2014, August). What Do People Desire in Their Leaders? The Role of Leadership Level on Trait Desirability. *The Leadership Quarterly, 25*(4), 711–729. doi:10.1016/j.leaqua.2014.04.001

Northouse, P. G. (2018). *Leadership: Theory and Practice* (8th ed.). Sage Publications.

Oakleaf L. (2016). *Trait theory of leadership*. Organization and Administration in Recreation, Sport and Leisure Management, Missouri Western.

Pacanowsky, M. E., & O'Donnell-Trujillo, N. (1983). Organizational communication as cultural performance. *Communication Monographs, 50*(2), 126–147. doi:10.1080/03637758309390158

Pauchant, T. C., & Mitroff, I. I. (1992). *Transforming the crisis-prone organization: Preventing individual, organizational, and environmental tragedies*. Jossey-Bass.

Ruben, B. D., De Lisi, R., & Gigliotti, R. A. (2021). *A guide for leaders in higher education: Core concepts, competencies, and tools* (2nd ed.). Stylus.

Ruben, B. D., & Gigliotti, R. A. (2017). Are higher education institutions and their leadership needs unique? Vertical and horizontal perspectives. *Higher Education Review*, *49*(3), 2752.

Seeger, M. W., & Ulmer, R. R. (2001). Virtuous responses to organizational crisis: Aaron Feuerstein and Milt Cole. *Journal of Business Ethics*, *31*(4), 369–376. doi:10.1023/A:1010759319845

Sellnow, T. L., & Seeger, M. W. (2020). *Theorizing crisis communication* (2nd ed.). Wiley- Blackwell.

Smircich, L., & Morgan, G. (1982). Leadership: The management of meaning. *The Journal of Applied Behavioral Science*, *18*(3), 257–273. doi:10.1177/002188638201800303 PMID:10260212

Solnit, R. (2020). *'The way we get through this is together': The rise of mutual aid under coronavirus.* Available at: https://www.theguardian.com/world/2020/may/14/mutual-aid-coronavirus-pandemic-rebecca-solnit

Sorensen, P., & Yaeger, T. (2018). *Theory X and Theory Y – Management.* Oxford Bibliographies.

Spector, B. (2016). Carlyle, Freud, and the Great Man Theory More Fully Considered. *Leadership*, *12*(2), 250–260. doi:10.1177/1742715015571392

Spector, B. (2019). *Constructing crisis: Leaders, crises and claims of urgency*. Cambridge University Press. doi:10.1017/9781108551663

Sutin, S. E., & Jacob, W. J. (2016). *Strategic transformation of higher education: Challenges and solutions in a global economy*. Rowman and Littlefield.

Ulmer, R. R., Sellnow, T. L., & Seeger, M. W. (2018). *Effective crisis communication: Moving from crisis to opportunity* (4th ed.). SAGE.

Varma, T. M. (2011). Crisis communication in higher education: The use of "negotiation" as a strategy to manage crisis. *Public Relations Review*, *37*(4), 373–375. doi:10.1016/j.pubrev.2011.08.006

Vroom, V. H., Yetton, P., & Jago, A. (1973). A decision tree model of leader behavior: An extension of the contingency approach. *Human Relations*, *16*(4), 451–467.

Wallace, J. D., Burton, B. G., Chandler, R. C., & Darby, D. G. (Eds.). (2020). Special issue for suddenly online: Considerations of theory, research, and practice. *The Journal of Literacy and Technology*, *21*(2), 1–148.

Wango, G. (2009). *School Administration and Management: Quality Assurance and Standards in Schools.* Jomo Kenyatta Foundations.

Winn, Z. (2017, April 13). *The Virginia Tech shooting's impact on emergency preparedness.* https://www.campussafetymagazine.com/clery/virginia_tech_shooting_anniversary_emergency_preparedness/

ADDITIONAL READING

Bass, B. M. (1985). *Leadership and Performance beyond Expectations*. Free Press.

Bass, B. M., & Avolio, B. J. (1989). *Manual for the Multifactor Leadership Questionnaire*. Consulting Psychologists Press.

Brennan, J. A., & Stern, E. K. (2017). Leading a campus through crisis: The role of college and university presidents. *Journal of Education Advancement and Marketing*, *2*(2), 120–134.

Fairhurst, G. T., & Grant, D. (2010). The Social Construction of Leadership: A Sailing Guide. *Management Communication Quarterly*, *24*(2), 171–210. doi:10.1177/0893318909359697

Hendrickson, R. M., Lane, J. E., Harris, J. T., & Dorman, R. (2012). Academic leadership and governance of higher education: A guide for trustees, leaders, and aspiring leaders of two- and four-year institutions. *Stylus (Rio de Janeiro)*.

Moerschell, L., & Novak, S. S. (2019). Managing crisis in a university setting: The challenge of alignment. *Journal of Contingencies and Crisis Management*, *28*(2), 30–40.

Seeger, M. W., & Ulmer, R. R. (2001). Virtuous responses to organizational crisis: Aaron Feuerstein and Milt Cole. *Journal of Business Ethics*, *31*(4), 369–376. doi:10.1023/A:1010759319845

Stern, E. K. (2009). Crisis navigation: Lessons from history for the crisis manager in chief. *Governance: An International Journal of Policy, Administration and Institutions*, *22*(2), 189–202. doi:10.1111/j.1468-0491.2009.01431.x

KEY TERMS AND DEFINITIONS

Approach: Style or way being chosen by the leader to lead an educational institution.

Blended: This is an approach to educational leadership that combines all workable leadership approaches (classical and modern) to steer an institution successfully in exceptional circumstances.

Classical: It refers to long-established earlier or traditional forms or styles of leadership.

Crisis: An erratic occurrence that jeopardizes significant prospects of shareholders and might utterly affect an institution's productivity and engender undesirable results (Coombs,2015).

Higher Education: It ebodies all organized learning and training activities at the tertiary level. This includes conventional universities (i.e., those with conventional arts, humanities, and science faculties) as well as specialized universities (like institutions specializing in agriculture, engineering, science, and technology). The concept also includes conventional post-secondary institutions (like Polytechnics, colleges of education, and vocations). Under the umbrella of 'higher education' come all forms of professional institutions (UNESCO 1994).

Leadership: One's ability to guide and influence the staff in an educational institution to attain their set and intended goals.

Transformative: Leadership blends various leadership approaches, visualizes the bigger picture, communicates the vision, engages stakeholders collaboratively, inspires creativity, engages adaptability, and consolidates collective knowledge and skills from all stakeholders to thrive (Mitroff, 2004; Bass and Avolio,1994).

Chapter 12
Tales of the Unexpected:
Post-Crisis Transformational Leadership

Catherine Hayes
University of Sunderland, UK

ABSTRACT

This chapter explores the potential for strategic organizational frameworks and applied methodological perspectives to aid workforce contexts in harnessing the transferable agency of those equipped experientially for leadership in times of unprecedented crisis. Academic research and scholarship must be needs-led not methods-driven, and as the recent pandemic demonstrated, authentic leadership had never been more important. Transitional change through the crisis of the global pandemic led to shifts to greater acknowledgement of the need for a different type of knowledge creation and replacing the contexts of validity and reliability in empirical research with those of trustworthiness and authenticity. In providing some of these organizational frameworks and methodological perspectives in an accessible manner, many can be systematically applied to the context of everyday strategic planning and institutional management settings so that tangible target outcomes are visible, achievable, and perhaps, most importantly, remain person-centered.

INTRODUCTION

"Education breeds confidence. Confidence breeds hope. Hope breeds peace."

(Confucius 551—479 B.C.E.)

Business in the context of the 21st Century knowledge economy is driven by the dynamics of policy, practice and the institutions and organisations which drive their capacity to function and develop professionally and in the context of applied research amidst emerging crises (Bogoviz, 2019). It is within these contexts that the emergence of Professional Doctorate programmes, the Doctorate of Business Administration, in particular, which has forged a landscape of the need to address the professionalisation of knowledge, to acknowledge the agency that applied knowledge equips personnel with and how more traditional mechanisms of doctoral education are less suited to the application of theory to practice and

DOI: 10.4018/979-8-3693-1926-0.ch012

more suited to theoretical emergence and academic contexts such as education (Cardoso et al, 2020). The gap between perceptions of usefulness and purposefulness of the two though, has narrowed in recent years (Aarnikoivu, 2021). The prospect of responding reflexively and adaptively to new crisis events and key epiphanies such as the COVID-19 pandemic has ensured the visible and tangible impact of professional doctorate programmes in practice, reflecting a shift to greater respect for a different type of knowledge creation and replacing the contexts of validity and reliability in empirical research with those of trustworthiness and authenticity in applied praxis environments such as the workplace (Dirks and de Jong, 2021).The objectives of this chapter are fourfold in a) providing a theoretical basis for the facilitation of knowledge creation in work based settings and its translation into practice via optimal leadership in crisis b) Framing the translation of doctoral knowledge in crisis by mid-career professionals; c) The consideration of the complex ambiguity surrounding knowledge creation in crisis and post-crisis from a methodological perspective and d) Introducing transformative learning theory as a lens through which the need for cognitive, metacognitive and epistemic perspectives can be acknowledged and used to drive positive action in workplace crises.

Background

Reliance on transdisciplinary, interdisciplinary and the multiple perspective they bring to a civic society is also significant. In positioning professional doctorate pathways as key indicators of both situational and axiological change in the latter parts of the 20th and early parts of the 21st Century, there ought also to be a consideration of how flexible these programmes are in equipping middle management to deal with the complex ambiguity of crisis scenarios, skills of reflexive responsiveness and the capacity to engage with reflection on key aspects of crises. More importantly, however, is how lessons might best be learned for the future of industries and sectors where knowledge is the harness and staff are the key drivers of responsiveness in practice (Hancock, 2020). Alongside these contextual considerations, there is also the need to consider the situated nature in terms of the context and setting of the academic delivery of Professional Doctorate programmes in Continuing Professional Development practice (Karas et al, 2020). The vast array of professional doctorates now available to mid-career professionals in Western education, is demonstrative of the established need for professionals to move beyond just the context of advanced scholarship and knowledge acquisition that characterises doctoral programmes and this is something which has a high degree of potential transferability to other contexts and settings, recovering from crisis and in need of new strategic leaders and academics. In attuning educational delivery to needs led provision, rather than the creation of new knowledge for creation's sake, as is often the case in traditional PhD's, there has been a whole raft of debates on the concept of fitness for purpose in relation to 'real life' application in the world. One mechanism by which this has been evaluated is the way mid-career professional have influenced key aspects of economic generation, social change or influenced spheres of practice within or beyond their own disciplinary professional identities. Consequently, interdisciplinary professional doctorate programmes have afforded a generation of mid-career professionals the opportunity to challenge assumptions and key presuppositions which frame their current disciplinary and interdisciplinary praxis and to legitimise how their negotiation of complex ambiguity is credible and rooted in systematic and responsive change to the dynamic landscapes of research-based practice. Defining exactly what a knowledge economy is, is challenging for several reasons, especially pertaining to doctoral level education. (Bolisani and Bratianu, 2018). Therefore, there are as many different versions as there are national and global economies. For the purposes of this chapter, however, the term knowledge economy will refer to an economy via which in the creation, harnessing and exploitation of knowledge is a significant indicator of consequent wealth, creativity, and innovation. Since the term now largely characterises most of organisational culture in global workforces, the term transcends the strategic and influences every aspect of operational practice in contemporary societies, with a new and focused culture on the accumulation of wealth and the expected norm of competition.

The impact of this definition on Higher Education over the last two decades is also reflective of the Neoliberalist approaches to business organisational infrastructure and human capital which enables and facilitates them. As a direct consequence of this the relevance of doctoral education and training has been hauled to account and has seen a shift in perception of professional doctorate candidates in comparison to their counterparts undertaking traditional Doctor of Philosophy programmes. This shift is reflective of the additional mechanisms of knowledge production and creation in praxis and their facilitation within the context of these programmes. The term 'real world' has also become a key differentiator between the type of knowledge creation when comparing the two doctoral pathway outcomes, with the work-based research focus of professional doctorates reflecting the shift away from theoretical hypothesising to interventional approaches within now more commonly executed research methodologies such as participatory action research, case study, theory of change, and ethnography. Within academic institutions delivering doctoral programmes the progression of knowledge-based economies has also driven the far clearer differentiation between academic impact and research impact. Alongside this the exponential increase in digital technologies have fuelled the progressive application of knowledge to practice and the necessary re-shaping and refining of workforce skills and development (de Araujo et al, 2021). Because of this, professional doctorates have become progressively intersectional, embracing, and incorporating interprofessional practice as an integral part of multi and interprofessional delivery mechanisms, where signature pedagogies and disciplines can be embraced at an epistemological level.

Integrating Academic and Professional Practice in Crisis

It has been argued that the intersectionality between academic disciplines and signature pedagogies in the creation and application of knowledge characterises its major strengths. A historical overview of the emergence of professional doctorate pathways provides an insight into how this has become operationalized in practice across disciplines which predominantly surround education, business, medicine, and allied healthcare alongside contemporary additions in engineering and computing. Stemming from the need to extend boundaries within and between professional disciplines and the emergence of more student, person, patient, and client centred approaches in the context of multi-disciplinary teamwork across all professions, accounts for the rise in popularity of programmes with a specific focus on knowledge creation and implementation in practice. At a macro level, onus on the requirement for needs led knowledge creation was introduced by governments whose work with educational agencies and employers had examined the concept of fitness to practice, upon completion of academic education and training programmes throughout Western education. Consequently, emphasis was placed on bridging the theory practice gap and how academic institutions could directly contribute towards providing students with the skills to work in competitive, productive organisations, where knowledge translation into practice at the front line of international workforces could have tangible impact and evidence of contribution to changing 'real world' policy and practice in times of crisis and post crisis recovery.

Transformative Learning in Crises Leadership and Management

The concept of transformative learning as a singular metatheory has direct relevance to time, context, and culture, all of which are relevant and influence contribution to the knowledge at the front line of work-based praxis. All concepts of transformative learning can be aligned with the complex ambiguity of the workplace, since they begin with the recognition of disorienting dilemmas that characterise the 'real world'. Since being able to challenge presupposition and longstanding assumptions that are directly associated with having worked within the same professional context as a mid-career professional entails leads to the opportunity for a new perspective or reference frame, this also provides an opportunity to contemplate and negotiate the need for change. Reconciling meaning from lived experience enables new frames of reference to be established of direct relevance to multifaceted variables that frame human interaction, dialogue, and communication alongside additional complexities of the workplace. Within the context of a cycle of ongoing monitoring and evaluation, when applied this leads to the further identification of disorienting dilemmas and the opportunity to reframe multiple perspectives. In the context of acute times of crises where all these things can be temporally exacerbated and the speed of decision making is crucial to optimal outcomes, then transformative learning becomes a tool in the armory of professional leaders and managers.

In the clear delineation between the concept of transformation and transformative learning as a process it is critical introspection as part of reflective practice which can lead to profound change on both an individual and collective contribution to professional practice. Critical introspection engages both conscious and unconscious parts of the psyche and frames ontological positioning of worldview and consequent ways of knowing and making meaning of experience in practice-based settings. From the pedagogic perspective of teaching and facilitating generic skills of transformative learning, it is imperative that educators delineate between the perspective transformation described seminally by Mezirow and transformative learning as a metatheory, within which several different approaches can be subsumed. The contextual significance of education and training for mid-career professional often engages them in being proactive learners, via various teaching methodologies such as problem-based learning approaches, which are essentially transformative in that they permit the introduction of complex ambiguity and encourage higher order critical thinking and, consequently, incorporate higher order learning outcomes as part of discernment in processes of decision making.

Transformative learning as a metatheory provides educators with a lens through which they can view student learning and ascertain the need for both functional and emotional change. Central to this is the role of reflection which consequently develops into a capacity for critical reflexivity.Transformative learning is desirable on several counts within the context of management and leadership in practice, particularly in relation to the domain specificity and conscientisation of education for applied practice. Learning that has the capacity to transform frames of reference and assumptions to make them more discriminating, open, reflexive and emotionally able to change. This offers a deeper structural or paradigmatic shift that alters capacity for negotiating the world and forever altering perceptions of it. In this sense it is important to consider the structural and agentic properties of teaching and learning as a process.

Identities of Professionalism in Reflexive Action

How core concepts of performativity, transformative learning, and the potential to measure the impact of knowledge translation at the front line of person-centred care can all influence crisis management continues to be of particular significance throughout the Global Covid-19 Pandemic. The potential to educate staff capable of leading through the complex ambiguity and series of unknown outcomes that need to be negotiated in such circumstances is the central focus of how best crises can and ought to be managed in practice-based settings such as the workplace and across wider society. Indeed, it is this complex ambiguity of the world, which ensures work- based research undertaken inside professions by those with culturally and temporally marked identities of their own professionalism, ought always to be acknowledged in relation to the concepts of both truth and verisimilitude.

Framing History and Contexts of Historical Development

The relatively expedited process of relationship building in the needs led knowledge base of the Nineteen Nineties ensured a lasting and dynamic change was imposed on institutional relationship development between stakeholders, academic institutions, and the need to equip personnel with the skills of systematic research were identified as paramount to definitions of success (Peters, 2020a). With prevailing politics and the consequent polemic change which occurred as a direct consequence, Neoliberalism prevailed across Higher Education where existing criticism of ‘ivory tower’ teaching, learning and research was highlighted and shunned. The PhD was posited as having little relevance to the ‘real world’ identification of gaps in research and the ability to bridge the theory-practice gap was high on the agenda for progressive change. Being able to critically evaluate the current and extant literature in the process of informed clinical decision making and deliberative policy analysis became a core skill in the armoury of all professionals and within contexts such as the UK National Health Service also served as a means of being able to prioritise resources, with the face of healthcare management being totally modernised so that management versus leadership came to the fore in terms of how progression, development and sustainability of the NHS might best be achieved. Of course, this has specific relevance to how the same culture can continue to be shaped in times of crises.

Emphasis focused on the new genesis of professionals who could actively recognise the need for different modes of learning and consequently gave credence to the ever-increasing number of curriculum designers who embraced the concept of stakeholder input in the initial justification of formal academic curricula at doctoral level. As a means of addressing the ‘real-world’ issues faced within the context of crisis this has ensured a very specific focus on the situational contexts of doctoral level learning, the need for critical reflexivity in relation to epistemic bias and cognition and a direct contribution to the emergent knowledge economy of the 21st Century. The liminal shift between what has historically constituted education versus training developed into widespread debate, with knowledge translation from academic learning to applied praxis, featuring heavily within it. Consequently, definitions of knowledge and in particular the knowledge underpinning applied practice at the front line of professions across society became a new agenda, at a point when traditional hierarchies of business were themselves being challenged in practice, as a means of ensuring that interprofessional working could be optimally delivered. Alongside optimal delivery, the polemic impact of these considerations also raised a whole array of accompanying political and feminist debates in relation to women in business environments and professions across modern society. Whilst it is not the purpose of this chapter to explore this facet of the knowledge economy emergence, it remains an important legacy of the time and it is now that these issues are being addressed directly through equality, diversity, and inclusion initiatives across Western society.

Operationalising Strategic Management in Crisis

In terms of strategic management in crises situations and being reflective of the modern economic demands exacerbated by the COVID-19 Pandemic, the emergence of digital technology and technology enhanced learning has shaped the face of HEI educational delivery and commitment to progressive change. (Herbst et al, 2021). This has seen an accompanying series of investments in equipment which has served to further separate and create systemic inequalities in the knowledge production capacity and capability on a global level, making the gap between developing countries and their Western counterparts even wider. In congruence with this HEIs have taken advantage of this somewhat enforced paradigmatic shift in the creation of knowledge and in keeping with neoliberalist perspectives have designed, developed, and sustained a competitive stance in the process of developing and brokering knowledge across an array of post-crisis stricken Covid economies.

Empowerment Application for Crises in Action

Table 1. Empowerment for crisis in action: Doctoral learning pathways

	Doctoral Learning Pathways
Generic Skill Sets for Crisis Management	• Develops the ability to address complex management problems and to present advanced work-based solutions • Develops the ability to develop effective leadership capacity in the context of professional management fields • Ability to develop a critical perspective in fields of professional management practice
Situational Specificity and Contextual Relevance in Crisis Situations	• Ability to make ethically sound judgements within social, professional and legal frameworks • Ability to personally manage reflective practice within ethical and moral standards frameworks • Ability to frame decisions and make judgements in ways which recognise and respond to concerns for social and environmental sustainability • Negotiating complex ambiguity and navigating uncertainty • Implementation and management of responsive change
Higher Level Learning Outcomes in Crisis	• Ability to apply knowledge and diagnose complex and paradoxical situations • Ability to address and solve problems using well developed skills, competencies, capabilities, and knowledge • Ability to formulate and communicate complex solutions in simple practical ways • Ability to address strategic and operational implications of proposed solutions
Bridging the Theory and Practice Nexus in Crisis	• Ability to select, analyse and synthesise data at an advanced conceptual level • Ability to establish new ideas. Build conceptual models, clarify understandings, and determine new arenas of thought of an intrinsically practical nature • Ability to demonstrate critical, imaginative, integrative, and analytical thinking
Communities of Practice and Collective Intersectionality in Crisis	• Ability to work efficiently and co-operatively as a member of a community of scholars to contribute to the growth of knowledge of others and to develop group learning skills • Ability to foster a learning environment that motivates research progress, stimulates thinking and helps to create new and responsive ideas
Critical Reflexivity and Challenging of Presupposition and Assumption in Crisis	• Conceptual frameworks to practice and research processes and to inform and develop enhanced 'deep' management knowledge • Ability to reflect on lived experience

Temporal Perspectives of Knowledge Economies in Crisis Management

The relatively expedited process of relationship building in the needs led knowledge that came from the late nineties, had an impact on the overall dynamic between stakeholders, academic institutions, and the need to equip personnel with the skills of systematic research (Peters, 2020a). Amidst this polemic change and the emergence of a neoliberalist system of Higher Education, there had been longstanding criticism of the relevance of traditional, theoretical doctoral programmes such as the PhD, which bore little relevance to the bridging of the theory-practice nexus. Next, the emphasis on developing the concept of the 'researching professional' versus the 'professional researcher' was compounded in the context of the allied health professions as a legacy of the 1980's preliminary introduction of 'evidence-based practice', arguably the forefather of the rationale for all degree professionals across the sector. This need to be able to critically evaluate the current and extant literature in the process of informed clinical decision making and deliberative policy analysis became a core skill in the armoury of healthcare providers and was also arguably used as a means of rationing resources and eliminating the anecdotal evidence base of several allied health professions in the UK. As a direct consequence of these cultural changes Higher Education Institutions (HEIs) have responded with the provision of Continuing Professional Development (CPD) pathways, aimed specifically at mid-career professionals for whom knowledge and skills enhancement became the norm (Nancarrow and Borthwick, 2021). The new genesis of practitioners who could actively recognise the need for different modes of learning, gave credence to the ever-increasing number of curriculum designers who embraced the concept of stakeholder input in the initial justification of formal academic curricula at doctoral level (Singe et al, 2021). As a means of addressing the 'real-world' issues faced within the context of mid-career professionals this brought a specific focus to the situational contexts of doctoral level learning, the need for critical reflexivity in relation to epistemic bias and cognition and a direct contribution to the emergent knowledge economy. The liminal shift between what constituted education versus training became a widespread debate, with knowledge translation from academic learning to applied praxis, featuring heavily within it. As a consequence, definitions of knowledge and in particular the knowledge underpinning applied practice at the front line of patient centred care became a new agenda, at a point when traditional hierarchies of care were themselves being challenged in practice, as a means of ensuring that multi-disciplinary team approaches could be optimally delivered and the concept of 'patient centred care' became less tokenistic and as a consequence the need to move away from traditional pyramidal knowledge relationships within and between allied health professions and medical staff became apparent. Indeed, it was from this stage of knowledge economy emergence that allied health professions became titled as such instead of the old moniker, the professions allied to medicine, where the pyramidal organisational structures within which healthcare was delivered, largely determined the capacity of allied health professionals to have an input into clinical decision making or discussion when a member of medical staff was present. This raised a whole array of accompanying political and feminist debates in relation to female dominated professions such as nursing and allied health, versus medicine, which was then, largely male dominated. Whilst it is not the purpose of this chapter to explore this facet of the knowledge economy emergence, it remains an important legacy of the time and it is now that these issues are being addressed directly through equality, diversity and inclusion initiatives across both allied health and medical care professions.

Leadership Credibility and Adaptive Crisis Management Skills

In terms of strategic development and reflective of the modern economic demands exacerbated by the COVID-19 Pandemic, the emergence of digital technology and technology enhanced learning is shaping the educational delivery mechanisms of HEIs (Herbst et al, 2021). This has seen an accompanying series of investments in equipment which has served to further separate and create systemic inequalities in the knowledge production capacity and capability on a global level, making the gap between developing countries and their Western counterparts even wider. In congruence with this HEIs have taken advantage of this paradigmatic shift in the creation of knowledge and in keeping with neoliberalist perspectives have designed, developed, and sustained a competitive stance in the process of developing and brokering knowledge across an array of economies (Bielak et al, 2008). In relation to doctoral level learning, this has ensured that the specialised nature of doctoral learning pathways for the individuals partaking in them has impacted too, on the need to create communities of practice and a broadscale acknowledgement of learning 'in situ'. Historically, research undertaken about mid-career professionals was about elevating the status of the academic institutions from where the research came, rather than impacting on person centred care at the front line of healthcare (Pols, 2014). The changing perspectives and accompanying narratives surrounding doctoral level learning, have changed this dynamic and correspondingly doctorates have become a mechanism of equipping organisational learning, rather than paving a way out for the most academically capable and often the most clinically capable mid-career professionals (Thomas et al, 2014).

Knowledge Translation for Crisis Management

Recent adaptations to Higher Education doctoral provision now recognise the need to reframe a society's dominant methodological position in terms of their paradigmatic insufficiency and their potential for the breakdown of barriers. Seminal delineations of Mode 1 and Mode 2 knowledge still serves as a useful point of departure in terms of the distinction between theoretical and situated research (Rigolot et al, 2021). They posited that Mode 1 knowledge provided discipline specificity and explanatory methodological approaches which aimed to address the development of knowledge without specific stakeholder need – rather as an approach of doing research for research's sake. The alternative, Mode 2 knowledge, was context specific, designed specifically to solve a problem linked to the workplace and characterised by critical reflexivity and an awareness of the knowledge economy within which it is produced (Gibbons & Nowotny, 2001). This work served to underpin the whole basis of design for Western professional doctorate programmes in the latter part of the 20th Century and still pervades in the new emergent peri-pandemic world we currently exist within. It is more an issue of strategic relevance than resonance which pervades the academic debates surrounding which is 'better', a traditional PhD or a Prof Doc in practice. Contemporary lenses now frame the two as being equal but different and having an inherent degree of overlap, which before now was largely ignored or overlooked. This is usually linked to misconceived perceptions of relative quality of the two programmes with the Prof Doc often being denigrated as a poor runner up (Becton et al, 2020). This originally stemmed from the difficulty HEIs had in establishing the metric evaluation of systematised knowledge and the academic rigour underpinning it. The legitimacy of knowledge has also been contested in relation to situation and context with a longstanding tradition that knowledge had to emanate from universities to have any degree of credibility. Often though, the so termed knowledge produced in the 'ivory towers' of academic was not transferable into practice since its legitimacy lacks the degree of authenticity and trustworthiness that can only be reconciled from purposive experiential evidence (Chapman, Colvin &., & Cosentino, 2020). Professional Doctorates and interdisciplinary professional doctorates have afforded the difference between purist knowledge and scholarship to be legitimised within the context of clinical and professional practice (Coates et al, 2020).

Performativity in Crisis

The term performativity has particular significance to the overall execution and delivery of Professional Doctorate education and how the programme is perceived, relative to its traditional counterpart, the Doctor of Philosophy. Performative research is that which integrates the need for address of issues in practice, with the process of undertaking that change as an integral part of the process (Aguilar & Guénette, 2021; Østern et al, 2021). It is here that the greatest potential confusion for professional doctorate candidates occurs, since the complex ambiguity of research context, research question or issue and the researcher are blurred by the lens of professional practice they are examined through. The integration of knowledge production and applied praxis also has implications for the concept of tacit knowledge, which is historically challenging to metrify or make tangible, yet has a profound impact on clinical mastery and the notion of proficiency in clinical and professional practice (Ding, Aoyama & Choi, 2020). Whilst traditional narratives highlighting this have been subsumed into the Neoliberalist knowledge economies of European and Western universities, there still remains a tension between how the quality and rigor of AHP professional doctorates, where knowledge is produced at the front line of person-centred care and integrated back into practice as a process, reflexively and often intuitively, rather than just as a posited emergent theoretical outcome (Peters, 2020b). Within this degree and level of performativity, the concept of transdisciplinarity also comes to the fore. It is here where so many aspects of disciplinary praxis merge into one at an epistemological level that the gradual silos of education and training are pragmatically redundant and the transcendence of disciplinarity ensures that knowledge produced in action, from all manner of academic signature pedagogies and academic disciplines becomes one at the centre of person-centred care (Hayes & Smith, 2020). The scope of the interdisciplinary doctorate acts as a catalyst for optimal and intuitive clinical decision making at the heart of patient centred care but is heavily reliant on the experiential learning of practitioners within their everyday role and the critical reflexivity they apply to these lived experiences, critical epiphanies and everyday interactions (Graff & Von Wehrden, 2021).

Responsive Disciplinarily in Crisis

It is institutionalised attitudes towards investigation and analysis which are modernist in nature. The inaccessible lexicon of the fields of disciplines and studies has ensured they have remained largely inaccessible to researcher with a broader perspective on positionality. Signature pedagogies and academic disciplines, rather than contexts of interprofessional working are often seen as the point of departure in doctoral education pathways (Shulman, 2005). The emphasis placed on interdisciplinary professional doctorate study focuses on research which contributes to a community of practice, is methodologically systematic and which is theoretically underpinned. In the context of the workplaces within which mid-career professionals work, it also affords doctoral candidates to move beyond reductionist approaches to binary answers or the testability of knowledge to the unpicking of the complex ambiguity with people, which is often a basis of contention. An embedded part of this complex ambiguity is the disciplinary tradition that frames cultural contexts and practice and therefore can often overshadow the potential for epistemological congruence. Hayes and Smith (2020) highlighted the reconnection of academic disciplines and signature pedagogies at an epistemological level. This has implications for both clinical and professional practice within mid-career professionals since their practice has a fundamentally uniquely applied origin and consequently so do DProfs originating in this field. This contrasts with traditional PhD study and has become as source of contention for those stalwarts of PhD delivery for whom a clear operational definition of rigour is paramount within the context of practice (work-based) research initiatives. It is the values, attitudes, and behavioural responses that professionals of different disciplinary career trajectories project, which can unduly frame collective and individual professional identity. Whereas a contingent framework or lens through which professional practice can be regarded is important, the analysis of individual disciplines is something which has temporal resonance alongside cultural longevity. It is here that disciplinary expertise serves to delineate and provide liminal shifts within and between professional disciplines, but which also frames the challenges of interdisciplinary and multidisciplinary education. The notion of performativity is central to the capacity of being able to bridge disciplines via work-based praxis (Macfarlane, 2021). Alongside functional tasks, AHP candidates are also afforded the opportunity to use creativity as a bridge to innovation in patient centred care in providing solution focused responses to key areas of research in their respective disciplinary fields of healthcare. Moving away from the taught element of Professional Doctorate programmes through to the supervisory phase, necessitates educationalists moving away from operationalising practice as facilitators but move to the role of supervisor, which changes the dynamic of the learning relationship between doctoral candidate and educationalist to one of project manager and critical friend.

Translational Research in The Workplace

Characterised by postmodernism facets of hermeneutics, deconstruction and social constructivist and constructionist perspectives now frame translational research. This enables epistemic standpoints or perspectives in the world of the workplace to be considered, whilst at the same time acknowledging epistemic bias and researcher positionality. Translational research in terms of operational definition, focuses on the way research findings can be applied to practice. It has a direct impact on the phases of research design and methodological approaches that are integrated into the process of implementation science (Lundgren-Resenterra & Kahn, 2019). The actual research-based innovations that emanate because of translational research, stem from the successful connectivity of these differing elements of the research process and ultimately, within the professionalism context, how these either directly or indirectly contribute to research for patient benefit. Usually within the context of professional settings, these are collaborative projects with an interdisciplinary focus, led by people from within the representative professions and managed formally via steering groups and committees who monitor the quality and progress of research funded by the public purse or charitable donations. As part of the process of inclusion and an ethos of patient centred care, there has been an increase in the incorporation of the 'patient voice' into the earliest stages of research design and methodology. This has been an attempt to ensure that beyond tokenism, patient experience and the critical reflexivity that occur because of being a patient can be integrated into areas for illumination and responsive action. This has also been echoed across strategic curriculum development within HEIs where tripartite relationships in curriculum design

Figure 1. Constructive alignment in epistemological stance

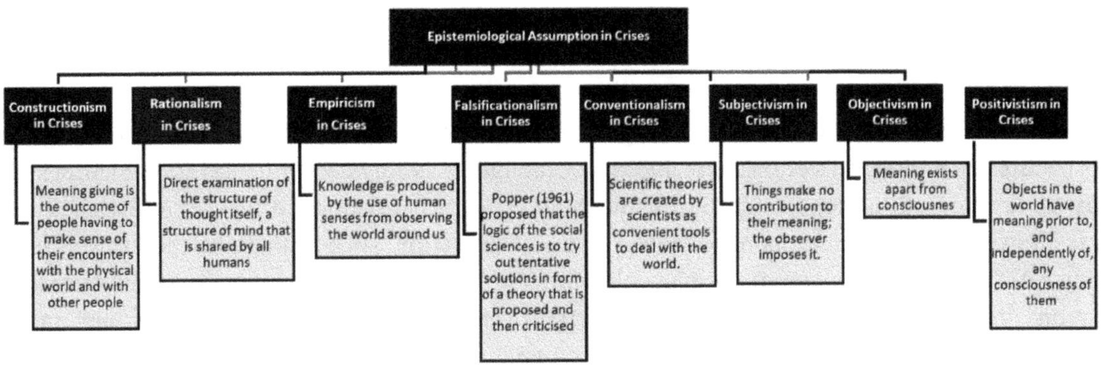

and justification the processual norm are now. this is a clear and fundamental acknowledgement that alongside systematic scientific knowledge of processes in allied healthcare provision, that the experience of patients, too has a value in terms of need and focus of what people have lived through in their healthcare trajectories.

Framing Human Knowledge in Crisis

The transdisciplinary reach of Professional Doctorate study crosses philosophy, social science, medical science, education, and includes and incorporates performativity, memory, language, imagination, emotion, intuition, reasoning, perception and experience. Aristotle, (350 BCE) posited, 'knowledge is an achievement that involves reaching the truth through cognitive ability', (Greco, 2020; Sosa, 2007). Defining knowledge beyond the context of abstract philosophy for applied praxis remains challenging, even. At a philosophical level, the nature of man's search for 'what is knowledge' is the epistemological debate has endured if human curiosity, from which all rationality for research emanates. The complexity of knowledge is one which transcends disciplinary perspectives, and which is contextualised within experiential learning and work-based praxis. Perspectives from the Chicago School of functionalism and instrumentalism on the logic and theory of knowledge was developed by the pragmatist philosopher and psychologist Dewey, whose work was at the time in complete contrast to traditional forms of understanding truth and knowledge. (Hook 2008). Within practice-based settings, the concept of epistemology and epistemic cognition is the basis upon which the disruptive transformative learning can best be implemented (Hayes & Smith, 2020). In challenging assumptions about the situated nature of work-based practice, this begins with the proposition of two questions, firstly, 'What is knowledge?' and secondly, 'What is it possible to know?' in the context of making meaning and sense making in the world. On an individualised basis and in the construction of areas for address via practice-based research, facilitation and discourse enable engagement with higher order thinking. This develops and facilitates skills of reflection and an accompanying capacity for reflexive practice (Frick, 2021). This then leads to recognition of and claim to a fundamentally unique contributions to the existing knowledge base.

Refection and Reflexive Praxis

Understanding reflection (and the multitude of definitions appertaining to it) is a core threshold concept in transformative learning, many researching professionals have minimal experience of sustainable reflective practice, and in my experience do not place value on the process and the valuable impact of reflection. The starting point for this learning with mid-career professional students is them cognising introspection, how they reflect on their innermost thoughts' feelings and memories, sometimes characterised as informal reflection. Once this threshold concept is understood the disruptive facilitation moves along the continuum to more advanced forms of reflection, such as that of Schön's (1983) work linking reflection to the context of them as mid-career professionals in the workplace with 'reflection-in-action' and 'reflection-on-action'. Dewey's (1939) reflective thought, the bridge of meaning that connects one experience to the next, gives direction and impetus to growth, moving the learner from a disturbing state of perplexity (referred to as disequilibrium) to a harmonious state of settledness (equilibrium), this links to Piaget's (1977) theory of cognitive development discussed above. Rogers (2002) provided a contemporary interpretation on Dewey's phases of reflection in relation to the process of 'intellectualisation' or 'locating the problem'. Positing that this stage is of fundamental significance in facilitating the reflector's move from their impressionistic formulation of the problem to the articulation of an idea which can provide a sense of relief. This relief stems from the process of reflection and hence processes of sense making and meaning making beginning. Even so, this has been laid open to critique in relation to the complex ambiguity of terminology surrounding the two constructs, particularly the degree of insecurity that lies within the construction of reflection-in-action (Moon, 1999). More recently geopolitics and the stage of Neoliberalism have largely addressed some of these issues but Schön's work remains a vital threshold concept to enable the professional development of mid-career professionals from a diverse array of healthcare backgrounds (Rasi et al, 2014).

A Transformative Reflection and Reflexive Model proposed by Lawson et al (2014) was derived from Kolb's (1984) experiential learning cycle, itself founded on, (1) concrete experience; (2) reflective observations; (3) abstract conceptualisation; (4) active experimentation, and which forms a conceptual self-reflective process for use in the facilitation of transformative learning in professional practice. The process was designed to focus in on the reflection stage and extend it into a liminal space within which critical experiences can be critically reflected on from alternative perspectives. This provides greater insight and an increased understanding of the underlying assumptions held in relation to how sense and meaning from experience is made.

Mid-career professional students' cognisance of their uncritically acquired tacit knowledge performs an important role in the professional workplace, it is the instantly accessed often relied on repository of knowledge of others that is liable to be biased (discussed above) and self-confirming. Smith (2001) posits that tacit knowledge is created through two opposing processes, the first is routinisation, where explicit procedural knowledge is converted to tacit knowledge through repetition. The second is in Kolb's model of reflection, deriving explicit knowledge through reflection that would have otherwise remained in episodic memory to be used tacitly. Cognition of tacit knowledge is a key threshold concept in working with mid-career professional students engaged in disruptive transformative learning, it is the automatic innate ability for facilitators to draw on critically assimilated experience, Smith (2001), which suggests this is an unconscious skill. The disruptive pedagogical framework focusses on developing and elaborating this unconscious skill that mid-career professional students already possess, this is the same pool of untapped self-knowledge that the transformative coaching model elicits with mid-career professional students. The 'reflective judgement model' posited by King and Kitchener (2004a) is applied within teaching and facilitation to support student's intellectual development through questioning their beliefs regarding the nature and certainty of knowledge, through leveraging their fluid intelligence. King and Kitchener's (2004a, 2004b) work on reflective judgement has extended the body of evidence in cognitive psychology, and the understanding of Mezirow's epistemic assumptions about knowledge and knowing, it is what they describe as '*complex reasoning*', King and Kitchener (2004b). Their reflective judgement model is grounded in the seminal work of John Dewey (1933), and the cognitive-development tradition from Piaget's (1965) schema theory, previously discussed above.

Interdisciplinary Relevance to Transformation in Crisis

The integration of work in the brain, and its capacity for transformative impact, is largely dependent on bringing together the fields of psychology and neuropsychology. Operationally defining this is challenging, with an array of published literature framing this subtly but significantly differently—for example MacLean (1990) described anxious and curious brain states and Taylor and Marienau (2016) reflective and reflexive practice. Sims (2017), Etherington (2004), and Pollner (1991) underpinned the concept of reflective judgement, King and Kitchener (2004) critical self-reflection. Cattell (1963) posited meta-cognitive processes, performance, and self-monitoring whereas Stankov (2000, 2003; Stankov et al, 2006) provided key insights into curriculum justification and design. These can be clearly constructively aligned to the seminal work of Mezirow's on transformation through the 'ten phases of transformative learning' and epistemic questioning (Mezirow, 1990, 2000, Mezirow & Taylor, 2009).

It is during the intellectual processes of thinking about what is known, that critical judgement, Connolly (2000), can be identified as one of the mental processes used to decision make, this may involve thoughts and emotions, (either or both), depending on philosophical beliefs. This links to the work of Hogarth (1987) who offered a conceptual model of judgement occurring within a system composed of three elements: (1) the person; (2) the task and the environment; and (3) the action that results from the judgements and which can subsequently affect both the person and the task environment, and by definition this was normally the starting point for the persons learning journey as they were sitting in front of me having volunteered or been volunteered for an education program with me. The judgement occurs in the so-called task environment, the operations that lead to the judgement are acquisition of information with; the processing of information; and the output which feeds back into the schema and then the action, and outcome. The question of how equilibration, heuristics and bias can have an effect in operation the acquisition, is an important element, and may be the point at which transformation could occur.

In relation to the disciplinary context of teaching and learning, as Mezirow and Taylor (2009) indicate, this transformation can be epochal with a sudden and dramatic reorienting insight or incremental, involving a progressive series of transformations and related points of view that culminate in a transformation in habit of mind (Carvalho & Cardoso, 2020). This provides an explanatory basis to the process of acquisition, processing and outputs of decision making. Overall, this provides a foundational metaphorical platform for the application of Mezirow's theory in the support of critical reflective thinking and consequently change.

Mezirow challenged traditional established psychologists who regarded adult learning through the lens of predominantly behavioural or psychoanalytical perspectives and the whole notion of experientialism. Modern criticism of Mezirow posits that that his theory ought to incorporate far more social context and be fundamentally aligned to the needs of the individual. Mezirow may have tacitly taken for granted that this was at the core of his purpose by referring to himself as a social action educator. This is also supported by his early life experiences as an adult educator, where his early fostering of democratic social action through community development and adult literacy programs in the USA and abroad in developing countries, became a core characteristic of his contribution to professional practice. Merriam and Brockett (2011) debated whether one must already be at a mature level of cognitive functioning to engage in any transformational learning process, which had not been questioned previously. As they posited, for transformational learning to occur, one must be able to critically reflect and engage in rational discourse; and both activities are characteristic of higher levels of cognitive functioning.

MEASURING IMPACT IN TRANSLATIONAL RESEARCH

The Doctoral Impact Translational Research Assessor Framework has been developed to support a tripartite evaluation of research impact with institutional adopters of research, academic institutions, and researchers. The framework focuses around eight key areas of research activity undertaken within the context of Professional Doctoral study, incorporating, and embedding core principles of equality, diversity and inclusion for all representative mid-career professionals and their professional members.

The eight areas of designated research activity focused on within the framework are:

1. Interdisciplinary leadership, governance, and accountability.
2. Membership and professional registration of the UK Health and Care Professions Council and their global equivalents.
3. Capacity for wider dissemination, capacity building and uptake of knowledge creation in multidisciplinary and interprofessional knowledge sharing.
4. Impact on processes, mechanisms and methodologies, initial education, training and teaching of mid-career professionals in professional and clinical practice.
5. External, peer reviewed recognition for contribution to professional or clinical practice attributable to the research contribution in work-based praxis.
6. Potential for accreditation of education and training, either as Continuing Professional Development or as an addition or enhancement to student learning at recognised AHP provider institutions.
7. Evidence of influence on AHP outreach and formal engagement initiatives.
8. Tangible Contribution to opportunities for employment, the career development, and trajectories of existing mid-career professionals and positive influence on the attraction, recruitment, and retention capability of future mid- career professionals.

The five accompanying levels of overarching good practice are:

- Level 1: Transfer and Adoption of Principles of Work Based Practice.
- Level 2: Development and Integration of Initiative into Work Based Practice.
- Level 3: Societal Engagement
- Level 4: Transformation Potential
- Level 5: Sustainability Potential

To complete the impact evaluation framework, those organisations where the research has been implemented in practice are asked to self-assess their ongoing progress across each of the eight designated categories on a straightforward Excel database. The range of self-assessment scores are an opportunity to grade the level of impact within the organisation from 0 – 10 where 0 is least impactful and 10 is of maximal impact. This makes it possible for all the outcomes from participating organisations and the mid-career professions represented to be combined in establishing shared perspectives or differences in experience of the implementation in practice (Laver-Fawcett & Cox,2021).

Appleby and Pilkington's (2014) conceptual framework is utilised with mid-career professional students to examine their own critical professionalism within their work environment, communities of professional practice, and identity. The disruptive pedagogical facilitation supports mid-career professional students to understand how their meaning perspectives, frames of reference and habits of mind are constructed and de-constructed, during the critical-reflection process.

Critical Professional Conceptual Framework

Appleby and Pilkington identified reflection as being one of the fundamental components of professional learning, drawing on the work of Dewey's (1938) sense making, and Habermas' (1974), insight into knowledge generation through reflection. They explicate the work of Brookfield (1985) who posited that professionals need to move 'beyond self-referential approaches' to reflection and widen their view to include the perspectives of others and the external environment. They make an important distinction in the role of a dialogic approach to reflection in communities a professional practice, where there is a role for 'the peer', 'the facilitator', 'the mentor', and 'the critical friend'. This aligns with my praxis and Mezirow's discourse, which links to the development of dialectical and critically reflective thinking. The facilitator and critical friend enable a critical approach to the co-construction of knowledge and understanding with the mid-career professional students. In praxis, mid-career professional students are challenged to recognise and examine the precursors of critical professionalism and demonstrate the required thinking and knowledge to demonstrate their own praxis. This is followed by them determining how they identify and then share their assumptions about their own critically professional behaviours. In the disruptive pedagogical facilitation, this framework aligns the theoretical concepts from the first four sections of this chapter an integrates them visually and verbally, illuminating to the mid-career professional students the challenges of perspective transformation when viewed through the lens of their professional behaviours. Critical self-awareness of a mid-career professional student's own cognitive performance and emotional intelligence can impact on their ability to engage in the affective aspect of perspective transformation and is a key threshold concept for mid-career professional students to engage with (Tyndall et al, 2021).

Practice-Based Research Action Research

One of the most popular methodological approaches rationalized by Professional Doctorate students is Action Research. The whole ethos of the approach is to make a functional and pragmatic difference to the challenges of work-based praxis, which can consequently be studied through the lens of experiential learning and the resultant outcomes of experience in practice. It has been posited as a liminal boundary between theory and practice in terms of its academic rigour.

Insider Research and Epistemic Positionality

Action Research, for most work-based practitioners affords them the chance to build and iteratively develop their practice, whilst still working at the front line of patient centred care in the mid-career professionals. This may pertain to the progressive development of their role or form a fundamental part of their progressive development. Action Research has been implemented in the context of social justice as a mechanism of addressing systemic inequalities in organisations and the agency they afford their employees, whether deliberately or subconsciously. The duality of Action Research as being a valuable source of organisational improvement and a mechanism of ensuring that the concept of epistemic positionality is not of detriment to the research process but acknowledged as an embedded part of it. The overall sense of operational analysis and address that action research facilitates is one of the factors which provides a means of acknowledging and addressing power imbalances both on a personal and a professional level.

Complexity Theory in Translational Action Research

Pluralist stances towards situated methodological development are actively encouraged. Organisations currently are encouraged to wholly reject the long held prevailing paradigm of mechanistic efficiency driven hierarchical command and control organisations. It also ought to be noted that recent philosophy defines the natural world as being definitively different to the social world. Consequently, complexity theories were designed to fit the natural world rather than the social world although language and discourse have triggered)0relevant insights into the behaviour of the social world (Brister, Frodeman, & Briggle, 2020). We are far less complex than the complexity of the universe and therefore can only ever view or experience reality through categorical man-made frameworks that allow us to make sense and meaning of the world humanity occupies and experiences. Complexity theories do see boundaries that divide the hard from the soft and the natural from the social in terms of what they really are – artificial and arbitrary.

Complexity based thinking is not a one size fix all and the importance of question led research rather than methods driven approaches remain paramount if research is to remain authentic and trustworthy in the context of the workplace – nor does it denigrate the perspectives of empirical methodologies, which ought to be characterised by validity and reliability.

What the professional doctorate offers allied health professionals is the opportunity to undertake and develop context-specific challenges to core questions of issue from their practice and the overlapping disciplinary contexts of their colleagues from interprofessional working and multidisciplinary teamwork.

Complexity theory is one lens through which the emergence of the knowledge economy within the context of mid-career professionals can be regarded. Both intrinsic and extrinsic factors impinging on the effectiveness of organisational hierarchies to share common goals within patient centred care can be examined in terms of their parameters and scope of individual professional and clinical practice. By definition the organisations within AHP practice is undertaken are subject to the need for complicated adaptation in unpredictable and ambiguous contexts. The interrelationships of professional working within healthcare settings necessitate that each individual clinical discipline and hence collective rather than individual professional identity of each worker, is such that it can be largely self-organised and responsive to need as and when required in person centred care (Mavri, Ioannou & Loizides, 2021).

Mid-career professional doctorate research is often characterised by a strong emphasis on participatory access and engagement with multiple stakeholders, with whom the candidate often works in the context of their mid-career professional practice (Armsby, Costley, & Weller, 2021). The broad capacity that this context provides for progressive evaluation and feedback, makes it an ideal means of achieving strategic and operational practice. Workplace environments are often influenced by local, regional, and national politics at macro, meso and micro levels where polemic and cultural impacts are most tangible in practice. Tailoring evidence-based practice alongside the demographic profiling of the area is central to participatory implementation across local, regional and national workforce providers (Kelly et al, 2021). This facilitates organisational capability and capacity building as a central by-product and the development of a contextualised, authentic, and trustworthy evidence base.

Within the context of practice based doctoral research, there is a need to consider models of research utilisation and their overlapping relationship with complex ambiguity in professional practice. Translational research is an operational necessity to achieve a substantive level of impact with research interventions or investigations which have occurred, often at a local level. There are fundamental challenges and barriers to the effective identification and introduction of evaluative frameworks which can evaluate the effectiveness or impact of translational research in practice-based settings.

An integrative framework capturing the effective implementation and translation of research into practice for allied healthcare professional doctorate candidates can be achieved by the adaptation of the Ottowa Model of Research Utilisation (OMRU). The model affords context specific adaptation and recognition of the situated nature of practice-based research. This specifically adapted doctoral knowledge translation model also ensures that processes of knowledge translation can be actively guided so that evidence from the extant literature can become an embedded part of the process alongside the acknowledgement, consolidation or refuting of the potential for implementation of research findings evaluated to date. The following suggested framework makes two additions to the six current recommended outcomes of the formal OMRU framework. These have been added to accommodate the individualised situational specificity and signature disciplines or pedagogies of professional identities represented in the context of workplace professional practice (Shulman, 2005).

Some of these barriers may be common across all contexts of AHP doctoral study. Unlike empirically based research, work-based research is situated in the complex ambiguity of non- controlled environments where the unpredictability of unfolding events in cultures, contexts and is subject to political turbulence and change at macro, meso and micro levels. This can correspondingly impact on the processes of research, design, iterative developmental progression, implementation and consequently the outcomes of research and its impact in practice. Whereas practice-based research is stringently methodologically designed, it cannot be framed on the premises of linear causation, eliminate epistemic bias or function purely in a reductionist sense. Attributing causality is therefore not something that falls within the remit of professional doctorate research in the context of AHP education since the complexity and multifaceted basis of embedded research issues are such that they cannot be split into causal variables, since they are so interdependent and meaning making would be lost in an attempt to extract and singularly study or account for them.

As the systematic framework (Table 1: The Crisis Management Complexity Framework - Transformative Dimensions & Characteristics) emphasises translation of knowledge to academic and clinical communities within the context of client centred care, it has been straightforwardly titled 'The Crisis Management Framework', since implementation is largely dependent on the collation of complex ambiguity, multi levels of causation, and characterised by spontaneous change. The framework emphasises an incremental, capacity building approach which integrates numerous qualitative and mixed methods approaches in the context of mid-career professional research (Berreta, 2021).

Table 2. The crisis management complexity framework - transformative dimensions & characteristics

TRANSFORMATIVE DIMENSIONS	CHARACTERISTICS
Adopter, Public and Private Stakeholder Dimensions	These complementary dimensions incorporate those organisations and organisational staff, i.e., the human capital, who will adopt and implement research-based practice at the front line of societal and civic contribution.
Adoption Related Dimensions	These dimensions frame the potential use and transferability of research-based practice and practice-based research 'in situ' and the modifications that may need to be made as a consequence, which in turn may impact on both the intended and unintended integration of research.
Contextual, Ergonomic, Ecological and Environmental Dimensions	These dimensions actively incorporate core facets of societal service provision, including the impact of structure and agency within organisational environments where transformative knowledge can be translated into practice.
Implementation Dimensions	Linking directly to the process of implementation, these dimensions account for mechanisms of operational knowledge transfer, which can ensure transformative knowledge from research-based practice can be effectively implemented beyond the point of theorising.
Situationally Specific Dimensions	These extends beyond context and environmental dimensions to the individual and context based individual experiences of the workers involved. These are impacted upon by the level of assumption and epistemic bias with which people operate and frame transformational knowledge in practice.
Signature Disciplinary Identity Dimensions	Linking to Shulman's (2005) seminal recognition of signature pedagogies and disciplinary boundaries, this dimension considers the active delineation of collective and individual professional identities.
Research Implementation or Intervention Dimensions	This set of dimensions links directly to the nature and design of the research intervention and implementation process. Functional by design, this dimension incorporates quality management, functional operationalisation of the research project and the mechanisms and impact of engagement this has with the rest of the organisational dimensions (Reynolds, 2021).
Outcome Related Dimensions	In terms of an outcomes-based approach to the integration and translation of transformation in practice, outcome related dimensions are aligned with those resulting directly from implementation of the research project intervention or implementation process (Sin, Soares & Tavares, 2020).

CONCLUSION

In considering the challenges faced by HEIs across the globe in providing doctoral research pathways that are both fit for purpose, and which address the longstanding capacity for the knowledge they generate to be applied at the forefront of crisis, lack of capacity to evaluate impact has been a distinct issue. This chapter has considered the various factors impacting on these processes and made some preliminary suggestions as to how both transformative learning and impact evaluation might be managed and implemented in practice. Interdisciplinary learning opportunities extend both the pragmatic and academic reach of mid-career professionals within clinical and professional praxis, building further their capacity and capability and their perception as credible and dependable researchers, 'in situ'. The chapter has emphasised the challenges of working in cultures and contexts with longstanding issues stemming largely from misplaced assumption and pre-supposition, which when aligned with the published extant evidence based shows the liminal tension between theory, praxis and illuminates the potential to address anecdotal experience and to provide a degree of credibility for a systematic and analytical insight. Optimal crisis responses and their consequent outcomes are undoubtedly dependent upon it.

REFERENCES

Aarnikoivu, M. (2021). The spatiotemporal dimension of doctoral education: A way forward. *Studies in Higher Education, 46*(11), 2431–2443. doi:10.1080/03075079.2020.1723530

Aguilar, R. P., & Guénette, M. F. (Eds.). (2021). *Situatedness and Performativity: Translation and Interpreting Practice Revisited*. Leuven University Press. doi:10.11116/9789461663863

Appleby, Y., & Pilkington, R. (2014). *Developing critical professional practice in education*. Niace.

Armsby, P., Costley, C., & Weller, G. (2021). The role of practice in doctoral degrees. *Research in Post-Compulsory Education, 26*(3), 257–273. doi:10.1080/13596748.2021.1920258

Becton, Y. J., Bogiages, C., D'Amico, L., Lilly, T., Currin, E., Jeffries, R., & Tamim, S. (2020). An emerging framework for the EdD activist. *Impacting Education: Journal on Transforming Professional Practice, 5*(2), 43–54. doi:10.5195/ie.2020.131

Bielak, A. T., Campbell, A., Pope, S., Schaefer, K., & Shaxson, L. (2008). From science communication to knowledge brokering: the shift from 'science push'to 'policy pull. In *Communicating science in social contexts* (pp. 201–226). Springer. doi:10.1007/978-1-4020-8598-7_12

Bogoviz, A. V. (2019). Industry 4.0 as a new vector of growth and development of knowledge economy. In *Industry 4.0: Industrial Revolution of the 21st Century* (pp. 85–91). Springer. doi:10.1007/978-3-319-94310-7_8

Bolisani, E., & Bratianu, C. (2018). The emergence of knowledge management. In *Emergent knowledge strategies* (pp. 23–47). Springer. doi:10.1007/978-3-319-60657-6_2

Boud, D., Costley, C., Marshall, S., & Sutton, B. (2021). Impacts of a professional practice doctorate: A collaborative enquiry. *Higher Education Research & Development, 40*(3), 431–445. doi:10.1080/07294360.2020.1765744

Brister, E., Frodeman, R., & Briggle, A. (2020). The Future of Field Philosophy: Lessons Learned and Next Steps. In *A Guide to Field Philosophy* (pp. 363–368). Routledge. doi:10.4324/9781351169080-24

Brookfield, S. (1985). 'Sources in self-directed learning theory and practice'. Wiley Online Library. *New Directions for Adult and Continuing Education, 1985*(25), 87–90. doi:10.1002/ace.36719852510

Cardoso, S., Tavares, O., Sin, C., & Carvalho, T. (Eds.). (2020). *Structural and Institutional Transformations in Doctoral Education: Social, Political and Student Expectations*. Springer Nature. doi:10.1007/978-3-030-38046-5

Carvalho, T., & Cardoso, S. (2020). Conclusion: The transformations in doctoral education—A comprehensive and critical approach. *Structural and institutional transformations in doctoral education: Social, political and student expectations. Issues in Higher Education*, 375-389.

Cattell, R. B. (1963). Theory of fluid and crystallized intelligence: A critical experiment. *Journal of Educational Psychology, 54*(1), 1.

Chapman, S., Colvin, L. E., & Cosentino, S. (2020). Translational aspects of the multidisciplinary study of metacognition. *Translational Issues in Psychological Science*, *6*(1), 26–31. doi:10.1037/tps0000224 PMID:33728364

Coates, H., Croucher, G., Moore, K., Weerakkody, U., Dollinger, M., Grosemans, I., Bexley, E., & Kelly, P. (2020). Contemporary perspectives on the Australian doctorate: Framing insights to guide development. *Higher Education Research & Development*, *39*(6), 1122–1139. doi:10.1080/07294360.2019.1706451

Connolly, T. (2000). *General Introduction*. Judgement and Decision Making.

de Araujo, L. M., Priadana, S., Paramarta, V., & Sunarsi, D. (2021). Digital leadership in business organizations. *International Journal of Educational Administration, Management, and Leadership*, 45-56.

Dewey, J. (1938). *1997. Experience and education*. Touchstone.

Dewey, J. (2010). *How we think*. Book Tree.

Ding, W., Aoyama, A., & Choi, E. (2020). An approach to the diffusion of tacit knowledge: Learning from imitation to creation. *International Journal of Knowledge Management Studies*, *11*(2), 103–116. doi:10.1504/IJKMS.2020.106276

Dirks, K. T., & de Jong, B. (2021). Trust within the workplace: A review of two waves of research and a glimpse of the third. *Annual Review of Organizational Psychology and Organizational Behavior*, 9.

Etherington, K. (2004). *Becoming a reflexive researcher: Using ourselves in research*. Jessica Kingsley Publishers.

Festinger, L. (1962). Cognitive dissonance. *Scientific American*, *207*(4), 93–106. doi:10.1038/scientificamerican1062-93 PMID:13892642

Frick, L. (2021). Doctoral Creativity as an Epistemological Force in Saving and/or Destroying the World. *The PhD at the End of the World: Provocations for the Doctorate and a Future Contested*, 143-153.

Gibbons, M., & Nowotny, H. (2001). The potential of transdisciplinarity. In *Transdisciplinarity: joint problem solving among science, technology, and society* (pp. 67–80). Birkhäuser. doi:10.1007/978-3-0348-8419-8_7

Goleman, D. (1996). Emotional intelligence. Why it can matter more than IQ. *Learning*, *24*(6), 49–50.

Graff, H., & Von Wehrden, H. (2021). Discourses of Boundary Crossing. *Beyond Interdisciplinarity: Boundary Work, Communication, and Collaboration*, 36.

Greco, J. (2020). *Knowledge-Producing Abilities*. Cambridge University Press. doi:10.1017/9781108666404.006

Habermas, J. (1974). On social identity. *Telos*, *1974*(19), 91–103. doi:10.3817/0374019091

Hancock, S. (2019). A future in the knowledge economy? Analysing the career strategies of doctoral scientists through the principles of game theory. *Higher Education*, *78*(1), 33–49. doi:10.1007/s10734-018-0329-z

Hayes, C., & Smith, P. (2020). Defining and framing epistemology. In *Beyond Disciplinarity* (pp. 6–22). Routledge.

Herbst, P., Crespo, S., Matthews, P. G., & Lichtenstein, E. K. (2021). Dissertating through disruptions: COVID-19 and the need for a research infrastructure. *Journal for Research in Mathematics Education*, *52*(2), 110–116. doi:10.5951/jresematheduc-2020-0300

Hogarth, R. M. (1987). *Judgement and choice: the psychology of decision* (2nd ed.). Wiley.

Hook, S. (2008). *John Dewey: an intellectual portrait*. Cosimo, Inc.

Israelit, M. (1996). Nathan Rosen: 1909–1995. *Foundations of Physics Letters*, *9*(2), 105–108. doi:10.1007/BF02186252

Kahneman, D. (2011). *Thinking, fast and slow*. Macmillan.

Kahneman, D. (2013). Thinking, fast and slow. Farrar, Straus and Giroux.

Kahneman, D., Slovic, P., & Tversky, A. (1982). *Judgment under uncertainty: heuristics and biases*. Cambridge University Press. doi:10.1017/CBO9780511809477

Karas, M., Sheen, N. J., North, R. V., Ryan, B., & Bullock, A. (2020). Continuing professional development requirements for UK health professionals: A scoping review. *BMJ Open*, *10*(3), e032781. doi:10.1136/bmjopen-2019-032781 PMID:32161156

Kelly, S., Nicholson, J., Johnston, P., Duty, D., & Brennan, R. (2021). Experienced professionals and doctoral study: A performative agenda. *Industrial Marketing Management*, *92*, 232–243. doi:10.1016/j.indmarman.2020.02.018

King, P. M. & Kitchener, K. S. (2004a). *Judgment model: Twenty years of research on epistemic cognition*. Erlbaum.

King, P. M. & Kitchener, K. S. (2004b). *Reflective judgment: Theory and research on the development of epistemic assumptions through adulthood*. Taylor & Francis.

Kolb, D. A. (1984). *Experiential Learning*. Prentice-Hall.

Laver-Fawcett, A., & Cox, D. L. (2021). *Principles of Assessment and Outcome Measurement for Allied Health Professionals: Practice, Research and Development*. John Wiley & Sons.

Lundgren-Resenterra, M., & Kahn, P. E. (2019). The organisational impact of undertaking a professional doctorate: Forming critical leaders. *British Educational Research Journal*, *45*(2), 407–424. doi:10.1002/berj.3503

Macfarlane, B. (2021). The neoliberal academic: Illustrating shifting academic norms in an age of hyper-performativity. *Educational Philosophy and Theory*, *53*(5), 459–468. doi:10.1080/00131857.2019.1684262

Mather, M., Shafir, E., & Johnson, M. K. (2003). Remembering chosen and assigned options. *Memory & Cognition*, *31*(3), 422–433. doi:10.3758/BF03194400 PMID:12795484

Mavri, A., Ioannou, A., & Loizides, F. (2021). Cross-organisational Communities of Practice: Enhancing creativity and epistemic cognition in higher education. *The Internet and Higher Education, 49*, 100792. doi:10.1016/j.iheduc.2021.100792

Meissner, P., Cottler, L. B., & Michener, J. L. (2020). Engagement science: The core of dissemination, implementation, and translational research science. *Journal of Clinical and Translational Science, 4*(3), 216–218. doi:10.1017/cts.2020.8 PMID:32695491

Merriam, S. B., & Brockett, R. G. (2011). *The profession and practice of adult education: An introduction.* John Wiley & Sons.

Mezirow, J. (1990). *Fostering critical reflection in adulthood.* Jossey-Bass Publishers San Francisco.

Mezirow, J. (2000). *Learning as transformation: Critical perspectives on a theory in progress.* Jossey-Bass.

Mezirow, J., & Taylor, E. W. (2009). *Transformative learning in practice: Insights from community, workplace, and higher education.* John Wiley & Sons.

Mezirow, J., & Taylor, E. W. (Eds.). (2009). *Transformative learning in practice: Insights from community, workplace, and higher education.* John Wiley & Sons.

Moon, J. (1999). *'Describing higher education: some conflicts and conclusions'. Psychology Press.* Benchmarking and Threshold Standards in Higher Education.

Nancarrow, S., & Borthwick, A. (2021). *The allied health professions: a sociological perspective.* Policy Press.

Østern, T. P., Jusslin, S., Nødtvedt Knudsen, K., Maapalo, P., & Bjørkøy, I. (2021). A performative paradigm for post-qualitative inquiry. *Qualitative Research.*

Peters, M. A. (2020a). *Digital socialism or knowledge capitalism?* Taylor and Francis. doi:10.1080/00131857.2019.1593033

Peters, M. A. (2020b). Towards a theory of knowledge socialism: Cognitive capitalism and the fourth knowledge revolution. In *Knowledge Socialism* (pp. 15–31). Springer. doi:10.1007/978-981-13-8126-3_2

Piaget, J. (1977). The role of action in the development of thinking. In *Knowledge and development* (pp. 17–42). Springer. doi:10.1007/978-1-4684-2547-5_2

Pollner, M. (1991). Left of ethnomethodology: The rise and decline of radical reflexivity. *American Sociological Review, 56*(3), 370–380. doi:10.2307/2096110

Pols, J. (2014). Knowing patients: Turning patient knowledge into science. *Science, Technology & Human Values, 39*(1), 73–97. doi:10.1177/0162243913504306

Rasi, L., Dopita, M., & Cervinkova, H. (2021). Neoliberalism's Paradoxical Effect and European Doctoral Education Reforms in Post-socialist Europe. *Critical Education, 12*(9).

Reed, K., Reed, B., Bailey, J., Beattie, K., Lynch, E., Thompson, J., Vines, R., Wong, K. C., McCrossin, T., & Wilson, R. (2021). Interprofessional education in the rural environment to enhance multidisciplinary care in future practice: Breaking down silos in tertiary health education. *The Australian Journal of Rural Health, 29*(2), 127–136. doi:10.1111/ajr.12733 PMID:33982852

Reynolds, S. S., Howard, V., Uzarski, D., Granger, B. B., Fuchs, M. A., Mason, L., & Broome, M. E. (2021). An innovative DNP post-doctorate program to improve quality improvement and implementation science skills. *Journal of Professional Nursing, 37*(1), 48–52. doi:10.1016/j.profnurs.2020.12.005 PMID:33674108

Rigolot, C. (2020). Quantum theory as a source of insights to close the gap between Mode 1 and Mode 2 transdisciplinarity: Potentialities, pitfalls and a possible way forward. *Sustainability Science, 15*(2), 663–669. doi:10.1007/s11625-019-00730-8

Schön, D. A. (1983). *The Reflective Practitioner: How Professionals Think in Action*. Ashgate.

Shulman, L. S. (2005). Signature pedagogies in the professions. *Daedalus, 134*(3), 52–59. doi:10.1162/0011526054622015

Sin, C., Soares, D., & Tavares, O. (2020). Coursework in industrial doctorates: A worthwhile contribution to students' training? *Higher Education Research & Development*, 1–15.

Singe, S. M., Sheldon, L., Rynkiewicz, K., Manning, C., Filep, E., Zuk, E., & Hargrave, C. (2021). Mentorship Experiences of Doctoral Students: Understanding Desired Attributes of Doctoral Student Mentors. *The Internet Journal of Allied Health Sciences and Practice, 19*(1), 13. doi:10.46743/1540-580X/2021.1967

Smith, E. A. (2001). *The role of tacit and explicit knowledge in the workplace*. MCB UP Ltd.

Sosa, E. (2007). *A virtue epistemology: Apt belief and reflective knowledge* (Vol. I). OUP Oxford. doi:10.1093/acprof:oso/9780199297023.001.0001

Stankov, L. (2000). *Complexity, metacognition, and fluid intelligence*. Elsevier.

Stankov, L. (2003). *Complexity in human intelligence*. American Psychological Association.

Stankov, L., Danthiir, V., Williams, L. M., Pallier, G., Roberts, R. D., & Gordon, E. (2006). Intelligence and the tuning-in of brain networks. *Learning and Individual Differences, 16*(3), 217–233. doi:10.1016/j.lindif.2004.12.003

Thomas, A., Menon, A., Boruff, J., Rodriguez, A. M., & Ahmed, S. (2014). Applications of social constructivist learning theories in knowledge translation for healthcare professionals: A scoping review. *Implementation Science : IS, 9*(1), 1–20. doi:10.1186/1748-5908-9-54 PMID:24885925

Tversky, A. & Kahneman, D. (1974). Judgment under uncertainty: Heuristics and biases. *Science, 185*(4157), 1124–1131.

Tyndall, D. E., Flinchbaugh, K. B., Caswell, N. I., & Scott, E. S. (2021). Troublesome knowledge for entry-level PhD nursing students: Threshold concepts essential for the research-focused doctorate. *Journal of Professional Nursing, 37*(3), 572–577. doi:10.1016/j.profnurs.2021.03.006 PMID:34016316

ADDITIONAL READING

Annarelli, A., & Nonino, F. (2016). Strategic and operational management of organizational resilience: Current state of research and future directions. *Omega*, *62*, 1–18. doi:10.1016/j.omega.2015.08.004

Creaton, J. (2021). The impact of professional doctorates in the workplace: Evidence from the criminal justice sector. *Research in Post-Compulsory Education*, *26*(3), 274–289. doi:10.1080/13596748.2021.1920259

Cruickshank, N. (2020). He who defends everything, defends nothing: Proactivity in organizational resilience. *Transnational Corporations Review*, *12*(2), 94–105. doi:10.1080/19186444.2020.1764326

Herrera, A. E. B. (2022). Pandemic Lessons. *Mexican Journal of Medical Research ICSA*, *10*(19), 1–2. doi:10.29057/mjmr.v10i19.8369

Kantur, D., & İşeri-Say, A. (2012). Organizational resilience: A conceptual integrative framework. *Journal of Management & Organization*, *18*(6), 762–773. doi:10.5172/jmo.2012.18.6.762

Kusa, R., Duda, J., & Suder, M. (2022). How to sustain company growth in times of crisis: The mitigating role of entrepreneurial management. *Journal of Business Research*, *142*, 377–386. doi:10.1016/j.jbusres.2021.12.081

Lalonde, C., & Roux-Dufort, C. (2013). Challenges in teaching crisis management: Connecting theories, skills, and reflexivity. *Journal of Management Education*, *37*(1), 21–50. doi:10.1177/1052562912456144

Miceli, A., Hagen, B., Riccardi, M. P., Sotti, F., & Settembre-Blundo, D. (2021). Thriving, not just surviving in changing times: How sustainability, agility and digitalization intertwine with organizational resilience. *Sustainability (Basel)*, *13*(4), 2052. doi:10.3390/su13042052

Sedlack, R. (2021). Crisis Management in a Pandemic: Voices from the Field. *Issues of Educational Leadership: Crisis Management During Challenging Times*, 21.

KEY TERMS AND DEFINITIONS

Action Research: Also, often and interchangeably, termed Participatory Action Research (PAR), co-operative enquiry and action learning is a research approach focused on the systematic improvement and positive change of the structure and agency afforded to people within context specific settings.

Disruptive Innovation: Refers to the innovation that transforms previously inaccessible products and ensures their availability to wider more generalised populations.

Impact: Something that has a marked effect or influence.

Knowledge Transfer: Is a diverse range of activities used in the support of mutually beneficial collaborations within and between universities, businesses and the public sector for the civic benefit of society.

Performativity: A philosophical means of describing the power of language to effect change in the world.

Transformative Learning: A process of individually or collectively changing perspectives, which has three distinguishable dimensions of psychological response, convictional attitude, and behavioural change.

Chapter 13
Meeting the Future:
Identifying and Preparing the Next Generation of Leaders for Crisis Management

Sherihan Radi
https://orcid.org/0000-0002-0749-3800
Brandenburg University of Technology Cottbus–Senftenberg, Germany

ABSTRACT

The purpose of this research is to investigate the effects of crises on higher education systems while examining the moderating effect of crisis management and its impacts on crises and higher education systems. In order to achieve the research objectives, the researcher collected data through an online survey. The study revealed that crises have a negative impact on higher education systems, while crisis management has a negative effect on crises and a positive impact on higher education systems. The study also found out that crisis management moderates the relationship between crises and higher education systems. Based on the results, the researcher presented recommendations for identifying and proactively preparing the next generation of leaders in education institutions to be able to enhance resilience and deal and react efficiently with any crisis that may occur.

INTRODUCTION

Colleges and universities have been challenged with several crises over the past few years. One of these challenges is the coronavirus disease 2019 (COVID-19) pandemic which crossed national borders and has affected people without regard to their nationality, income or level of education creating several consequences that did not leave education as an exception.

According to Schleicher (2020), the COVID-19 pandemic had severe impacts when the lockdowns interrupted schooling in schools, colleges and universities all over the world. It has exposed several inequities and inadequacies in education systems when it switched to online learning by using access to broadbands and computers. Students had to rely more on their own resources, in order to be able to

DOI: 10.4018/979-8-3693-1926-0.ch013

learn remotely. Teachers and professors as well had to adapt to new modes of teaching and pedagogical concepts for which they have not been trained before making them face the risk of falling behind. This crisis also raised questions about the value offered by higher education in colleges and universities such as the education content and networking. Furthermore, crises are not only limited to pandemics or health emergencies. They also include natural disasters, financial setbacks, economic as well as political and environmental disorders.

Consequently, disruption caused by crises threaten the purpose and mission of colleges and universities. During these times leaders in the higher education sector are pressured to deal with the characteristics of urgency and complexity of crises. Therefore, the importance and relevance of crisis management arises to serve the purpose of higher education and to deal with the wide range of crises and the diversity of stakeholders. Moreover, the topic of crisis management became more essential and relevant for practitioners and scholars too as mentioned by Gigliotti (2022).

Problem Statement

In accordance with Leishangthem (2017), higher education is considered one of the key indicators of development in any country.

The foundation of colleges and universities is based on a set of core values and principles coming naturally to their members during periods of stability providing guidance as by Gigliotti (2022).

Based on HKUST (n.d.), these core values offer principles that guide colleges and universities towards achieving the advancement of their visions and missions.

On the other hand, during times of crisis, members of colleges and universities are obliged to respond to the emerging urgent demands. As a result, these principles and values decrease and diminish as described by Gigliotti (2022).

This study discusses various types of crises that may arise and influence higher education systems, while exploring the importance of crisis management. It also highlights the impact of crisis management on both crises and higher education systems. It also identifies the moderating impact of crisis management on the relationship between crises and higher education systems.

Aim of the Study

The study investigates the impact of crises on higher education systems and explores the importance, value and effect of crisis management on higher education systems and on crises. It also seeks to examine the moderating role and influence of crisis management. Furthermore, it presents recommendations on the effective implementation of crisis management in higher education systems where the next generation of leaders are identified and prepared for effectively and efficiently dealing with crises.

Research Objectives

The research objectives are to examine the impact of crises on higher education systems, in addition to seeking to accomplish the following objectives:

- To identify the influence of crisis management on higher education systems.
- To recognize the impact of crisis management on crises.

- To examine the moderating effect of crisis management on the relationship between crises and higher education systems.
- To explore the importance and value of crisis management in higher education systems.
- To provide recommendations for the effective implementation of crisis management to enable leaders to deal with crises on efficient and effective levels.

Research Questions

The study tends to answer the following questions:

- What is the impact of crises on higher education systems?
- What is the influence of crisis management on higher education systems?
- What is the effect of crisis management on crises?
- What is the moderating effect of crisis management on the relationship between crises and higher education systems?
- What is the importance and value of crisis management in higher education systems?
- How to prepare the next generation of leaders to effectively and efficiently deal with crises?

Research Hypotheses

H1: Crises have a significant negative impact on higher education systems.
H2a: Crisis management has a significant positive impact on higher education systems.
H2b: Crisis management has a significant negative impact on crises.
H3: Crisis management has a moderating effect on the relationship between crises and higher education systems.

SIGNIFICANCE OF THE STUDY

Theoretical Importance

To enhance contribution to the field of dealing with crises, changes and transformations faced by higher education systems. Additionally, the importance of this study lies in filling the gap by examining the topic of crisis management in relation to its value in identifying and preparing a new generation of leaders who are concerned with higher education systems. Moreover, this study will provide support to scholars and academicians contributing additional insight and knowledge on the studied subject.

Practical Importance

The way numerous crises affecting higher education are of complex nature and therefore, colleges and universities are required to have the ability to deal with any emerging crisis. They are also required to adapt to any changes resulting from the emergence of crises, in order to be able to further continue their mission and not to be influenced by any damage or negativities occurring as consequences of different crises.

These colleges and universities have to be always updated with regard to any crises or changes that occur worldwide. Therefore, they are in great need of having a crisis management that assists them in successfully achieving their mission and goals.

Thus, implementing a crisis management which enables leaders who are working in the field of higher education to successfully deal with any unexpected events and crises, have proper planning and resolve problems is crucial. Furthermore, this study presents recommendations which may be useful for colleges, universities and individuals dealing with higher education systems. It will ensure that crisis management practices are properly implemented and well understood to successfully accomplish missions and goals as well as to achieve effectiveness. This will also assist in forming a new generation consisting of successful leaders who are able to get through crises and efficiently deal with any contradictions and challenges.

Literature Review

This section discusses the literature related to the concept and types of crises, concept of crisis management and higher education. It also describes the importance, value, stages and principles of crisis management.

Concept of Crisis

There are several definitions for the term crisis. As discussed by Johansson & Pihl (2023), the interconnected society in recent times resulted into the magnitude effect of crises globally and the definition of crisis became wider. Crisis has been defined as a threat which has negative impacts on operations if it has not been handled on a proper way.

Bouncken et al. (2022) referred to crises as intense changes which have not been foreseen, are not short incidents and remain over a period of time. They become more serious over time leading to the creation of fear in organizations and individuals. Hence, solidarity is reduced and extra challenges are created. The success of actions taken to adapt with the crises can also be limited.

As expressed by Sayed et al. (2021), the various definitions of a crisis include a serious threat to the norms and values of a social system creating the necessity of making critical decisions, due to the existence of uncertain circumstances and time pressure. A crisis can be characterized as a life-threatening uncertain situation which requires an instant response. It can also threaten high-priority objectives of decision-makers, as it occurs surprisingly and limits the time for response.

Pontes (2020) stated that a crisis is a wide word referring to a series of various situations that prevent the normal process evolution leading to significant damages.

Zamoum & Gorpe (2018) described the characteristics of a crisis as urgent, significant, deceptive, high pressure and rare. Crises create threats of reputation, financial loss as well as public safety. They also threaten the values of an organization due to their discontinuity period.

A crisis is also a critical incident involving threat, injury, death, disruption, damage to property or to the environment as defined by the Pittsburg State University (n.d.).

Another definition for crisis is negative change, a period or event which is characterized by uncertainty, unexpectedness and a threat. Crises cause dangerous and unstable situations influencing individuals, groups or the whole society. They can emerge all of a sudden without or with little warning as described in Wikipedia (n.d.).

In sum, a crisis can be defined as an uncertain, unexpected and negative situation causing unstable consequences and requires a prompt response.

Types of Crises

Crises can occur in various types. In accordance with C2 (2022), there is no way of predicting when a crisis will strike, but when it occurs, it could influence the whole operation and its continuity in a serious way.

Samawi (2021) conveyed that crises include several phases. They start with the occurrence of the crisis which requires the recognition of the crisis and reaching solutions. When the decision-maker fails to face the crisis, the second phase occurs which includes the expansion of the crisis and the increasing of its strength. The third phase is involved with the crisis becoming extremely destructive. Afterwards, the crisis shrinks until it loses its strength and starts to fade and disappear. At this final stage, the society begins its crisis recovery.

According to Sayed et al. (2021), different types of crises include environmental, global goods consumption, political and health crises. Environmental crises include climate change which is happening instantly, as well as global warming described as a climate crisis. Other environmental crises involve floods, tornadoes, tropical cyclones heat waves caused by extreme weather conditions. Other examples of environmental crises are biodiversity loss and deforestation. All these extreme events cause losses of life, displacement and may lead to major economic effects. Further, there are crises of global goods consumption where access to basic needs for instance access to food, water and electricity is challenging. Political crises such as war result into safety, social and economic effects. With regard to health outbreaks affecting populations, examples are Ebola, diphtheria, HIV-AIDS and recently the emergence of COVID-19 pandemic.

In brief, it is not always possible to predict when a crisis will happen, as it has many different types. For instance, it includes environmental, health, political and global good consumption crises.

Concept of Crisis Management

Crisis management is defined as a process that deals with unfavorable situations which could damage an organization in a serious way as by Norton (2022).

Crisis management can also be defined as a guide on actions that should be taken during the occurrence of any crisis as per Rodsevic (2022).

Maghdid et al. (2022) added that crisis management is a continuous comprehensive and integrated effort in an organization. It aims at understanding, preventing and managing the crisis while considering the stakeholders' interests.

Rodsevich (2022) mentioned another definition for crisis management as the whole coordination of an organization response to a crisis including prevention, preparation, media training, pre- and post-crisis.

Another definition of crisis management is that it is a management process responding to a crisis situation strategically through skilled and well-trained managers to reduce losses as described by Samawi (2021).

Nteka (2021) cited that crisis management is the process and technique by which crises are resolved, reduced or prevented. It aims at controlling unpleasant situations through minimizing uncertainty and risk.

As mentioned by Pontes (2020), crisis management consists of measures aiming at reacting to crises and maintaining a balanced situation. Its purpose is also to restrain the damage of a crisis.

Crisis management requires numerous types of knowledge such as communication, legal and social knowledge besides expertise and various skills in order to reach solution as stated by Zamoum & Gorpe (2018).

Based on Juneja (n.d.), crisis management is the art of dealing with unexpected and sudden events.

Summarily, crisis management can be defined as a process that responds to and deals strategically with crises and sudden unfavorable situations by reducing risk and uncertainty.

Importance and Value of Crisis Management

Johansson & Pihl (2023) pointed out that the hyperconnected society, networks and interconnected systems led to the increase in global disturbances. Consequently, crises have been spreading at a wider space across the whole world. Adapting to a world full of complexities, uncertainties and rapid changes has become a necessity. Therefore, crisis management has become an essential subject over the last decade.

As specified by Rodsevic (2022), crisis management prepares institutions for crises by allowing them to devise strategies that enables them to competently and confidently deal with crises and alleviate damage. On top of that, it protects the institution's financial well-being and reputation.

Likewise, not only does crisis management reduce the threat of a crisis, it also assists in managing the crisis impact on the institution reputation, as clarified by Crises Control (2022).

Samawi (2021) pointed out that the importance of crisis management as one of the modern independent sciences increased and is considered as one of the future sciences. It adapts to changes and is connected to all humanities.

On the other hand, crisis management consists of several rules which were discussed by Norton (2022). These include the avoiding of and the preparation for a crisis.

Marker (2020) highlighted that with the speedy escalation of crises. crisis management is not anymore optional and neglecting it is very risky.

Since crisis management has numerous benefits, it has become crucial. As explained by Juneja (n.d.), not only does it assist the different institutions and organizations in properly adjusting to sudden changes, it also helps them understand the causes of crises. Additionally, it enables them to cope with crises in the best possible ways and take the required precautions.

In summation, crisis management' s value and importance became significant amidst today's rapidly changing world that is filled with uncertainties and complexities. Crisis management doesn't only prepare institutions to properly cope and deal with crises, it also protects their reputation and financial well-being.

Crisis Management Stages

Crisis management contains several stages. As demonstrated by AlertMedia (2023), the first stage is the pre-crisis stage, followed by the crisis stage, the response stage and finally the post-crisis stage as follows:

Stage 1: Pre-Crisis Stage

Based on AlertMedia (2023), some crises can occur at any time without any warning such as natural disasters. At the same time there are other crises which happen with prior warnings for example pre-weather warning that forecasts extreme weather. This first stage happens before the occurrence of the crisis where the pre-warning signs are identified and the organization actively observes any potentials threats and includes ways for dealing with the emergency.

As indicated by the London Premier Centre (2023), the importance of this stage is derived from its ability to decrease the negative effects and outcomes of crises. It is capable of reducing or preventing serious problems in the future through data analysis that can predict what could be happening and what is required to be done. In this case, this stage deals with possible repercussions and scenarios.

Goh (2021) added that this stage is involved with prevention and preparation. It is also focusing on decreasing the known risks which could lead to the appearance of a crisis as per Noggin (2021).

Stage 2: Crisis Stage

This is the stage when the crisis already happened and can no longer be prevented. During this stage, the situation is assessed, a plan is activated and the members of the organization are alerted as indicated by AlertMedia (2023).

Stage 3: Response Stage

According to AlertMedia (2023), in this stage, the required resources are used to deal with the crisis effects. Prompt decisions are taken to efficiently manage the situation and decrease any risks.

Goh (2021) conveyed that this stage comes after the acute crisis phase has passed and the organization focuses on reassuming its normal functions. It is about the actual response to an actual crisis as by Noggin (2021).

Stage 4: Post-Crisis Stage

As indicated by Noggin (2021), all crises end but they could also reoccur, particularly if the organizations don't complete this final stage. In accordance with AlertMedia (2023), the period of a crisis varies according to its type. During this stage, any damages are assessed in order for the organizations to resume its operations. This final stage could also include considering the well-being of the staff who may have been affected by the crisis negative impacts. It is also concerned with the better preparation for the coming crises as expressed by Goh (2021). Additionally, based on Noggin (2021), organizations begin to also reflect and fulfill the commitments they made during the occurrence of the crisis.

In short, crisis management consists of a pre-crisis stage, crisis stage, a response stage and finally a post-crisis stage.

Principles of Crisis Management

There are several key principles of an effective crisis management as described by García (2021). Some of them are as follows:

Flexibility

García (2021) referred to flexibility as the ability to focus, learn and make decisions during uncertain periods. It is also about the capability to use different solutions and approaches.

As per the Management Training Institute (2020), flexibility is considered the most important skill during a crisis. The reason for that is that it makes organizations more responsive and quickly adaptable to change and complicated situation without any hesitation. Flexibility also assists leaders in guiding the staff through challenges. This makes the staff feel supported and more confident and also builds their trust in their leaders.

Communication

Based on Reputation Management (2023), effective communication is a key in efficiently dealing with crises. It is characterized by empathy, honesty and transparency. It is about keeping communication in a regular and ongoing way. The way of responding to the crisis as well as the persons who will be communicating with stakeholders is also determined.

Zemke (2020) expressed that communication is considered fundamental as it offers stabilization through providing guidance, clarity and trust. It assists organizations to continue and recover during times of uncertainty created by crises to achieve success.

As clarified by Rodsevich (2022), clear, quick and accurate communication provides the required knowledge enabling an organization to make the right decisions during a crisis.

Leadership

García (2021) mentioned that leadership is concerned with the ability to deal with complexities and changes. Based on Deloitte (2019), leadership during a crisis is different than usual as requirements and demands become greater. This principle requires the existence of strategic support and long-term planning. It is also related to the implementation of the right process as well as a critical oversight.

Transparency

It is related to being transparent about the whole situation, providing updates and establishing trust with stakeholders as per Reputation Management (2023).

Bermude & Izquierdo (2020) added that sharing updated, timely and transparent information is crucial during times of crises where unsteadiness and uncertainty exist.

Resourcefulness

As indicated by Goldenberg (2022), resourcefulness refers to the ability to restore the remaining resources and making full use of it. It comprises of using the resources for any crises that could happen in the future while making the most effective use of the available existing resources. On the other hand, it is about being openminded to new possibilities, learning new things and understanding problems. Additionally, resourcefulness also includes making the situation better by doing things in a different way.

García (2021) pointed out that resourcefulness is about the ability to develop structures, knowledge, plans and information.

Team Work

According to Schlimm (2020), teamwork happens across different hierarchies and departments and involves performance.

Based on Queensland Health (2018), teamwork uses the different individual strengths and skills towards specific goals dealing with responding and resolving a crisis. It aims at giving up individual goals and achieving common goals through the alignment of individuals. In addition to what is mentioned, teamwork is essential with regard to crisis management, due to the nature of crises which is conflicting information, limited timeframe, increased stress and unclear aims.

Speed and Urgency

Quick decisions and actions are taken immediately and are not postponed or ignored as discussed by García (2021). It is concerned with acting fast and acknowledging the crisis as by Reputation Management (2023).

Focus

Not being distracted by other matters, on the contrary to focus on fulfilling a particular mission and objectives as described by García (2021).

To sum up, an efficient crisis management is composed of several principles. Among them are flexibility, communication, leadership, transparency, resourcefulness, team work, speed & urgency and focus.

Concept of Higher Education

As referred to by the Lis Education Network (2022), higher education is defined as the advanced education level that is beyond primary and secondary levels. It involves colleges, universities and other institutions providing specific, professional and specialized programs. It provides various degrees for instance Bachelor's, Master's and Doctorate degrees.

Accordant with the UNESCO (2023), higher education is a rich asset scientifically and culturally, due to its ability to create personal development and a stable future for students. It also has the ability to promote innovation and research. At the same time, higher education boosts social, technological and economic change.

Based on the Lis Education Network (2022), higher education is a transformational phase in the students' education empowering their advanced learning, expertise, personal growth and thinking. It prepares them for their careers and future challenges by equipping them with the required qualifications and skills. By doing that, these students are able to pursue leadership roles and high-level positions. It encourages their intellectual curiosity while promoting a research culture. Higher education also enables them to contribute to the whole world and to their communities. Not only does higher education shape the individuals' future, it also shapes the society.

Concisely, higher education is defined as the advanced education level beyond primary and secondary levels. Besides promoting innovation and research, it promotes students' personal growth, skills, advanced learning, expertise and qualifications. It shapes their future and enables them to pursue high-level positions and leadership roles. On the other hand, it shapes communities and the whole society.

Conceptual Framework

Figure 1 below was constructed according to the study's objectives.

Figure 1. Conceptual model
Source: By Researcher

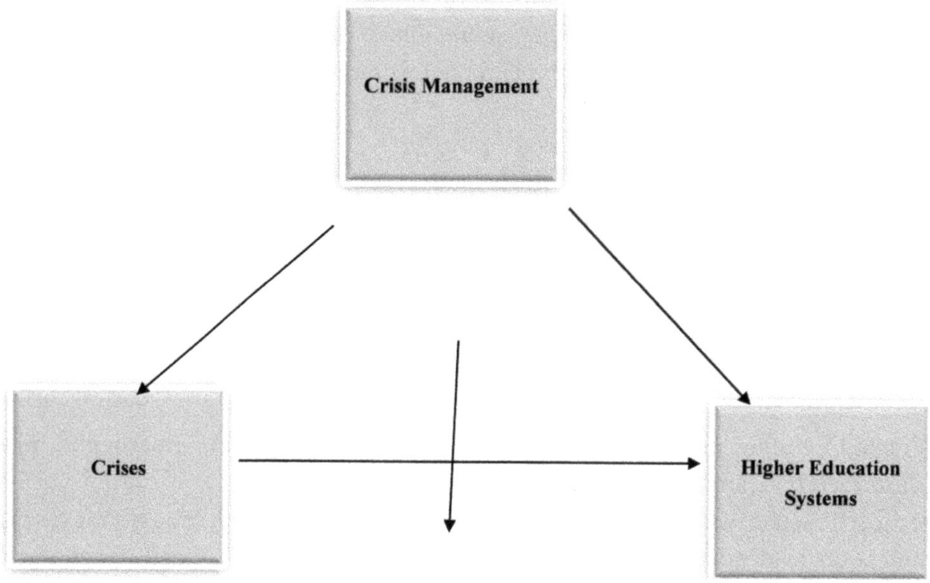

The conceptual framework of this research model shows the impact of crises on higher education systems. It also illustrates the impact of crisis management on crises individually and on higher education systems individually. The framework also demonstrates the moderating effect of crisis management on the relationship between crises and higher education systems

The Impact of Crises on Higher Education Systems

As conveyed by Bouncken et al. (2022), since crises are complex, they result into negative impacts. The higher their complexity, the harder it is to estimate their effects and causes and the more difficult it becomes to achieve solutions. Any inactivity or incorrect actions may lead to the escalation of the process related to dealing with the crises.

Based on Maghdid et al. (2022), no matter what the type of crisis occurred, its long- or short-term consequences are damaging and can be acute. So, regardless of the type or size of the crisis, the whole system could be brought down, as stated by C2 (2022).

Samawi (2021) expressed that crises create a state of disturbance and imbalance, which results into an interruption in the educational system. It also obstructs the attention of the members from achieving the educational goals and performing their work. Therefore, immediate measures are required, in order to return to normal status and to prevent the aggravation of the situation.

As implied by Queensland Health (2018), due to the sudden and rapid changing nature of crises, they are difficult to handle. They represent a challenge and don't allow enough time to analyze the crisis or collect information about it. Even if some information is collected, it could be incomplete or in conflict with other information. Consequently, immediate attention and properly analyzing this information is required, in order to be able to make the right decisions. Crises also create stress and time pressure and require appropriate response besides efforts to restore the situation. Stress caused by crises also leads to anxiety, fear, mental block and even panic.

Different crises have different effects on higher education systems. The impact of some types of crises which are considered by the researcher as common crises are discussed further.

For instance, economic crises could change higher education at micro and macro levels, as referred to by Tsiligiris (2012). On the micro level, the consequences could lead to the reduction of students' numbers on expensive programs which forces colleges and universities to reduce the number of staff. On the macro level, the impact could be long-term with regard to macro policy and planning which transforms higher education into a form of a marketized service-provider for customers where students receive value for the money they paid focusing less on research. Consequently, colleges and universities would create alternative income sources while stopping non-competitive programs.

With regard to the impact of health crises, the global outbreak of the recent highly infectious disease (COVID-19) pandemic will be discussed further as an example.

According to Luchian (2021), COVID-19 exposed the weakness in education systems related to direct learning which was implemented where both teachers and students should be present at the same place and time. Isolation measures made direct learning and face to face studies impossible and forced colleges and universities to look for other forms of teaching methods. It was not possible to continue with the applied inter-university exchanges related to student exchange programs, the admission of international students besides international research collaboration. On the other hand, financial problems were evident, due to the loss of tuition fees which were paid by international students. Moreover, research activities were interrupted, due to the non or limited access to field investigations and laboratories. Additionally, several scholarships have been cancelled or delayed. Learning opportunities were limited and were switched to online access, due to the conditions of social distancing and isolation implemented in the whole world. Switching to the use of smartphone, computers, internet and information technology was an essential necessity. Colleges and universities had no other options than transiting shifting to distance learning. Online education and digital communication generated several challenges. Communication between students and other students, as well as between students, teachers and professors was limited. Interpersonal contacts were lost and mentoring possibilities were decreased. Online courses and teaching materials required additional expenses in a form of the implementation of IT services, devices to be used for remote work by professors and teachers.

On the other hand, climate change is an example of environment crises.

As stated by Concern Worldwide US (2022), one of the consequences of climate change is lower education levels. Closures related to climate crises threaten the quality of education, as professors and teachers together with students won't be able to catch up on the missed educational program. Besides, educational institutions are destroyed by typhoons and education is interrupted.

Further, crises seriously affect the teaching process, as well as the life order, as they inflict mental or physical harm on both teachers and students as mentioned by Zhen & Bian (2015).

To put it briefly, crises have negative effects on higher education systems. Their consequences can be acute and damaging. Even so, they can lead to a state of imbalance and disturbance interrupting the educational system. As a result, the attention of the members of these systems can be shifted from performing their work and achieving the educational goals.

The Impact of Crisis Management on Higher Education Systems

Bouncken et al. (2022) indicated that management in some institutions might underestimate or ignore the threat and complication of crises. They may take simple actions or postpone essential decisions. A lack of a crisis management puts the whole organization at risk of fully recovering from crises as by C2 (2022).

Based on the Blue Moon Consulting Group (n.d.), crises in general happen very fast and are difficult to manage. Thus, chaos in higher education systems is not acceptable and the management is required to have a defined process to achieve an effective response. Several institutions look at crises from the point of trying to find an answer for what they will tell stakeholders when they start to make inquiries. Here lies the danger for these institutions in the way they would deal with crises and if they would manage them poorly, their reputation could be significantly damaged and their response would be remembered. On the other hand, crisis management seeks to respond to questions related to determining what should be done and who should be involved. It is involved with defining the process that would manage the consequences and the impacts from a broader range for all these issues and events that occurred. It doesn't only respond to crises; it also manages potentially unexpected events and impacts. Consequently, providing a successful management of crises, improves the reputation and the leadership of these institutions.

In accordance with Samawi (2021), crisis management is concerned with the prediction of possible crises and dealing with them at the best level of professionalism while using the available resources. The purpose is to ensure the achievement of the least degree of damage to the educational systems with the lowest cost.

In addition, as conveyed by Hayes (2022), crisis management identifies threats to both an organization and its stakeholders with the purpose of effectively responding to these threats. It is concerned with reacting to negative events both during and after their occurrence.

In relation to higher education systems, when facing crises, they must also deal with the time constraints and pressure from the media, parents, staff, local community groups, partners and others as conveyed by the Blue Moon Consulting Group (n.d.). Adding further that the response of these systems is critical and is a key determinant in forming the way these systems and their leaderships will be perceived. Therefore, the response should be compassionate, credible and consistent. Hence, crisis management is necessary in this case and its influence is highlighted. Based on the Blue Moon Consulting Group (n.d.), crisis management has the appropriate and correct organizational mindset which embodies reputational risk into the process of making decisions in an active way.

Moreover, based on the Change Leader, Inc. (n.d.), through crisis management, higher education systems can create new policies and changes allowing them to reach new objectives. They can also perform with integrity and a culture that welcomes new change.

If we take the COVID-19 worldwide pandemic as an example for a recent global crisis, we would recognize the importance of crisis management, particularly in higher education systems.

In line with the above-mentioned, crisis management courses are essential for leaders in higher education systems to assist them in dealing with any unpredictable and complicated challenges in their institutions as mentioned by AI and the LinkedIn community (n.d.).

To summarize, crisis management has a significant impact on higher education systems. Whenever there is a lack of a crisis management, the whole systems are at risk of completely recovering from crises. Moreover, crisis management enables higher education systems to effectively manage and respond to crises and also improves the reputation of these systems and their leaderships. Nevertheless, crisis management identifies threats and ensures the least degree of damage to the educational systems with the lowest cost.

The Impact of Crisis Management on Crises

The significance and impact of crisis management on crises was discussed by Samawi (2021), who mentioned that it appears through improving the available skills and resources and providing practical abilities and skills. The aim is to decrease the crises consequences besides determining recovery actions to return to the normal status.

Over and above that, based on the Institute for Public Relations (2007), crisis management is able to constantly handle crises, as it protects both organizations and stakeholders and reduces the consequences of crises. At the same time, it is also capable of lessening or preventing the damage a crisis can impose on organizations and stakeholders. This is due to the way it is designed which composes of pre-crisis, crisis response and post-crisis stages. It is concerned with preventing crises or preparing to face them. On the other hand, it responds to crises and searches for methods to be better prepared for any forthcoming crises. During the post-crisis stage, repair continues.

Briefly, crisis management's impact on crises is significant as it responds to crises, prevents or lessens its negative consequences and the damage they can impose.

Crisis Management as a Moderator in the Relationship Between Crises and Higher Education Systems

Crisis management plays a role as a moderator between crises and higher education systems as explained further.

Based on the London Premier Centre (2023), managing crises which is one of the major methods of crisis management reduces the negative and harmful effects and consequences of crises before beginning to aggravate. It improves the capacity to withstand challenges.

As a further matter, crisis management enables higher education systems to recover after facing crises.

As stated by C2 (2022), crisis management decreases the potential negative effects of unexpected events and risks and helps in understanding the required steps that should be taken, in order to respond to certain threats.

Crisis management also assists in transforming to new levels and reduces the impact of crises as added by Alkhawlania et al. (2019).

Apart from that, Zhen & Bian (2015) expressed that recovery after crises is an essential part of crisis management in higher education systems. It involves assessing and learning from the crises which leads to the promotion of crisis management functions in the future. It is also concerned with aftercare which includes ways to help professors and students to recover on the psychological level from the crises.

In addition to the above-mentioned, Wang & Hutchins (2010) cited that crisis management is able to expose at any time any weaknesses in the systems and establishes capabilities for these systems to learn from previous experiences. Not only does crisis management act as an emergency plan preparing for particular crises but also as a program that plans for a wide range of various types of crises with efficiency and speed.

Table 1. Participants demographic information

Items	N	Minimum	Maximum	Mean	Standard Deviation
Gender	169	01	02	1.5089	.50141
Age	169	25	59	1.8994	.61378
Civil Status	169	01	02	1.6450	.47994
Educational Level	169	10	16	2.6568	.47619
Professional Experience	169	05	30	1.6331	.65126
Job Level	169	01	03	1.6213	.72277

Source: By Researcher

To sum up, crisis management plays a role as a moderator between crises and higher education systems. It decreases the potential negative effects of crises, assists higher education systems in responding to crises and recovering.

Methodology

Primary data was collected from 169 respondents through an online questionnaire. A five-point Likert scale was used ranging from 1: strongly disagree to 5: strongly agree. The Cronbach alpha reliability of the scale measuring crisis management is 0.92 while the Cronbach alpha reliability of the scale measuring crises is 0.923 and the one measuring higher education systems is 0.93. Secondary data was collected by using information from other researchers and studies.

Data Analysis and Findings

Table 1 below shows the demographic information of the 169 participants in the online survey as well as the mean and standard deviation which have been calculated.

Table 2 below demonstrates the correlation result.

Based on Table 2, correlation was used to examine the relationship between the variables; crises, higher education systems and crisis management. Additionally, crises and higher education systems are negatively correlated. Further to that, the table also shows that crisis management is positively correlated

Table 2. Correlation result

Item	C	CM	HES	CXHES
C	1			
CM **HES**	-.341** .000 -.235**	1 .285	1	
	.002	.001		
CXCM	.892**	.110	.391**	1
	.000	.002	.000	

Source: By Researcher

Table 3. Regression analysis

	Model	B	Un-Standardized Coefficients Standard Error	t-Value	Sig
C	-.196		.040	-4.902	.000
CM	-.321		.062	-5.137	.000
CXCM	.218		.009	25.552	.000

R^2 = .185, Adj. R^2=.175, F-Value= 18.805, p value= 0.000
R^2 = .240, Adj. R^2=.226, R^2 Change= .055, F-Value= 11.990, p value= 0.001
Source: By Researcher

to higher education systems and negatively correlated to crises. On the other hand, crisis management has a moderation influence on the variables; crises and higher education systems.

With regard to regression analysis, it was demonstrated in Table 3 below.

The impact of crises and crisis management besides the effect of crisis management as a moderator was assessed by the use of regression analysis. Table 3 above illustrates that the standardized coefficient of regression for crises which is (-.365) shows that it is significantly and negatively related with the dependent variable.

Crisis management with (-.383) as the standardized coefficient of regression has significant negative correlation with the variable. The results revealed that crisis management acts as a moderator for the relationship between crises and higher education systems. Coefficient (R2= .185) shows that the predictors explained 18.5% of variations while the F-value shows that the model is significant ($p<0.05$). The R2=.240 and the R2change=.055 demonstrate that crisis management is moderating the relationship between crises and higher education systems. This is due to the variation after the incorporation of the moderating variable which is 24% with the R2 change value which is 5.5% showing that the model is significant.

In conclusion and in accordance with the above-mentioned, the results of this study are consistent with the secondary sources collected from other authors, scholars and researchers. They discussed the various effects of crises on higher education systems in section (3.1) as well as the impact of crisis management on higher educations systems in section (3.2). The numerous influences of crisis management on crises in section (3.3) in addition to its moderating effect in the relationship between crises and higher education systems in section (3.4) have also been highlighted.

Regardless of the different impacts explained in sections (3.1, 3.2, 3.3 and 3.4) they all confirm that crises have a significant negative impact on higher education systems while crisis management has a significant positive impact on higher education systems and a significant negative impact on crises. They also confirm the moderating effect of crisis management in the relationship between crises and higher education systems.

Based on the aforementioned, it is concluded that all the research hypotheses (H1, H2a, H2b and H3) are confirmed and none of them is rejected.

Future Study

Crisis Management is an expanded topic which opens a broad range of several subjects for future research. This research mainly focuses on the impact of crises and crisis management on higher education systems and the impact of crisis management on crises. It also discusses the moderating effect of crisis management in the relation between both crises and crisis management.

Since the study is focused only on higher education systems, it is recommended that further studies could be extended to other sectors.

CONCLUSION AND RECOMMENDATIONS

Colleges and universities have been challenged with numerous crises over the past few years from natural disasters, disease outbreaks, economic, political and environmental challenges. These crises create instant chaos and generate devastating and severe impacts on all life aspects and higher education systems are not an exception. These systems are also susceptible to various types of crises and their negative consequences can't be underestimated, as they can fundamentally change systems.

The study revealed the significant negative impact of crises on higher education systems as well as the negative impact of crisis management on crises. On the other hand, it highlighted the significant positive effect of crisis management on higher education systems.

It also showed that crisis management plays an important role as a moderator between crises and higher education systems. Additionally, crisis management is essential for higher education institutions and lacking it, leads to the inability of being prepared when unexpected crises and challenges appear. When institutions are properly prepared, long-term damages chances become less. Despite of the fact that it is not possible for any high education system to be fully prepared for all types of crises, they can still benefit from the role of crisis management. This is reflected in updating and developing a crisis management plan that can adapt to changes.

Consequently, based on the aforementioned, the researcher recommends that higher education institutions should adopt a proper crisis management plan that enables them to respond effectively to any crisis. This requires several steps:

- The formation of a diverse and innovative crisis team in higher education institutions assigned with different roles aiming at achieving one common goal; being in charge of promptly dealing with and responding to any crisis.
- Constantly carrying out trainings and development courses related to suggesting ways of responding to crises. These courses should not only be for crisis team members but also for members of staff in these high education institutions which include administrative as well as teaching staff.
- Having the necessary resources, updated emergency equipment and technological devices to respond early to crises while being able to restore any damage.
- Frequently updating the adopted crisis management plan, in order to have a general idea of any potential threats. The crisis management plan should be easily accessible and understood by all members. Further to that, it should include risk analysis, an emergency contact list and a chart that contains the names and responsibilities of the crisis management team.
- The identification of coordinating individuals in higher education institutions who shall be in charge of notifying and conveying the news of the occurrence of crises. They should also be able to act as a transparent communication central point with other members with regard to any vital information regarding the emerging crisis.

- Creating means of effective communication in an institution between its members to overcome the crisis as well as with stakeholders.

In conclusion, crisis management is essential for higher education systems to survive the complexities and challenges of the 21st century. Without giving attention to the implementation of a proper crisis management, it would be difficult to cope with all challenges and the higher education systems would find it impossible to adapt to changes or prevent negative consequences and damages.

REFERENCES

AI and the LinkedIn community. (n.d.). *What are the most common mistakes that higher education leaders make in crisis management?* Retrieved October 21, 2023, from www.linkedin.com: https://www.linkedin.com/advice/3/what-most-common-mistakes-higher-education

AlertMedia. (2023, August 10). *The 4 Stages of Crisis and How to Manage Them.* Retrieved 27 October, 2023, from www.alertmedia.com: https://www.alertmedia.com/blog/stages-of-crisis/

Alkhawlania, M. A., Bohari, A. b., & Shamsuddin, J. b. (2019). actors that Influence Crisis Management in Yemeni Organizations: The Mediating Effect of Decision- making Styles. *International Journal of Innovation. Creativity and Change*, *5*(2), 11–34.

Bermude, H., & Izquierdo, A. (2020, July-August). *Trust and Transparency in Times of Crisis.* Retrieved October 29, 2023, from www.prsa.org: https://www.prsa.org/article/trust-and-transparency-in-times-of-crisis

Blue Moon Consulting Group. (n.d.). *Higher Education: Crisis Management A Proactive & Strategic Approach to Managing Reputational Risk.* Blue Moon Consulting Group. Retrieved from https://bluemoonconsultinggroup.com/wp-content/uploads/2022/03/BMCG-Higher-Ed-1-Crisis-Management-Whitepaper-2-8-22.pdf

Bouncken, R. B., Kraus, S., & Ancillo, A. (2022). Management in times of crises: Reflections on characteristics, avoiding pitfalls, and pathways out. *Review of Managerial Science*, *16*(7), 2035–2046. doi:10.1007/s11846-022-00580-2

C2. (2022, November 11). *Crisis Management Team: Function, Roles & Responsibilities.* Retrieved October 25, 2023, from https://continuity2.com/blog/crisis-management-team

Change Leader, Inc. (n.d.). *Crisis Management and Transformation.* Retrieved October 13, 2023, from https://changinghighered.com/higher-education-consulting/crisis-management-transformation/

Concern Worldwide US. (2022, August 3). *Climate change is one of the biggest threats to education — and growing.* Retrieved October 12, 2023, from ww,concernusa.org: https://concernusa.org/news/education-and-climate-change/

Crises Control. (2022, September 14). *What is Crisis Management and Why are Crisis Management Strategies Important?* Retrieved October 11, 2023, from https://www.crises-control.com/blogs/what-is-crisis-management-and-why-is-it-important/

Deloitte. (2019). *Stepping in: The board's role in crisis management.* Deloitte. Retrieved from www2.deloitte.com: https://www2.deloitte.com/content/dam/Deloitte/uk/Documents/risk/deloitte-uk-risk-global-on-the-boards-agenda-crisis-management.pdf

García, J. E. (2021, January 26). *15 Principles of Crisis Management.* Retrieved October 28, 2023, from www.linkedin.com: https://www.linkedin.com/pulse/15-principles-crisis-management-jorge-eduardo-garc%C3%ADa

Gigliotti, R. A. (2022, February 24). *Crisis Leadership in Higher Education: Historical Overview, Organizational Considerations, and Implications*. Retrieved September 30, 2023, from https://oxfordre.com/politics/display/10.1093/acrefore/9780190228637.001.0001/acrefore-9780190228637-e-2045

Goh, M. (2021, June 28). *Identify the Stages of a Crisis.* Retrieved October 27, 2023, from https://blog.bcm-institute.org/crisis-management/identify-the-stages-of-a-crisis

Goldenberg, E. (2022, April 4). *How To Lead Through A Crisis.* Retrieved October 30, 2023, from https://www.jobillico.com/blog/en/how-to-lead-through-a-crisis/

Hayes, A. (2022, March 22). *Crisis Management: Definition, How It Works, Types, and Example.* Retrieved October 8, 2023, from https://www.investopedia.com/terms/c/crisis-management.asp

HKUST. (n.d.). *Mission & Vision.* Retrieved October 1, 2023, from https://hkust.edu.hk/about/mission-vision

Institute for Public Relations. (2007, October 30). *Crisis Management and Communications.* Retrieved October 23, 2023, from https://instituteforpr.org/crisis-management-and-communications/

Johansson, L., & Pihl, L. (2023). *Crisis Management in Theory and Practice: How and Why Organizations Work the Way They Do*. Jönköping International Business School.

Juneja, P. (n.d.). *Crisis Management - Meaning, Need and its Features*. Retrieved October 10, 2023, from https://www.managementstudyguide.com/crisis-management.htm

Leishangthem, L. D. (2017, March). Political Instability and Its Influence on Higher Education: A Study of Students' Perceptions in Manipur University. *International Journal of Humanities & Social Science Studies*, *III*(V), 300–307. https://oaji.net/articles/2017/1115-1491480171.pdf

Lis Education Network. (2022, January 23). *Meaning, Purpose & Objectives of of Higher Education.* Retrieved October 27, 2023, from https://www.lisedunetwork.com/higher-education/

London Premier Centre. (2023, August 7). *The role of proactive crisis management in protecting institutions and communities.* Retrieved October 30, 2023, from https://www.lpcentre.com/articles/the-role-of-proactive-crisis-management-in-protecting-institutions-and-communities

Luchian, I. (2021). The impact of the covid-19 crisis on higher education. *Geopolitical. Social Security and Freedom Journal*, *4*(1), 15–26. doi:10.2478/gssfj-2021-0002

Maghdid, R. S., Othman, B. A., & Omer, A. K. (2022). From Crisis to Crisis Management; Causes and Impacts of Crises in the Public Sector. *Polytechnic Journal of Humanities and Social Sciences*, *3*(1), 65–74. doi:10.25156/ptjhss.v3n1y2022.pp65-74

Management Training Institute. (2020, November 24). *Five Benefits of Being Flexible During a Crisis.* Retrieved 29 October, 2023, from https://managementtraininginstitute.com/five-benefits-of-being-flexible-during-a-crisis/

Marker, A. (2020, August 31). *The Essential Guide to Crisis Management.* Retrieved October 27, 2023, from https://www.smartsheet.com/content/crisis-management-guide

Noggin. (2021, July 5). *A Guide to Crisis Management.* Retrieved October 27, 2023, from https://www.noggin.io/blog/the-stages-of-crisis-understanding-the-crisis-management-lifecycle

Norton, C. (2022, January 7). *What Is Crisis Management And Why Its Important?* Retrieved October 25, 2023, from https://prohibitionpr.co.uk/public-relations/crisis-management/what-is-crisis-management-and-why-its-important/

Nteka, N. (2021). Crises analysis and management. *Entrepreneurship, 9*(1), 64–77. doi:10.37708/ep.swu.v9i1.6

Pittsburg State University. (n.d.). *Definition of a Crisis.* Retrieved October 6, 2023, from https://www.pittstate.edu/president/policies/definition-of-a-crisis.html

Pontes, M. (2020, March 30). Crisis Management Principles in the war against COVID-19. Retrieved October 28, 2023, from https://www.aberje.com.br/crisis-management-principles-in-the-war-against-covid-19/

Queensland Health. (2018). *Rural Perioperative Team Training Program eamwork and Crisis Resource Management (CRM) Principles.* Clinical Excellence Division, Department of Health. the State of Queensland: the State of Queensland (Queensland Health). Retrieved from https://clinicalexcellence.qld.gov.au/sites/default/files/2018-12/RPTTP%20CRM%20Principles.pdf

Reputation Management. (2023, May 30). *5 Key Principles of Crisis Management in ORM.* Retrieved October 28, 2023, from https://www.reputationmgmt.net/crisis-management/5-key-principles-of-crisis-management-in-orm/

Rodsevic, M. (2022a, November 15). *What is Crisis Management and why is it important?* Retrieved October 11, 2023, from https://prlab.co/blog/what-is-crisis-management-and-why-is-it-important/

Rodsevich, M. (2022b, November 10). *What is Crisis Communication and why is it important?* Retrieved October 29, 2023, from https://prlab.co/blog/what-is-crisis-communication-and-why-is-it-important/

Samawi, F. (2021). Educational crisis management requirements and its relation to using distance learning approach: a cross-sectional survey secondary stage schools in Al-Balqa'a governorate during COVID-19 outbreak from the perspectives of teachers. *Turkish Online Journal of Distance Education-TOJDE, 22*(3), 196-212.

Sayed, Y., Singh, M., Pesambili, J. C., Bulgrin, E., & Mindano, G. (2021). *Crises and the COVID-19 pandemic: education responses and choices during times of disruptions: A global synthesis and Sub-Saharan African country analysis: Cabo Verde, Burkina Faso.* Education International.

Schleicher, A. (2020). The impact of COVID-19 on education - Insights from education at a glance 2020. OECD.

Schlimm, D. (2020, April 9). *Teamwork in Times of Crisis.* Retrieved October 28, 2023, from https://www.jenoir.com/blogs/blog/440593-teamwork-in-times-of-crisis

Sekaran, U. (2008). *Research Methods for Business.* John Wiley and Sons.

Tsiligiris, V. (2012, March 18). *The impact of the economic crisis on higher education.* Retrieved 12 October, 2023, from https://www.universityworldnews.com/post.php?story=20120314055536933

UNESCO. (2023, April 20). *What you need to know about higher education.* Retrieved October 27, 2023, from https://www.unesco.org/en/higher-education/need-know

Wang, J., & Hutchins, H. M. (2010). Crisis Management in Higher Education: What Have We Learned From Virginia Tech? *Advances in Developing Human Resources, 12*(5). Advance online publication. doi:10.1177/152342231039443

Wikipedia. (n.d.). *Crisis.* Retrieved October 6, 2023, from https://en.wikipedia.org/wiki/Crisis

Zamoum, K., & Gorpe, T. S. (2018, June 27). *Crisis Management: A Historical and Conceptual Approach for a Better Understanding of Today's Crises.* Retrieved October 27, 2023, from https://www.intechopen.com/chapters/60813

Zemke, A. (2020, June 13). *Role of Strategic Communication in Crisis Management and Business Continuity.* Retrieved October 29, 2023, from https://beehivepr.biz/strategic-communication-in-crisis/

Zhen, Y., & Bian, S. (2015). Study on Crisis Management Strategy for Universities. *International Conference on Education, Management, Commerce and Society (EMCS)*, 273-276. 10.2991/emcs-15.2015.56

Chapter 14
Rebuilding Higher Education Systems Impacted by Crises:
The Impact of Wars and Conflicts on Higher Education

Mulatedzi Calvin Rammbuda
https://orcid.org/0000-0002-1230-4621
University of South Africa, South Africa

ABSTRACT

This chapter aims to compare and contrast the experiences and responses of three universities in Africa that have faced different forms of wars and conflicts. The chapter collected data through textual analysis using a qualitative comparative case study methodology. The study compared and contrasted the experiences and reactions of three African institutions that have dealt with various types of wars and conflicts using Savo Heleta's idea of higher education in post-conflict situations. The chapter also discovered that by promoting dialogue, critical thinking, social cohesion, and respect for human rights in the participating countries, higher education has significantly contributed to post-conflict reconstruction and peacebuilding. This chapter seeks to add to the body of knowledge on higher education in conflict-affected contexts and offer suggestions for improving the adaptability and value of higher education in these situations. As suggested in this chapter, governments should work together with a variety of stakeholders.

INTRODUCTION

Higher education is often seen as an important driver of social and economic development, especially in low- and middle-income countries. However, in contexts affected by war and conflict, higher education also faces numerous challenges and risks that can affect its quality, relevance and accessibility. Yevdokymova (2023) highlights that the quality and relevance of education is declining due to disruptions in curricula, outdated teaching methods and politicised assessment systems. Higher educational institutions are often targets of violence, suffer from deprivation of resources and opportunities, and are uprooted

DOI: 10.4018/979-8-3693-1926-0.ch014

(Mlaba, 2023). Raqib (2017) and Taka (2021) agree that wars and conflicts destroy infrastructure, affect the way education is delivered, endanger students and teachers, cause trauma and mistrust, and much more. In many African countries, wars and conflicts have led to lower enrolment and attendance rates, brain drain, educational inequality and challenges in teaching and learning. The African higher education system has struggled with issues of funding, accountability, autonomy, quality control and responsiveness to societal demands (Mba, 2017). This book chapter aims to conduct a comparative study of three universities in Africa that have experienced or are still experiencing wars and conflicts. The University of Cape Town (UCT) in South Africa, Makerere University in Uganda and the University of Juba in South Sudan are the three African universities where this study investigated how wars and conflicts affect higher education. These universities represent different types and phases of conflicts as well as different historical, cultural and institutional contexts. These universities have also experienced conflicts of different types and intensities, such as colonial oppression, apartheid, civil war, political instability and social unrest. This chapter has examined not only how these universities have promoted knowledge, democracy and human rights in their respective circumstances, but also how these institutions have responded to the opportunities and challenges of war and conflict. The study aims to contribute to the existing literature on higher education and conflict, which is still limited and fragmented, particularly in the African context, and also to provide policy recommendations and practical implications for universities, governments, donors and civil society actors interested in supporting and improving higher education in conflict-affected contexts. The study hopes to stimulate further research and dialogue on the challenges and opportunities of higher education in conflict and peacebuilding.

BACKGROUND

The impact of war and conflict on higher education is a complex and multifaceted issue that has gained increasing attention in recent years. Higher education plays a crucial role in the development of human capital, social cohesion and democratic values in any society, but especially in countries affected by violence and instability. However, higher education in conflict-affected contexts also faces many challenges and risks, such as physical destruction, human loss, displacement, violations of academic freedom, barriers to access and gender equality, and psychosocial trauma (Milton, 2018; Mlaba, 2023). The disruption of academic operations, damage to infrastructure, displacement of staff and students, and threats to the security of the university community at Makerere University are all consequences of war and conflict (Head, 2022). Higher education can also contribute to the causes and dynamics of conflict by reproducing inequalities, mobilising grievances and promoting divisive ideologies. Moja and Cloete (2021) note that in the case of South Africa's historical black universities and technical and vocational education institutions, conflict has arisen due to problems with the management, administration and restructuring of education. In addition, Camera (2023) shows that political and social tensions in society, such as the Gaza war, can have an impact on some conflicts.

In addition, UCT has been impacted by both the post-apartheid transition, which brought social and political instability and violence, and the apartheid system (1948–1994), which mandated racial segregation and discrimination in education. These past and present tensions have impacted on the accessibility, equity and quality of higher education, as well as the university's identity, curriculum, governance and culture (Du Plessis, Jansen van Vuuren, Simons, Frantz, Roman & Andipatin, 2022). Despite these difficulties, UCT has maintained its position as a leading university in Africa and is considered one of the best in the world. The University of Juba was affected by the Sudanese Civil War (1983–2005), which led to the independence of South Sudan in 2011, and the South Sudanese Civil War (2013–2020), which led to massive devastation and displacement. These wars and conflicts have severely affected the relevance and quality of higher education as well as the growth, operation and sustainability of the university.

However, by engaging in peace-building initiatives that promote social justice, human rights, dialogue and reconciliation, universities seek to reduce the impact of war and conflict (Kester, Abura, Sohn & Rho, 2022, Mlaba, 2023). The creation of social values that support a culture of peace, harmony and non-violence can be greatly enhanced by higher education (Pherali, 2022). Education can be used to promote positive social change and peace-building by cultivating the attitudes and beliefs of groups (Johnson, 2013; Sahar & Kaunert, 2021). Insufficient funding for education systems, a lack of qualified lecturers affected by the conflict or hired by other parties, and the need to change curricula and strategies to address the root causes of the conflict and promote peace and development are some of the other difficulties associated with post-conflict reconstruction of higher education. Pellini (2021) also points out that if post-conflict financial resources are prioritised for critical services, staff and students may die or be forced to relocate, undermining institutions. Although this is rapidly changing, post-conflict recovery of higher education systems is often not prioritised. Higher education institutions continue to be affected by wars and conflicts, even though there are methods to end them. The chapter aims to provide answers to the following research questions:

- How have wars and conflicts affected the access, quality and relevance of three African universities in their countries?
- What are the common and unique challenges and opportunities faced by the three African universities affected by war and conflict?
- What contribution have three African universities made to peacebuilding and development in their countries?

The objectives are:

- To analyse the impact of war and conflict on the accessibility, quality and relevance of three African universities in their countries;
- To compare and contrast the challenges and opportunities faced by three African universities affected by war and conflict; and
- To examine how three African universities have contributed to peacebuilding and development in their countries

LITERATURE REVIEW

Definitions of the Key Words

Access: The ability or opportunity to use or benefit from something, e.g. a service, resource or right.

Impact: The results or consequences of an action, event or phenomenon.

Human rights: The fundamental rights and freedoms to which everyone in the world is entitled, regardless of nationality, race, religion, gender or other status3. Human rights are based on the principles of dignity, equality and justice.

Mitigate: To make something less severe, harmful or damaging or to take action to reduce the negative impact of something.

Peacebuilding: The process of creating and maintaining conditions for lasting peace, which includes addressing the root causes of conflict and violence, promoting reconciliation and social cohesion, and strengthening institutions and governance.

Reconstruction: The process of rebuilding or restoring something that has been damaged or destroyed, such as physical infrastructure, social services or economic systems.

Social cohesion: The degree of trust, cooperation and solidarity between members of a society that reflects their sense of belonging, identification and mutual support.

Social justice: The fair and equitable distribution of resources, opportunities and benefits in society, as well as the protection of human rights and the elimination of discrimination and oppression.

Transformation: The process of fundamentally or radically changing or improving something, such as a system, structure or relationship.

Impact of Conflicts and Wars on Three African Universities

During Idi Amin's rule in the 1970s and the civil wars that followed his overthrow, Makerere University was destroyed and neglected. However, it was never closed and is still an ongoing research and teaching university with 3,000 graduate students and about 30,000 undergraduate students (Kuyok, 2017). Due to the war, the University of Juba was relocated to Khartoum in the late 1980s; after the peace agreement, it was brought back to Juba in 2010. It struggled with problems such as lack of staff, poor infrastructure and limited budget (Mlaba, 2023). The resistance movements of the 20th century and the apartheid government had an impact on UCT in South Africa (Milton, 2018). Wars and conflicts can have devastating effects on academic staff, students, curriculum, infrastructure and institutional resources in devastated areas. Wars and conflicts often result in the destruction or damage of educational facilities, as well as the death, injury, displacement or abduction of academic staff and students (Milton, 2018). As a result, survivors can suffer severe psychological and emotional distress, and their chances of education and employment can be severely compromised. Groves and Kasangaki (2011) point out that the MakCHS has been involved in addressing the challenges of providing health services

The Role and Contribution of Three African Universities in Promoting Peacebuilding, Stability, and Development in Their Communities and Countries

Makerere University has played an important role as a partner and actor in peacebuilding in Africa through its outreach, teaching, research and learning initiatives (McNamee & Muyangwa, 2021). It has proven that it is able and willing to support the principles of security, respect, equality and peace for all. Universities can contribute to promoting a culture of non-violence and peace through their outreach initiatives, research projects and academic programmes (Mwamki, 2017). A Centre for Peace and Development Studies (CPDS) has been established at the University of Juba to conduct research, provide training and advocate for peace, conflict and development issues in and around South Sudan (Nyuykonge & Zondi, 2017). The ideals and principles of peace, such as respect, tolerance, communication, cooperation and solidarity, can be taught to students (Enaifoghe, 2022). According to Millican, Kasumagić-Kafedžić, Masabo and Monica (2021), higher education can promote a culture of peace, dialogue and cooperation by providing values, information and skills that can be used to prevent, mitigate or resolve conflicts. In addition, they can conduct studies that provide information and evidence for conflict prevention and peacebuilding. They can work with their stakeholders and communities to share knowledge and resources and support peace and development projects in the neighbourhood.

Makerere University has made a significant contribution to promoting development, stability and peacebuilding in the countries and communities it serves — both in Uganda and across the continent. Groves and Kasangaki (2011) point out that Makerere University has been involved in addressing health challenges in Uganda. Numerous influential politicians, academics, activists and professionals who have influenced the political, economic and social development of their countries and regions have come from Makerere University of Uganda, including the founding president of Tanzania, Julius Nyerere, the presidents of Uganda, Milton Obote and Yoweri Museveni, the environmentalist and Nobel Peace Prize winner Wangari Maathai and the former Archbishop of York, John Sentamu (Khadiagala, 2021). They can also encourage staff and students to take on leadership roles and develop an entrepreneurial spirit, as well as develop and implement ideas that benefit society as a whole. To create knowledge and promote the application of research findings in policy and practise, Makerere University has entered into partnerships with local, regional and global partners. For example, it has received funding from the African Union, the European Union and SIDA to conduct studies on peacebuilding and post-conflict reconstruction (Khadiagala, 2021). They can collaborate on initiatives and programmes by forming partnerships and sharing best practises, experiences and lessons learned (Kester et al., 2022). They can also support and influence decisions and policies that impact peace and development at a variety of levels (Kester et al., 2022). On the other hand, according to Enaifoghe (2022), UCT has been engaged in peace and conflict research, human rights, democracy, governance, development and social justice. The Centre for Conflict Resolution, the Institute for Democracy, Citizenship and Public Policy in Africa, the Centre for Social Science Research and the [Centre for African Studies] are just some of the facilities and institutes at UCT that deal with these issues.

Sudan has a long history of conflict and peace negotiations, with more than sixty peace agreements under the rule of Omar al-Bashir (1989–2019). The African Conflict and Peacebuilding Review (ACPR) journal focuses on the Juba 2020 peace agreement for Sudan and the prospects for peacebuilding in Darfur (De Guttry, Ndiloseh & Amoroso, 2022). It includes interdisciplinary contributions from scholars and practitioners dealing with the Sudanese and African context of peace and conflict. The University of Juba and the Rotary Peace Centre offer various academic programmes for peace and development leaders from or in Africa (Enaifoghe, 2022). These programmes aim to promote social reform, human rights, indigenous engagement and conflict resolution skills. They also promote the 'method-free' approach to peacebuilding, which empowers communities to participate in peace grantees' social change initiatives. According to Ruby (2021), Rotary Peace Centres offer specially designed courses to prepare people to engage in peacebuilding and conflict resolution wherever they settle. On the other hand, Tuhereze (2021) asserts that students can be trained to analyse, prevent and resolve conflicts at all levels of society through the Postgraduate Diploma in Peacebuilding and Conflict Transformation. In collaboration with the Rotary International Peace Centre, the University offers a one-year programme called the Postgraduate Diploma in Peace Building and Conflict Transformation. The target group for this programme are professionals with at least five years of relevant experience in peace or development work.

The Coping Strategies and Actions of Three African Universities in Response to the Impact of War and Conflict

According to Barnett, Deng and Yoasa (2008), some of the strategies and measures adopted by the University of Juba to cope with the effects of war and conflict include the following: Relocating the University from Khartoum to Juba in 2006, following the signing of the Comprehensive Peace Agreement, to re-establish its presence and identity in the South; implementing an English language policy as part of the Government of South Sudan's vision to promote national unity, international communication and resistance to Arabization; and addressing the language challenges of students and staff through the provision of intensive English language courses, language support services and bilingual materials. Berke and Sell (2018) also believe that new curricula and programmes should be developed in collaboration with local and international partners to meet the needs and aspirations of the post-conflict society, and that the quality and relevance of education should be improved by enhancing the university's academic standards, infrastructure, facilities and resources, that access to education and equity in education should be improved by enrolling and retaining more students, especially women and disadvantaged groups, and by offering scholarships and financial aid, and that the role of the university as a catalyst for peace, development and social change should be promoted through research, community service and advocacy.

UCT has implemented a number of strategies and measures to deal with and recover from the effects of war and conflict by creating and enforcing transformation plans and policies that promote diversity, equity and inclusion among staff and students and prohibit unfair discrimination on any grounds; conducting urban water management research that explores ways to improve social equity and resilience in Cape Town, particularly in relation to droughts, flooding and sanitation issues; creating a strategic plan that outlines UCT's goals and vision for the period 2020–2025 and addresses key opportunities and challenges for the university; and establishing the African Climate and Development Initiative (ACDI) as a strategic initiative to facilitate and expand climate change research (University of Cape Town, 2020). To strengthen its position as a centre of world knowledge in the heart of Africa, Makerere University has created a strategic plan for the years 2020–2030 (Nabatte, 2020; Makerere University, 2021). The university has implemented a number of strategies and measures to address and recover from the effects of war and conflict. These include promoting an inclusive, diverse and respectful culture within the university community, strengthening the university's financial sustainability and mobilising resources to ensure adequate funding for its core missions, and strengthening institutional capacity for peace and conflict resolution through research, teaching and outreach.

THEORETICAL FRAMEWORK

Using Heleta's (2020) concept of higher education in post-conflict contexts, the researcher investigated how wars and conflicts affect higher education at three African universities. This concept describes the period after a formal ceasefire, when society begins to recover and rebuild the damaged or destroyed institutions, relationships and systems. The main focus in this area is on the reconstruction and modernization of higher education systems and institutions, which Heleta (2020) states are crucial for the recovery and development of war-torn societies. Heleta (2020) also explains how higher education can play a strategic role in stabilising and promoting the post-conflict reconstruction of communities and states by providing quality education, strengthening local capacity, fostering creativity and supporting the peace and reconciliation process. However, it also recognises the challenges and issues facing post-conflict higher education, such as infrastructure reconstruction, academic capacity development, curriculum reform, quality assurance, governance and funding. Wars and conflicts have a catastrophic impact on higher education as the opposing sides usually attack, destroy or damage universities. The consequences include loss of life, uprooting of staff and students, disruption of academic operations, deterioration of academic standards and disconnection from academic communities around the world. In addition, war and conflict can also affect curricula and teaching practise in higher education, as they can promote or perpetuate prejudice and preconceived ideas among students and staff, as well as racial, religious or political differences. Higher education in post-conflict environments must therefore address the psychological and social trauma, the legacy of violence and hatred, and the institutional and physical damage caused by the conflict (Heleta, 2020).

Furthermore, the researcher explores the policies and strategies that these universities adopt to deal with the negative consequences of violence and instability on their academic outcomes and activities by using the concept of decolonial standpoint introduced by Chasi (2021). This concept is about breaking down power structures and concepts of knowledge that support the perpetuation of geopolitical, racial and gendered hierarchies. It is not only about changing the location and biography of knowledge, but also about changing one's role from that of a primary recipient to that of a creator of internationalisation practises and knowledge. Chasi (2021) bases her decolonial perspective on the notion that decolonization should be actively incorporated into the internationalisation of higher education in this country as one of the main concerns of the South African higher education system. In this sense, decolonization means resisting colonial and Eurocentric influences on the creation and exchange of information and promoting the inclusion and recognition of regional and indigenous languages and knowledge systems.

Wars and conflicts affect higher education in different ways. They can damage its resources and create new barriers to equity and access (Chasi, 2021). They can also exacerbate the unequal and exclusionary effects of globalisation on higher education, particularly for the Global South (Milton, 2018). However, wars and conflicts can also provide opportunities for innovation and reform in higher education, enabling it to resist global homogenization and respond to local needs and realities (Pellini, 2021; Yevdokymova, 2023). Furthermore, wars and conflicts can affect the role and responsibility of higher education in promoting social change and peacebuilding in the post-conflict and post-conflict period (Kester et al., 2022). This requires careful reflection on how higher education can address the roots and effects of violence and injustice and how it can promote peace, tolerance and dialogue between diverse and divided groups (Kester et al., 2022). Heleta's (2020) definition of post-conflict higher education and Chasi's (2021) concept of decolonial perspective are useful tools for understanding and assessing the impact of war and conflict on higher education.

Methodology

This interpretive study with a qualitative design examined how war and conflict affected the administration of higher education in three African universities through a comparative case study methodology. This methodology enabled the researcher to find patterns and differences in the way wars and conflicts have affected the administration of higher education in different settings. The researcher selected three universities in Africa located in different countries affected by conflict, namely UCT in the Republic of South Africa, the University of Juba in South Sudan and Makerere University in Uganda. The higher education systems in these countries face different opportunities and challenges for peace-building and development as well as different forms and intensities of conflict. The following questions were asked:

- How have wars and conflicts affected the access, quality and relevance of three African universities in their countries?
- What are the common and unique challenges and opportunities faced by three universities in Africa that have been affected by war and conflict?
- How have higher education institutions contributed to peacebuilding and development in these countries?

Through the use of a qualitative comparative case study methodology, the researcher explored the similarities and differences between the three cases and uncovered the factors and circumstances influencing each outcome. The chapter collected data through textual analysis using a qualitative comparative case study methodology. The data came from a variety of sources, including academic articles, books, news articles, media reports, strategies, reports, document analysis, lesson plans, curricula, and testimonies. To better understand the opportunities and challenges that war and conflict bring to higher education, this chapter examined three African universities and the strategies and measures these institutions had put in place to deal with the negative impact of instability and violence on their academic programmes and outcomes.

Findings and Discussion

Impact of Wars and Conflicts on Access, Quality, and Relevance of Three African Universities in Their Countries

This section answers the research question of how wars and conflicts have affected the access, quality and relevance of three African universities in their countries. Heleta (2020) and Milton (2021) point out that higher education in conflict-affected regions can face many challenges, such as deterioration of access, quality and equity, disruption of learning processes, damage to infrastructure and resources, insecurity and violence, and politicisation and polarisation of the academic community. According to Heleta (2020), Makerere University also faced severe financial constraints, academic interference and government repression that undermined its autonomy and quality. On the other hand, Mlaba (2023) argues that the University of Juba was relocated to Khartoum in the late 1980s, after the peace agreement, but faced problems such as staff shortages, poor infrastructure and limited budget. The move presented the university with numerous challenges, such as the lack of infrastructure, facilities and security in Juba, the disruption of academic operations and the difficulty of retaining and recruiting qualified staff. The university also had to contend with the outbreak of a new civil war in South Sudan in 2013, which forced it to close the university for several months and displaced many students and staff. Pelini (2021) pointed out that the university has tried to resume its work and contribute to peace and nation building in South Sudan, but still faces many obstacles and uncertainties.

In addition to changing the location and biography of knowledge, it is also about changing one's own role from that of a primary recipient to that of a creator of internationalisation practises and knowledge. In her decolonial perspective, Chasi (2021) assumes that decolonization as one of the main problems of the South African higher education system should be actively included in the internationalisation of higher education in this country. These past and present tensions have impacted on the accessibility, equity and quality of higher education, as well as the identity, curriculum, governance and culture of UCT University. Du Plessis, Jansen van Vuuren, Simons, Frantz, Roman and Andipatin (2022) state that past and present tensions have impacted on the accessibility, equity and quality of higher education as well as the identity, curriculum, governance and culture of UCT University. According to the theoretical framework used by Heleta (2020), higher education can play a strategic role in stabilising and promoting the post-conflict reconstruction of communities and states by providing quality education, strengthening local capacity, fostering creativity and supporting the peace and reconciliation process.

Common Challenges of African Universities Affected by War and Conflict

This section answers the second research question on the common and unique challenges and opportunities faced by three African universities affected by war and conflict: Makerere University, University of Juba and University of Cape Town. This chapter highlights some of the common challenges faced by these African universities, as reported by Milton (2018), Heleta (2020), Kester (2022) and Hima (2023). Wars and conflicts destroyed or damaged the physical infrastructure of these universities, such as buildings, equipment, libraries and laboratories. This affected the quality and availability of teaching and learning facilities and hindered the development of research and innovation. For example, Heleta (2020) reports that during the civil war in South Sudan, the University of Juba lost most of its infrastructure and resources and had to move to a makeshift campus with limited facilities. The loss of physical infrastructure not only affects the quality of education, but also hinders the reconstruction and development of post-conflict societies. Another common challenge was the displacement, death or traumatization of academic staff and students, resulting in reduced capacity, quality and diversity of academic staff, violence or political interference, lack of funding, resources and support from local and international actors who often neglect higher education in post-conflict situations difficulties in maintaining academic standards, accreditation and quality assurance amidst rapid expansion and diversification of higher education providers.

The Unique Challenges of the Three African Universities

This section answers the second research question on the common and unique challenges and opportunities faced by three African universities affected by war and conflict. Heleta (2015) points out that Makerere University in Uganda faced repeated closures and strikes due to political unrest, administrative problems and financial mismanagement. It has also struggled to cope with the influx of refugees from neighbouring countries such as South Sudan and the Democratic Republic of Congo, who have different educational backgrounds and needs. The University of Juba in South Sudan is suffering from the ongoing civil war that broke out in 2013, forcing the university to relocate several times and work in difficult conditions. Other problems include ethnic polarisation, corruption and a lack of qualified staff and students. The University of Cape Town in South Africa has struggled with the legacy of apartheid and colonialism, which has led to inequalities, injustices and exclusions in the higher education system. It has also faced protests and demands from students and staff for decolonization, transformation and social justice. Heleta's (2020) definition of post-conflict higher education is relevant to the discussion of the impact of war and conflict on higher education as it highlights the difficulties and importance of rebuilding and reshaping higher education after violent conflict.

The Common and Unique Opportunities of the Three African Universities

This section answers the second research question on the common and unique challenges and opportunities faced by three African universities affected by war and conflict. The findings show the potential to contribute to peacebuilding, reconciliation and development in their societies through the provision of education, research and innovation that address the causes and consequences of conflict. Hima (2023) demonstrates the opportunity to collaborate and network with other higher education institutions locally and internationally to share experiences, resources and best practise. The opportunity to diversify and expand their academic offerings, teaching methods and target audiences to meet the changing needs and demands of their societies. This chapter highlights some of the unique opportunities available to the three universities. Makerere University in Uganda has the opportunity to leverage its reputation and history as one of Africa's oldest and most prestigious universities to attract and retain talented staff and students and influence policy and practise in the region. The University of Juba in South Sudan has the opportunity to shape the future of the world's youngest nation by educating the next generation of leaders, professionals and citizens. The University of Cape Town in South Africa has the opportunity to lead the transformation and decolonization of higher education in Africa by promoting diversity, inclusion and social justice in its curriculum, research and culture.

The Engagement of Three African Universities in Peacebuilding, Democracy, and Social Justice

This section answers the research question of what contribution universities have made to peacebuilding and development in three African countries, how three universities have contributed to peacebuilding and development in their countries. The findings show that both Makerere University, the University of Juba and the University of Cape Town have played a role in promoting peace, democracy and social justice. McNamee and Muyangwa (2021) agree that Makerere University has played an important role as a partner and actor in African peacebuilding through its outreach, teaching, research and learning initiatives. It has proven that it is able and willing to support the principles of security, respect, equality and peace for all. According to Anigbogu (2023) and Clarke-Habibi (2021), Makerere University has been involved in a number of programmes aimed at addressing Africa's problems and development needs, including regional integration, education for social change, research and innovation, and peace and conflict resolution. Makerere University also offers courses and programmes on conflict and peace studies, human rights and democracy, social justice and ethics, and gender and women's studies. Makerere University has also produced notable alumni who have shaped the political and social landscape of Uganda and the region, such as Milton Obote, Yoweri Museveni, Julius Nyerere and Wangari Maatha.

On the other hand, according to Enaifoghe (2022), UCT is committed to peace and conflict research, human rights, democracy, governance, development and social justice. The Centre for Conflict Resolution, the Institute for Democracy, Citizenship and Public Policy in Africa, the Centre for Social Science Research and the [Centre for African Studies] are just some of the facilities and institutes at UCT that deal with these issues. Analysis of the documents shows that UCT is committed to social justice, equality and change. It strives to redress historical injustices, promote equal opportunities for all, represent the demographics of South Africa, protect human rights, uphold the inherent dignity of every human being and meet the needs of the country's evolving democracy. Sadiq et al. (2019) report that UCT also conducts research and imparts knowledge on a range of topics related to social justice, democracy and peace, including politics, prejudice, gender and race. UCT also offers courses and programmes on human rights, transitional justice, social justice and peace studies, among others. UCT has also produced notable alumni who have played an important role in promoting peace and democracy in South Africa and on the continent, such as Nelson Mandela, Desmond Tutu, Thabo Mbeki and Nkosazana Dlamini-Zuma.

Sudan has a long history of conflict and peace negotiations. De Guttry, Ndiloseh and Amoroso (2022) point out that the African Conflict and Peacebuilding Review (ACPR) brings together a diverse group of scholars and practitioners from Africa and other parts of the world who study Sudan and the Darfur conflict in depth. The APCR is dedicated exclusively to the Juba 2020 peace agreement for Sudan and the prospects for peacebuilding in Darfur. The Rotary Peace Centre, on the other hand, was established to offer postgraduate degree programmes for peace and development leaders who are from or have worked in Africa. According to Ruby (2021), Rotary Peace Centres offer specially designed courses to prepare people to engage in peacebuilding and conflict resolution wherever they settle. On the other hand, Tuhereze (2021) asserts that students can be trained to analyse, prevent and resolve conflicts at all levels of society through the Postgraduate Diploma in Peacebuilding and Conflict Transformation. In collaboration with the Rotary International Peace Centre, the University offers a one-year programme called the Postgraduate Diploma in Peace Building and Conflict Transformation. The University of Juba is also home to the Centre for Peace and Development Studies, which is dedicated to researching and promoting peacebuilding, conflict transformation and development issues in South Sudan and the Horn of Africa. The University of Juba has also produced influential alumni who have contributed to the peace process and democratic transition in South Sudan, such as Riek Machar, Rebecca Nyandeng and Taban Deng Gai. From the look of things, Chasi's (2021) concept of the decolonial perspective can be seen as a pertinent and helpful framework for analysing and evaluating the impact of wars and conflicts on higher education, as well as for imagining and implementing alternative, emancipatory higher education models that are more inclusive, egalitarian and transformative.

CONCLUSION, RECOMMENDATION, AND IMPLICATIONS

In this chapter, three African universities were used to analyse how war and conflict affect higher education in different ways. It has shown how higher education can suffer from or contribute to violence and how it can support peace and resilience through its functions and activities. Higher education should promote peace and conflict sensitivity by integrating it into its curricula, pedagogy and assessment. It should also support and promote research and innovation that is relevant, rigorous and responsive to the needs of conflict-affected societies. In addition, it should promote inclusivity, diversity and equity and address the causes and effects of conflict such as poverty, marginalisation, discrimination and human rights violations. Furthermore, it should improve the quality and relevance of its teaching, research and services and focus them on peacebuilding and sustainable development. This chapter has two implications: one theoretical and one practical. Theoretically, it adds to the literature on higher education and conflict by comparing and contextualising three African cases. It also provides a framework that links conflict, violence, peace and resilience and shows how they relate to the structural, institutional and individual aspects of higher education. In practical terms, it provides insights and lessons for improving higher education policy and practise in conflict-affected contexts by highlighting challenges and opportunities and proposing recommendations and actions. It also identifies some areas for further research and investigation, such as how higher education impacts on conflict, how higher education contributes to post-conflict recovery and reconciliation, and how to assess the role of higher education in peacebuilding and resilience.

Competing Interests

The author of this publication declares that there are no competing interests

Funding

This research received no specific grants from public, commercial or charitable funding bodies. Funding for this research was provided by the author of this book chapter.

REFERENCES

Anigbogu, V. C. (2023). *The role of universities in responding to Africa's problems and development needs*. Makerere University.

Azis, A. A. (2017). The Contribution of Post-Conflict Education to Social Transformation and Sustainable Development. *Indonesian Journal of International Studies*, *2*(1), 63–70. doi:10.22146/globalsouth.28853

Barnett, J., Deng, R., & Yoasa, N. B. (2008). Post-conflict language issues at the University of Juba, Southern Sudan: Policies, proficiencies and practicalities. *ARAS*, *29*(1&2), 92–103.

Berke, T., & Sell, J. (2018). Educational challenges in South Sudan. In J. Sell & T. Berke (Eds.), *Strategies, policies, and directions for refugee education* (pp. 121–138). Springer. doi:10.1108/S2055-364120180000013005

Camera, L. (2023). The war on college campuses. *U.S. News & World Report*, *163*(2), 36–40.

Chasi, S. (2021). South Africa's Policy Framework for Higher Education Internationalisation: Decolonial Perspective. *Think (London, England)*, *87*, 34–39.

Clarke-Habibi, S. (2021). *Higher education can do more to transform conflict and build peace*. University World News.

De Guttry, A., Ndiloseh, M., & Amoroso, A. M. (2022). The 2020 Juba Agreement for Peace in Sudan and Prospects for Peacebuilding in Darfur. *African Conflict and Peacebuilding Review*, *12*(2), 1–4. doi:10.2979/africonfpeacrevi.12.2.01

Du Plessis, M., Jansen van Vuuren, C. D., Simons, A., Frantz, J., Roman, N., & Andipatin, M. (2022). South African higher education institutions at the beginning of the Covid-19 pandemic: Sense-making and lessons learnt. *Frontiers in Education*, *6*, 740016. doi:10.3389/feduc.2021.740016

Enaifoghe, A. (2022). South Africa's Efforts and Mechanisms in Maintaining Peace and Stability in Southern African Development Community Subregion. In E. Spiegel, G. Mutalemwa, C. Liu, & L. R. Kurtz (Eds.), *Peace Studies for Sustainable Development in Africa. Advances in African Economic, Social and Political Development*. Springer. doi:10.1007/978-3-030-92474-4_47

Ezati, B. A. (2014). Addressing Pedagogical Training Needs of Teaching Staff: Lessons from Makerere University Short Professional Development Programs 2006 – 2010. *American Journal of Educational Research, 2*(12), 1190–1198. doi:10.12691/education-2-12-9

Groves, S & Kasangaki, A. (2011). *Makerere University College of Health Sciences' role in addressing challenges in health service provision at Mulago National Referral Hospital.* Academic Press.

Head, M. (2022). The impact of COVID-19 on mental health: A systematic review. *Journal of Clinical Psychology, 78*(1), 1–34. doi:10.1002/jclp.23087

Heleta, S. (2015). *How to rebuild higher education in countries torn apart by war.* The Conversation. Retrieved from https://theconversation.com/how-to-rebuild-higher-education-in-countries-torn-apart-by-war-48791

Heleta, S. (2020). Higher Education in Post-Conflict Settings. In The International Encyclopedia of Higher Education Systems and Institutions. Springer. doi:10.1007/978-94-017-8905-9_220

Hima, H. (2023). Learning on the move: Resetting the agenda for education and learning in conflict-affected settings. *International Journal of Educational Development, 83*, 102413. doi:10.1016/j

Johnson, A. T. (2013). University agency in peacebuilding: Perspectives on conflict and development in Kenya. *Prospects, 43*(3), 329–345. doi:10.1007/s11125-013-9276-8

Kester, K. (2020). *The United Nations and Higher Education: Peacebuilding, Social Justice and Global Cooperation for the 21st Century*. Information Age.

Kester, K., Abura, M., Sohn, C., & Rho, E. (2022). Higher education peacebuilding in conflict-affected societies: Beyond the good/bad binary. *International Journal of Comparative Education and Development, 24*(3/4), 160–176. doi:10.1108/IJCED-04-2022-0027

Khadiagala, G. M. (2021). The African Union in Peacebuilding in Africa. In T. McNamee & M. Muyangwa (Eds.), *The State of Peacebuilding in Africa*. Palgrave Macmillan. doi:10.1007/978-3-030-46636-7_12

Kuyok, K. A. (2017). Higher Education in South Sudan: Living with Challenges. *Industry and Higher Education*, (89), 16–18. doi:10.6017/ihe.2017.89.9839

Makerere University. (2021). *Strategic plan 2020 – 2030: Unlocking the knowledge hub in the heart of Africa.* https://www.mak.ac.ug/sites/default/files/Strategic%20Plan%202020-2030.pdf

Mba, J. C. (2017). *Challenges and prospects of Africa's higher education.* Global Partnership for Education.

McNamee, T & Muyangwa, M. (2021). *The State of Peacebuilding in Africa Lessons Learned for Policymakers and Practitioners: Lessons Learned for Policymakers and Practitioners.* doi:10.1007/978-3-030-46636-7

Millican, J., Kasumagić-Kafedžić, L., Masabo, F., & Almanza, M. (2021). Pedagogies for peacebuilding in higher education: How and why should higher education institutions get involved in teaching for peace? *International Review of Education, 67*(5), 569–590. doi:10.1007/s11159-021-09907-9 PMID:34219800

Milton, S. (2018). Rebuilding and Reforming Higher Education. In *Higher Education and Post-Conflict Recovery*. Palgrave Macmillan. doi:10.1007/978-3-319-65349-5_4

Mlaba, K. (2023). *How do war and conflict impact education?* Penguin.

Moja, T., & Cloete, N. (2021). Transforming higher education in South Africa: A new approach to governance. *African Issues*, *29*(1), 44–58.

Nabatte, P. (2020). African universities urged on developing comprehensive safeguarding policies with critical elements of safety & protection. *Journal of Higher Education in Africa*, *18*(2), 45–60. doi:10.1234/5678

Nyuykonge, C., & Zondi, S. (2017). South African Peacebuilding Approaches: Evolution and Lessons. In C. Call & C. de Coning (Eds.), *Rising Powers and Peacebuilding*. Rethinking Peace and Conflict Studies. Palgrave Macmillan. doi:10.1007/978-3-319-60621-7_6

Pellini, A. (2021). *War crashes higher education systems – countries like Somalia need a system-wide reboot*. ODI: Think change. Retrieved from https://odi.org/en/insights/war-crashes-higher-education-systems-countries-like-somalia-need-a-system-wide-reboot/

Pherali, T. (2022). *Conflict, Education and Peace in Nepal: Rebuilding Education for Peace and Development*. Bloomsbury. doi:10.5040/9781350028784

Raqib, S. (2017). Far-reaching consequences of wartime attacks on education. *International Journal of Educational Development*, *55*, 12–19. PMID:31024119

Ruby, J. (2021). The seven centers of peace. *Journal of Peace Studies*, *12*(3), 45–56.

Sadiq, H., Barnes, K. I., Price, M., Gumedze, F., & Morrell, R. G. (2019). Academic promotions at a South African university: Questions of bias, politics and transformation. *Higher Education*, *78*(3), 423–442. doi:10.1007/s10734-018-0350-2

Sahar, A., & Kaunert, C. (2021). Higher education as a catalyst of peacebuilding in violence and conflict-affected contexts: The case of Afghanistan. *Peacebuilding*, *9*(1), 57–78. doi:10.1080/21647259.2020.1731123

Taka, M. (2021, May 17). *Conflict kills education: Rwandan experiences show how lost years can be recovered.* The Conversation. https://theconversation.com/conflict-kills-education-rwandan-experiences-show-how-lost-years-can-be-recovered-158314

Tuhereze, E. (2021). Postgraduate diploma in peace building and conflict transformation. *Journal of Peace Studies*, *18*(2), 45–60. doi:10.1080/12345678.2021.0000000

University of Cape Town. (2020a). *Health and safety policy*. University of Cape Town. https://www.uct.ac.za/main/explore-uct/policies/health-and-safety-policy

University of Cape Town. (2020b). *History introduction.* https://www.uct.ac.za/main/explore-uct/history-introduction

YevdokymovaN. (2023). Resilience and Reform: The Impact of War and Conflict on Innovation in Higher Education. doi:10.2139/ssrn.4602827

Chapter 15
The Global Citizens in Global (Un)certainties:
Southern Higher Education Students in the North

Md. Tariqul Islam
https://orcid.org/0000-0002-4165-5594
Hiroshima University, Japan

ABSTRACT

In the era of globalization, neoliberal free trade regards higher education as a private good that facilitates international mobility for and of higher education. Similarly, HEIs establish an economic image of overseas higher education that encourages young people to invest in global mobility to pursue higher education. In this context, following a qualitative research methodology, the chapter analyzes the (un)certainties of 18 physically mobile young Bangladeshi (Southern) students. It discusses their consciousness issues and confidence level regarding consuming higher education in Australia (North). To analyze their perceptions, they were considered global citizens (i.e., economic and moral actors in the globalized world). Based on the empirical data, the identified features of young Bangladeshi students and their areas of concern provide the scope to identify them as conscious and confident neoliberal consumers in the global education market who came to Australian universities with neoliberal aspirations.

INTRODUCTION

In the post-World War II era, economic and societal development in most nations was mainly guided by Keynesian economic theory with its emphasis on the welfare and state planning of the economy. Following the introduction of neoliberalism as a political project in the 1980s and the fall of the Soviet Union at the end of that decade, a significant shift in societies, their economies, institutions, and education sectors (Rizvi & Lingard, 2000) was observed as a result of the rapid expansion of global free market ideas. Both developmental and critical literature use the term 'globalization' to describe this

DOI: 10.4018/979-8-3693-1926-0.ch015

shift. Davies (2006, p. 9) observes that "globalization is both a threat and an opportunity, in terms of the impact of trade, technology, media, social organization, and cultures." From a development standpoint, globalization impacts nearly every element of our lives (Rizvi, 2007). It hastens the formation of persistent networks, resulting in the emergence of international interdependencies. In this framing, the main discussions around globalization focus on economic development and the significant role of transnational companies (Osler & Vincent, 2002). However, we see globalization as more than a form of economic imperialism and a way to create a single global market economy (Shultz, 2007). Because of the uneven characteristics of globalization, various types of inclusion and exclusion are visible through complex and dynamic sets of international, national, and local relationships.

We can find the causes and impacts of globalization in political, technological, cultural, and economic practices, which have undergone significant changes in recent decades. Thus, globalization is interconnected with international commerce, virtual economies, electronic capital transfers, labor and production, information, ecology, legal and administrative systems, culture, and civil society (Osler & Vincent, 2002). In other words, the impacts of globalization extend beyond the realm of public policy. Cultural practices may be influenced, potentially to a greater extent than economic and political factors. Similarly, Urry (2003b) sees the role of social science as reporting on the transformation of local, regional, national, environmental, and cultural aspects within globalization's new robust capitalist structure. In other words, globalization is also viewed "in relation to the various ways in which global processes now work – through the flows of people, capital and information, as well as ideas and images found in the media" (Rizvi, 2008, p. 18) and to how political, social, technological, economic, and cultural changes are initiated (Rizvi & Lingard, 2000), what does this imply for contemporary citizenship? If, as Held and Moore (2007) contend, these transformations are also occurring at the individual level and are shaping values, aspirations, self-understandings, and identities, what might this imply for how individuals perceive such dynamics, including new demands and challenges on us as individuals?

For this reason, Bauman (2009, 2013a) supports using education for global citizenship to develop young people's individual and social skills to engage them in critical dialogue with others as they engage with this globalized world and make meaning of it. Similarly, Appadurai (2007) suggests that education could be mobilized to address the capabilities of young people to assist them in navigating cultural maps where their future aspirations can be formulated to promote an understanding of their objectives concerning global perspectives and norms. More broadly, the United Nations (2015) has set targets through SDG 4 to 'ensure inclusive and equitable quality education and promote lifelong learning opportunities for all' is realized by 2030. Specifically, this goal aims to develop the required knowledge and skills for global citizenship through target 4.7, enabling all learners to promote sustainable development.

Very little literature can be found on international student mobility for higher education and the student's educational and lived experience of globalized mobility (Islam, 2021, 2022, 2023d, 2023a, 2023b). In other words, a small amount of the literature focuses on the micro aspects of globalization discussed above, such that the individuals' lived experiences remain hidden. Most of the research in this field continues to focus on the macro themes that characterize globalization debates. Thus, the influence of globalization on young students in international mobility pursuing higher education in shaping their characteristics and developing them as conscious and confident global citizens in the global (un)certainties is well unexplored. Young students' citizenship experiences in the form of international mobility are a micro aspect of globalization that needs to be considered as part of the discussions surrounding the global higher education market. For this reason, higher educational institutions must be encouraged to consider the characteristics of young students' mobility motivations and experiences to prepare them as future global citizens and ensure their survival and success in the competitive global free market of education. In this context, adopting the conceptual model of education for global citizenship (Islam, 2019) and theorization of Southern students in the North (Islam, 2023c), in this chapter, I describe and discuss some of the micro or individual (Urry, 2003a) aspects of globalization, such as mobility, identity, competence, and consciousness by examining some of the relationships that individuals have with macro issues, such as economics, education, culture, and citizenship. Here, I emphasize identifying globalized young people's distinguishing characteristics to understand the new form of citizenship they are experiencing within the complex economic and moral dilemmas of globalized mobility. This discussion will contribute towards a further understanding of the influence of globalization on the micro level, especially on young students in international mobility pursuing higher education.

Globalized Young People

In this age of globalization, many young people aspire to a better, more prosperous, and just world that seems increasingly borderless. Simultaneously, globalization increases anxieties and fears about rising inequalities and economic uncertainties. Young individuals on a global scale are frequently anticipated to cultivate a comprehension of cultural diversity, progressively integrating it into their daily encounters. Moreover, they must have the skills to compete with workers around the globe for jobs to survive in the world economy (Dill, 2013). We can say that global citizens require two significant characteristics. Firstly, they ought to become *economic actors* in the global labor market. Secondly, they must become *moral actors* responsible for their actions impacting others and the environment. In the following section, a literature review identifies some of the significant characteristics of global citizens that are often contested, contradictory, or overlapping.

Global Mobility

In the globalized world, Urry (2003b) considers physical, imaginative, and virtual mobilities as the rights of people. Rizvi and Beech (2017) observe that in contemporary society, the phenomenon of global mobilities has become a prominent force that is reshaping the social environments in which individuals currently reside, engage in employment, and pursue education. Similarly, mobility is considered one of the most popular values (Bauman, 1998) for young people. According to Skrbis et al. (2014), in the majority of instances, individuals form anticipations and ambitions regarding future mobility throughout the early stages of their lives. However, it is important to note that these goals are subject to continuous evolution and alteration during the course of one's life. These shifts primarily occur due to discernible demographic or structural factors. International mobility, however, has become one of the defining aspirations of young people (Rizvi, 2008). Dolby and Rizvi (2008) demonstrated the diverse range of potential movements in which young individuals can engage. Firstly, individuals relocate internationally as immigrants due to economic or political motivations, as well as for educational pursuits. Furthermore, they actively strive to establish novel living situations by assuming the role of change agents. Ultimately, individuals virtually traverse global culture, media, cinema, fashion, and the Internet. Many mobile young people develop their identity and future aspirations within international mobility. However, the patterns of the movement of young people at present are different from those of previous generations, who used to transform the sites in which they lived. On the other hand, globalized young people have a multiple and mobile sense of belonging through occupying a very different space. As a result, the mobility of young people at the international level has an implicit impact on political ecology through the development of global citizens.

Global Identity and Relationships

At present, young people live in a world with fewer borders, as they can interact with any part of the world through developed communication technologies and transportation systems. Rizvi (2017) suggests that the current era need novel approaches to understanding global interconnectedness and interdependence that extend beyond the economic dimensions of globalization. This understanding should also encompass moral and intercultural considerations. In this situation, globalized young people might consider themselves global citizens rather than national citizens (Bostad & Ottersen, 2014). Similarly, Wood and Black (2018) argue that the conventional understanding of citizenship, which is characterized by fixed, territorial, and status-based attributes, does not adequately encompass the modern experiences of young individuals as citizens. In other words, the concept of citizenship has evolved to encompass more than just a legal status and the regulations that govern it (Walsh & Black, 2018). However, globalized people exist in a complex relationship to traditional, state-bound citizenship forms. According to Dill (2013), no global political body might confer citizenship status, rights, or responsibilities on global citizens. As a result, they live in a complicated situation, as they do not have any political home and ruling authority for this global citizenship; instead, they have a deep commitment to a broader economic and moral purpose and universality. Thus, young people are experiencing a new dimension of political identity, which is fluid and multiple compared with citizenships from rigid single nation-state contexts (Armstrong, 2006; Oxley & Morris, 2013). Many globalized young people also dream of a formal status of all human beings as global citizens. To fulfill this dream, they build relationships with the state and other political organizations through cosmopolitanism by promoting global governance and establishing a world state.

Informed and Competent About Global Culture

In the context of globalization, there is an expectation for young individuals to embody the qualities of being socially responsible and considerate towards both their immediate local communities and the broader global community. This expectation arises from recognizing that they inhabit a world where cultural boundaries are increasingly permeable (Dolby & Rizvi, 2008). Therefore, individuals must possess cultural and intellectual competence to effectively navigate the challenges brought about by the globalization of media and language facilitated by satellite television and the Internet. The importance of cultural competencies is increasing as a way to effectively navigate the diverse cultural landscape that exists within a nation. According to Walsh (2016), developing cultural competencies is crucial to establishing geopolitical and economic connections with individuals from different nations. Young individuals must comprehend various aspects of global society, including cultures, languages, religion, economics, science, and technology (Braskamp, 2008).

Additionally, they should have a heightened awareness and reverence for diverse cultural traditions. De Costa (2016) posits that developing intercultural communicative competence is pivotal in constructing global citizenship. Furthermore, the utilization of English as a Lingua Franca facilitates the cultivation of this skill. In the contemporary context, characterized by a highly interconnected and boundary-less global landscape, young individuals are compelled to engage in cross-cultural collaboration as an integral component of the global labor force (Dill, 2013). In order to thrive in the era of globalization, young individuals must possess a keen understanding of cross-cultural dynamics, as well as effective communication and collaboration abilities. This necessity arises from both economic and moral standpoints. In this setting, individuals must adopt an inclusive stance towards cultural diversity, cultivate a receptive mindset, and actively foster connections with others by enhancing their intercultural competence. These efforts are essential for attaining both economic and moral objectives.

Competence in Economic and Technological Aspects for Survival and Success

Economic transformations in the globalized world have a direct effect on the lives of young people. In this context, as global workers, young people are experiencing a world of intolerance and economic competitiveness. They must develop themselves as competent workers and consumers for the global capitalist society and rely on individualism to do so (Dill, 2013). From the capitalist point of view, globalized young people serve the intention of global economic integration (Folk, 1993). Young people play their role in the competitive free market economy as labor, human capital, and consumers for economic growth and international development following the ideologies of neo-liberalism and capitalism (Oxley & Morris, 2013). According to Walsh (2016), as the lifestyles of young people become challenging and they face new problems for participating in the contemporary workforce, it is essential to promote the necessary literacies. Young people facing the realities of increasing virtual and physical mobility and no longer bound to a single place or nation-state need to accept the importance of local knowledge and think about and engage with the practicalities of global employment requirements.

Globalized young people should also have some economic and technological competencies to flourish and thrive in the neoliberal world. They need to achieve the knowledge, skills, and competencies to compete as human capital to succeed in this new world of challenging (and at times confronting) economic realities. This global citizenship encompasses a combination of technical proficiencies, logical problem-solving abilities, and appropriate sensitivities for individuals who own agency in expressing themselves freely (Dill, 2013). Young people must be equipped with digital literacy and soft skills for working (Walsh, 2016). These skills are essential for young people to adapt to the profound structural changes and the world of insecure work influenced by globalization in specific countries and internationally.

Consciousness of Global Issues and Events

Globalized young people see themselves coming from a long tradition of 'local' communities and normative perspectives to create a better world (Folk, 1993). In the cosmopolitan age, globalized young people regard the global community as one without boundaries where all members are concerned with and responsible for each other and the planet (Dill, 2013). From a moral point of view, young people could be considered members of a single community where the relationships of individuals and groups are based on a universally accepted notion of human rights (Oxley & Morris, 2013). According to Osler and Vincent (2002), globalized young people should respect and value others and their experiences from local and global perspectives.

Moreover, they should appreciate diversity and be interested in taking action for the betterment of others. According to Dill (2013), global consciousness is an essential element for globalized young people, which they develop with the combination of consciousness about others' perspectives, consideration of humanity as the base of the community, and moral conscience for the advancement of the world. For this reason, only recognizing differences across countries does not express the concept of developing a global perspective; it is instead about integrating all people from all races, cultures, and religious backgrounds. Globalized young people should not limit their perspectives to cultural differences historically associated with nations and countries. Moreover, they should understand and respect justice, equity, fairness, and equal opportunities as virtues and values relevant to a global community. Nevertheless, it is crucial to acknowledge that these ideals and virtues should not be regarded as universally accepted truths, but rather as significant and debatable objectives that are integral to our debates, which also embrace other traditions (Braskamp, 2008). According to Oxley and Morris (2013), some globalized young people also see their role from a spiritual point of view based on love, caring, and cultural symbolism available in some religious teachings, faith-based meditations, and humanism.

Furthermore, for the sake of human survival, young people should develop attitudes and skills regarding energy, resources, environment, and political choices to maintain a sense of fairness and uphold human rights (Folk, 1993). This type of globalized young person acts as a transnational activist to develop a global civil society as concerns class, ethnicity, gender, disability, sexuality, or any other social context. Globalized, conscious young people advocate for changes in the actions of humans, governments, global civil society, and the corporate world. They emphasize the importance of nature, human rights, future generations' security, and Indigenous people's rights when seeking to fulfill the aspiration of sustainable development (Oxley & Morris, 2013). In addition, young people need to understand a sustainable future where political aspects are emphasized along with sustainability's environmental, social and economic aspects. Political literacy provides them with the knowledge and understanding of political systems' working procedures and languages and advocacy and mobilization skills to participate in effective change.

In the context of globalization, the younger generation must possess the necessary skills and knowledge to effectively address and overcome obstacles that arise both within their local communities and on a worldwide scale (Osler & Vincent, 2002). Critical thinking is a prominent attribute among young individuals in a globalized society. According to Oxley and Morris (2013), individuals with critical thinking skills prefer to scrutinize the motives, procedures, and objectives underlying various occurrences within the globalized society. Moreover, they reflect thoughtfully on the potential ramifications of these events. They examine the emergence of new identities and connections and the necessary competencies for survival. The younger generation critically examines moral and cultural practices while also considering the potential for social mobility within the context of globalization. They adopt a dual perspective, acknowledging the experiences and perspectives of both those who hold power and those who are marginalized.

METHODOLOGY

In the chapter, a qualitative research methodology was employed to conduct a study involving a sample of 18 Bangladeshi students who are currently enrolled in educational programs at two universities in Australia.

Research Site and Participants

I deliberately chose two universities located in Melbourne, Australia. To safeguard the confidentiality of the designated Australian universities, I used the pseudonyms Eastern University (EU) and Southern University (SU). To begin with, the chosen universities exhibit a substantial enrollment of international students, including those from Bangladesh, compared to other Australian universities. Additionally, both selected universities have committed to fostering the development of students as global citizens, as seen by the inclusion of this objective in their graduate learning outcomes and expected traits. I recruited a specified quantity of research participants who could furnish information regarding the topic and fundamental phenomena of the investigation using snowball sampling (Creswell, 2013).

Methods

During the initial data collection phase, an online pre-interview survey (Collinson, 2004) was sent to the participants. This survey aimed to obtain pertinent general background information regarding their mobility and ability to pursue a degree at an Australian institution. The pre-interview survey was designed, and data was collected online using Qualtrics, as described by Snow et al. (2013). The survey's primary objective is not to quantify data but to provide a comprehensive understanding and justification for the study to encourage participant involvement. Additionally, the survey aims to collect essential general information that will be used to analyze qualitative data. Furthermore, I conducted individual interviews with these young mobile students to collect qualitative data regarding their attitudes and educational experiences in Australian universities, allowing them to express their thoughts and perspectives in their own words. The inclusion of the perspectives of young students is of utmost importance in this chapter, given the significant variations in their individual experiences within the context of globalized mobility. In the present setting, the utilization of semi-structured interviews is seen as essential due to its capacity to provide comprehensive insights into the lives of young pupils. This study aims to ascertain the perspectives of mobile students in their youth regarding Australian higher education within the context of globalization. The interview approach seeks to address the issue of invisibility by thoroughly elucidating the concealed discourses present in the life narratives of young mobile students within a global framework (Brinkmann, 2013; Elliott & Urry, 2010).

Data Analysis

The data for this study was analyzed using a combination of a pre-interview survey, interviews, and document analysis to address the research objectives. A figure was constructed based on the data obtained from the pre-interview survey, which was subsequently analyzed in conjunction with the qualitative data gathered during the interviews. Within the confines of this chapter, I have employed the theme analysis approach, wherein I have scrutinized every novel piece of data concerning the overarching objective of the research.

RESULTS AND DISCUSSION: THE MOBILE MERCHANTS

In this section, I will analyze the educational aspirations of physically mobile young Bangladeshi people and discuss their issues of consciousness and level of confidence regarding consuming higher education in Australia. To explore their perceptions, I will consider them economic and moral actors in the globalized world. Regarding university selection, all of my participants expressed their desire to study in a higher-ranked university. At the same time, they also considered the country, city, cost, and credibility of the certificate of that university. Many choose a double major or double degree (Harpur, 2010) through an academic program. Most importantly, all of my participants confidently depend on the Internet for all their decisions related to higher education. These features of young Bangladeshi students and their areas of concern allow me to identify them as conscious and confident neoliberal consumers in the global education market who came to Australian universities with neoliberal aspirations.

The Desire for Higher Ranked University

By and large, neoliberal consumers aspire to consume well-known brand products that provide an extended warranty period for a competitive price, prestige in society (Marginson, 2006), and individual satisfaction through consumption (Maringe, 2006). Similarly, my participants articulate their desire to study in a higher-ranked university in one of the most livable cities of a developed country. Moreover, they decided to enroll after comparing the cost of education and the credibility of academic certificates in the global job market among several higher-ranked universities. The issue of ranking the two universities that I selected as my research sites is reflected in Fazi's account from a young Bangladeshi student's perspective:

Southern University and Eastern University these are the uni[versities] to go especially when you are from Bangladesh and you hear that these universities are in the top 500 in the world. Obviously, you would not think about anything else. Like ok, these universities are the best. (Fazi)

Like Fazi, Mridul rationalizes his choice of destination country for higher education based on the quality and cost of each. However, he places more weight on the educational expenses related to studying in Australia:

Australian universities have a good reputation compared to Canada and the UK. I would say Australia is better in terms of the quality of education and most importantly, they have cheaper option compared to that of the UK and Canada. (Mridul)

Further, Jake and Sif emphasized the importance of a livable city when selecting a higher educational institution:

So when I thought of Australia. I tried going to universities that were ranked and affordable. I liked Melbourne at that time because Melbourne is supposed to be the seventh consecutive year for it to be the most livable city. When I was researching on Melbourne, Southern University seems the best at that time. (Jake)

I was always interested in Melbourne as Melbourne already has some good reputation in terms of the world's most livable city. If you want to choose Melbourne, you do not have many options. Suppose, if you want to go for a group of eight, you just have two options in Melbourne. Regarding the feasible solutions, even Eastern University is expensive, and it is not cheap, but it is affordable [for me]. (Sif)

Though Miridul, Sif, and Jake mentioned the factors of quality of education, country or city in university selection, the common area of their concern is the cost. Similar to them, Shuvo and Beatrix mentioned their calculations to find a balance between their budget and university brand:

Actually, I was thinking about the financial issue, but I got an international Merit Scholarship from the university, and my tuition fees were deducted a bit. Comparing the other universities, it is almost the same when I got my scholarship. (Shuvo)

My father gave me a budget that you should be under this amount. And I wanted a good ranking. So I thought about Southern University. It matched to my requirements. (Beatrix)

In this context, Sif explained young students' decision about how they select a suitable Australian university through a financial analysis of its ranking and fees:

Eastern University was one of the convenient solutions for me. For us convenient means the cost. If you want to go for something, very expensive that is not possible and if you want to go for something very cheap, even that is not feasible. So, you have to choose something in the middle. (Sif)

In addition to ranking and fees, young students consider the duration and structure of programs, the accommodation facilities and costs in the nearby areas, and the availability of part-time work for students in the host country. For instance, Imti explains that through a three-year bachelor program, he will invest less time and obtain his degree more quickly. According to him:

It [the bachelor program] is three years of length so it was a short one and I took it. Actually, the study takes away time, and I do not want to waste time. I have a plan for doing business, especially in IT. (Imti)

Besides, Fazi compares the program delivery structures between two of my research sites. He explains his preference for the trimester system rather than the semester system because of the provision of a one-month break after each trimester. According to him:

I feel Eastern University is more complicated than Southern University. Cause, Eastern University has a semester system, and their system is pretty fees forward. Trimesters, three months a unit and one month off. It is very easier to understand what is going on. When I got to know that Southern University is not complex than Eastern University, it attracted me. (Fazi)

Fazi likes Southern University's trimester system and the one-month break after each trimester, as this opportunity gives him more time to earn through part-time work to pay study-related fees. In other words, his preference is similar to those of consumers who like to purchase products in installments. In this installment system, consumers can use the product, and at the same time, they have the time to earn money to pay for the product. The importance of the break after each trimester to survive in Australia is also echoed by Irti:

Since I had to study overseas, I had to look for a few certain things. Such as, what would be the accommodation, how much time I would be allowed to work, how much money I will be making, how would be the study pressure and everything. Since I will be a full-time student, I also have to work part-time, because the accommodation is not always easy. (Irti)

Figure 1. Participants' current area of study

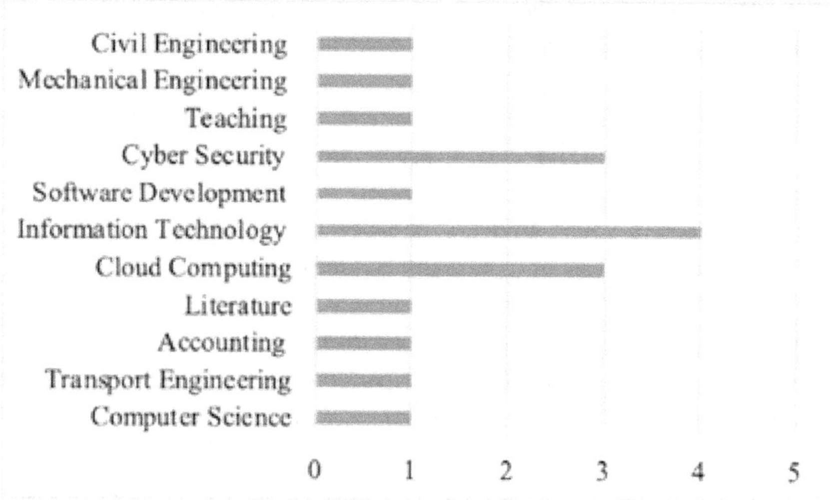

Like most of my participants, part-time work is vital for Irti to contribute financially to his overseas education or accommodation. As a Southern University student, Irti can use the trimester break to work part-time without any study pressure.

Cost and Quality of and Return from Education

In my participant selection criteria, I did not specify their current area of study. Still, I opened it to all disciplines to see the trend of Bangladeshi students regarding their area of study selection. Among my 18 participants, I found Bangladeshi students tend to choose their discipline or first major within eleven areas of study:

Figure 1 shows my participants' current study area; most are from the Faculty of Sciences and Engineering, two are from the Faculty of Social Sciences, Arts and Education, and only one is from the Faculty of Business Studies. However, many of them have a second major area of study that they were not asked about during the pre-interview survey.

Figure 2 displays Bangladeshi students' reasons behind choosing their area of study. Most of my participants selected their area of study based on career aspirations. Among my eighteen participants, fourteen hope their selected study area offers them employment or career opportunities. Many of them also gave importance to the university ranking for their disciplines. Similarly, interview data supports the pre-interview survey data regarding young students' career concerns for discipline selection. The statements below from Anisul and Sif reflected how IT or IT-related subjects such as Computer Science, Software Development, Cloud Computing, and Cyber Security address the career aspirations of young students. According to them:

Cyber Security is the core unit of IT. If you do research online, you will see there are many vacancies in cybersecurity. In 2020, the cybersecurity issue will be the main issue in the IT sector. (Anisul)

Figure 2. Bangladeshi students' reasons for choosing a particular area of study in Australian universities

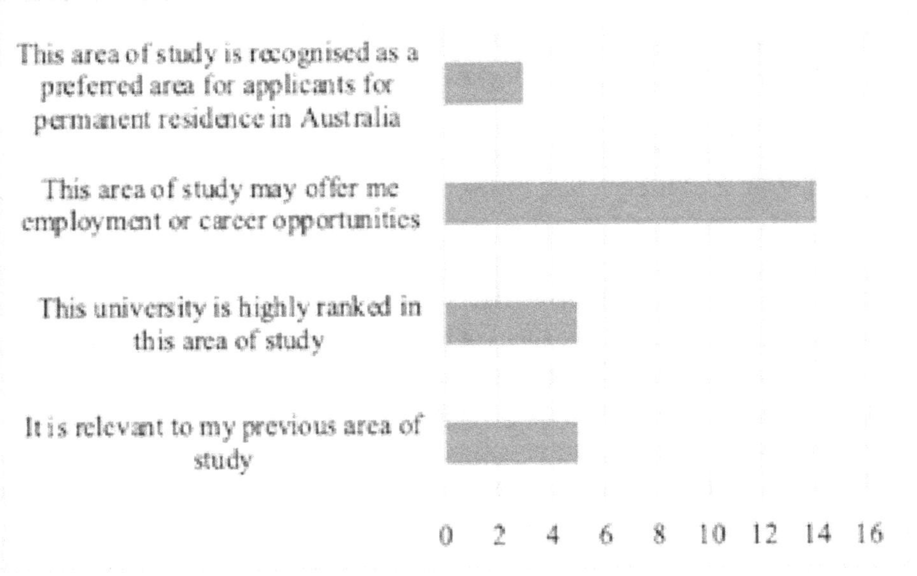

The entire industry of IT is growing. I think everything regarding IT will stop at a point. When you stop there, there will be one point starting again that is secure those devices. Nowadays, everything is going to the cloud. Whenever you have data in the cloud, you have to have security to secure the cloud. So it is a growing industry, and network security is an industry inside an industry. I think that is a better [career] option. (Sif)

Tamal is studying Civil Engineering at Southern University. He explained the global job prospects of his discipline and rationalized his choice of area of study based on an economic perspective:

I found that in Australia and Bangladesh as well as in another developed country there are a lot of things happening related to Civil Engineering. So, its job opportunities much higher than the other sectors. (Tamal)

I mentioned earlier that many of my participants are doing a double major. The selection of two different areas of study can be seen as the outcome of a risk factors analysis regarding future career prospects. For example, Beatrix is doing her bachelor's in literature while she takes Psychology as her second major. She rationalized her decision concerning career strengths, weaknesses, opportunities, and threats at home or abroad. According to her:

I am doing my double major. One is in literature, another is in Psychology, and Psychology is my backup. I think it has better career prospects at least in here [Australia]. If ever I plan to stay back [in Bangladesh], I already have a major in literature, and I can go back and get a job as a teacher in Bangladesh. (Beatrix)

The acceptability of the certificate in the global job market is a vital factor for young Bangladeshi students in choosing their study destination. The acceptability of an Australian degree boosts confidence among my participants concerning future job prospects worldwide. Prince illustrates young students' perception of the value of certificates from Bangladeshi universities and Australian universities:

Australian certificate is acceptable all over the world. Like our [developing] country's bachelor [degree] from a few universities not even acceptable. You have to do the bachelor again [to get a job overseas]. (Prince)

In addition to showing an appreciation for the credibility of an Australian certificate, many young Bangladeshi students deliberately check the status of their selected discipline in the lists of eligible skilled occupations for PR in Australia. Figure 1 reflects that some of my participants' area of study selection is linked with Australia's migration policies. Young Bangladeshi students are physically mobile to Australian universities due to the attractiveness of this migration policy-controlled labor market (Riaño et al., 2018; Rizvi, 2008; Wei, 2013) rather than their interest in becoming an expert in a particular discipline. For instance, in 2007-08, under various visa class applications, 21,421 international students were granted residency and full work rights in Australia (AE, 2009; Gothe-Snape, 2018). The link between the future possibility of living permanently in Australia and the selection of an academic area of study is reflected in Fazi's excitement:

I saw the benefits of IT, and it is on the list of PR as well. When I chose IT, and I saw it on that list, it makes me happier. If I had an 80% interest in IT, it made it 90%. (Fazi)

From the above discussion, I argue that, to face what Bauman (2013d) identifies as the postmodern challenges, most young Bangladeshi students are studying professional programs and doing double majors in Australian universities. In other words, the neoliberal imagination of globalization makes Bangladeshi elite young students choose dominant fields of study to ensure financial and social benefits (Banks et al., 2007; Rizvi, 2011). Based on the selected discipline and credibility of Australian certificates, Jake showed his capacity to aspire (Appadurai, 2007) to a boundaryless career that is generated by the neoliberal imagination:

I did not say that I would only apply for a job only in Australia. My final aim is to move on to wherever the world provides me with a better career. It [Australian certificate] is going to be accepted [worldwide]. Australia is not my final destination. My final destination is a career. (Jake)

In this context, Anisul provided a mathematical calculation of how he may receive economic returns (Eder et al., 2010; Maringe, 2006; Varghese, 2008) from his financial investment in higher education in Australia:

I am dreaming of working overseas like America or Australia. I do not want to go back and start a job in my country. Because I am spending here lots of money and if I go back and work in my country, I will not be able to earn that much money, not even half of that. I am spending around 80 thousand dollars, and if I get a job in Australia, the minimum salary will be 50 thousand or more at the starting level. So, it will take like 2 years or 3 years to get the whole money back. (Anisul)

According to Anisul, only an opportunity to work in a developed country can ensure a return from investing in overseas higher education. In the above quote, he uses the example of the Australian job market, from where he expects to return the whole invested amount within two to three years if he obtains a job in his field. In opposition to Anisul's view, very few of my participants expressed their interest in contributing (Tran & Vu, 2017) to Bangladesh after graduating from an Australian university. For instance, Topu expressed his desire to work in the telecommunication sector in Bangladesh:

I saw that the telecom industries in Bangladesh are growing and I have an interest in Computer Network and Security as well as Software Development. I thought that maybe I could contribute back to my home. (Topu)

From the above-discussed data, I claim that by analyzing present benefits and future outcomes, young Bangladeshi students consciously select their university and area of study in Australia. In other words, most of them show their confidence in choosing a product according to its resale value and calculating the rate of economic return from their investment in overseas education. Their aspirations from this investment in global higher education mainly include obtaining a job in a developed country, living permanently in one of the world's most livable cities, and becoming part of the Western lifestyle.

Dependency on the Internet

Now, I will discuss the process followed by young Bangladeshi students in making a conscious decision and the aspects they consider to be confident about overseas higher education. After analyzing the interview data, I argue that my participants' dependency on the Internet is behind their conscious decision about international mobility for higher education and their confidence about their future career, which Bauman (1998) refers to as a secondary and derivative sense. The mediation of the Internet as a strong media to create a porous cultural border in the globalized world motivates young Bangladeshi students to be internationally mobile and pursue Australian higher education (Islam, 2022). Australian universities present their features in attractive ways through their websites, YouTube channels, and social media to develop trustworthiness among young students about their offers. For instance, Anisul and Pikachu became motivated by the keyword 'worldly' (SU, 2013) and the program named 'Industry-based Learning' (EU, 2018), respectively. Like conscious techno-reliant consumers, young Bangladeshi students analyze university and discipline rankings, explore customers' reviews about program structure, cost, and quality of education, and examine the possible return from investment in global higher education through online marketplaces. The role of the Internet in developing young students' confidence about rankings, course curriculum, and career prospects is reflected by Shuvo, Anisul, and Tamal:

The ranking [of Eastern University] is really good, and you can see their course curriculum [on the Internet], the way they designed the program is really good. (Shuvo)

Yeah, when I was in my country, I research online about a career in IT or Cyber Security. (Anisul)

When I finish my college, I started to research on further education and the universities in other countries like who are doing good. I did my research on the Internet. I looked for what the people think about the university and what the people think who are studying in universities. So what I remember is some people were saying like if you study at Southern University, there is a high chance that you can get a good job. (Tamal)

From the above quotes, I observe that in the Internet era, Australian universities need not worry about education agents' power of persuasion (Bauman, 2013b) to enroll international students.

Critical Thinking

In addition to these initiatives to make young students succeed in the labor market and academic courses, teachers also introduce examples to promote critical thinking (Oxley & Morris, 2013; Walsh, 2016) among students. In business studies, the discussion of social inequality and injustice is considered by Mridul as an effort by the lecturer to make the students aware of what Appadurai (2013) suggests is essential for informed and critical citizenship. According to him:

In management [course], we have learnt about how Adidas is reaping off the employees that they hire most in Africa. They are paying them low wages. A woman from Africa said that just to make you look fashionable, I have to live on two dollars. It is like Adidas is paying somebody two dollars equivalent in African currency for an hour of labour. But their products they are producing are basically very expensive. I think because they [teachers] bring up the cases we know the types of the problem being faced by different developing countries. I think this way they are trying to make us more aware of the situation on what is going on. (Mridul)

Mridul's experience can be described as an example of the ethics of possibility (Appadurai, 2013) practiced by a faculty member in an academic course at his Australian university. Appadurai identifies the ethics of possibility as the aspiring capacities that develop informed, creative, and critical global citizenship. Additionally, multidisciplinary courses create an opportunity to know logic from other academic backgrounds and develop a critically aware consciousness. Sif mentions the moral learning opportunities that multidisciplinary courses develop other than economic competencies for the job market:

The thing Eastern University is doing they are mixing up some specific courses for us. Suppose we study our core subjects, but regardless of those subjects, they are putting some subjects those are a mix of some other discipline like law, accounting something like that. I mean that is actually an opportunity to a mix-up with people from another discipline, different thinking and different logic. (Sif)

Along with academic and professional care, Australian universities ensure multidimensional and multicultural support (Bek, 2017) through several clubs, associations, and sports facilities. These non-academic organizations also enrich the Universities' customer care systems. The multicultural environment of Australia is one of the crucial motivating factors for young Bangladeshi students to choose this country as their study destination (Islam, 2021). The table shows the number of active clubs and associations in two Australian universities to develop young students as global citizens.

In addition to different clubs and associations, my research sites have sports centers to encourage students to keep fit and help them balance study, social activities, training, and competition in a conflicting situation (SU, 2018). Indeed, Topu finds Eastern University's sports facilities essential to maintaining balance in his life by supporting his mental, physical, and social development:

Table 1. Number of available clubs and societies at two research sites

Category	Eastern University	Southern University
Academic	41	7
Cultural	18	8
Sports	41	8
Religious/spiritual	7	3
Social/social welfare	8	9
Halls of residence	4	-
Performing arts	4	-
Political	2	-
Special interest	24	-
Total	149	35

Source: (EU, 2018b, 2018e; SU, 2018a)

Other than that [academic learning], they [Australian universities] have excellent sports facilities, that is very important for your mental health. The sports center [at EU] is helping me to keep fit both mentally and physically which is affecting my academic results, which in turn will help me in future [career]. Here, you can meet people as well in social sports, so it is helping me to maintain a balanced lifestyle. (Topu)

Moreover, Topu linked this support system to his future career as he connects sports with academic results. Like Topu, many of my participants consider these non-academic arrangements significant from a career perspective. For instance, Dina mentions the intention of these various clubs to promote diversity, while Anisul uses this diversity to expand his network, and Mridul imagines his career success in this multicultural learning and networking process:

They [clubs] actually try to make you embrace diversity. (Dina)

I work in SUSA and SUSA organizing 5-7 parties in the O-Week, and they organize tours, some training sessions like first aid training. If I go on a tour, I am meeting there new people, so I am making friends there, and I am just enlarging my network. (Anisul)

I would say graduates from Southern University might work in China, Bangladesh or India. These different countries have different cultures. It is better if you have a knowledge earlier from undergrad level. So, in future when you are going into the job, you will be successful. (Mridul)

Similarly, Sif and Pikachu shared their experience of academic and non-academic support systems at Eastern University. They understand that faculty and association members are trying to encourage them to learn beyond educational content to meet the requirements of their future professional fields. As stated by Sif, the student associations of his university inform them about employers' requirements:

Eastern University Student Association and the Eastern University Postgraduate Association are doing monthly events. They are always motivating us and coming up with some career aspects every semester.

They are always explaining things to us what we will need in the industry, what we will require in the industry. Time to time, we are getting used to with the present industry requirements. (Sif)

Correspondingly, Pikachu's lecturers encouraged her to become involved with some social activities that she will be able to include on her CV. According to her:

I think I got an idea of what else I should be doing other than my studying from lectures. because I heard it a lot in my lectures, you should be thinking of doing something else, and you need to make your resume stronger, there is not so much education can take you. They [employers] would also want to see other qualifications in your resume. They want to see you are involved in social work than just studying. (Pikachu)

Universities should organize These non-academic arrangements to develop students' consciousness and responsibility about the global community, human rights, cultural diversity, democracy, tolerance, political, environmental, social, and economic sustainability, and justice. These areas of consciousness and responsibility are seen as the moral aspects of global citizenship, which have been recognized in the literature (e.g., Dill, 2013; Oxley & Morris, 2013; Tran & Vu, 2017; Walsh, 2016). As I discussed above, some of my participants are becoming more consciously aware through these initiatives promoting cultural diversity and critical thinking. However, at this stage, I argue that Australian universities' academic and non-academic support systems may not fulfill their goal of promoting knowledge and the moral aspects of education for global citizenship. This is because these universities always seem to emphasize to young students that they need to develop competence for the future labor market through mobility and connectedness opportunities (Moskal & Schweisfurth, 2018; Tran & Gomes, 2017) and only rarely promote the importance of a balanced life and citizenship. Moreover, these universities might be upgrading their facilities, but it can be argued that the primary purpose of this is to attract young students; in other words, they promote different cultural and social organizations and events as a means of marketing and branding rather than being primarily for students' social, physical and mental well-being. Most importantly, these facilities and services are mainly arranged by these neoliberal universities to ensure their improved position in the global university ranking systems (Rizvi, 2011) by increasing graduate employability (Blackmore et al., 2014; Clarke, 2018; Huang & Turner, 2018). As Rizvi (2017) argues, the universities in question have adopted neoliberal imaginaries that have had a detrimental impact on the marginalization of moral and cultural discussions pertaining to the fundamental objectives of education.

In the same way, young Bangladeshi students use these learning opportunities and support systems to understand people from different cultures, enlarge their global networks (Nghia, 2015), and enrich their CVs for their future careers rather than emphasizing the moral aspects of global citizenship. In other words, they acquire multicultural skills to succeed in the labor market but not to accumulate ethical skills and social benefits from a shared life (Bauman, 2009, 2013a). These academic and non-academic ethics of care (Abdullah et al., 2017) and multivocality (Bauman, 2013c) mainly benefit students as they prepare for the global society as labor market actors. Thus, young Bangladeshi students are welcoming the Western nature of education (Waters, 2006), i.e., all of the academic and non-academic support systems (Nghia, 2015), arrangements and services (Rizvi, 2011) of Australian universities. From young Bangladeshi students' point of view, they find that these initiatives will benefit them in terms of their aspired future careers and economic gains.

CONCLUSION

From the perspective of economic imperialism alone, globalization functions as a mechanism to establish a unified global market economy. Due to globalization, various manifestations of exclusion and inclusion have emerged, each discernible through complex and dynamic networks of domestic, international, and global affiliations. This study explores the characteristics of global citizens by examining the experience and aspirations of Bangladeshi higher education students to pursue their studies in a university of higher ranking in a developed country known for its good quality of living. Findings suggest that the individuals decided to enroll after thoroughly evaluating the tuition fees and the perceived value of academic credentials in the international employment sector across multiple prestigious higher education institutions. The desire expressed by individuals represents neoliberal consumers' desires to consume widely recognized brand products that offer a longer warranty duration at a competitive price. This desire is driven by the pursuit of social prestige and personal fulfillment derived from consumption.

In response to the contemporary challenges posed by postmodernism, a significant number of young students from Bangladesh are pursuing professional degrees and engaging in the pursuit of dual majors in Australian universities. In essence, the conceptualization of globalization within the neoliberal framework influences the decision-making of privileged young students in Bangladesh, leading them to opt for fields of study that are perceived as dominating to secure economic and societal advantages. Some research participants are experiencing increased conscious awareness due to engaging in efforts that promote cultural diversity and critical thinking. Nevertheless, Australian universities' academic and non-academic support systems may not effectively achieve their objective of fostering knowledge acquisition and the ethical dimensions of education for global citizenship. This phenomenon can be attributed to the universities' consistent prioritization of promoting competency among young students for the prospective labor market, mostly through providing mobility and connectedness opportunities. However, these institutions seldom emphasize the significance of leading a well-rounded life and cultivating a sense of citizenship. Furthermore, it can be contended that the main objective behind upgrading facilities in these universities is to appeal to a younger demographic. In principle, they utilize various cultural and social organizations and events as a marketing strategy to enhance their brand image rather than prioritizing their students' social, physical, and mental well-being. Neoliberal universities mainly organize these institutions and services to improve their standing in global university rankings by promoting graduate employability.

In the current era marked by globalization, a considerable proportion of young individuals possess hopes for a global community that is more egalitarian, prosperous, and integrated. This vision entails a gradual reduction in the significance of boundaries. Simultaneously, globalization brings anxieties and apprehensions over individuals' increasing inequities and economic uncertainties. In the present-day context, there is a growing imperative for young folks across the globe to cultivate a holistic comprehension of cultural diversity, given its progressively pervasive integration into their everyday encounters. Moreover, individuals must possess the requisite skills and abilities to effectively contend with workers from different geographical areas worldwide, ensuring their lives' sustainability within the global economic landscape. It is believed that the attributes mentioned above of globalized youth are a fusion of economic and moral dimensions. The viability and attainment of young individuals are contingent upon acquiring economic competencies and fulfilling requisite moral considerations.

REFERENCES

Abdullah, D., Abd Aziz, M. I., & Mohd Ibrahim, A. L. (2017). The stories they tell: Understanding international student mobility through higher education policy. *Journal of Studies in International Education, 21*(5), 450–466. doi:10.1177/1028315317720766

AE. (2009). *The Australian education sector and the economic contribution of international students.* Australian Council for Private Education and Training. http://www.acpet.edu.au/uploads/files/Reports_Submissions/International_Students_14MAY2009__FINAL_.pdf

Appadurai, A. (2007). The capacity to aspire: Culture and the terms of recognition. In D. Held, H. L. Moore, & K. Young (Eds.), *Cultural politics in a global age: Uncertainty, solidarity and innovation* (pp. 29–35). Oneworld Publications.

Appadurai, A. (2013). *The future as cultural fact: Essays on the global condition.* Verso.

Armstrong, C. (2006). Citizenship in global perspective. *Policy & Practice-A Development Education Review, 3*, 14–25.

Banks, M., Olsen, A., & Pearce, D. (2007). *Global student mobility: An Australian perspective five years on.* IDP Education Pty Ltd.

Bauman, Z. (1998). *Globalization: The human consequences.* Columbia University Press.

Bauman, Z. (2009). *Does ethics have a chance in a world of consumers?* Harvard University Press.

Bauman, Z. (2013a). *Community: Seeking safety in an insecure world.* John Wiley & Sons.

Bauman, Z. (2013b). *Consuming life.* John Wiley & Sons.

Bauman, Z. (2013c). *The individualized society.* John Wiley & Sons.

Bek, H. (2017). Understanding the effect of loneliness on academic participation and success among international university students. *Journal of Education and Practice, 8*(14), 46–50.

Blackmore, J., Gribble, C., Farrell, L., Rahimi, M., Arber, R., & Devlin, M. (2014). *Australian international graduates and the transition to employment.* Deakin University.

Bostad, I., & Ottersen, O. P. (2014). Global presence, global responsibility and the global citizen. In A. B. Sterri (Ed.), *Global citizen: Challenges and responsibility in an interconnected world* (pp. 1–3). Sense Publishers. doi:10.1007/978-94-6209-929-6_1

Braskamp, L. A. (2008). Developing global citizens. *Journal of College and Character, 10*(1), 1–5. doi:10.2202/1940-1639.1058

Brinkmann, S. (2013). *Qualitative interviewing.* Oxford University Press. doi:10.1093/acprof:oso bl/9780199861392.001.0001

Clarke, M. (2018). Rethinking graduate employability: The role of capital, individual attributes and context. *Studies in Higher Education, 43*(11), 1923–1937. doi:10.1080/03075079.2017.1294152

Collinson, V. (2004). Learning to share, sharing to learn: Fostering organizational learning through teachers' dissemination of knowledge. *Journal of Educational Administration*, *42*(3), 312–332. doi:10.1108/09578230410534658

Creswell, J. W. (2013). *Qualitative inquiry and research design: Choosing among five approaches*. Sage publications.

Davies, L. (2006). Global citizenship: Abstraction or framework for action? *Educational Review*, *58*(1), 5–25. doi:10.1080/00131910500352523

De Costa, P. I. (2016). Constructing the global citizen: An ELF perspective. *Journal of Asian Pacific Communication*, *26*(2), 238–259. doi:10.1075/japc.26.2.04dec

Dill, J. S. (2013). The longings and limits of global citizenship education: The moral pedagogy of schooling in a cosmopolitan age. In *Routledge advances in sociology* (Vol. 109). Routledge.

Dolby, N., & Rizvi, F. (2008). *Youth moves: Identities and education in global perspective*. Routledge. doi:10.4324/9780203938836

Eder, J., Smith, W. W., & Pitts, R. E. (2010). Exploring factors influencing student study abroad destination choice. *Journal of Teaching in Travel & Tourism*, *10*(3), 232–250. doi:10.1080/15313220.2010.503534

Elliott, A., & Urry, J. (2010). *Mobile lives*. Routledge. doi:10.4324/9780203887042

EU. (2018). *Industry-based learning (IBL) program*. https://www.monash.edu/it/future-students/industry-experience/industry-based-learning

Folk, R. (1993). The making of global citizenship. In J. Brecher, J. B. Childs, & J. Cutler (Eds.), *Global visions: Beyond the new world order* (pp. 39–52). South End Press.

Gothe-Snape, J. (2018). *Record number of international students sticking around on visas with full work rights*. ABC News. https://www.abc.net.au/news/2018-07-27/temporary-graduate-visa-485-boom/10035390

Harpur, J. (2010). *Innovation, profit and the common good in higher education*. Palgrave Macmillan. doi:10.1057/9780230274624

Held, D., & Moore, H. L. (2007). Introduction: Cultural futures. In V. Rao & M. Walton (Eds.), *The capacity to aspire: Culture and the Terms of Recognition* (pp. 1–16). Oneworld Publications.

Huang, R., & Turner, R. (2018). International experience, universities support and graduate employability – perceptions of Chinese international students studying in UK universities. *Journal of Education and Work*, *31*(2), 175–189. doi:10.1080/13639080.2018.1436751

Islam, M. T. (2019). (Re)searching for the development of a conceptual model of education for citizenship in the context of young people's globalised mobility in higher education. *Globalisation, Societies and Education*, *17*(2), 194–207. doi:10.1080/14767724.2019.1583093

Islam, M. T. (2021). Cultural cultivation for the mobility of Bangladeshi higher education students in Australia. *Higher Education Research & Development*, *41*(5), 1572–1585. doi:10.1080/07294360.2021.1902950

Islam, M. T. (2022). Media mediation on academic advancement for youths' international mobility pursuing higher education. In S. Bansal, V. Ahuja, V. Chaturvedi, & V. Jain (Eds.), *Multidisciplinary Perspectives Towards Building a Digitally Competent Society* (pp. 208–229). IGI Global. doi:10.4018/978-1-6684-5274-5.ch012

Islam, M. T. (2023a). (Un)employability and (un)employment issues towards international mobility for higher education. In P. O. de Pablos, X. Zhang, & M. N. Almunawar (Eds.), *Education Institutions, Skills, and Jobs in the Digital Era: Toward a More Inclusive and Resilient Society*. IGI Global.

Islam, M. T. (2023b). Bangladeshi students' family fertilization for pursuing higher education in Australia. In G. Morris & L. Li (Eds.), *Handbook of Research on Developments and Future Trends in Transnational Higher Education* (pp. 247–264). IGI Global. doi:10.4018/978-1-6684-5226-4.ch013

Islam, M. T. (2023c). Bauman or Appadurai? In theorising mobilities of the Southern higher education students in the North. In K. Jermsittiparsert, P. Suanpang, A. Nuemaihom, & P. Siriattaku (Eds.), *Handbook of Research on Current Trends in Asian Education, Psychology, and Sociology* (pp. 1–25). IGI Global.

Islam, M. T. (2023d). International students' perception and experience of higher education for global citizenship. In J. Keengwe (Ed.), *Handbook of Research on Innovative Digital Practices and Globalization in Higher Education*. IGI Global. doi:10.4018/978-1-6684-6339-0.ch015

Marginson, S. (2006). Dynamics of national and global competition in higher education. *Higher Education, 52*(1), 1–39. doi:10.1007/s10734-004-7649-x

Maringe, F. (2006). University and course choice: Implications for positioning, recruitment and marketing. *International Journal of Educational Management, 20*(6), 466–479. doi:10.1108/09513540610683711

Moskal, M., & Schweisfurth, M. (2018). Learning, using and exchanging global competence in the context of international postgraduate mobility. *Globalisation, Societies and Education, 16*(1), 93–105. doi:10.1080/14767724.2017.1387768

Nghia, T. L. H. (2015). Factors influencing prospective international students' motivation for overseas study and selection of host countries and institutions: The case of Vietnamese students. In C. Gomes, & S. Chang (Eds.), *26th ISANA International Education Association Conference Proceedings*. ISANA International Education Association Inc.

Osler, A., & Vincent, K. (2002). *Citizenship and the challenge of global education*. Trentham Books Limited.

Oxley, L., & Morris, P. (2013). Global citizenship: A typology for distinguishing its multiple conceptions. *British Journal of Educational Studies, 61*(3), 301–325. doi:10.1080/00071005.2013.798393

Riaño, Y., Van Mol, C., & Raghuram, P. (2018). New directions in studying policies of international student mobility and migration. *Globalisation, Societies and Education, 16*(3), 283–294. doi:10.1080/14767724.2018.1478721

Rizvi, F. (2007). Postcolonialism and globalization in education. *Cultural Studies ↔ Critical Methodologies, 7*(3), 256–263.

Rizvi, F. (2008). Epistemic virtues and cosmopolitan learning. *Australian Educational Researcher, 35*(1), 17–35. doi:10.1007/BF03216873

Rizvi, F. (2011). Theorizing student mobility in an era of globalization. *Teachers and Teaching, 17*(6), 693–701. doi:10.1080/13540602.2011.625145

Rizvi, F. (2017). Globalization and the neoliberal imaginary of educational reform. *Education Research and Foresight Series, 20*, 1–13.

Rizvi, F., & Beech, J. (2017). Global mobilities and the possibilities of a cosmopolitan curriculum. *Curriculum Inquiry, 47*(1), 125–134. doi:10.1080/03626784.2016.1254500

Rizvi, F., & Lingard, B. (2000). Globalization and education: Complexities and contingencies. *Educational Theory, 50*(4), 419–426. doi:10.1111/j.1741-5446.2000.00419.x

Shultz, L. (2007). Educating for global citizenship: Conflicting agendas and understandings. *The Alberta Journal of Educational Research, 53*(3), 248.

Skrbis, Z., Woodward, I., & Bean, C. (2014). Seeds of cosmopolitan future? Young people and their aspirations for future mobility. *Journal of Youth Studies, 17*(5), 614–625. doi:10.1080/13676261.2013.834314

Snow, J., Mann, M., Page, T., & Nash, M. (2013). *Qualtrics survey software: Handbook for research professionals*. Qualtrics Labs, Incorporated.

SU. (2013). *Choose wisely. choose worldly. choose Southern*. https://www.youtube.com/watch?v=O4I1uRjC1Ds

SU. (2018). *Sport and gyms*. https://www.deakin.edu.au/life-at-deakin/student-life/sports-and-gyms

Tran, L. T., & Gomes, C. (2017). Student mobility, connectedness and identity. In L. T. Tran & C. Gomes (Eds.), *International student connectedness and identity: Transnational perspectives* (pp. 1–11). Springer. doi:10.1007/978-981-10-2601-0_1

Tran, L. T., & Vu, T. T. P. (2017). 'Responsibility in mobility': International students and social responsibility. *Globalisation, Societies and Education, 15*(5), 561–575. doi:10.1080/14767724.2016.1195729

United Nations. (2015). *Transforming our world: 2030 agenda for sustainable development*. United Nations. https://sustainabledevelopment.un.org/content/documents/21252030 Agenda for Sustainable Development web.pdf

Urry, J. (2003a). Global complexities. Distant proximities: Dynamics beyond Globalization. By Rosenau James N.Princeton. *International Studies Review, 5*(2), 250–252. doi:10.1111/1521-9488.502010

Urry, J. (2003b). Global complexity. *Polity*.

Varghese, N. V. (2008). *Globalization of higher education and cross-border student mobility. Unesco*. International Institute for Educational Planning Paris.

Walsh, L. (2016). *Educating Generation Next: Young People, Teachers and Schooling in Transition*. Springer. doi:10.1057/9781137526403

Walsh, L., & Black, R. (2018). *Rethinking youth citizenship after the age of entitlement.* Bloomsbury Academic.

Waters, J. L. (2006). Geographies of cultural capital: Education, international migration and family strategies between Hong Kong and Canada. *Transactions of the Institute of British Geographers*, *31*(2), 179–192. doi:10.1111/j.1475-5661.2006.00202.x

Wei, H. (2013). An empirical study on the determinants of international student mobility: A global perspective. *Higher Education*, *66*(1), 105–122. doi:10.1007/s10734-012-9593-5

Wood, B. E., & Black, R. (2018). Globalisation, cosmopolitanism and diaspora: What are the implications for understanding citizenship? *International Studies in Sociology of Education*, *27*(2–3), 184–199. doi:10.1080/09620214.2017.1415161

KEY TERMS AND DEFINITIONS

Education Marketing: Education marketing is a specialized field within the broader marketing discipline. It encompasses a range of methodologies, tactics, and instruments employed to systematically examine the market, tailor a school's offerings to align with the requirements and aspirations of prospective customers, and effectively communicate the unique value proposition of the educational institution.

Global Citizenship: The concept of global citizenship refers to the collective efforts made by individuals and communities to address social, environmental, and economic issues, acknowledging the universal citizenship of every individual.

Globalization: Globalization is a conceptual framework employed to elucidate the process by which trade and technology have engendered heightened interconnectedness and interdependence within the global community. Globalization encompasses the economic and societal transformations that have ensued as a consequence.

International Mobility: International mobility is developing educational, professional, social, and intercultural competencies that enable individuals to navigate and adapt to foreign travel and relocation experiences effectively.

Neoliberalism: Neoliberalism is a term commonly employed in contemporary discourse to denote reform policies prioritizing market orientation. These policies encompass the elimination of price restrictions, the deregulation of capital markets, the reduction of trade barriers, and diminished state power in the economy, mainly through privatization and austerity measures.

University Ranking: University rankings are compilations that seek to evaluate the caliber and effectiveness of universities by considering a range of criteria. Various performance factors determine university rankings, including research productivity, academic standing, global engagement, and student satisfaction. These rankings aim to offer an assessment of the overall excellence of universities.

Chapter 16
The Impact of War on Higher Education:
The Context of Wars in Syria, Iraq, Ukraine, and Afghanistan

Mustafa Kayyali
https://orcid.org/0000-0003-3300-262X
HE Higher Education Ranking, Syria

ABSTRACT

This chapter investigates the far-reaching consequences of conflict on higher education systems in Syria, Iraq, Ukraine, and Afghanistan. It dives into the issues that these countries face, including the destruction of academic infrastructure, displacement of professors and students, curriculum changes, and gender inequities. Despite these adversities, professors and students demonstrate tremendous perseverance. International collaborations and support play an essential role in protecting education opportunities. The chapter underlines the continuing importance of education in building stability, peacemaking, and growth in war-torn communities, offering hope for a brighter future.

INTRODUCTION

War, the essence of human struggle, inflicts indelible wounds on communities, transcending the immediate physical devastation to penetrate the very fabric of civilization. Beyond the terrible tales of battlefields and geopolitical conflicts, war leaves in its wake a path of deep disturbances that extend well beyond the frontlines. As the globe continues to cope with the complexity of modern warfare and its global repercussions, it becomes important to study one of its often-overlooked casualties: higher education. The quest of knowledge and the promotion of intellectualism have been important pillars of human progress throughout history. Higher education institutions have acted as sanctuaries of enlightenment, where the torch of wisdom is transmitted from one generation to the next. They are locations where individuals transcend boundaries of culture, creed, and situation to stimulate innovation, critical

DOI: 10.4018/979-8-3693-1926-0.ch016

thinking, and societal growth. Yet, in places embroiled by protracted conflicts, the sanctity of these intellectual havens is persistently tested, and the entire character of higher education comes under threat.

This chapter continues on a comprehensive investigation of the profound and complex repercussions of war on higher education, with a special focus on the war-torn landscapes of Syria, Iraq, Ukraine, and Afghanistan. Each of these nations has endured persistent instability, lengthy bloodshed, and social upheaval, and their higher education systems have borne the burden of these chaotic circumstances. As we delve into the various challenges and shared experiences encountered by these four distinct war-affected regions, it is vital to note that the implications of war on higher education are not confined to the destruction of physical infrastructure alone. The corrosive impacts of conflict pervade every part of academia, resulting to the relocation of researchers and students, the erosion of curriculum coherence, the stifling of research and innovation, and the weakening of international academic collaboration. We sincerely think that education, even in the most terrible conditions, contains the promise of transformation. It is a beacon of hope that can illuminate the darkest of times, and it remains the basis upon which the future of every nation is constructed. Therefore, while we negotiate the labyrinthine environment of war's influence on higher education, we renew our commitment to understanding and advocating for the resilience and rebirth of intellectual pursuits in the face of hardship. The journey may be long and riddled with problems, but the destination—a brighter, more educated, and wealthy future—is one worth working for.

Higher Education in Crisis: The Syrian Perspective

The Syrian crisis, now entrenched in its second decade, has not only devastated the nation physically but has also inflicted significant wounds on its once-thriving higher education system. What was once a beacon of academic brilliance in the region has become a heartbreaking saga of destruction, relocation, and resilience. The impact of war on higher education in Syria is a horrific tale that unfolds in chapters of decaying infrastructure, shattered academic standards, and a dogged quest of knowledge despite all difficulties.

Syria's higher education institutions, long acclaimed for their brilliance and diversity, now stand as sad reminders of the toll conflict exacts on the pillars of intellectual advancement (Al Azmeh, 2021). The physical architecture of universities and colleges, once bustling magnets of learning, sits in ruin. The war has cruelly turned campuses to rubble, leaving classrooms as nothing more than echoes of what they once were, and libraries burnt or pillaged, wiping decades, often centuries, of scholarly heritage. Beyond the tangible wreckage, the intangible loss is immense. The vibrant intellectual environment that formerly permeated these institutions has been replaced by an eerie silence, broken only by the noises of warfare. The devastation not only deprives current and future students of a conducive learning environment but also severs the historical continuity of Syrian intellectual excellence. The conflict's catastrophic result is the migration of both academics and students. Thousands of researchers, forced out by violence or seeking sanctuary in more stable academic environments abroad, have left behind a hole in Syria's higher education scene (Kayyali, 2022). This depletion of intellectual capital impedes the growth of critical research and creativity and robs pupils of the supervision of experienced educators. Students, too, have been unknowing casualties of relocation. The persistent violence has disrupted their education, leaving many with stopped or indefinitely delayed studies as students traverse the risks of conflict and seek safety in safer locations. For these pupils, access to educational materials, including textbooks and computers, has become a luxury—an inconceivable contrast to the pre-conflict past (Mualla, 2022).

The protracted conflict has ushered in an era of uncertainty surrounding academic standards in Syria. The usual operation of educational institutions has been interrupted, leading to inconsistencies in curricula and evaluation systems. The lack of access to up-to-date resources and the experience of faculty members has limited students' capacity to acquire a well-rounded education. Moreover, the degradation of academic standards has ramifications beyond the borders of Syria. The validity of degrees and certificates from Syrian institutions is increasingly questioned, risking the worldwide acceptance of Syrian graduates. The quest for educational excellence, long a source of national pride, now hangs in the balance.

Rebuilding Syria's higher education system in the post-conflict age is a mammoth challenge, but it is one that has tremendous promise. International organizations, NGOs, and governments must play a crucial role in helping the rehabilitation of educational institutions, the repatriation of academics, and the provision of scholarships for displaced students. Efforts should be multidimensional, covering not only the physical infrastructure but also the intellectual and cultural aspects of higher education. The safety of educational workers and facilities must be ensured, producing a climate conducive to learning and research. Collaborative activities between international and local stakeholders can pave the way for a complete recovery. Syria's higher education system, long a source of national pride, today serves as a striking symbol of the lingering wreckage wrought by conflict. Yet, it is also a sign of endurance, determination, and the everlasting belief in the transforming potential of education. As we turn to the future, it is our joint responsibility to assist the revival of higher education in Syria, to restore the broken academic institutions, and to empower the next generation of Syrians to rebuild their homeland through knowledge and innovation. In the face of adversity, the spirit of higher education persists in Syria, lighting the path towards a brighter future—a future where the impact of war on higher education is overshadowed by the enduring pursuit of knowledge, the restoration of academic excellence, and the rebuilding of a nation scarred but not defeated. The path to recovery is long and tough, but with concentrated efforts and worldwide solidarity, it is a journey that bears the potential of redemption for Syria's higher education system and, by extension, the nation itself.

War's Toll on Iraqi Higher Education: Challenges and Resilience

Iraq, with its great heritage of scholarship and intellectual achievement, has navigated a difficult route in recent decades. Prolonged hostilities, like the Iran-Iraq War, the Gulf War, and the aftermath of the 2003 war, have left indelible wounds on the country's higher education system. Despite the persisting problems, Iraqi educators, students, and institutions have showed amazing resilience in the face of adversity, embarking on a journey of rehabilitation and rebirth.

Decades of conflict have taken a significant toll on Iraq's higher education system, formerly recognised as among the most advanced in the Middle East. The Iran-Iraq War (1980-1988) launched the era of unrest, causing considerable damage to educational facilities. The Gulf War (1990-1991) further strained the system, disrupting academic life and creating major brain drain as intellectuals sought sanctuary overseas. However, it was the aftermath of the 2003 war that delivered the most serious blow to Iraq's higher education (Bedan, 2022). Campuses were plundered, staff personnel were attacked, and academic standards dropped. This time saw the emigration of many Iraqi intellectuals, leaving a hole that has been tough to replace. One of the most significant issues encountered by Iraqi higher education has been the brain drain of its intellectual elite (Ameen, 2019). Professors and scholars departed the country to escape violence, persecution, and instability (Jafar, 2021). This huge migration deprived Iraq of its intellectual capital, leaving universities scrambling to maintain academic standards and research productivity.

In addition to the brain drain, millions of Iraqis were displaced internally or sought safety abroad, affecting the education of an entire generation of pupils. Educational resources, including libraries, laboratories, and research institutions, were ravaged throughout the battle, further impeding academic advancement. Despite these great hurdles, Iraq's higher education industry has exhibited an unflinching dedication to recovery. With the growing stabilization of the security situation, some displaced academics have returned to their posts, contributing to the rebuilding process. Universities have concentrated on rebuilding infrastructure and restoring key facilities. International organizations and foreign governments have played a vital role in aiding the revival of Iraqi higher education. Initiatives to give scholarships for Iraqi students to study abroad have been essential in preserving access to education for talented individuals during the turmoil. Additionally, agreements between foreign universities and Iraqi institutions have fostered knowledge sharing and faculty growth (Jafar, 2022). It is worth mentioning that international cooperation is not restricted to reconstructing physical infrastructure; it also extends to knowledge-sharing and research partnership. Such relationships have the potential to revive academic life in Iraq and help it restore its standing as a center of intellectual brilliance.

While there have been substantial achievements in restoring Iraqi higher education, significant obstacles persist. Ongoing security concerns, political uncertainty, and economic problems pose obstacles to the sustainable revival of educational institutions. The need for continuing investment in infrastructure, faculty development, and research capacity is crucial. Moreover, there is a pressing need to address the core causes of the brain drain and build an atmosphere that attracts and maintains academic talent within Iraq. Ensuring the safety and security of professors and students is important to rebuilding trust in the higher education system (Jameel, 2020).

In the midst of adversity, Iraqi higher education stands as a beacon of hope. The commitment of professors and students to maintain their academic history and restore their institutions is a monument to the lasting power of knowledge. As Iraq navigates the hard path to recovery, it is vital that the international community continues to contribute assistance and resources to enable Iraqi higher education reclaim its place on the global arena. The obstacles faced by Iraqi higher education are not insurmountable, and the tenacity displayed by instructors and students alike is a source of inspiration. As the nation rebuilds and stabilizes, higher education will play a crucial role in supporting development, reconciliation, and intellectual advancement. In the face of adversity, Iraq's higher education sector is determined to rise, to reclaim its status, and to contribute to the nation's brighter future—a future where the impact of war on higher education becomes a story of challenges met with resilience, knowledge overcoming conflict, and the enduring pursuit of academic excellence.

The problems encountered by Iraqi higher education are great, but the tenacity displayed by instructors, students, and the larger community illuminates a route toward recovery. As Iraq navigates the hard process of rebuilding, the restoration of its higher education system is not simply a pragmatic requirement but a symbolic triumph against adversity. Through international support, targeted initiatives, and a commitment to inclusivity, Iraqi higher education can reclaim its standing as a regional hub of learning, contributing to a brighter future for the nation and inspiring other conflict-affected regions to overcome their own educational challenges.

Ukraine's Struggle for Academic Normalcy: Navigating Conflict and Education

Ukraine, a nation positioned at the crossroads of Europe, has found itself involved in geopolitical tensions and conflict in the 21st century. Amidst these obstacles, the higher education system has become a war of its own, testing the resilience of professors, students, and institutions. The impact of war and political upheaval on Ukraine's academic environment is a complicated narrative of adaptation, survival, and determination. Ukraine's spiral into the turbulence of conflict began with the annexation of Crimea by Russia in 2014, a decision that sent shockwaves through the nation. Subsequently, the commencement of the conflict in Eastern Ukraine, focused in Donetsk and Luhansk, unfolded, changing not only the country's political landscape but also having important ramifications for higher education. Universities and colleges in the impacted regions found themselves on the front lines, with campuses becoming literal battlegrounds. Academic life was upended, and staff and students alike found themselves caught in the centre of turmoil, fighting to maintain a semblance of normalcy while confronting uncertainty and risk. The conflict in Eastern Ukraine brought about considerable obstacles for curricular continuity (Sokol, 2022). Universities in affected areas had to alter their educational programs to accommodate the immediate requirements of students while grappling with the harsh reality of armed war. Some colleges even made the difficult decision to move their activities to safer cities, compelling students and professors to migrate in pursuit of knowledge despite the upheaval. The polarization of Ukraine's political scene further compounded issues.

Amidst the turbulence, Ukraine's higher education industry appealed to the global community for support. Collaboration with foreign universities and institutes became a lifeline, supporting the exchange of knowledge, faculty growth, and research cooperation (Oleksiyenko, 2023). Scholarships and programs that foster academic mobility have played a significant role in sustaining opportunities for students and scholars to continue their study and research (Atamanyuk, 2021). International aid has been crucial in helping Ukrainian institutions modernise their facilities and teaching approaches to meet global standards. These connections have not only helped Ukraine maintain its place as a hub of learning but also emphasised the relevance of international academic participation during times of war. The prolonged conflict has hindered academic mobility both within Ukraine and beyond its borders. Students and professors from affected regions have faced challenges in pursuing their studies or teaching duties in other parts of the country (Sokolova, 2023). Safety concerns, logistical hurdles, and administrative obstacles have restricted the free flow of intellectual talent. International mobility, too, has experienced difficulties.

In the face of adversity, Ukrainian higher education institutions, lecturers, and students have displayed amazing perseverance. Despite the hurdles, many colleges have continued to function, adapt, and innovate. Students have showed extraordinary commitment to pursue their education, even under the most trying circumstances. Ukraine's higher education sector faces a future where it must negotiate continuing wars and political instability (Debych, 2023). Reconciliation and repairing divisions within academics will be important for its progress. Upholding academic independence, cultivating critical thinking, and promoting inclusion will be vital in protecting the integrity of higher education in Ukraine. As the nation attempts to achieve stability and resolve problems, higher education may play a vital role in moulding the future of Ukraine. Education is a great tool for building understanding, tolerance, and social solidarity, offering a road towards reconciliation and sustainable development. The perseverance demonstrated by Ukraine's higher education community is a tribute to the enduring worth of knowledge in the face of tragedy. Ukraine's rich cultural legacy, firmly entwined with its higher education system, plays a significant role in the endurance of academic institutions (Hnydiuk, 2023). Efforts to preserve and promote Ukraine's cultural identity within the academic sector might contribute to a sense of continuity and national pride. Universities can become focal sites for cultural preservation, incorporating heritage studies into their courses and partnering with international organizations to safeguard and document cultural assets. This merging of cultural resilience with academic pursuits not only protects Ukraine's unique identity but also develops a sense of belonging and purpose within the higher education community.

As Ukraine seeks to overcome the legacy of conflicts and construct a more stable future, higher education appears as a linchpin for national unity. Educational institutions can provide as neutral ground for debate, encouraging understanding, and promoting togetherness among varied cultures. International organizations, foreign governments, and educational institutions worldwide have a responsibility to support and uplift their Ukrainian counterparts. This support extends beyond financial aid; it involves encouraging collaborative research initiatives, giving opportunities for academic interaction, and advocating for the relevance of education in post-conflict healing. The global community can play a critical role in ensuring that Ukraine's higher education not only survives but thrives in the face of adversity.

Ukraine's quest for academic normalcy under conflict is a multilayered story of persistence, adaptation, and the eternal pursuit of knowledge. The challenges faced are enormous, but the dedication to protecting and promoting higher education in Ukraine is constant. As the nation tries to establish a more stable and peaceful future, the role of higher education in moulding that future cannot be understated. In the face of uncertainty, Ukraine's higher education community remains determined in its purpose to overcome problems and contribute to the nation's intellectual and social advancement. The global community, in turn, has a key role to play in aiding Ukraine on this revolutionary path.

Education Under Fire: Afghanistan's Journey to Revive Higher Learning

Afghanistan, a country continually scarred by warfare and political instability, has wrestled with the severe impact of war on its higher education system. Over the decades, the country has endured a series of conflicts, invasions, and revolutions that have wrecked its educational infrastructure and hampered the pursuit of knowledge. The tale of higher education in Afghanistan is a monument to the endurance and determination of professors and students in the face of continuous adversity.

Afghanistan's higher education system has a long history, with universities like Kabul University created in the 20th century. However, the country's route to prosperity was delayed by decades of violence, beginning with the Soviet invasion in 1979 and followed by years of civil war, and the succeeding US-led war in 2001. Each period of battle left its mark on the educational landscape. Universities and colleges in Afghanistan have suffered the brunt of the violence. Campuses were damaged or destroyed, libraries were robbed, and educational infrastructure deteriorated (Azimi, 2020). The flight of intellectuals and professors during the conflict generated a serious brain drain, reducing the intellectual resources required for sustaining higher education.

The issues addressed by Afghan universities are multifaceted. The deterioration of academic quality owing to continuous conflict has damaged the integrity of degrees and certifications. Curriculum coherence has been disrupted, resulting to gaps in students' education and a lack of consistency with worldwide norms. Amidst the chaos, Afghan educators and children have demonstrated extraordinary perseverance. Universities have operated under improvised venues, and faculty members have continued to educate despite the hard circumstances. The passion for knowledge remains undiminished among Afghan youth, many of whom have overcome security threats to seek higher education. Furthermore, the Afghan diaspora has played a key role in financing higher education in their motherland (Kester, 2021). Afghan academics and professionals residing overseas have launched programs and collaborations to provide remote learning opportunities, mentorship, and scholarships for Afghan students and educators. International organizations, governments, and non-governmental organizations (NGOs) have been essential in providing help to resuscitate higher education in Afghanistan. Scholarships for Afghan students to study abroad, funding for educational initiatives, and cooperation between Afghan and foreign universities have helped sustain higher education in the country. These collaborations have not only focused on restoring infrastructure but also on capacity building, faculty development, and curriculum enhancement. They have underlined the necessity of international cooperation in sustaining the educational legacy of Afghanistan. The route to recovery for Afghanistan's higher education sector is filled with hurdles. Ongoing security concerns, political uncertainty, and economic challenges continue to hamper the progress of educational institutions. The need for ongoing investment in infrastructure, faculty development, and research capacity is crucial. Additionally, addressing gender imbalances and promoting inclusivity for underrepresented communities, such as internally displaced persons, is vital. Education should be a tool for social reform, developing tolerance, understanding, and reconciliation in a society devastated by violence. Higher education in Afghanistan possesses the potential to be a catalyst for reconstructing the nation. It can prepare the Afghan young with the skills, knowledge, and critical thinking talents needed to confront the difficulties of post-conflict rebuilding, governance, and development.

Moreover, higher education can play a crucial role in forging national unity and healing. It can give a venue for communication, understanding, and cooperation among many communities and factions inside Afghanistan (Welch, 2020). As Afghanistan navigates its uncertain future, the importance of higher education in determining that future cannot be emphasised. Education is not only a dream; it is a beacon of hope for a nation weary of conflict. The tenacity and determination displayed by Afghanistan's educators and students are a monument to the enduring worth of education in the face of relentless struggle.

In the face of the obstacles provided by conflict, Afghan higher education institutions have exhibited extraordinary ingenuity in teaching and learning. Embracing technology has become a fundamental component of this adaptability. Online learning platforms, virtual classrooms, and digital resources have emerged as lifelines for students and instructors, enabling them to pursue academic endeavours despite the physical limits. The integration of vocational training programs that match with the needs of the post-conflict economy has gained traction. These programs not only give practical skills to students but also aid to the reconstruction efforts by supplying a skilled workforce. As Afghanistan seeks to overcome the legacy of conflicts and construct a more stable future, higher education appears as a linchpin for national reconciliation. Educational institutions can provide as neutral ground for debate, encouraging understanding, and promoting togetherness among varied cultures. Initiatives that promote tolerance, diversity, and mutual respect within institutions can extend beyond campus bounds, impacting broader social attitudes. By establishing ideals of tolerance and collaboration, higher education becomes a strong weapon for mending the wounds of the past and creating bridges toward a harmonious future.

Afghanistan's struggle to resuscitate higher learning under the shadow of conflict is a varied story of challenges faced with fortitude, of knowledge overcoming turmoil, and of the persistent pursuit of academic greatness. The challenges are enormous, but the commitment to protecting and expanding higher education in Afghanistan remains unshaken. In the face of adversity, Afghanistan's higher education community stands ready to contribute to the nation's intellectual and social progress, offering hope for a brighter future where the impact of war on higher education is overshadowed by the enduring pursuit of knowledge and the rebuilding of a nation scarred but not defeated. The global community, in turn, has a key role to play in aiding Afghanistan on this transformative path.

CONCLUSION

The pages of this chapter have served as a window into the volatile world of higher education in the midst of conflict. Through a critical analysis of the diverse repercussions of war on academics in Syria, Iraq, Ukraine, and Afghanistan, we have sought to unearth the intricate web of difficulties and opportunities that emerge in the wake of protracted conflicts. As we near the conclusion of our voyage, it is abundantly obvious that the repercussions of war on higher education are substantial, enduring, and far-reaching. The ravages of conflict reach well beyond the physical ruin of campuses and infrastructure; they infect the thoughts and aspirations of professors, students, and entire communities. The displacement of professors, the disruption of curriculum, the stifling of research and innovation, and the weakening of international academic relationships are merely a few of the challenges that higher education institutions must cope with in conflict-ridden countries. Amidst this grim evaluation, there shines a light of hope—a monument to the tenacity of the human spirit and the ongoing relevance of education. It is obvious that even in the most severe situations, higher education endures, frequently against all odds. The stories of educators who risk their lives to impart information, of students who resist the turmoil to pursue learning, and of communities that prioritize rebuilding their intellectual roots are witness to the indestructible spirit of education.

It is vital to note that the impacts of conflict on higher education are not uniform across these locations. While each country faces its own distinct set of issues, there are also commonalities that unite them. The decline of academic quality, the loss of human capital, the disruption of educational systems, and the persistent trauma on individuals and communities are threads that weave through this intricate tapestry of war's aftermath. But if we reflect on the tenacity and determination demonstrated by educators, students, and communities, we see a glimmer of promise. The rebuilding and regeneration of higher education in war-torn places is a difficult yet possible task. The power of education to generate stability, peace, and sustainable development remains undiminished, and it is this potential that must be used to accelerate change.

As we complete our investigation of the impact of war on higher education in Syria, Iraq, Ukraine, and Afghanistan, let us not lose sight of the enduring relevance of education in repairing fragmented societies. The road ahead may be tough, but it is a road worth going. By investing in the regeneration and resilience of higher education, we invest in the future of these nations and, eventually, in the goal of global peace and prosperity. In the face of hardship, we are reminded that education, like a phoenix, can emerge from the ashes of strife, bringing with it the promise of a brighter, more educated, and more harmonious future. The chapters that lie ahead for Syria, Iraq, Ukraine, and Afghanistan may be unwritten, but they are brimming with possibilities. It is our collective responsibility to guarantee that the story of higher education in these war-affected countries continues to be one of resilience, hope, and development. In closing, let us reaffirm our commitment to the enduring power of education, for it is through education that societies can emerge from the darkest of times and chart a course towards a better tomorrow—a tomorrow where the impact of war on higher education becomes a story of triumph over adversity, of knowledge overcoming chaos, and of the human spirit rising above conflict's shadow.

REFERENCES

Al Azmeh, Z., Dillabough, J., Fimyar, O., McLaughlin, C., Abdullateef, S., Aloklah, W. A., Mamo, A. R., Abdulhafiz, A. H., Al Abdullah, S., Al Husien, Y., Al Mohamad Al Ibrahim, A., Al Ibrahim, Z., Barmu, T., Farzat, A., & Kadan, B. (2021). Cultural trauma and the politics of access to higher education in Syria. *Discourse (Abingdon)*, *42*(4), 528–543. doi:10.1080/01596306.2020.1715922

Ameen, N., Willis, R., Abdullah, M. N., & Shah, M. (2019). Towards the successful integration of e-learning systems in higher education in Iraq: A student perspective. *British Journal of Educational Technology*, *50*(3), 1434–1446. doi:10.1111/bjet.12651

Atamanyuk, S., Semenikhina, O., & Shyshenko, I. (2021). Theoretical fundamentals of innovation of higher education in Ukraine. *Pedagogy and Education Management Review*, (2), 30-36.

Azimi, M. N., & Balakarzai, A. T. (2020). Nation Building Through Higher Education System Resurgence in Afghanistan. Handbook of Education Systems in South Asia, 1-22. doi:10.1007/978-981-13-3309-5_56-1

Bedan, S. (2022). Higher Education in Iraq between the Past and the Present. *Journal of Positive School Psychology*, 4718–4731.

Debych, M. (2023). Internationalization of higher education in the context of recovery of Ukraine: Goals and challenges. *Continuing Professional Education: Theory and Practice*, *2*(75), 7–15.

Hnydiuk, I., Hataullina, E., Marshuk, L., Timoshenko, N., & Cherneha, V. (2023). Formation of the system of financial support of higher education institutions in Ukraine. *Amazonia Investiga*, *12*(65), 141–148. doi:10.34069/AI/2023.65.05.14

Jafar, H. (2021). *Internationalization of Higher Education and Imported Universities in Post-conflict Iraq through a Policy Borrowing Lens*. University of Toronto.

Jafar, H., & Sabzalieva, E. (2022). Faculty Experiences of Higher Education Internationalization in Post-Conflict Iraq and Tajikistan. *Journal of Comparative and International Higher Education*, *14*(2), 47–65. doi:10.32674/jcihe.v14i2.3453

Jameel, A. S., & Ahmad, A. R. (2020). Factors impacting research productivity of academic staff at the Iraqi higher education system. *International Business Education Journal*, *13*(1), 108–126. doi:10.37134/ibej.vol13.1.9.2020

Kayyali, M. (2022). *Quality Assurance*. Internationalization, and Ranking of Higher Education Institutions in the Context of the Syrian War.

Kester, K. (2021). Toward a conflict-sensitive approach to higher education pedagogy: Lessons from Afghanistan and Somaliland. *Teaching in Higher Education*, 1–20. doi:10.1080/13562517.2021.2015754

Mualla, W., & Mualla, K. J. (2022). Higher Education in Syria Post-war and Pandemic: Challenges and Opportunities. *Higher Education in the Arab World: New Priorities in the Post COVID-19 Era*, 283-309.

Oleksiyenko, A., Shchepetylnykova, I., & Furiv, U. (2023). Internationalization of higher education in tumultuous times: Transformative powers and problems in embattled Ukraine. *Higher Education Research & Development*, *42*(5), 1103–1118. doi:10.1080/07294360.2023.2193727

Sokol, T., & Melko, L. (2022). Current challenges of higher education in Ukraine. *Pedagogy and Education Management Review,* (1), 49-53.

Sokolova, E. (2023). Improvement of innovative forms of public administration in the higher education system of Ukraine. Publishing House "Baltija Publishing".

Welch, A., & Wahidyar, A. (2020). Quality assurance in Afghan higher education: Achievements and challenges. *Asian Education and Development Studies*, *9*(4), 479–493. doi:10.1108/AEDS-09-2018-0146

Chapter 17
Impact of Armed Conflicts on Higher Education Northwestern Syria Context

Mahdi Khaled Alkol
Idlib University, Syria

ABSTRACT

The conflict in Syria is almost entering its 14th year, and its negative effects have been clearly reflected in all economic, educational, health, service, and other sectors across Syria, including the education and higher education sectors, where many young people were deprived of completing their university education as a result of repeated displacement and instability. Some negative effects of the conflict on the higher education were the migration of competencies, experiences, minds, and those with advanced degrees.

INTRODUCTION

The effects of armed conflict extend to many sectors and aspects of life in the countries that suffered from these scourges and wars. war and armed conflicts displace the population, disrupt schooling, and cause health crises, which cause death, almost 750,000 people are killed as a result of the conflict each year. (Patricia Justino, 2010). these sectors have suffered major infrastructure damage, especially the higher education sector and universities, which had a large part of These damages. Scott Gates concludes that many factors are associated with a higher risk of conflict (Erik Melander, 2005).

Nearly thirteen years after the start of the conflict and fighting in Syria, the war has cast its shadow on all fields of economic, and political life in Syria. Higher education is one of the pillars of development in societies and an essential factor for achieving progress and prosperity in any society. These conflicts lead to the destruction and dispersion of university infrastructure and negatively affect the quality and capabilities of education.

Concept of Armed Conflict

Non-international armed conflict is a protracted armed confrontation that occurs in many countries between the forces of armed groups and government forces (International Committee of the Red Cross

DOI: 10.4018/979-8-3693-1926-0.ch017

(ICRC)). There is another kind of armed conflict that happens between two or more countries, which is called international conflict, regardless of the cause or severity of the conflict. Many international or non-international conflicts occur around the world, in some countries such as Syria, Iraq, Palestine, Yemen, Ukraine, Libya, Brazil, Myanmar, Nigeria, Somalia, and Congo. These armed conflicts are primarily governed by international humanitarian law, which is a set of rules – either stipulated in treaties or recognized through custom – that limit the permissible behavior of parties to a conflict (www.amnesty.org, 2014).

Armed conflict refers to a state of tension or conflict that involves the use of military force and weapons in an attempt to achieve certain objectives or protect certain interests. Armed conflict can have a strong impact on the security, social, and economic stability of the areas concerned. The main features of armed conflict include Using military force, whether by a state's formal armed forces or informal armed groups, the presence of two or more parties, which can be states, armed groups, or any entity that uses military force to achieve its goals. Armed conflict also includes Using advanced weapons, such as guns, cannons, aircraft, and missiles, and is characterized by physical violence and the possible destruction of infrastructure. In addition to Specific objectives, parties involved in a conflict have certain goals to achieve, and these goals can be related to controlling territory, changing the political system, or achieving religious or ideological goals. Armed conflict often has a significant impact on innocent civilians, whether through direct attacks or isolation and lack of basic services.

It can be said that armed conflict is often accompanied by the lack of respect for human rights, such as forced displacement, and targeting of civilians. It may be linked to the control of natural resources, such as oil or water, and may lead to economic or separatist conflict. Armed conflict can be violent and complex, and identifying causes and effects requires a deep understanding of the circumstances and dynamics surrounding it. Amnesty International tries to document in general violations of international law during armed conflicts and campaigns against these violations. Regardless of who commits it or where it occurs, supporting the demands of victims to achieve justice starts from national authorities to international institutions. but for the sorrow, it doesn't contribute effectively to protecting civilians and vulnerable groups, it cannot hold aggressors against others and violators of the law accountable.

Higher Education During Crises and Emergencies

The term "higher education in emergencies" refers to higher education opportunities that provide international protection to young refugees and displaced people living in areas affected by armed conflict, natural disasters, or violence. and improve youth affected by crises' access to inclusive, equitable education and learning outcomes, by protecting and increasing external funding, working together in a spirit of international cooperation to build crisis-resilient education systems. and that are more resilient and meet needs in the face of conflict conditions, social unrest, and natural hazards - while ensuring the continued availability of education in emergencies post-conflict. Crisis education is the set of activities that need to be undertaken to achieve continuity of regular learning in times of acute crises or long-term instability (Alshami, Alzanfaly,2021). Young people living in conflict-affected areas are particularly vulnerable because they are a group most likely to be recruited, forcefully or otherwise, into either national armies or armed groups. This causes a huge loss of talent on the part of countries affected by the crisis, the same talent that is essential to rebuilding these countries.

Impediments to a Return to Higher Education

Several groups in Syrian society are affected by the conflict, one of these groups is the young people, where many of them had to spend several years away from the classroom, they feel embarrassed when they decide to return to education, corresponding to their academic level. Many of them become too old and they can't sit on the study desks at the university, Also, they feel by the marginalization from members of the community (Cervantes-Duarte, Fernández-Cano, 2016).

Impact of Armed Conflict Upon Higher Education

Many countries have witnessed long years of conflicts that have affected human development. (Al-Munaifi, Al-Eryani, 2021). Conflict disrupts access to and provision of education, and can affect human capital stocks in the long term (Umueva, Musa, Hatch, 2018). Armed conflicts usually cause the educational process and all academic activities to stop in regions and countries that witness these conflicts. As well as the inability of students to continue their studies and attend their lectures, and complete their research. Armed conflicts also cause a loss of human capital, with teaching and administrative staff and students exposed to forced displacement, killing, and migration.

Despite all the challenges facing higher education, there have been efforts to adapt and rebuild the higher education system in Syria. However, the road to recovery and rebuilding higher education in Syria remains long and challenging, requiring international support, resources, and a sustained commitment to education as a fundamental human right. The impact of armed conflicts on the quality of education is not limited to attacks on the lives of students and teachers not only, it constitutes a violation of the right to education, as achieving good quality education helps build peace and reduces, opportunities for conflict in the country and raising awareness to reduce violence. In light of conflicts, the necessary skills cannot be achieved to contribute to peaceful integration, reconstruction, and sustainable development within societies (Raheema, 2017).

The effects of armed conflicts on higher education can be classified according to the following classification system (Cervantes-Duarte & Fernández-Cano, 2016)

1- Damaging or destroying the educational infrastructure.
2- Loss of the academic community, experiences, and skills.
3- The teaching staff in the university is unqualified.

Armed conflict in Syria had main effects on higher education, higher education faced numerous challenges in Syria after 2011 that could significantly impact its functioning and quality. Crises can take various forms, including armed conflicts and natural disasters. Here are some key challenges and considerations related to higher education in conflicts, these challenges can be defined as follows:

Destruction of university infrastructure, when the ongoing conflict in Syria intensified and its pace and intensity increased, universities and educational institutions suffered massive damage. The university buildings, classrooms, libraries, administrative offices, laboratories, and research facilities were damaged partially or completely, also Theft or destruction of educational equipment and advanced technology, in addition to Loss of academic documents and vital information. Some challenges appear in the loss of human resources, knowledge, and experiences, many students and academics were displaced leaving their native residence, as a result of the unstable and dangerous security situation, many students and faculty members were forced to leave their universities and search for safe places, the conflict in Syria led to migration of many students and academics, which led to the loss of new generations of learners and participants in the academic community, and effects on their education and academic careers in the long term. Safety and security concerns also emerged, the safety and security of students, faculty, and staff become paramount during crises. Universities may need to implement measures to ensure the well-being of the campus community. Also, there is difficulty in accessing the university and educational facilities, the escalation of security tensions and the destruction of infrastructure has made access to the universities a major challenge for many students -It poses a major obstacle to them completing their studies-, teaching staff, and administrators, especially for students living in affected areas and hot zones of conflict.

Crises often lead to the disruption of academic activities, including the closure of universities, cancellation of classes, and interruption of research projects. In addition to the previous challenges the lack of scientific contributions and declining quality of higher education, the ongoing conflicts affected negatively scientific research and innovation, reducing Syria's contribution to global scientific research, Lack of resources and academic fragmentation contributed to the reduction of the quality of higher education in Syria. Students suffer from a lack of materials and practical opportunities to acquire practical skills, Collaborative international programs, research projects, and student exchanges may be disrupted. Also, the conflict caused Limited or loss of communication with international academic bodies, which led the Syrian universities to lose international communication and opportunities for academic exchange, which limits the ability of entities working in higher education in Syria to communicate with the international community, international network of education and research. Which deprives them of benefiting from global scientific and cultural developments, affecting the global engagement of universities.

In addition to the previously mentioned challenges, other challenges arise, such as psychological impact on civilians including(students-faculty-academics), where the Syrian conflict caused high levels of health problems and psychological effects and problems such as fear, depression, anxiety, and post-traumatic stress disorder on students, faculty, and staff. Many Syrians witnessed traumatic events such as bombing, shooting, and forced displacement, which contributed to these problems. Universities often face financial challenges during crises, including reduced funding, and increased demand for financial aid from affected students, which have high tuition costs and other study-related expenses. In northern Syria, annual fees range between USD 150 and USD 500 at semi-private universities, while some private universities charge up to USD 1800 per year. All universities depend on students as their main and only source of funding (Alkol, 2023). The conflicts caused the devaluation of the currency and the deterioration of the Syrian economy. This has led to a reduction in funding allocated to higher education, which has negatively affected the ability of universities to provide the infrastructure and resources necessary to provide high-quality education. The dispute could affect the reputation of the university and its attractiveness to international students and academics.

Medical education has been particularly affected, as providing health care and practical training can be difficult challenges under existing conditions. University libraries and research centers were at risk of destruction, resulting in the loss of valuable documents and research. In addition to all previous, educational opportunities for women and girls have been significantly affected by the conflict, and rates of early marriage and involvement in domestic work have increased, the early marriage prevented many girls from pursuing their education, and many factors contribute to the practice of child marriage, some of these factors are gender inequality, poverty, exclusion, Marginalization and lack of security, especially among displaced and conflict-affected communities. The rates of child marriage are usually accompanied by an increase in An increase in fertility rates, which may harm the mother's health And the child's in the short and long term, and undermines the ability to make decisions related to the life of the child and her family (ESCWA,2019).

Impact of Conflict on Some Segments of Society

Conflict and difficult situations can lead to multiple phenomena that affect some segments of society, as the graduates from the universities, are facing major challenges in searching for job opportunities due to the devastated economic structure and lack of jobs, this is what prompted them to migrate abroad to the neighboring countries as Turkia and some Europian countries in search of work and to escape the ongoing conflict between the fighting parties. The high cost of education at its various levels has deprived many people of continuing their education and academic achievement, especially the poor and needy who cannot meet the high costs of studying for their children, the high costs of education constitute a great burden on poor families, especially in light of the lack of sufficient support for education by the government and the concerned authorities. School and university tuition fees are high compared to the income and economic situation of many families, especially the displaced who live in difficult living conditions, especially those who live in camps. many students resort to borrowing money to pay the costs of the studies, which leads to the accumulation of significant student debt and may affect their financial future. Another segment of society that was affected by the conflict was girls, the deterioration of the situation of girls and young women, leads to child marriage and affects the social and economic situation of families and society as a whole, among these phenomena is the effects of early marriage on girls. Here are some possible effects in some aspects:

1. Education and Health

Early marriage may lead to a girl's education being interrupted, as she may be forced to leave school to devote herself to married life and family. Girls who marry young may be exposed to greater health risks, including risks of pregnancy at an early age that may affect the health of the mother and child.

2. Psychological Development and Family Economy

Early marriage can have a significant impact on a girl's psychological development, as she may face psychological and social pressures as a result of the responsibilities of married life at an early age. Early marriage can have a significant impact on the family economy, as marrying at an early age may challenge a girl to engage in the labor market and achieve financial independence.

3. Legal Rights and Social Life

In some cases, early marriage can result in a girl losing some of her legal rights, as she may be unable to make independent decisions based on marriage and family laws in some societies. Early marriage may have an impact on a girl's social life, as she may miss the opportunity to interact with her peers normally.

4. Health Behavior and Personal Development Opportunities

Early marriage may cause changes in health behavior, as pregnancy and motherhood at an early age can have an impact on a girl's overall health. Girls who marry young may have fewer opportunities to develop their skills and gain education and personal experiences.

Higher Education and Post-Conflict Recovery

Higher education has an essential role in the post-conflict recovery process, contributing to rebuilding societies, fostering stability, and laying the foundation for sustainable development. Post-crisis, universities face the challenge of rebuilding infrastructure, restoring academic programs, and supporting the recovery of the campus community. Here are key aspects of the relationship between education in the university stage and post-conflict recovery:

1. Human Capital Development and Rebuilding Social Fabric

Higher education is instrumental in developing human capital by providing advanced skills, knowledge, and expertise. This educated workforce is essential for reconstruction efforts and the overall development of a post-conflict society. Universities serve as hubs for intellectual and cultural exchange. The re-establishment of the institutions of higher education helps rebuild the social fabric by bringing people together, fostering dialogue, and promoting tolerance and understanding.

2. Peacebuilding and Reconciliation, Economic Reconstruction

Higher education contributes to peacebuilding and reconciliation by promoting dialogue, critical thinking, and understanding. It provides a platform for addressing historical grievances and fostering a sense of shared identity among diverse communities. Educated individuals contribute significantly to economic reconstruction. Higher education equips people with the skills needed for entrepreneurship, innovation, and participation in various sectors, which is vital for rebuilding a post-conflict economy.

3. Innovation and Research, Cultural Preservation

Universities are hubs of innovation and research. Post-conflict societies benefit from research activities that address challenges and contribute to sustainable solutions. Innovation and technology can play a main role in the recovery process. The institutions of higher education often play a role in preserving cultural heritage. The continuity of academic and cultural activities helps safeguard the identity and traditions of a society that may have been affected during the conflict.

4. Social Inclusion and Equality, Capacity Building

Higher education promotes social inclusion by providing equal education opportunities. It can address gender disparities, ensure access for marginalized groups, and contribute to a more equitable and just post-conflict society. Post-conflict nations often face a shortage of skilled professionals. Higher education institutions play a vital role in capacity building by training a new generation of leaders, educators, and professionals.

5. International Collaboration, Community Engagement

Higher education facilitates international collaboration and partnerships. Engaging with the global academic community provides post-conflict societies with resources, knowledge exchange, and opportunities for collaborative research. Universities can actively engage with local communities to address their needs. Community-oriented research and outreach programs contribute to the social and economic development of the regions affected by conflict. The institutions of higher education can play a vital role in the overall recovery of communities affected by crises by contributing expertise, research, and community engagement.

6. Psychosocial Support

Higher education institutions can provide psychosocial support services to help individuals cope with the trauma of conflict. Counseling services and mental health programs contribute to the overall well-being of students and staff. Despite the challenges, universities can play a vital role in reconstruction processes by providing academic and research expertise. In conclusion, higher education serves as a linchpin for post-conflict recovery, offering a pathway to stability, socio-economic development, and lasting peace. International support, investment, and collaboration are essential to harness the full potential of higher education in the rebuilding process.

Rebuilding and Reforming Higher Education Systems in The Aftermath of Conflict

It can be said that rebuilding higher education in post-conflict countries is a difficult and complex task that contains many tensions. Such as physical reconstruction, working to restore the established higher education system, and developing capabilities and skills. As well as issues related to reforming the public sector, reforming the higher education system, supporting the public sector versus the private sector, and moving towards independence (Milton, 2018).

The recovery process in the higher education sector in Syria after the war is an enormous challenge given the significant turnover that the country has witnessed. Factors that play a role in the recovery process include humanitarian, economic, and political aspects. Here are some items that can affect the recovery process:

❖ Infrastructure rebuilding and providing funding and support:

Recovery requires rebuilding the university infrastructure that was damaged or destroyed during the conflict. This includes rebuilding academic buildings and other facilities. The educational sector needs significant funding and international support to rebuild universities and provide the necessary academic and technical capabilities.

❖ Securing academic staff and enhancing security and stability:

Qualified academic personnel and researchers must be attracted and training programs provided to compensate for the loss of academic personnel during the conflict. Also, recovery requires stable security to ensure educational activities proceed without disruption.

❖ Restoring cultural and educational identity and redirecting research priorities:

Universities can play an important role in restoring the cultural and educational identity of the entire society. Also, universities should consider redirecting research priorities to meet society's current needs and challenges.

❖ Promoting digital education and encouraging innovation and entrepreneurship:

Shifting to online education and digital technologies can be a way to deliver academic courses even under difficult circumstances. Encouraging innovation and entrepreneurship in scientific research and technological applications can be part of a recovery strategy.

❖ Achieving integration with the local community and promoting vocational and technical education:

Universities must be integrated with the local community and better meet its needs. Career and technical education can play a large role in preparing students to participate in the reconstruction process. Despite the significant challenges, focusing on these elements can contribute to accelerating the recovery process in higher education in Syria.

Armed Conflict in Syria and Reasons for the Outbreak of War

Popular demonstrations in Syria in March 2011 sparked what later became an ongoing armed conflict in Syria that affected a large portion of the Syrian population and subjected them to displacement either inside the country or outside it. During the outbreak of war, many people lost their lives due to armed violence, and others became detainees or missing persons.

The Impact of the Armed Conflict in Syria on Higher Education

The armed conflict that broke out in Syria in 2011 is considered one of the biggest challenges facing the country, and one of the sectors that has been greatly affected is higher education. The conflict represents the total destruction of university infrastructure and poses enormous challenges to students and academics alike. This chapter tries to explore the impact of the armed conflict in Syria on the higher education system, from deliberations of the economic and social context to the unique challenges faced by the academic community.

Historical Context

The history of conflict in Syria extends back a decade, with pro-democracy protests beginning in 2011 and developing into an armed conflict. Over time, the opposition factions fragmented, and regional and international powers intervened, resulting in a deterioration of the security and economic conditions.

- The impact of the destruction on university infrastructure

Among the most affected aspects of the conflict context is the severe destruction of university infrastructure. It included academic buildings, libraries, and laboratories, resulting in a significant loss of educational and cultural resources.

- Migration of academic staff and students

One of the main challenges facing higher education in Syria is the migration of academic staff and students. Due to security threats and instability, many of them have migrated to other countries in search of safety and better educational opportunities. Many people and researchers may ask how the Syrian civil war started. The Syrian civil war is a tragic conflict that has lasted 11 years. Though some observers argue about what caused the Syrian Civil War, it's clear that it started in the background of the Arab Spring, a series of popular uprisings against despotic governments across the Middle East. Though Assad (Syria's president) was confident that he would be spared, deep political and economic tensions plagued the country (De Gree, Cheprasov, 2023).

The violence escalated in Syria after the outbreak of the revolution, and a civil war began. many armed and rebel groups sprung up, the conflict became more than a fighting among Syrians (between a supporter of the Assad regime and an opponent of it). The foreign powers began to support sides of the conflict, by sending the money, and weaponry, as the chaos gave rise to organizations that worsened extremist jihadist organizations, such as the Islamic State (IS) group and al-Qaeda, became involved. which formed a source of concern to the international community which saw them as a big threat(bbc.com, 2023).

Good governance requires legal frameworks and the protection of human rights, particularly the minorities' rights. A country with the rule of law tends to accept non-violent mechanisms to end the conflict, it supports peaceful responses to civilian demands. Since the rule of law requires the guarantee of individuals' rights, it is also important to recognize its strong links with the acceptance of the rights of others. such the rule of law requires both formal institutions and a culture supporting them (Karim, Mousavi Shafaee,2018).

CONCLUSION

The impact of the armed conflict on higher education in Syria is clearly visible, as most educational facilities have been destroyed, resulting in a significant loss of resources and technologies that contribute to quality education, also the conflict led to the migration of academic personnel, leaving universities without the necessary expertise and affecting the ability to provide high-quality education, an education gap appeared due to the academic courses stopped, which affects students' progress and weakens the academic and research dimensions of universities. This conflict caused a psychological impact on students and academics, as they suffer from psychological stress and loss of security, which affects their academic performance and emotional stability, Employment opportunities for graduates have been significantly affected by the weak economy and lack of career opportunities, educational opportunities for women and girls have been significantly affected by the conflict, and rates of early marriage and involvement in domestic work have increased, taking on the responsibilities of the husband, children, and home at an early at a very young age under 18 years. Early marriage prevented many girls from pursuing their education.

Financing and reconstruction challenges appeared, and the educational sector in Syria faces major challenges in securing the necessary funding to rebuild infrastructure and provide educational services. there is an urgent need for international support to rebuild higher education in Syria, including supporting academic personnel and providing financial and technical resources. The conflict reflects its impact on the identity and future of Syrian academics, as they face enormous challenges in surviving and escaping difficult circumstances. Rebuilding higher education in Syria requires effective integration with the international community and humanitarian actors to achieve development and stability in the region. These conclusions reflect the challenges of higher education in Syria and highlight the urgent need for sustained efforts and international cooperation to rebuild the educational sector and secure a sustainable educational future, these points bring together the extent of the impact of the conflict on the educational system in Syria and highlight the major challenges that this system faces in the post-conflict phase.

REFERENCES

Alkol, M.(2023).Challenges and Difficulties of University Study in Northwestern Syria. *International Higher Education, 11.*

AlMunifi, A., & Aleryani, A. (2016). Internal efficiency of Higher education system in armed conflict-affected countries-Yemen case. *International Journal of Educational Development, 83.*

Alshami, S., & Alzanfaly, A. (2021). Future directions for planning education in times of crises: the Corona pandemic as an example. Regional Center for Educational Planning, Sultan Kabus University.

Cervantes-Duarte, L., & Fernández-Cano, A. (2016). Impact of Armed Conflicts on Education and Educational Agents: A Multivocal Review. *Revista Electrónica Educare (Educare Electronic Journal), 20*(3).

De Gree, A., & Cheprasov, A. (2023). *Syrian Civil War Timeline.* Impact & Outcome.

Economic and Social Commission for Western Asia. (2019). *Development challenges in conflict settings, The impact on child marriage and adolescent fertility.* E/ESCWA/ECRI/2019/2.

International Committee of the Red Cross (ICRC). (2008). *How is the Term "Armed Conflict" Define in International Humanitarian Law?* Opinion Paper, March.

Justino, P. (2010). How Does Violent Conflict Impact on Individual Educational Outcomes, The Evidence So Far. UNESCO. *Education for All Global Monitoring Report, 2011*, 2.

Karimi, M., & Mousavi Shafae, S. M. (2018). Poor Governance and Civil War in Syria. *Turkish Journal of Middle Eastern Studies, Vol, 5*(1), 49–71. doi:10.26513/tocd.383377

Melander, E. (2005). Gender Equality and Intrastate Armed Conflict. *International Studies Association Quarterly, 49*(4).

Milton, S. (2018). *Higher Education and Post-Conflict Recovery.* Academic Press.

Omoeva, C., & Moussa, W. (2018). *The Effects of Armed Conflict on Educational Attainment and Inequality.* Education Policy and Data Center, EPDC Research Paper No. 18-03.

Raheema, N. (2017). The impact of armed conflicts on quality Education in Iraq. *Al-Mostansiriyah Journal for Arab and International Studies, 14*(57).

KEY TERMS AND DEFINITIONS

Armed Conflict: This means a conflict between regime forces and armed groups that poses a threat to the lives of civilians, and causes massive displacement of the people, and violations of human rights and international humanitarian law.

Capacity Building: The improvement of the capability of the individual or organization to produce, and perform.

Civil War: War between organized groups within the same country. The aim may be to control the country or a region, to achieve independence for a region, or to change government policies.

Crisis: An unstable situation in the country, whether in the economic, social, political, or security sphere, which leads to the disruption of most aspects and fields of work in it, and causes many problems such as unemployment, crime, and others.

Cultural Preservation: The protection of cultural heritage, to ensure its survival for future generations. It involves the preservation of cultural artifacts, traditions, languages, and practices that are considered valuable and significant.

Economic Reconstruction: A process for creating a proactive vision of economic change.

Higher Education: Higher education is a stage of formal learning that comes after the completion of secondary education.

Northwestern Syria: Northwestern Syria is the region around Aleppo, Syria. Home to the UNESCO World Heritage Site the Dead Cities and countless archaeological remains, it is as of 2023 contested by the Syrian government, Turkish-backed rebels, and Kurdish rebels. Travel in this region is both difficult and dangerous. Northwestern Syria contains many cities such as Efrin, Aleppo, Idlib, Maarat al-Numan, Saraqib.

Post-Conflict Recovery: The improvement of the security, military, and the restoration of law and order, political and governance, economic and development, economic reform, and improvement of prevailing social conditions.

Chapter 18
Stuck in the Limbo:
Syrian Higher Education After 13 Years of War

Mustafa Kayyali
https://orcid.org/0000-0003-3300-262X
HE Higher Education Ranking, Syria

ABSTRACT

This chapter explores the unstable situation of Syrian higher education in the aftermath of 13 years of ongoing conflict. As the nation grapples with the ramifications of war, its educational institutions have been significantly damaged, leading to a condition of limbo for students, professors, and administrators alike. Drawing on an in-depth examination of the obstacles faced, including infrastructure destruction, brain drain, and ideological upheavals, this chapter examines the complex processes determining the contemporary status of Syrian higher education. Furthermore, it discusses alternative pathways for rebuilding and rejuvenating the industry, highlighting the important significance of international collaboration and innovative concepts in navigating the path forward.

INTRODUCTION

The Syrian conflict, now running over thirteen years, has caused severe and far-reaching devastation upon all elements of society, with higher education emerging as one of its most grievously impacted sectors. Once hailed as a bulwark of intellectual progress and enlightenment, Syrian universities and colleges today stand as shattered ruins of a bygone period, their halls resounding with the melancholy sounds of a lost generation. In this comprehensive examination, we seek to unravel the intricate web of issues besieging Syrian higher education, highlighting the devastating repercussions wreaked by years of conflict and displacement. As the fighting progressed, so too did the toll on educational infrastructure. Campuses previously bustling with the interchange of ideas and the quest for knowledge now lie in ruins, their once-majestic buildings reduced to ashes by the unrelenting onslaught of conflict. Libraries, once repositories of wisdom and enlightenment, have been turned to ashes, their treasured volumes lost

DOI: 10.4018/979-8-3693-1926-0.ch018

to the annals of history. Laboratories, where revolutionary research previously flourished, now lie idle, their equipment looted or destroyed in the chaos of conflict. The physical devastation of educational institutions has not only disturbed the academic pursuits of students and teachers but has also inflicted a debilitating blow to the basic underpinnings of knowledge transfer and scholarly inquiry.

The migration of educated professionals, generally referred to as brain drain, has further aggravated the problem facing Syrian higher education. Faced with depleting resources, persistent insecurity, and restricted prospects for professional growth, many academics and researchers have been driven to depart their homeland in search of safety and stability. The loss of this intellectual capital has left a gaping void within the higher education sector, depriving universities of the expertise and institutional memory needed to negotiate the challenges of rebuilding and renewal. The outflow of talent has not only delayed efforts to recover but has also maintained a cycle of stagnation and deterioration, further increasing the sense of dread and hopelessness gripping the academic community. Beyond the physical destruction and human displacement, the Syrian crisis has created fundamental ideological differences within intellectual circles (Prosyukova et al, 2017). Universities, once bastions of free thought and open research, have become battlegrounds for conflicting ideologies, with dissenting voices repressed and alternative perspectives marginalized. The stifling of academic freedom and the suppression of dissent have produced a climate of fear and intimidation, where professors and students alike are obliged to self-censor their views and ideas for fear of retaliation. The degradation of intellectual freedom not only weakens the integrity of higher education but also threatens to perpetuate the cycle of violence and extremism engulfing Syrian society.

In addition to the devastation of infrastructure and brain drain, the conflict has wrought havoc upon the accessibility of education for millions of Syrians, particularly women and girls. With entire communities uprooted and displaced, access to higher education has become a faraway dream for many, with financial restraints, security concerns, and cultural barriers standing in the way of academic advancement. Refugee camps, while giving a semblance of shelter and security, lack the resources and infrastructure needed to enable meaningful educational opportunities, keeping countless young Syrians locked in a cycle of poverty and despair. For women and girls, traditional gender norms and cultural expectations sometimes serve as insurmountable impediments to education, with many compelled to abandon their goals of higher learning in favor of early marriage or domestic obligations. Despite these severe hurdles, there is a ray of hope for the future of Syrian higher education. International solidarity and collaboration have emerged as beacons of resilience, with governments, NGOs, and educational institutions coming together to support the repair and regeneration of Syrian universities and colleges. From supporting scholarships and academic exchange programs to giving technical aid and capacity-building support, the international community has played a critical role in helping Syrian higher education weather the storm of conflict and displacement. Through creative ideas and collaborative alliances, stakeholders both inside and outside Syria's borders are working diligently to repair what has been lost and set the groundwork for a brighter future.

By leveraging the transforming power of education, we may transcend the scars of conflict, bridge the divides of ideology and identity, and spread the seeds of stability and prosperity for future generations. As we continue on this road of rebuilding and renewal, let us not lose sight of the strength and drive of the Syrian people, whose thirst for knowledge and quest for a better tomorrow will ultimately prevail over the darkness of violence and despair.

The Impact of War on Syrian Higher Education

The Syrian conflict, a protracted and challenging disaster now extending over thirteen years, has left an unmistakable effect on every element of Syrian society, including its higher education system. Once acclaimed as a symbol of intellectualism and growth in the area, Syrian universities and colleges now find themselves coping with a series of diverse issues arising from the relentless onslaught of war, displacement, and ideological upheaval. In this wide inquiry, we start with a thorough exploration of the numerous problems encountered by Syrian higher education, probing the deep web of ramifications produced by years of violence and unrest (Tozan, 2023).

At the forefront of the crisis lies the widespread damage to physical infrastructure. Syrian campuses, once busy hubs of scholarship and creativity, now lie in ruins, their once-grand edifices reduced to ashes by the ravages of war. Classrooms, labs, and libraries, once alive with the interchange of ideas and the pursuit of knowledge, now remain as haunting reminders of the toll exacted by bloodshed and destruction. The disappearance of these crucial educational venues not only disturbs the academic pursuits of students and staff but also deals a significant blow to the basic underpinnings of scholarly inquiry and intellectual growth. The protracted nature of the conflict has led to the enormous displacement of millions of Syrians, resulting in a major interruption to higher education access and participation (Abedtalas et al, 2020). Many students have been forced to abandon their studies as they flee conflict and seek safety in safer zones, while others confront insurmountable barriers to obtaining educational opportunities in the chaos of relocation. The displacement crisis has upended the lives of countless students and academics, robbing them of the ability to pursue their academic dreams and leaving a generation of Syrians bereft of the transformational power of education. Compounding these issues is the phenomenon of brain drain, which has seen a major outflow of educated professionals, intellectuals, and students leave the conflict-ridden country in pursuit of safety and opportunity abroad. Faced with limited chances for professional growth and pervasive insecurity at home, many professors and researchers have been obliged to seek asylum in foreign places, leaving behind a gap in expertise and institutional knowledge within Syrian higher education institutions. The loss of this intellectual capital not only hinders efforts at rehabilitation and renewal but also perpetuates a cycle of decline and stagnation within the sector.

Beyond the physical destruction and human displacement, the Syrian conflict has produced fundamental ideological splits among academic circles, further worsening the problem facing higher education. Universities, once bastions of free thought and open research, have become battlegrounds for competing narratives, with dissenting voices repressed and alternative perspectives marginalized. The restriction of academic independence and the imposition of ideological orthodoxy have hampered intellectual discourse and innovation, hampering the development of critical thinking skills and constraining the intellectual horizons of students and scholars alike. The conflict has had a terrible impact on the mental health and well-being of students and teachers, increasing previous traumas and generating new sources of psychological suffering. The persistent exposure to violence, dislocation, and uncertainty has taken a terrible toll on the psychological resilience of those within the academic community, leading to widespread anxiety, despair, and post-traumatic stress disorder (Al Azmeh et al, 2021). Addressing the mental health needs of learners and professors is therefore essential in any comprehensive effort to rebuild and revitalize Syrian higher education, ensuring that individuals can heal from the psychological wounds of war and resume their academic pursuits with renewed vigor and resilience.

As Syria stands at a crossroads, the rehabilitation of higher education will be important to the country's long-term prosperity and security. It will require not only enormous financial investment but also a commitment to nurturing a culture of academic independence, critical inquiry, and creativity. Only through the combined efforts of Syrians and the world at large can the promise of higher education be restored to the students and scholars of Syria, ensuring that they have the expertise and abilities they need to rebuild their country and shape a brighter future for generations to come. In light of the issues facing Syrian higher education, restoration efforts must focus on the needs and aspirations of students and teachers. This includes investing in the physical facilities of universities and colleges, giving access to mental health services, and establishing chances for professional development and scholarly exchange. Additionally, attempts to reconstruct higher education must be guided by values of inclusivity and diversity, ensuring that all members of society have equal access to educational possibilities regardless of gender, ethnicity, or financial background.

The restoration of higher education allows for reinventing the role of universities and colleges in Syrian society. By encouraging a culture of innovation and entrepreneurship, higher education institutions may play a major role in advancing economic development and social advancement. Through cooperation with business, government, and civil society, universities can leverage the creativity and inventiveness of students and staff to address important challenges facing Syrian society, from reconstructing infrastructure to supporting sustainable development. In the aftermath of violence, higher education also has a key role to play in encouraging reconciliation and peacebuilding. By promoting dialogue, tolerance, and mutual understanding, colleges can help bridge the gaps that have ripped Syrian society apart. Through initiatives such as interfaith conversation programs, peace studies courses, and community involvement projects, higher education institutions can enable students to become agents of positive change in their communities and beyond. The reconstruction of Syrian higher education is not simply about rebuilding facilities and restoring academic programs—it is about forging a brighter future for Syria and its people. By investing in education, we invest in the next generation of leaders, inventors, and peacemakers who will determine the fate of their nation. As Syrians seek to rebuild their country from the ashes of conflict, higher education will be a beacon of hope and a catalyst for change, lighting the way toward a more prosperous, inclusive, and peaceful future for all (Pherali & Millican, 2020).

The challenges facing Syrian higher education in the wake of the crisis are substantial and diverse, but not insurmountable. By tackling the devastation of physical infrastructure, the displacement problem, the brain drain phenomena, and the ideological conflicts within academic circles, Syrians and the international community may work together to rebuild and revive higher education in Syria. Through investment in education, promotion of academic freedom and critical inquiry, and nurturing of creativity and entrepreneurship, higher education may once again become a cornerstone of Syrian society, empowering individuals to rebuild their lives and their nation in the aftermath of war.

Challenges Faced by Higher Education Institutions

The Syrian crisis, now ongoing for over a decade, has left an indelible effect on every element of Syrian society, including its higher education system. Once acclaimed as a symbol of intellectualism and growth in the area, Syrian universities and colleges now find themselves coping with a series of diverse issues arising from the relentless onslaught of war, displacement, and ideological upheaval. In this wide inquiry, we start with a thorough exploration of the numerous problems encountered by Syrian higher education, probing the deep web of ramifications produced by years of violence and unrest. At the forefront of the crisis lies the widespread damage to physical infrastructure. Syrian campuses, once vibrant hubs of learning and creativity, now lie in ruins, their classrooms, labs, and libraries reduced to ashes by the ravages of war. The destruction of infrastructure not only disrupts the academic calendar and impedes the delivery of educational services but also constitutes a substantial impediment to the eventual rehabilitation and regeneration of the sector (Watenpaugh et al, 2014).

In addition to the devastation of physical infrastructure, higher education institutions in Syria are challenged with a significant scarcity of skilled instructors and personnel. The brain drain issue, spurred by pervasive insecurity and restricted possibilities, has led to a large departure of skilled individuals, including professors, researchers, and administrators, who have sought sanctuary abroad. This loss of human capital not only deprives universities of vital knowledge and institutional memory but also impairs their ability to deliver quality education and conduct meaningful research (Jesry, 2022). The prolonged violence has exacerbated existing inequities in access to higher education, particularly for disadvantaged and vulnerable communities. Displaced persons, refugees, and internally displaced individuals confront considerable impediments to obtaining educational opportunities, including limited financial resources, bureaucratic hurdles, and a lack of infrastructure. As a result, many young Syrians find themselves unable to pursue their academic dreams, creating cycles of poverty and exclusion. The politicization of higher education has arisen as a serious concern facing institutions in Syria, with universities and colleges becoming battlegrounds for contending ideological ideologies. The restriction of academic freedom, censorship of dissenting voices, and imposition of ideological conformity harm the integrity and credibility of academic institutions, inhibiting critical research and intellectual discussion. As a result, higher education in Syria risks becoming an instrument of political indoctrination rather than a bulwark of knowledge and enlightenment.

Despite these enormous hurdles, there remains hope for the future of higher education in Syria. The perseverance and drive of students, educators, and administrators, together with the outpouring of international sympathy and support, offer a ray of optimism amidst the gloom of conflict. By addressing the root causes of the crisis, putting resources into the rehabilitation of infrastructure, and fostering a culture of liberty in academia and innovation, Syrians and the world at large can work together to ensure that higher education remains a beacon of optimism and potential for generations to come. In reaction to the loss of physical infrastructure, efforts must be made to reconstruct and rehabilitate university campuses and facilities across Syria. This will require major financial investment, technical skill, and coordination among numerous parties, including government agencies, international organizations, and academic institutions. Priority should be given to repairing important infrastructure such as classrooms, labs, and libraries, as well as ensuring that buildings are outfitted with modern amenities and facilities to facilitate teaching, learning, and research activities (Mourtada, 2019).

Initiatives should be initiated to address the dearth of skilled teachers and staff in Syrian higher education institutions. This may involve recruiting and training new faculty members, motivating competent professionals to return to Syria, and developing relationships with international universities and organizations to support knowledge transfer and capacity-building. Efforts should also be made to support early-career researchers and academics, providing them with chances for professional growth, mentorship, and networking to help them thrive in the academic environment.

In addition to physical and human resources, attention must be paid to overcoming the structural barriers that restrict access to higher education for disadvantaged and vulnerable communities in Syria. This may include creating policies and initiatives to support students from disadvantaged backgrounds, such as scholarships, financial aid, and academic support services. Efforts should also be made to remove bureaucratic impediments and expedite administrative processes to make it easier for students to enroll in and complete their courses. Steps should be taken to defend academic freedom and encourage intellectual diversity inside Syrian higher education institutions. This may require implementing legislation to defend academic freedom, ensuring that universities have autonomy in their governance and decision-making processes, and creating a culture of open research and discussion (Mualla, 2020). Efforts should also be taken to counteract censorship and ideological compulsion, allowing students and faculty to openly express their opinions and engage in rigorous intellectual discourse without fear of retaliation.

The issues facing higher education institutions in Syria are complicated and numerous, but they are not insurmountable. By addressing the damage to physical infrastructure, rebuilding human capacity, increasing access to education, and encouraging academic freedom and intellectual diversity, Syrians and the international community can work together to rebuild and revive higher education in Syria (Shaban, 2020). In doing so, they help ensure that higher education remains a cornerstone of Syria's recovery and reconstruction, offering hope, opportunity, and a brighter future for generations to come. In light of the issues facing Syrian higher education, restoration efforts must focus on the needs and aspirations of students and teachers. This includes investing in the physical infrastructure of universities and colleges, giving access to mental health services, and establishing chances for professional development and scholarly exchange. Additionally, attempts to reconstruct higher education must be guided by values of inclusivity and diversity, ensuring that all members of society have equal access to learning possibilities regardless of gender, ethnicity, or financial background. The restoration of higher education allows for reinventing the role of universities and colleges in Syrian society. By encouraging a culture of innovation and entrepreneurship, higher education institutions may play a major role in advancing economic development and social advancement. Through cooperation with business, government, and civil society, universities can leverage the creativity and inventiveness of students and staff to address important challenges facing Syrian society, from reconstructing infrastructure to supporting sustainable development.

In the aftermath of violence, higher education also has a key role to play in encouraging reconciliation and peacebuilding. By promoting dialogue, tolerance, and mutual understanding, colleges can help bridge the gaps that have ripped Syrian society apart. Through initiatives such as interfaith conversation programs, peace studies courses, and community involvement projects, higher education institutions can enable students to become agents of positive change in their communities and beyond. Ultimately, the rehabilitation of Syrian higher education is not simply about rebuilding facilities and reinstating academic programs—it is about constructing a brighter future for Syria and its people. By investing in education, we invest in the next generation of leaders, inventors, and peacemakers who will determine the fate of their nation.

As Syria stands at a crossroads, the rehabilitation of higher education will be important to the country's long-term stability and prosperity. It will require not only enormous financial investment but also a commitment to nurturing a culture of academic independence, critical inquiry, and creativity. Only through the combined efforts of Syrians and the international community can the promise of higher education be restored to the students and scholars of Syria, ensuring that they have the expertise and abilities they need to rebuild their country and shape a brighter future for generations to come. The obstacles facing Syrian higher education are great, but with commitment, ingenuity, and international help, they may be addressed.

Brain Drain and Its Implications

The discussion of brain drain has emerged as one of the most significant obstacles facing higher education institutions in Syria in the aftermath of almost a decade of fighting. Defined as the emigration of trained individuals, including professors, researchers, and students, brain drain has had far-reaching ramifications for the quality of education, the sustainability of academic institutions, and the chances for post-war reconstruction. One of the most immediate repercussions of the brain drain on higher education in Syria has been the loss of significant human capital. Skilled professionals, including seasoned educators, inventive researchers, and talented administrators, have been driven to abandon the country in pursuit of safety and opportunity abroad, leaving behind a paucity of expertise and institutional memory. This loss not only lowers the quality of education and research undertaken within Syrian universities and colleges but also offers serious difficulties to the long-term viability and sustainability of the sector.

The migration of competent professionals has worsened existing shortages of qualified teachers and staff within higher education institutions in Syria. With many educators and researchers departing the country in search of better prospects elsewhere, universities and colleges are left unable to fill vacancies, resulting in greater workloads, reduced morale, and a deterioration in academic quality. This, in turn, impairs the ability of institutions to attract and retain students, further prolonging cycles of decline and stagnation. Brain drain has had serious repercussions for the future of scientific research and innovation in Syria. With many of the country's most skilled researchers and innovators now residing abroad, the capacity of Syrian universities and research organizations to undertake cutting-edge research and discover innovative solutions to critical societal concerns has been significantly hampered. This not only complicates efforts at post-war reconstruction and growth in the economy but also reinforces cycles of dependency and underdevelopment inside the country (Mualla & Mualla, 2022).

Beyond its immediate impacts on higher education institutions and the scientific community, brain drain has broader socio-economic ramifications for Syrian society as a whole. The loss of competent professionals deprives the country of significant human capital, impeding efforts at economic recovery and social progress. Moreover, brain drain exacerbates existing disparities within Syrian society, as those with the means to depart seek opportunities overseas, leaving behind those who are less fortunate to suffer the burden of the conflict's effects. The repercussions of brain drain are numerous and intricately connected with the issues facing Syria as it strives to rebuild and recover from years of turmoil. In addition to the loss of expertise and institutional capacity, brain drain also affects efforts to build a culture of innovation and entrepreneurship within the country. Many of Syria's greatest minds, confronted with restricted prospects and a lack of resources, are opting to take their talents elsewhere, robbing the country of the creative energy and inventiveness needed to drive progress and development.

Despite these daunting hurdles, there remains potential for tackling the phenomena of brain drain and its ramifications for higher education in Syria. By investing in initiatives and schemes aimed at attracting and keeping skilled professionals, fostering a culture of entrepreneurship and creative thinking, and rebuilding trust in the country's higher education system, Syrians and the rest of the world can work together to ensure that the promise of education remains within reach for future generations. One potential path for combating brain drain is through the development of scholarship programs and research grants targeted at motivating skilled workers to remain in Syria or return to the nation after seeking opportunities abroad. By offering financial support and professional development opportunities, such programs can help retain talent inside the country and strengthen the capacity of higher education institutions to perform cutting-edge research and deliver quality education to students.

In addition to financial incentives, initiatives to counteract the brain drain must also focus on building an enabling environment for innovation and entrepreneurship within Syria. This includes investing in research infrastructure, facilitating collaboration between academics, industry, and government, and fostering a culture of risk-taking and experimentation. By offering chances for Syrian researchers and innovators to develop and market their ideas within the country, governments may assist to stem the flow of brain drain and harness the imaginative abilities of Syria's youth. Rebuilding trust in Syria's higher education system will be vital for attracting and maintaining qualified professionals in the years to come. This would need resolving concerns like as corruption, political influence, and lack of academic freedom, which have damaged faith in the integrity and credibility of the country's institutions and colleges. By fostering openness, accountability, and meritocracy within higher education institutions, policymakers can assist in restoring faith in the system and encourage talented individuals to contribute their talents and expertise to the rebuilding of Syria.

The phenomenon of brain drain poses substantial challenges for higher education in Syria, compromising the capacity of institutions to deliver excellent education and conduct cutting-edge research. However, by investing in programs and initiatives focused on retaining talent, stimulating innovation, and rebuilding trust in the higher education system, Syrians and the international community can work together to overcome these obstacles and construct a brighter future for higher education in Syria. Only through concerted action and persistent dedication can we hope to address the core causes of brain drain and create an environment where all Syrians have the opportunity to fulfill their potential and contribute to the country's growth. Brain drain is not a new problem, but its impact on Syria's higher education system has been particularly severe in the aftermath of the extended conflict that has consumed the country. Before the onset of the conflict, Syria possessed a rather healthy higher education industry, with institutions and colleges attracting students and academics from around the region and beyond. However, the outbreak of conflict in 2011 has led to a tremendous migration of talent, as experienced professionals seek sanctuary overseas in pursuit of safety and opportunity.

The loss of competent professionals has had a rippling effect throughout Syria's higher education system, increasing existing issues and impeding efforts at rehabilitation and reconstruction. One of the most apparent impacts of brain drain has been the depletion of academics and staff within universities and colleges across the country. Many experienced educators and researchers have been forced to quit the nation, leaving behind a paucity of trained professionals and a lack of institutional continuity. The impact of brain drain on the quality of education and research within Syrian institutions and colleges cannot be emphasized. With fewer skilled faculty members available to educate and guide students, academic standards have decreased, and the general quality of education has worsened. Moreover, the loss of experienced researchers has hampered efforts to conduct important research and innovation, thus diminishing the capacity of higher education institutions to contribute to the country's development. In addition to its immediate effect on higher education institutions, brain drain also has broader socio-economic ramifications for Syrian society as a whole. The loss of competent professionals deprives the country of significant human capital, impeding efforts at economic recovery and social progress. Moreover, brain drain exacerbates existing disparities within Syrian society, as those with the means to depart seek opportunities overseas, leaving behind those who are less fortunate to suffer the burden of the conflict's effects.

Addressing the core causes of brain drain will require a diversified approach that addresses both the push and pull forces causing skilled workers to leave Syria. On the one hand, efforts must be made to improve conditions within the country, including resolving security concerns, offering economic opportunities, and ensuring respect for human rights and the rule of law. On the other side, measures should be taken to entice qualified people back to Syria, including granting financial incentives, providing support for research and innovation, and creating a suitable climate for professional growth and development. Investing in higher education is crucial to tackling brain drain and restoring Syria's human capital base.

By supporting universities and colleges, offering scholarships and research grants, and promoting collaboration between academics, industry, and government, policymakers may help retain talent inside the country and create chances for professional growth and development. Efforts should be made to encourage a culture of innovation and entrepreneurship within Syria, offering assistance for startups and small enterprises and generating chances for Syrian scholars and innovators to contribute to the country's development. Brain drain poses substantial issues for higher education in Syria, diminishing the capacity of institutions to deliver excellent education and conduct cutting-edge research. However, by addressing the core causes of brain drain and investing in higher education, Syrians and the international community can work together to achieve a brighter future for higher education in Syria. Only through joint action and persistent dedication can we hope to overcome the problems posed by brain drain and establish an environment where all Syrians have the opportunity to fulfill their full potential and contribute to the country's growth.

Ideological Shifts and Academic Freedom

The Syrian crisis has wreaked terrible consequences on several fronts, foremost among them being the deterioration of academic freedom within the country's higher education institutions. Once regarded for their position as centers of intellectual inquiry and diversity, Syrian universities and institutions have fallen prey to the chaotic forces of conflict, becoming battlegrounds for competing ideological agendas and silencing dissent in the process. This transition has major ramifications not only for the pursuit of knowledge within Syria but also for the larger landscape of higher education in the region. At the heart of the matter lies the erosion of academic freedom, a basic concept that underpins the integrity and legitimacy of any educational institution. In Syria, scholars, educators, and students alike find themselves traversing a perilous landscape where expressing dissenting viewpoints or engaging in critical inquiry can have dire implications. The politicization of academia has led to an environment of fear and intimidation, with censorship and surveillance becoming prevalent methods of control. As a result, the once-vibrant intellectual life of Syrian colleges has been repressed, giving place to a culture of conformity and silence.

One need only look at the experiences of intellectuals and educators within Syria to grasp the enormity of the obstacles they encounter. Many find themselves excluded and stigmatized for their ideas, compelled to self-censor, or fear prosecution by the authorities. The degradation of academic freedom has profound consequences for the quality of education and research undertaken within Syria's higher education institutions. Scholars and researchers, bound by ideological limitations and political pressure, find themselves unable to follow their academic interests freely or engage in meaningful scholarly inquiry (Kayyali, 2022). This not only inhibits attempts at knowledge production and dissemination but also weakens the trustworthiness and reliability of Syrian higher education on the world stage. As a result, Syrian institutions and colleges risk being left behind in an increasingly interconnected and competitive world.

The politicization of higher education has exacerbated existing differences within Syrian society, intensifying sectarian, ethnic, and ideological cleavages. Universities and colleges, once sanctuaries for intellectual discussion and dialogue, have become arenas for the spread of opposing ideologies, reinforcing cycles of mistrust and antagonism. This continues to undermine efforts at reconciliation and social cohesiveness, undermining the country's prospects for long-term stability and peace. In this way, the erosion of liberty in academia tends to deepen the very conflicts that have ripped Syria apart. Despite these significant hurdles, there remains hope for the restoration of academic freedom within Syrian higher education. The perseverance and drive of scholars, educators, and students, together with the solidarity and support of the world community, offer a ray of optimism in the gloom of conflict. By advocating for the protection of academic freedom, fostering spaces for free speech and debate, and supporting initiatives aimed at promoting diversity of thought and intellectual inquiry, Syrians and the wider world can work together to ensure that higher education remains a beacon of hope and enlightenment amid turmoil. One path for restoring academic freedom in Syria is through the formation of foreign relationships and cooperation. By creating partnerships with universities and academic institutions overseas, Syrian universities and colleges can access resources, expertise, and support networks that can help combat the impacts of ideological repression and censorship. Moreover, foreign relationships can create opportunities for Syrian scholars and students to connect with other perspectives and ideas, developing a culture of tolerance and openness inside higher education institutions.

In addition to foreign alliances, attempts to restore academic freedom in Syria must also focus on reforming institutions and capacity-building. This includes enhancing legislative frameworks and safeguards for academic freedom, encouraging transparency and accountability within higher education institutions, and providing training and assistance for faculty members and administrators. By empowering researchers and educators to protect their rights and campaign for academic freedom, Syrians may begin to regain the intellectual spaces that have been lost to ideological repression and political meddling. Initiatives aiming at fostering digital literacy and access to knowledge can help to reduce the impact of censorship and surveillance on academic freedom in Syria. By offering training and resources for students and academics to navigate online spaces with confidence, Syrians can continue to pursue their educational pursuits and engage in meaningful intellectual discourse, even in the face of government repression and censorship. The decline of academic freedom in Syria poses substantial obstacles to higher education institutions and the pursuit of knowledge. However, by advocating for the protection of academic freedom, fostering international collaboration, and promoting reforms to institutions and capacity-building, Syrians and the wider world can work together to ensure that higher education remains an anchor of intellectual inquiry and diversity amid conflict and turmoil. Only through collaborative activity and unflinching devotion can we hope to resist the ideological upheavals that threaten to destroy the basic foundations of academic life in Syria.

But beyond these immediate solutions, there is a need for a deeper social shift—a recommitment to the ideas of intellectual freedom and open inquiry that lay at the heart of higher education. This will require not only institutional reform but also a cultural shift—a recognition that variety of opinion and expression is not only desirable but crucial for the growth of knowledge and understanding. In this sense, the participation of civil society, particularly non-governmental organizations, advocacy groups, and grassroots movements, will be vital. By organizing public opinion and lobbying for change, these players can contribute to creating an atmosphere where academic freedom is valued and maintained. Efforts to restore academic freedom must be supplemented by actions to address the core causes of war and instability within Syria. This includes promoting discussion and reconciliation among the country's many communities, addressing grievances, and establishing a culture of tolerance and cooperation. Without addressing these underlying challenges, any attempt to restore academic independence will be partial and eventually unsustainable.

In the longer term, the restoration of academic freedom in Syria must be part of an overall strategy of rebuilding and reconstruction—a reinvention of higher education that embodies the values of democracy, pluralism, and human rights. This will require not only financial investment but also political will as well as dedication from all parties, including the Syrian government, the rest of the world, and civil society. By working together, Syrians can construct a higher education system that is resilient, inclusive, and sensitive to the needs of its people—a system that stands as a beacon of hope and inspiration for years to come.

Strategies for Rebuilding and Revitalizing Higher Education

In the wake of nearly a decade of conflict, the goal of rebuilding and rejuvenating higher education in Syria looms as a massive undertaking. The scars of war run deep, hurting every facet of the country's educational environment. From the destruction of physical infrastructure to the loss of trained experts and the degradation of academic freedom, Syrian universities and colleges face enormous barriers on the road to recovery. Yet, despite the debris and ruins, there lurks a glimmer of hope—a drive to rebuild, to restore, and to recapture the promise of higher education for future generations. The repair of physical infrastructure stands as a crucial priority in the quest to reinvigorate higher education in Syria. The war has left many institutions and colleges in a condition of disarray, with classrooms, libraries, and laboratories reduced to rubble. Rebuilding these institutions is not simply about erecting new buildings; it's about providing safe, conducive learning environments where students and faculty can thrive. This will require enormous financial commitment, technical skill, and logistical assistance, all of which must be provided by both domestic and international stakeholders. By prioritizing the rehabilitation of educational infrastructure, Syrians may begin to lay the framework for the rebirth of higher education and the restoration of normalcy to academic life.

Yet, physical rebuilding alone cannot address the multiple issues facing higher education in Syria. The migration of competent professionals—academics, researchers, and students—represents another major issue that must be confronted head-on. For years, Syria has witnessed a brain drain of historic proportions, as smart individuals abandon the nation in pursuit of safety and opportunities abroad. To stem this tide, concerted efforts must be undertaken to establish incentives for qualified professionals to return. This could involve the introduction of scholarship programs, research grants, and career development opportunities geared at attracting academics and researchers back to their hometowns. Moreover, measures that promote the reintegration of returning professionals into the higher education system—such as mentorship programs and networking events—can help to ensure their successful transfer and maximize their contributions to the rebuilding process. The restoration of academic freedom and the promotion of a variety of opinions inside Syrian institutions and colleges are critical for rejuvenating higher education in the country. The war has not only taken a toll on physical infrastructure but has also played havoc on the intellectual fabric of society. Scholars, educators, and students find themselves negotiating perilous ground, where expressing divergent viewpoints or engaging in critical inquiry can have fatal implications. The politicization of academia has led to restriction, surveillance, and intimidation, restricting intellectual debate and eroding the integrity of higher education institutions. To counteract this tendency, actions must be taken to abolish restrictive laws and regulations that hinder academic freedom, as well as to build institutional structures to preserve the rights of researchers, educators, and students. By providing a climate where free inquiry and open debate are appreciated and encouraged, Syrians may help to rejuvenate intellectual life inside higher education institutions and promote the pursuit of knowledge as a communal endeavor.

The international community has a significant role to play in assisting the restoration of higher education in Syria. The war has left the country's educational system in shambles, with few resources and the ability to accomplish significant transformation (Al Hessan et al, 2016). In this setting, overseas partners can provide crucial assistance, including financial support, technical experience, and logistical aid to bridge the gap between vision and reality. Moreover, initiatives aimed at promoting academic exchange and collaboration—such as joint research initiatives, faculty exchanges, and mobility for students programs—can help to foster a culture of creativity and outstanding performance within Syrian colleges and universities, thereby positioning them as centers of excellence in the region and beyond. The mission of reconstructing and reviving higher education in Syria is formidable, but not insurmountable. By prioritizing the repair of physical infrastructure, resolving the brain drain crisis, restoring academic freedom, and rallying foreign help, Syrians may begin to lay the framework for a brighter future. By investing in education, we invest in the future of Syria, empowering a new generation of leaders and innovators to rebuild their country and design a better tomorrow for themselves and their children.

The difficulties affecting higher education in Syria are diverse and require a comprehensive approach to address. In addition to reconstructing physical infrastructure and recruiting qualified professionals back to the country, efforts must also be taken to address the underlying causes of violence and instability, including economic disparity, political corruption, and social injustice. Only by addressing these core problems can Syrians hope to achieve a stable and prosperous future for their country and its educational institutions. The restoration of academic freedom and the promotion of a variety of thought inside higher education institutions are crucial for establishing a culture of innovation and excellence. This will require not only the removal of restrictive laws and regulations but also the promotion of democratic ideals, human rights, and the rule of law. By creating an environment where free inquiry and open debate are appreciated and encouraged, Syrians may help to restore their country's intellectual capital and position themselves as leaders in the global knowledge economy. Efforts to revive higher education in Syria must be supported by steps to promote inclusivity and diversity within the educational system. This involves ensuring equitable access to educational opportunities for all Syrians, regardless of their history or circumstances, as well as boosting the participation of women, minorities, and marginalized groups in higher education. By investing in the potential of all Syrians, regardless of their background, Syria can harness the talents and abilities of its whole people to promote innovation, economic growth, and social development.

The restoration of higher education in Syria is necessary for the country's longevity, prosperity, and progress. By prioritizing the reconstruction of physical facilities addressing the brain drain phenomenon, restoring freedom in academia, and promoting diversity and inclusion within the system of education, Syrians can begin to rebuild their country's educational institutions and lay the foundation for a brighter future. With the assistance of the international community and a dedication to democratic values and human rights, Syria may once again become a light of optimism and potential for its people and the world.

CONCLUSION

In conclusion, the journey throughout the changing scenery of Syrian higher education after 13 years of war reveals a complicated fabric of problems, perseverance, and promise. Despite the tremendous damage wreaked by violence, the spirit of study and intellectual endeavor perseveres, functioning as a beacon of resilience amidst the rubble of shattered universities and fractured communities. Throughout this chapter, we have observed the enormous toll exacted on Syrian higher education institutions, from the physical damage of campuses to the evacuation of talented experts and the degradation of academic freedom. The scars of conflict run deep, casting a long shadow over the future of education in Syria and leaving a generation of students and intellectuals caught in a state of uncertainty.

However, among the sorrow, there are glimmers of optimism that illuminate the route forward. International solidarity and collaboration have emerged as strong accelerators for change, overcoming differences, and forging collaborations aimed at rebuilding and rejuvenating higher education institutions. From scholarship schemes that provide opportunities for displaced students to pursue their studies abroad to initiatives that fund the rehabilitation of destroyed institutions, the global world has rallied behind the cause of Syrian higher education. The tenacity of Syrian students, educators, and administrators stands as a monument to the enduring ability of education to transcend suffering and inspire reform. Despite the formidable hurdles they face, they continue to persevere, employing limited resources and inventive techniques to ensure that the torch of knowledge continues to burn bright. As we reflect on the journey thus far, it becomes obvious that the reconstruction of Syrian higher education is not only a matter of reconstructing physical infrastructure but also of growing intellectual resilience and fostering a culture of inquiry and critical thought. It requires a coordinated effort to address the core causes of instability and conflict, including the regaining of freedom in academia and the promotion of a variety of thought.

Moving forward, the international community must stay strong in its support for Syrian higher education, providing both financial resources and moral support to those on the front lines of reconstruction. By investing in education, we invest in the future of Syria, inspiring a new generation of leaders and innovators to rebuild their country from the ashes of war. The tale of Syrian higher education after 13 years of conflict is not one of despair but of perseverance, determination, and hope. As we stand at the crossroads of reconstruction, let us grasp this chance to forge a brighter future for the students and scholars of Syria, one in which the pursuit of knowledge knows no limitations and the promise of peace shines bright. The journey towards restoring higher education in Syria is a hard one, but it is also a journey filled with hope and opportunity. By leveraging the strength, determination, and resourcefulness of its people, Syria can overcome the challenges that lie ahead and establish a brighter future for decades to come. With the backing of the international community and a dedication to the ideals of education, innovation, and peace, Syria may once again become a beacon of optimism and potential for its people and the globe.

But to properly comprehend the breadth and complexity of the challenges facing Syrian higher education, it is vital to go into the specific issues and impediments that must be addressed. One of the most important concerns is the physical damage to campuses and educational infrastructure. The war has left several institutions and colleges in collapse with classrooms, libraries, and laboratories devastated to rubble. Rebuilding these institutions is not only a matter of erecting new buildings; it's about providing safe, conducive learning environments where students and faculty can thrive. The migration of competent professionals has further worsened the issues facing higher education in Syria. Many scientists, researchers, and students have abandoned the country in pursuit of safety and opportunity overseas, leaving behind a dearth of expertise and institutional memory. This loss not only lowers the quality of education and research undertaken within Syrian universities and colleges but also offers serious difficulties to the long-term viability and sustainability of the sector. The decline of academic freedom has had severe ramifications for higher education in Syria. Scholars, educators, and students find themselves negotiating perilous ground, where expressing divergent viewpoints or engaging in critical inquiry can have fatal implications. The politicization of academia has led to restriction, surveillance, and intimidation, restricting intellectual debate and eroding the integrity of higher education institutions. Addressing these difficulties will require an integrated strategy that incorporates both short-term actions and long-term measures. In the moments following the conclusion of the conflict, efforts must be concentrated on reconstructing physical infrastructure, giving emergency aid to displaced students and scholars, and creating safe venues for learning and research. At the same time, efforts must also be taken to address the underlying challenges of academic independence, diversity, and inclusivity. This includes removing restrictive rules and regulations, developing a culture of tolerance and open debate, and increasing the inclusion of women, minorities, and marginalized populations in higher education. It will be vital to invest in the development of human capital, giving chances for professional growth, skills training, and entrepreneurship. By encouraging Syrians to take responsibility for their destinies, we can contribute to establishing a stronger, more inclusive, and successful society.

In conclusion, the obstacles facing higher education in Syria are severe, but not insurmountable. By collaborating, Syrians and the international community can rehabilitate the country's educational institutions and establish the groundwork for a brighter future. With drive, perseverance, and a commitment to the principles of education, creativity, and peace, we can ensure that higher education remains a beacon of optimism and potential for generations to come.

REFERENCES

Abedtalas, M., Alawak, A., Aloklah, W. A., Aljasem, A., & Sarmini, I. (2020). Syrian higher education and social capital in times of conflict. *Education and Conflict Review, 3*, 9–15.

Al Azmeh, Z., Dillabough, J., Fimyar, O., McLaughlin, C., Abdullateef, S., Aloklah, W. A., Mamo, A. R., Abdulhafiz, A. H., Al Abdullah, S., Al Husien, Y., Al Mohamad Al Ibrahim, A., Al Ibrahim, Z., Barmu, T., Farzat, A., & Kadan, B. (2021). Cultural trauma and the politics of access to higher education in Syria. *Discourse (Abingdon), 42*(4), 528–543. doi:10.1080/01596306.2020.1715922

Al Hessan, M., Bengtsson, S., & Kohlenberger, J. (2016). *Understanding the Syrian educational system in a context of crisis* (No. 09/2016). Vienna Institute of Demography Working Papers.

Jesry, M., Omar, F. A., Rashwani, A., Bark, I., Jammo, K., Ajam, S., & Kassab, Z. (2022). Exploring the value of a risk-management quality-assurance model to support delivery of quality higher education in the conflict-affected northwest of Syria. *International Journal of Educational Research Open, 3*, 100134. doi:10.1016/j.ijedro.2022.100134

Kayyali, M. (2022). *Quality Assurance*. Internationalization, and Ranking of Higher Education Institutions in the Context of the Syrian War.

Mourtada, H. (2019). Higher-Education Relevance in Post-War Syria. *Major Challenges Facing Higher Education in the Arab World: Quality Assurance and Relevance*, 353-362.

Mualla, W. (2020). The governance of higher education in post-war Syria. *Higher Education in the Arab World: Government and Governance*, 309-319.

Mualla, W., & Mualla, K. J. (2022). Higher Education in Syria Post-war and Pandemic: Challenges and Opportunities. *Higher Education in the Arab World: New Priorities in the Post COVID-19 Era*, 283-309.

Pherali, T., & Millican, J. (2020). Rebuilding Syrian higher education for a stable future. *Education and Conflict Review, 3*, 1–97.

Prosyukova, K., Babenko, O., Abrosimova, N., & Fominykh, A. (2017). Political crisis in Syria and its impact on education. In *INTED2017 Proceedings* (pp. 8602-8605). IATED. 10.21125/inted.2017.2036

Shaban, F. (2020). Rebuilding higher education in Northern Syria. *Education and Conflict Review*, *3*, 53–59.

Tozan, O. (2023). The evolution of the Syrian higher education sector 1918-2022: From a tool of independence to a tool of war. *Globalisation, Societies and Education*, 1–15. doi:10.1080/14767724.2023.2265854

Watenpaugh, K. D., Fricke, A. L., & King, J. R. (2014). *The War Follows Them. Syrian University Students and Scholars in Lebanon*. Institute of International Education.

Chapter 19
Mortality Awareness of Parents Affected by the Kahramanmaras-Centred Earthquake Disaster in Turkey

Ali Mazı
https://orcid.org/0000-0001-9188-8583
Ministry of National Education, Turkey

ABSTRACT

The current study determined the mortality awareness of mothers and fathers two months after the earthquake disaster that happened in the southern provinces of Turkey on 06.02.2023. The study employed the survey model, one of the qualitative research methods. The study group is comprised of 151 people. The data of the study were collected by using the multidimensional mortality awareness scale. In the analysis of the collected data, independent samples t-test and one-way ANOVA were used. The findings of the study revealed that the participants' mortality awareness levels vary significantly depending on the variables of gender, experienced losses, and status of psychological harm. It was concluded that the participants' mortality awareness decreased as a result of the psychological assistance received after the earthquake as activities aiming to reduce anxiety were conducted during the psychological assistance process. Conducting similar studies with different groups is suggested in order to gain more detailed information about mortality awareness.

INTRODUCTION

Occurrences that happen as a result of physical, natural and human interventions in such a way as to result in physical, economic and social losses in individuals' lives or to terminate their lives or activities or to stop them for a short time in that the society in which they live are negatively affected are called disasters (Ministry of National Education [MoNE], 2011). One of the disasters that most affect social life throughout the history of humanity is earthquakes. Five hundred thousand earthquakes occur in the

DOI: 10.4018/979-8-3693-1926-0.ch019

world annually, only one hundred thousand of them are felt and one hundred of them cause destructions in cities (Reuters, 2023).

When the effects of earthquakes on Turkey are considered, it is seen that 20288 earthquakes occurred in 2022, while the magnitude of 11418 earthquakes was measured above 4 on the Richter scale (Kandilli Observatory, 2023a). When an earthquake has a high magnitude, it does not only cause destruction in the region but also negatively affects human activities. The destruction caused by the 7.7 and 7.6 magnitude (Kandilli Observatory, 2023b) Kahramanmaraş earthquakes in Turkey on February 6, 2023 deeply affected the lives of approximately 13.5 million people in 11 provinces (Anadolu Agency, 2023). Any event that occurs at an unexpected time in a situation that cannot be prepared for and disrupts the physical and psychological integrity of individuals is defined as trauma (MoNE, 2011). Earthquakes can also be shown as one of the most important causes of trauma as they occur suddenly and threaten human life. It can be thought that earthquake which is one of the traumatic situations threatens human life and increases mortality awareness because it damages the integrity of the body.

According to the definition of mortality awareness, it is the individual's acceptance of death with all its dimensions, including fear or anxiety, and developing behaviors accordingly (Levasseur, McDermott & Lafreniere, 2015). The multidimensional nature of mortality awareness is not only limited to fear (Okcu, 2021). When the literature is reviewed, it is seen that studies on earthquakes have mostly been carried out in China and Japan (Fan, Zang, Yang, Mo & Liu, 2011; Honda et al., 2019; Nakajima, 2012). In addition, the sample of most of these studies consists of children.

There are various studies in the national and international literature on earthquake-related death anxiety, fear and awareness. Baris (2022) conducted a study on the perception of death. Fan et al. (2011) examined the symptoms of post-traumatic stress disorder, depression and anxiety in adolescents after the 2008 Wenchuan Earthquake in China. Guvercin (2006) examined the depression levels of earthquake survivor mothers and their children in terms of various variables. Honda et al. (2019) conducted a longitudinal study of the long-term psychological effects of the Great East Japan Earthquake (GEJE) on parents and their children. While Kadioglu (2022) examined death anxiety in young adults, Karakus (2013) determined the views of the students who experienced and did not experience the earthquake disaster by using the metaphor method. Nakajima (2012) conducted research on post-earthquake psychology. Ozcan (2000) worked on a group counselling program for the psychological treatment of adolescents affected by disasters. Ozel and Ozkan (2020) examined the concepts of loss and grief within a psychosocial approach. Pakjouei et al. (2021) conducted research on individuals with physical disabilities who were affected by the earthquake. Tanhan and Kardas (2014) determined the hopelessness and trauma levels of high school students after the Van earthquake. Tsai (2001) examined the opinions of students who experienced an earthquake in Taiwan.

The problem statement of the current study is "What is the mortality awareness of the mothers and fathers who were affected by the earthquake disaster that occurred in Kahramanmaraş and surrounding provinces on 06.02.2023?" In this connection, the sub-problem of the study is "Does the earthquake awareness of the mothers and fathers affected by the earthquake vary significantly depending gender, loss of family members, status of receiving physical treatment and psychological assistance, house damage and status of getting out of the wreckage?" This study aimed to reveal the negative effects of the concept of death on mothers and fathers after large-scale earthquakes by addressing mortality awareness, anxiety and fear, as well as all other constructs. The study is important in terms of revealing mortality awareness in individuals who experienced the disaster in such a short time as 2 months after the earthquake disaster. Moreover, it is thought that the results of the current study will provide guidance for further research on how to make it possible for people who experienced this event, which is defined as the biggest disaster of the 21st century, to reach more effective and efficient assistance.

BACKGROUND

Natural disasters are natural events that occur due to the movement of the plates in the depths of the earth or that usually develop suddenly and as a result cause serious socio-economic damages in human life as well as loss of life (Ozgen, Unaldi & Bindak, 2011). In addition, situations affecting human life such as global warming and erosion caused by climate change are defined as disasters (Erkal & Degerliyurt, 2009). Disasters that change the structure of nature and affect human life negatively are grouped according to the way they occur. The first of these groups is geological-morphological disasters. Earthquakes, tsunamis, volcanic movements caused by tectonic movements caused by the movements of land and oceanic plates in the depths of the earth can be given as examples to these disasters. The second group can be stated as disasters caused by climate change or meteorological changes. Examples of disasters in this group are floods, inundations, landslides, meteorological conditions such as extreme temperatures, extreme winter conditions, air pollution, drought, winds with destructive effects, acid rains, melting of glaciers. The third group can be called hydrographic disasters. Floods caused by excessive rainfall in rivers, pollution of rivers due to human activities, exceeding the carrying capacity of dams and their collapse, pollution of lake waters by various factors, sea swells are examples of this group. In the last group, biological disasters draw attention. They can be listed as forest fires, erosion, animal-borne epidemics, insect and locust invasions that occur after overheating or human activities (Sahin & Sipahioglu, 2002; Ozey, 2006; Ozgen, Unaldi & Bindak, 2011).

Earthquakes are one of the leading natural disasters of geological and geomorphological origin that adversely affect human activities (Öcal, 2007). Earthquake can also be defined as natural events that cause various tremors in the earth's surface due to the changes occurring in the depths of the earth's surface and seriously affect human life, but their effects can be catastrophic (Sür, 1993). The location of fault lines that will cause earthquakes in the earth is now recognised. However, the time and intensity of earthquakes are not known in advance. The science of seismology aims to predict the location, magnitude and time of occurrence of the earthquake and then to prevent loss of life and property by taking the necessary measures (Sunbul & Sunbul, 2018). In recent years, studies in the field of seismology have shown that the most important factor that prepares the ground for the occurrence of earthquakes is the "stress state" of faults (Steacy et al. 2005). Changes in the stress level of the fault are called as earthquake cycle. As a result of the movements of continental or oceanic plates before an earthquake or between two earthquakes (Interseismic), the intensity of stress increases in fault lines over time (Freed et al. 2007).

Earthquakes threaten human life in the areas where they occur and cause loss of both life and property. Faults, which form the fracture zone in the earth's surface, are explained by the damage they cause to the environment and building stock as well as the magnitude of the socio-economic losses they have caused throughout history (Ergunay, 2008). Since Turkey is located on active fault zones, earthquakes are frequently occurring. %99 of the population of the country, %98 of the production and industrial facilities and %75 of the power generation centres are located on active fault lines (Turkoğlu, 2001). The 17 August 1999 Gölcük earthquake has been reported to have an impact of 6.2 billion dollars on the GNP (DPT, 1999). Moreover, the economic cost of this earthquake is estimated to exceed 13 billion dollars including the public financing (Ozerdem & Barakat, 2000). Furthermore, earthquakes cause some psychological problems in the society as well as loss of life and property in the locations where they occur. According to the studies conducted after the Erzincan earthquake in Turkey, it was determined that the participants who experienced an earthquake were nervous and tense (Karanci et al, 1999). Besides, earthquakes also have some negative effects on people in the socioeconomic area. Most of the people who experienced the 17 August Gölcük earthquake lost their houses and belongings and became unable to make a living as a result of the destruction of their income sources. It has been determined that people living in the places where the earthquake occurred experienced changes in terms of belief and culture (Kasapoglu & Ecevit, 2001).

In order to minimise the damages of earthquakes and to raise awareness of the society, trainings to raise disaster awareness are important in preventing the increase in loss of life. Training to develop disaster awareness includes taking protective measures against all types of disasters. In earthquake education, it is defined as the education process in which the earthquake and the damages that may be caused by the earthquake can be determined based on the structure of the earth and then the ways of protection from these damages are taught (f, 2005). The vital effects of earthquakes in societies with socioeconomic and cultural differences should be effectively taught to all members of the society through education. For this reason, students who constitute the future of the society should go through a serious education process to develop earthquake awareness. A good transfer of an important subject such as natural disasters to the learners will ensure that the students are prepared for disasters, trained in subjects such as disaster protection and disaster management. The teaching of courses related to natural disasters in schools is rapidly becoming widespread worldwide (Ocal 2007). When the research studies were analysed, it was concluded that as the level of knowledge of individuals about earthquake increases, the rate of damage that can be caused by earthquake decreases (Ocal, 2007). Based on this result, it has been observed that individuals with a high level of knowledge about earthquakes are more likely to participate in activities to reduce the damages of earthquakes that may occur (Hurnen & McClure, 1997). All governments in the world are obliged to make efforts to reduce the disasters that may occur in their countries. Disasters can have a heavy impact on the human resources and economy of countries. The severity of the damage caused by this disaster is important in terms of showing the preparedness of the country and society exposed to the disaster and the extent to which disaster awareness is adopted. According to the widely accepted opinion, overcoming disasters successfully and recovering quickly by minimising the effects of disasters can only be ensured by effective use of the knowledge which is the basis of disaster trainings and learning at all levels (Varol, 2007).

The individuals affected by disasters face serious psychological problems. Besides, it has been determined that the number of individuals affected by disasters increases every year. Psychological traumas occurring after disasters negatively affect the daily life of individuals. This situation may impair the mental health of individuals throughout their lives. When they cannot receive the necessary psychological support, individuals may suffer from various mental disorders. As a result of this situation, it may also negatively affect the individuals around them affected by the disaster in this process. To give an example, psychological problems of the mother and father in the families affected by the disaster affect the psychology of the children affected by the disaster even more negatively (Dogan & Erdur, 2014). It can be stated that the concept of disaster is evaluated from many different perspectives in the field of mental health. The crisis state that occurs after the disaster is a situation that can affect the individual and the group (family, work environment, etc.) due to its negative consequences. Crisis situation is used for situations that require urgent psychological support after a disaster (Cornell & Sheras, 1998).

The post-disaster crisis situation causes psychological traumas in individuals. The effect of the traumas caused by this situation does not have the same level of psychological impact on each individual in the environment. Although the reactive behaviours that emerge after the trauma caused by the crisis environment vary from individual to individual, it has been observed that they are not permanent. However, the psychological impact after the trauma affects the quality of life negatively for a long time (Bullock, Haddow & Coppola, 2016). The determinant element of psychosomatic reactions in the post-disaster crisis environment is that it is related to the extent and magnitude of the impact of the disaster. In addition, it was determined that the reactions of individuals exposed to the disaster towards the crisis environment differed according to the time elapsed after the disaster. The process of reactional differentiation towards the crisis situation was evaluated in four parts (Karanci & Erdur, 2012). In the first psychological shock period that may emerge after the crisis, it may be the first twenty-four hours or longer after the occurrence of the disaster. Traumatic reactions that may occur in this period are stated as sudden physiological arousal, excessive sensitivity and feeling of restriction, inability to think logically and make decisions, problems in memory retention and concentration of attention, unreality of every apparent situation, petrification of emotions and short-term shock (Hacioglu et al. (2002). The second reaction period starts to be seen after two to six days. Traumatic reactions that may occur in this period are emotional confusion, anxiety, anger, irritability, blame, insecurity, fear of being alone. There are also various physical reactions. These reactions are basic reactions such as trembling, nausea, cardiac problems (palpitations etc.) and inability to stand still. The third reaction period is defined as the period of mental processing and thinking about what happened. This period is entered at the end of one week on average. Traumatic reactions that may occur in this period are defined as the disaster survivor's unwillingness to talk about the event, mourning for the lost relatives, experiencing strong emotions such as sadness and longing, having memory and focusing problems, experiencing anger and conflicts in interpersonal relations. The last period is called recovery and readjustment period. A lot of time has passed since the crisis situation created by the disaster. It is now necessary to adapt to daily life. The behavioural reactions that may occur in this period are slowing down to resist the events and behaviours of adaptation to life. In addition, disaster survivors start to feel good emotionally. Also, they start to make plans for the future. The disaster is now a part of the memories. The mind of the disaster survivor is not occupied with the disaster and bad situations as intensely as in the first times (Hacioglu et al. 2002). However, due to the magnitude and severity of the disaster, the post-disaster trauma period is not seen similarly in all individuals. It can also be seen that disaster survivors do not accept the disaster situation and show resistance towards recovery. As a result of the intense stress experienced, the crisis environment created by the disaster can be denied, suppressed or removed from the level of consciousness with different defence mechanisms. It is quite difficult for disaster survivors who react in this way to be included in daily life and it is seen as socially worrying (Yildiz, 2014).

All these psychological and physiological reactions caused by disasters are accompanied by individuals' encounters with the phenomenon of loss and mourning. People show a number of attitudes and behaviours in situations such as individuals' lives being in danger or experiencing such a situation, losing a relative, losing their financial resources. All of these emotional, behavioural and intellectual reactions are called grief reactions (Kukuoglu, 2018). The concept of grief is stated as the mental state experienced after the loss of a relative or the loss of anything to which attachment is developed. Grief reactions are shaped according to the environment, culture and crisis situation experienced (Yildiz, 2014). The mourning process is a natural process that occurs in the crisis situation after the disaster. It covers a time period between six months and two years on average. However, the mourning phenomenon which exceeds this period is seen as a harbinger of a psychopathological situation. In this case, traumatic grief, complicated grief, pathological grief, complex grief, unresolved grief, chronic grief stand out (Kukuoglu, 2018).

The most well-known theoretical model in the process of traumas and grief experienced by individuals after a disaster is Kubler-Ros's Five Stages of Grief Theory (Kubler-Ros, 1993). According to the perspective of this theory, individuals who mourn after a loss after a disaster experience the following process respectively. They complete the mourning process by denying the crisis situation and isolating themselves from the situation, going through anger process towards the crisis situation, bargaining to get rid of the crisis situation, experiencing depression due to the crisis situation and finally accepting the crisis situation (Kubler-Ros, 1993). Examples of the reactions given in this process include depression, hopelessness, anxiety, guilt, anger, loneliness as emotional reactions. Behavioural reactions include crying, withdrawal and burnout reactions. Cognitive reactions include constant thinking of the deceased, low self-confidence, thoughts of helplessness, difficulty in concentration and denial of the situation. In physiological reactions, substance use, loss of appetite, fatigue and somatic complaints are noteworthy.

The crisis situations experienced may also lead to the emergence of different psychopathologies, especially long-term (major) depression that may occur later in life. The negative effects caused by disasters can be evaluated in psychopathological dimension. In addition to this situation, it should also be considered that they may have very different effects on the society and family members. Large-scale events occurring in the physical environment, death of individuals themselves and their relatives, injuries and material losses in the post-disaster crisis environments negatively affect the mental health of individuals and their families. After the traumatic events occurring in the post-disaster crisis environment are announced to the whole world through the press, a large segment of the world is affected both psychologically and economically (Bullock, Haddow & Coppola, 2016). Families and undoubtedly children and adolescents can be affected by severe traumatic experiences such as disasters more than adults because of their lack of life experience and insufficient knowledge to gain life awareness. For this reason, it is an observable result that children and adolescents, who are still in need of care and attention of adults, need psychological support and necessary trainings the most after disasters (Yildiz, 2014). Especially in the aftermath of major disasters, sufficient psychological support should be provided to individuals, especially in the mourning process, with post-disaster Psychological First Aid services to eliminate the fear of death and loss of loved ones. Based on this idea, it is very important to provide post-traumatic psychological support and grief counselling services (Stroebe, Hansson, Stroebe, & Schut, 2001).

Method

Under this heading, the research model, study group, measurement tool, procedure and data analysis are explained.

Model

In the current study, survey model, one of the quantitative research methods, was used. This model allows the generalization of the results of the study, since data are collected from a large population (Fraenkel, Wallen & Hyun, 2012).

Table 1. Frequencies and percentages

Variable		N	Frequency	Percentage	Mean	Sd
Gender	Male	151	58	38.4	1.61	.48
	Female		93	61.6		
Loss of a Family Member	Yes	151	36	23.8	1.76	.42
	No		115	76.2		
Receiving Physical Treatment	Yes	151	8	5.3	1.94	.22
	No		143	94.7		
Receiving Psychological Assistance	Yes	151	10	6.6	1.93	.24
	No		141	93.4		
Damage to House	Minor and Moderate Damage	151	84	55.6	1.53	.66
	Heavy Damage		53	35.1		
	Ruined		14	9.3		
Going out of the Wreckage	Yes	151	10	6.6	1.93	.24
	No		141	93.4		

Study Group

The population of the current study consists of the mothers and fathers living in the earthquake-ridden regions. In this study, criterion sampling, one of the purposive sampling methods, was used. The criteria used in the selection of the participants were their being mother or father and their having experienced the earthquake. On the basis of these criteria, a total of 151 mothers and fathers who were earthquake victims were reached 2 months after the Kahramanmaraş earthquakes (between 01 and 15/04/2023) by two researchers who were also victims of the earthquake, first by phone and then they were sent the scale form created on the internet to complete. All the individuals reached for the administration of the scale lived in 11 provinces affected by the earthquake and participated in the study on a volunteer basis. Descriptive statistics of the study are presented in Table 1.

As seen in Table 1, a total of 151 mothers and fathers, 58 fathers (38.4%) and 93 mothers (61.6%) participated in the study. While there were 36 people (23.8%) with loss of family members, 115 people did not lose any of their family members. While 8 of the participants (5.3%) were treated in hospital, 143 people (94.7%) did not receive any physical treatment. While 10 people (6.6%) received psychological treatment after the earthquake, 141 people did not receive any psychological treatment. When the status of damage to the houses of the participants was examined, it was determined that there were 84 houses (55.6) with minor or moderate damage, 53 houses with heavy damage (35.1%) and 14 houses (9.3%) were completely ruined. It was determined that 10 of the participants (6.6%) were rescued from the wreckage and that 141 people (93.4%) were not trapped under the wreckage.

Measurement Tools

The measurement tools are explained under this heading.

Table 2. Results of the confirmatory factor analysis

Index	Bulut, McDermott, and McEwan (2017)	Findings of the Current Study	Result
/sd	1.45	1.33	Perfect Fit
CFI	.94	.92	Good Fit
TLI	.93	.91	Good Fit
IFI/SRMR	.94(IFI)	.07(SRMR)	Good Fit
RMSEA	.04	.04	Perfect Fit

Personal Information Form

It was developed by the researchers to collect information about the participants' gender, loss of close family members, damage to house, receiving physical treatment and psychological assistance and getting out of the wreckage.

Multidimensional Mortality Awareness Scale

The Multidimensional Mortality Awareness Scale was developed by Levasseur, McDermott and Lafreniere (2015) and adapted to Turkish by Bulut, McDermott and McEwan (2017). In the Turkish adaptation study, expert opinions were taken for language and culture compatibility of the scale. At this stage, the compatibility between the original scale and the adapted scale items was examined by 3 English language experts.

The scale developed by Levasseur, McDermott, and Lafreniere (2015) consists of 36 items and 5 factors. These factors were named as legacy, fearfulness, acceptance, disempowerment and disengagement. The scale is a 7-point Likert scale. The internal consistency coefficient Cronbach's Alpha value of the scale was found to be 0.79. The scale adapted to Turkish by Bulut, McDermott and McEwan (2017) consists of 30 items and 5 factors. A total of 317 people participated in the adaptation study. The scale is a 7-point Likert scale. The response options in the scale are "strongly disagree (1)", "disagree (2)", "a bit disagree (3)", "neither agree nor disagree (4)", "a bit agree (5)", "agree (6)", "strongly agree (7)". The internal consistency coefficient Cronbach's Alpha value of the scale was found to be 0.79. The Cronbach's Alpha value of the scale used in this study was found to be .867. It is understood that the scale is reliable (Dogan, 2019).

The sample size was checked before proceeding to confirmatory factor analysis. It was found that the study group was large enough for confirmatory factor analysis (Anderson & Gerbing, 1984; Boomsma, 1985; Goodwin, 1999; Muthén & Muthén, 2002; Stevens, 2002, Tanaka 1987). In order to ensure the construct validity of the current study, the results of the adaptation study conducted on the Multidimensional Mortality Awareness Scale by Bulut, McDermott and McEwan (2017) and the confirmatory factor analysis of the current study are given below. In the study by Bulut, McDermott and McEwan (2017), besides the model fit indices used in the current study, (IFI) index was also used. In the current study, SRMR value was used instead of (IFI).

When the model fit indices in Table 2 were examined, it was determined that the CFI, TLI values were at an acceptable level above 0.90. The /sd and the RMSEA indices were found to show a perfect fit. A RMSEA value below .05 indicates a perfect fit. In addition, a RMSEA value close to .08 indicates an acceptable fit (Browne & Cudeck, 1993). The scale used in the current study is seen to be valid and reliable.

Data Collection and Analysis

The data of the current study were collected by using the Multidimensional Mortality Awareness Scale to find an answer to the question "What is the death awareness of the mothers and fathers who were affected by the earthquake disaster that occurred in Kahramanmaraş and its surrounding provinces on 06.02.2023?".

The online scale prepared to obtain the data of the current study consists of two parts. In the first part, there are questions to elicit demographic information about the participants and in the second part, there is the Google Form consisting of the items of the Multidimensional Death Awareness Scale (Gosling and Mason, 2015). In addition, all the participants were contacted by phone and their consent was obtained. In the current study, in order to determine whether parametric or non-parametric test would be used in the data analysis of the Multidimensional Mortality Awareness Scale, first, the normality test was performed on the data. The results of the analysis conducted to determine the normality of the data are shown in Table 3.

As seen in Table 3, the Skewness and Kurtosis values of the data are between -1.5 and +1.5, indicating that the data collected with the scale are normally distributed (Tabachnick & Fidell, 2013). Before the normality test, the homogeneity of the data obtained with the scale was examined by using Levene's test. As a result of the test, it was seen that the scale items were homogeneous. Then, the normality test was conducted. The normality values of the scale were found to be insignificant (K-S, $p>.05$). When the results of the normality test conducted on the scale were examined, it was determined that the data showed a normal distribution (Morgan, Leech, Gloeckner & Barrett, 2004). In this context, the normality and homogeneity criteria required to use parametric tests were met (Koklu, Buyukozturk & Bokeoglu, 2007). As a result, independent samples t-test and one-way ANOVA test were used. In addition, Cohen's d values were checked to examine the effect size (Kilic, 2014). In the analysis of quantitative data collected in the current study, SPSS 26, JAMOVI and MPLUS 7 statistical packages were used.

Findings

The general mean score of the responses given by the participants to the items in the Multidimensional Mortality Awareness Scale administered to find an answer to the main research problem of the study "What is the mortality awareness of the fathers and mothers who were affected by the earthquake disaster?" is shown in Table 4.

Table 3. Normality analysis of the scale

Score	N	Mean	Sd	Skewness	Kurtosis	Kolmogorov-Smirnov (K-S)
Total	151	4.29	.96	.00	1.45	.09

As seen in Table 4, the mean mortality awareness score of the participants is 4.29. Based on this finding, it was determined that the death awareness of the participants was at the level of "a bit agree". This shows that the participants' earthquake-related mortality awareness started to form.

Results of the independent samples t-test conducted to determine whether the mortality awareness of the mothers and fathers who were affected by the earthquake varies significantly depending on their gender, loss of a family member, receiving physical treatment, receiving psychological assistance and getting out of the wreckage and the results are presented in Table 5.

As seen in Table 5, the mean mortality awareness score varied significantly depending on gender ($p<.05$). The effect size is seen to be small ($d=.44$). It was determined that the mortality awareness of the male and female participants after the earthquake varied significantly in favour of the female participants and this is thought to be due to the different natures of men and women. Second result of the analysis, it was found that the mean mortality awareness score of the participants varied significantly depending on the loss of a family member ($p<.05$). The effect size is seen to be small ($d=.39$). It was determined that the mortality awareness of the participants who lost a family member and who did not lose a family member after the earthquake varied significantly in favour of the participants who lost a family member. It is thought that loss of a family member increases mortality awareness due to the fact that they face death, witness the burials of their relatives at the time of the disaster and experience the pain of losing someone beloved.

Another result of the analysis, it was found that the mean mortality awareness score of the participants did not vary significantly depending on the status of receiving physical treatment ($p>.05$). This is thought to be because all health services needed by the participants are adequately provided. As a result of the analysis, it was found that the mean mortality awareness score of the participants varied signifi-

Table 4. General mean

Score	N	$\bar{x}$
Total	151	4.29

Table 5. Independent samples t-test

Score	Variable		N	$\bar{x}$	Sd	df	t	p	d
Total	Gender	Female	93	4.45	1.02	149	2.62	.01	.44
		Male	58	4.04	0.88				
Total	Loss of a Family Member	Yes	36	4.58	1.01	149	2.07	.04	.39
		No	115	4.21	0.92				
Total	Receiving Physical Treatment	Yes	8	3.41	1.69	7.2	-1.54	.16	-
		No	143	4.34	.88				
Total	Receiving Psychological Assistance	Yes	10	3.25	1.35	9.55	-2.55	0.2	.97
		No	141	4.36	.88				
Total	Getting out of Wreckage	Yes	10	4.61	1.61	9.04	.65	.52	-
		No	141	4.27	.90				

Table 6. Results of the one-way ANOVA

Score	Variable		N	$\bar{x}$	Groups	Ss	F	p	LSD	d
Total	Damage to House	Minor or Moderate	84	4.45	Between Groups	5.63	3.41	0.4	1-3	.53
		Heavy	54	4.15	Within Groups	132.62				
		Ruined	14	3.87	Total	138.25				

cantly depending on the status of receiving psychological assistance (p<.05). The effect size was found to be small (d=.39). It was determined that the mortality awareness of the participants who received psychological assistance and who did not receive psychological assistance after the earthquake varied significantly in favour of the participants who did not receive psychological assistance. This is thought to be because those who do not receive psychological assistance accept that death is a natural process. Last result of the analysis, it was found that the mean mortality awareness score of the participants did not vary significantly depending on the status of getting out of the wreckage (p>.05). This is thought to be because the destructiveness of the earthquake was quite high and that all the other participants witnessed wreckages around the city they lived in.

The results of the one-way ANOVA conducted to find an answer to the question "Does the mortality awareness of the mothers and fathers who were affected by the earthquake disaster vary significantly depending on the damage to their house?" are presented in Table 6.

As seen in Table 6, the mean mortality awareness score of the participants varied significantly depending on the status of damage to house (p<.05). This significant difference was found to be between the participants having houses with minor or moderate damage and the participants having houses with heavy damage in favour of the participants who have houses with minor or moderate damage. The effect size is seen to be medium (d=.53). This is thought to be because the mortality awareness the participants increased as the level of damage to their houses decreased as the participants living in houses with low and moderate damage helped rescue efforts.

RESULTS AND DISCUSSION

In the current study, it was aimed to determine the mortality awareness of mothers and fathers 2 months after the Kahramanmaraş-centred earthquake that occurred in the southern provinces of Turkey on 06.02.2023. The results of the study revealed that mean scores taken from the Multidimensional Mortality Awareness Scale by the participants varied significantly depending on the variables of gender, loss of a family member, receiving psychological assistance and damage to house. The results are discussed holistically below.

It was concluded that the female participants had higher mortality awareness after the earthquake. Kadioglu (2022) found that young adult women had significantly higher death anxiety levels than young adult men in his master's thesis on the relationship between death anxiety, five-factor personality traits and loneliness levels in young adults. This finding supports the results of the current study. In addition, Baris (2022) conducted a research on the perception of death in individuals in his master's thesis and revealed that the self-compassion and hope scores varied statistically significantly depending on the gender variable, the self-compassion scores depending on the age variable and the self-compassion scores depending on the status of having a child. Baris's study also supports the results regarding the female participants of the current study.

As those who lost a family member faced the reality of death, their mortality awareness increased more. Honda et al. (2019) investigated the post-earthquake problems of the individuals who experienced the loss of a family member. Honda et al. (2019) conducted a longitudinal study of the long-term psychological effects of the Great East Japan Earthquake (GEJE) on parents and their children. The purpose of the study was to monitor parents' post-traumatic stress disorder (PTSD) symptoms and behavioural problems in their children over time and analyse the long-term relationships among GEJE survivors. They used data from the GEJE Follow-up for Children study where children aged 4 to 6 years and their parents were recruited and followed-up immediately after GEJE in March 2011. Parental PTSD symptoms and children's behavioural problems decreased slightly over time and significant correlations were found between the questionnaires. Longitudinally, parental PTSD symptoms in 2012 showed a significant correlation with children's internalizing behavioural problems in 2014, after adjusting for children's behavioural problems in 2012 and parental PTSD symptoms in 2014. These results suggest that the effect of GEJE on parental PTSD symptoms and children's behavioural problems is long-term and lasts for at least 3 years. These possibilities should be carefully considered in mental health support for parents and their children in GEJE-affected areas.

It was concluded that post-earthquake physical injuries did not create mortality awareness of the participants. In addition, it was concluded that the mortality awareness of the participants decreased when activities were carried out to reduce anxiety about death in the psychological assistance received after the earthquake. The results obtained by Guvercin (2006) in his master's thesis named "Examination of depression levels of earthquake survivor mothers and their children in terms of various variables" support the findings of the current study. Accordingly, mothers who received psychological support after the earthquake had higher depression levels. Ozcan (2000) applied six-session group guidance to adolescents who were exposed to natural disasters in his master's thesis named a group counselling program on the psychological treatment of adolescents affected by disasters. According to the results of the study, a significant difference was found between the pre-test and post-test mean scores taken from the traumatic stress disorder scale, state anxiety scale and trait anxiety scale administered to the group who had experienced an earthquake. Fan, Zang, and Yang, Mo and Liu (2011) examined the Symptoms of Post-Traumatic Stress Disorder, Depression, and Anxiety in Adolescents after the 2008 Wenchuan Earthquake in China. The study examined the symptoms of post-traumatic stress disorder (PTSD), anxiety, and depression among 2,250 adolescents 6 months after the 2008 Wenchuan earthquake in China. The results showed that 15.8%, 40.5% and 24.5% of the participants reported clinical symptoms of PTSD, anxiety, and depression, respectively. Risk factors for the symptoms of these 3 disorders were determined as being female, advanced age and exposure to earthquake disasters.

It was concluded that being under the wreckage after the earthquake did not affect the mortality awareness of the participants. It was concluded that the mortality awareness of the participants whose houses were slightly or moderately damaged was higher. However, Guvercin (2006) stated that this occurred differently in children, and that depression levels were high in children who had been under the wreckage. This result does not support the results of the current study.

SUGGESTIONS AND RECOMMENDATIONS

- A qualitative research can be conducted on the death anxiety of mothers and their approaches to their children after the earthquake and it can be described what kind of problems they face.
- Strategies for coping with the fear of death and mourning for losses in the earthquake can be examined.
- Psychological support studies can be carried out by preparing a recovery program for the individuals affected by the Kahramanmaraş-centred earthquake.
- The experiences of individuals whose houses have been damaged can be described with hermeneutic research.
- In the current study, mothers and fathers were chosen as the target group. It is recommended to conduct similar studies on different groups in order to gain deeper insights into mortality awareness.

REFERENCES

Anadolu Agency. (2023). https://www.aa.com.tr/tr/gundem/depremler-10-ilde-yaklasik-13-5milyon-kisinin-yasadigi-bolgeyi-etkiledi/2808868#

Anderson, J. C., & Gerbing, D. W. (1984). The effect of sampling error on convergence, improper solutions, and goodness-of-fit indices for maximum likelihood confirmatory factor analysis. *Psychometrika, 49*(2), 155–173. doi:10.1007/BF02294170

Baris, S. (2022). *Perception of death in adult individual: A review according to self-compassion and hope variables* [Unpublished Master's Thesis]. Marmara University, Istanbul.

Boomsma, A. (1985). Nonconvergence, improper solutions, and starting values in LISREL maximum likelihood estimation. *Psychometrika, 50*(2), 229–242. doi:10.1007/BF02294248

Browne, M. W., & Cudeck, R. (1993). Alternative ways of assessing model fit. In Testing Structural Equation Models. SAGE Publications.

Bullock, J. A., Haddow, G., & Coppola, D. P. (2016). *Managing children in disasters: Planning for their unique needs*. CRC Press.

Bulut, M. B., McDermott, M., & McEwan, O. (2017). Adapting the multidimensional mortality awareness measure: A study of validity and reliability. *The Journal of International Social Research, 10*(49), 268–275. doi:10.17719/jisr.2017.1578

Cornell, D. G., & Sheras, P. L. (1998). Common errors in school crisis response: Learning from our mistakes. *Psychology in the Schools, 35*(3), 297–307. doi:10.1002/(SICI)1520-6807(199807)35:3<297::AID-PITS9>3.0.CO;2-K

Dogan, N. (2019). *Measurement and evaluation in education*. Pegem Academy.

Dogan, T., & Erdur, Ö. (2014). Disasters, traumas, crises and traumatic stress reactions. In Disasters, crises, traumas and psychological help. Turk PDR Dernegi.

DPT. (1999). *Economic and social effects of the earthquake, possible financing needs, measures to be taken in the short, medium and long term*. Ministry of Development. Ankara.

Ergunay, O. (1996). "What is disaster management? How should it be?" Search for solutions to Turkey's earthquake problems in the light of Erzincan and Dinar earthquakes. *Tubitak earthquake symposium proceedings book*, 263-272.

Erkal, T., & Degerliyurt, M. (2009). Disaster management of Turkey. *Eastern Geographical Review, 14*(22), 147-164. https://dergipark.org.tr/en/pub/ataunidcd/issue/2437/31194

Fan, F., Zhang, Y., Yang, Y., Mo, L., & Liu, X. (2011). Symptoms of posttraumatic stress disorder, depression, and anxiety among adolescents following the 2008 Wenchuan earthquake in China. *Journal of Traumatic Stress, 24*(1), 44–53. doi:10.1002/jts.20599 PMID:21351164

Fraenkel, J. R., Wallen, N. E., & Hyun, H. H. (2012). *How to design and evaluate research in education*. McGraw-Hill.

Freed, A., Syed, T., & Burgmann, R. (2007). Evolution of stress southern California for the past 200 years from coseismic, postseismic and interseismic processes. *Geophysical Journal International, 169*(3), 1164–1179. doi:10.1111/j.1365-246X.2007.03391.x

Goodwin, L. D. (1999). The role of factor analysis in the estimation of construct validity. *Measurement in Physical Education and Exercise Science, 3*(2), 85–100. doi:10.1207/s15327841mpee0302_2

Gosling, S. D., & Mason, W. (2015). Internet research in psychology. *Annual Review of Psychology, 66*(1), 877–902. doi:10.1146/annurev-psych-010814-015321 PMID:25251483

Guvercin, A. (2006). *Examining the depression levels of the mothers and their children who see an earthquake according to some variables*. Retrieved from https://www.proquest.com/dissertations-theses/depremzede-anneler-ve-çocuklarının-depresyon/docview/2564166661/se-2.

Hacioglu, M., Aker, T., Kutlar, T., & Yaman, M. (2002). Deprem tipi travma sonrasinda gelisen travma sonrasi stres bozuklugu belirtileri alt tipleri. *Dusunen Adam The Journal of Psychiatry and Neurological Sciences, 15*, 4–15.

Honda, Y., Fujiwara, T., Yagi, J., Homma, H., Mashiko, H., Nagao, K., Okuyama, M., Ono-Kihara, M., & Kihara, M. (2019). Long-term impact of parental post-traumatic stress disorder symptoms on mental health of their offspring after the great east japan earthquake. *Frontiers in Psychiatry, 10*, 496. Advance online publication. doi:10.3389/fpsyt.2019.00496 PMID:31404309

Hurnen, F., & Mcclure, J. (1997). The effect of increased earthquake knowledge on perceived preventability of earthquake damage. *Australasian Journal of Disaster and Trauma Studies*, 3.

Kadioglu, R. (2022). *Relationships among death anxiety, big five personality traits, and loneliness levels in young adults* [Unpublished Master's Thesis]. Necmettin Erbakan University, Konya.

Kandilli Observatory. (2023a). http://www.koeri.boun.edu.tr/sismo/2/deprem-verileri/yillik-deprem-haritalari/2022-yili-deprem-harita-grafik-ve-tablolari/

Kandilli Observatory. (2023b). http://www.koeri.boun.edu.tr/sismo/2/wp-content/uploads/2023/02/20230206_1024_KAHRAMANMARAS.pdf/

Karakus, U. (2013). The investigation of earthquake inexperienced and experienced students' perceptions by metaphor analysis. *Eastern Geographical Review, 18*(29), 97–116. doi:10.17295/ataunidcd.31309

Karanci, N., Aksit, B., Anafarta, M., Ogul, M., & Uner, G. (1999). *Let's Be Prepared Against Earthquakes.* Middle East University Disaster Management Application and Research Centre.

Kasapoglu, A., & Ecevit, M. (2001). *Sociological research of the earthquake.* Sosyoloji Dernegi.

Kilic, S. (2014). Effect size. *Journal of Mood Disorders, 4*(1), 44–46. doi:10.5455/jmood.20140228012836

Koklu, N., Buyukozturk, S., & Bokeoglu, C. O. (2007). *Statistic for Social Sciences.* Pegem.

Koroglu, A. (2018). Psychological traumas after natural disasters and sample psychological support education program. *Journal of Disaster and Risk, 1*(1), 39–52. doi:10.35341/afet.412005

Kubler-Ross, E. (1993). On death and dying. Academic Press.

Levasseur, O., McDermott, M. R., & Lafreniere, K. D. (2015). The multidimensional mortality awareness measure and model: Development and validation of a new self-report questionnaire and psychological framework. *Omega, 70*(3), 317–341. doi:10.1177/0030222815569440 PMID:26036058

Ministry of National Education. (2011). *Psychological support in disasters and traumatic events.* MEGEP. Retrieved from https://www.megep.meb.gov.tr/mte_program_modul/moduller_pdf/Afet%20Ve%20Travmalarda%20Psikolojik%20Destek.pdf

Morgan, G. A., Leech, N. L., Gloeckner, G. W., & Barrett, K. C. (2004). *SPSS for introductory statistics: Use and interpretation.* Psychology Press. doi:10.4324/9781410610539

Muthén, L. K., & Muthén, B. O. (2002). How to use a Monte Carlo study to decide on sample size and determine power. *Structural Equation Modeling, 9*(4), 599–620. doi:10.1207/S15328007SEM0904_8

Nakajima, Ş. (2012). Post-Earthqake psychology. *Okmeydanı Tıp Dergisi, 28*(2), 150–155. doi:10.5222/otd.supp2.2012.150

Ocal, A. (2005). The evaluation of earthquake education in the elementary school social studies courses. *Gazi Üniversitesi Gazi Eğitim Fakültesi Dergisi, 25*(1), 169-184. https://dergipark.org.tr/tr/pub/gefad/issue/6757/90874

Ocal, A. (2007). A research on elementary school preservıce teacher's earthquake knowledge level. *Mehmet Akif Ersoy University Journal of Education Faculty, 13*, 104-110. https://hdl.handle.net/20.500.12451/994

Okcu, C. (2021). Examination of the relationships between mortality awareness and mental health. *Türkiye Bütüncül Psikoterapi Dergisi, 4*(8), 43-62. Retrieved from https://dergipark.org.tr/tr/pub/bpd/issue/63116/958658

Ozcan, S. (2000). *Group counseling program about psychological wellbeing of adolescents* [Unpublished Master's Thesis]. Marmara University. İstanbul.

Özdoğan, S. (1993). The earthquake areas of Turkey. *Ankara University Journal of Turkey Cografyasi Arastirma ve Uygulama Merkezi, 2*, 53–68.

Ozel, Y., & Ozkan, B. (2020). Psychosocial approach to loss and mourning. *Psikiyatride Güncel Yaklasimlar, 12*(3), 352–367. doi:10.18863/pgy.652126

Ozerdem, A., & Barakat, S. (2000). After the Marmara earthquake: Lessons for avoiding short cuts to disasters. *Third World Quarterly, 21*(3), 425–439. doi:10.1080/713701047

Ozey, R. (2006). *Geography of disasters*. Istanbul: Aktif yayinevi.

Ozgen, N., Unaldi, Ü. E., & Bindak, R. (2011). Determining the teacher candidates' "Effective learning styles" on natural disasters. *Journal of Kirsehir Education Faculty, 12*(4), 303–323.

Pakjouei, S., Aryankhesal, A., Kamali, M., Seyedin, H., & Heidari, M. (2021). Earthquake and the educational needs of people with physical disabilities and associated groups: A qualitative study. *Archives of Rehabilitation, 22*(1), 48–65. doi:10.32598/RJ.22.1.3165.1

Reuters. (2023). https://www.ntv.com.tr/galeri/dunya/turkiye-ve-dunyada-yasanan-en-siddetli-depremler-dunyadaki-en-siddetli-deprem-hangi-ulkede-yasandi-dunyadaki-en-uzun-sureli-deprem-hangisidir,jpO-YIgJMqEK4Z6p0iKq16Q/uHBnLWlJPk2PDqI3GQpj9Q

Sahin, C., & Sipahioglu, S. (2002). Natural disasters and Turkey. Ankara: Gündüz Eg. ve Yay.

Steacy, S., Gomberg, J., & Cocco, M. (2005). Introduction to special section: Stress transfer, earthquake triggering and time dependant seismic hazard. *Journal of Geophysical Research, 110*(B5). Advance online publication. doi:10.1029/2005JB003692

Stevens, J. (2002). *Applied multivariate statistics for the social sciences*. Lawrence Erlbaurn Associates.

Stroebe, M. S., Hansson, R. O., Stroebe, W., & Schut, H. (2001). Introduction: Concepts and issues in contemporary research on bereavement. In M. S. Stroebe, R. O. Hansson, W. Stroebe, & H. Schut (Eds.), *Handbook of bereavement research: Consequences, coping, and care* (pp. 3–22). American Psychological Association. doi:10.1037/10436-031

Sunbul, F., & Sunbul, A. B. (2018). Evaluation of coulomb stress criteria in earthquake interactions: East Anatolian fault line. *Karaelmas Science and Engineering Jounal, 8*(2), 523–535.

Tabachnick, B. G., & Fidell, L. S. (2013). *Using multivariate statistics* (6th ed.). Pearson.

Tanaka, J. S. (1987). How big is big enough?": Sample size and goodness of fit in structural equation models with latent variables. *Child Development, 58*(1), 134–146. doi:10.2307/1130296

Tanhan, F., & Kardas, F. (2014). Investigating hopelessness and trauma levels of high school students exposed to the Van earthquake. *Sakarya University Journal of Education, 4*(1), 102–115. doi:10.19126/suje.52552

Tsai, C. C. (2001). Ideas about earthquakes after experiencing a natural disaster in Taiwan: An analysis of students' worldviews. *International Journal of Science Education, 23*(10), 1007–1016. doi:10.1080/09500690010016085

Turkoglu, N. (2001). Distribution of Turkey's area and population to earthquake regions. *Ankara University Journal of Turkey Cografyasi Arastirma ve Uygulama Merkezi*, *8*, 133–148.

Varol, N. (2007). Raising awareness of the society about natural and technological disasters and the role of "AFEM". *Proceedings book of TMMOB Disaster Symposium*, 127-131.

Wittkowski, J., & Paré, P. M. (2022). Introduction of the French version of the Multidimensional Orientation Toward Dying and Death Inventory (MODDI-F/fre). *Death Studies*, 1–9. doi:10.1080/07481187.2022.2108940 PMID:35950684

Yildiz, B. (2014). Children and adolescents. In T. Doğan & Ö. Erdur (Eds.), *Disasters, crises, traumas and psychological help* (pp. 25–43). Turk PDR Dernegi.

ADDITIONAL READING

Bulut, M. B., McDermott, M., & McEwan, O. (2017). Adapting the multidimensional mortality awareness measure: A study of validity and reliability. *The Journal of International Social Research*, *10*(49), 268–275. doi:10.17719/jisr.2017.1578

KEY TERMS AND DEFINITIONS

Awareness: It can be defined as the state of being aware.

Disasters: Each of the disasters such as floods, storms, earthquakes, etc. that cannot be prevented by human effort; catastrophe.

Earthquake: A tremor caused by the discharge of energy generated by movements in the earth's shell.

Education: Direct or indirect assistance, inside or outside of the school, for children and young people to acquire the knowledge, skills and understanding necessary for them to take their place in the life of society and to develop their personalities.

Mortality: It is defined as the state of being mortal.

Parents: Mother, father, and legal guardian.

Psychology: A group or an individual can be defined as a state of mind that dominates the thoughts and behaviour of a group or an individual.

Compilation of References

Aarnikoivu, M. (2021). The spatiotemporal dimension of doctoral education: A way forward. *Studies in Higher Education, 46*(11), 2431–2443. doi:10.1080/03075079.2020.1723530

Abbas, J. (2020). Service quality in higher education institutions: Qualitative evidence from the students' perspectives using Maslow hierarchy of needs. *International Journal of Quality and Service Sciences, 12*(3), 371–384. doi:10.1108/IJQSS-02-2020-0016

Abdullah, D., Abd Aziz, M. I., & Mohd Ibrahim, A. L. (2017). The stories they tell: Understanding international student mobility through higher education policy. *Journal of Studies in International Education, 21*(5), 450–466. doi:10.1177/1028315317720766

Abedtalas, M., Alawak, A., Aloklah, W. A., Aljasem, A., & Sarmini, I. (2020). Syrian higher education and social capital in times of conflict. *Education and Conflict Review, 3*, 9–15.

Abiddin, N. Z., Ibrahim, I., & Abdul Aziz, S. A. (2022). Non-governmental organisations (NGOs) and their part towards sustainable community development. *Sustainability (Basel), 14*(8), 4386. doi:10.3390/su14084386

Abiddin, N. Z., Ibrahim, I., & Aziz, S. A. A. (2022). The role of non-governmental organisations (NGOs) for community development: Way forward. *Journal of Management Information & Decision Sciences, 25*(2).

Aczel, B., Kovacs, M., van der Lippe, T., & Szaszi, B. (2021). Researchers working from home: Benefits and challenges. *PLoS One, 16*(3), e0249127. doi:10.1371/journal.pone.0249127 PMID:33765047

Adams, M. (2020). *What higher ed can learn from restaurants.* EAB. https://eab.com/insights/resource/strategy/higher-ed-lessons-restaurants-competition/

Adiyoso, W., & Kanegae, H. (2012). The effect of different disaster education programs on tsunami preparedness among schoolchildren in Aceh, Indonesia. *Disaster Mitigation of Cultural Heritage and Historic Cities, 6*(1), 165–172.

Adu-Baffoe, E., & Bonney, S. (2021). The Role of Non-Governmental Organizations in Basic Education Delivery in Ghana: Implications for Theory, Policy, and Practice. *International Education Studies, 14*(4), 35–47. doi:10.5539/ies.v14n4p35

AE. (2009). *The Australian education sector and the economic contribution of international students.* Australian Council for Private Education and Training. http://www.acpet.edu.au/uploads/files/Reports_Submissions/International_Students_14MAY2009__FINAL_.pdf

AFAD. (2023). *Disaster and Emergency Management Presidency Dictionary.* https://www.afad.gov.tr/aciklamali-afet-terimleri-sozlugu

Aguilar, R. P., & Guénette, M. F. (Eds.). (2021). *Situatedness and Performativity: Translation and Interpreting Practice Revisited.* Leuven University Press. doi:10.11116/9789461663863

Ahmed, Z., & Julis, S. H. (2015). Academic performance, resilience, depression, anxiety, and stress among women students. *Indian Journal of Positive Psychology*, *6*(4), 367–370.

AI and the LinkedIn community. (n.d.). *What are the most common mistakes that higher education leaders make in crisis management?* Retrieved October 21, 2023, from www.linkedin.com: https://www.linkedin.com/advice/3/what-most-common-mistakes-higher-education

Aitchison, C., Harper, R., Mirriahi, N., & Guerin, C. (2019). Tensions for educational developers in the digital university: Developing the person, developing the product. *Higher Education Research & Development*, *39*(2), 171–184. doi:10.1080/07294360.2019.1663155

Al Azmeh, Z., Dillabough, J., Fimyar, O., McLaughlin, C., Abdullateef, S., Aloklah, W. A., Mamo, A. R., Abdulhafiz, A. H., Al Abdullah, S., Al Husien, Y., Al Mohamad Al Ibrahim, A., Al Ibrahim, Z., Barmu, T., Farzat, A., & Kadan, B. (2021). Cultural trauma and the politics of access to higher education in Syria. *Discourse (Abingdon)*, *42*(4), 528–543. doi:10.1080/01596306.2020.1715922

Al Hessan, M., Bengtsson, S., & Kohlenberger, J. (2016). *Understanding the Syrian educational system in a context of crisis* (No. 09/2016). Vienna Institute of Demography Working Papers.

AlertMedia. (2023, August 10). *The 4 Stages of Crisis and How to Manage Them.* Retrieved 27 October, 2023, from www.alertmedia.com: https://www.alertmedia.com/blog/stages-of-crisis/

Alexander, B., Ashford-Rowe, K., Barajas-Murph, N., Dobbin, G., Knott, J., McCormack, M., Pomerantz, J., Seilhamer, R., & Weber, N. (2019). *Horizon report 2019 higher education edition.* EDUCAUSE. https://library.educause.edu/resources/2019/4/2019-horizon-report

Alkhawlania, M. A., Bohari, A. b., & Shamsuddin, J. b. (2019). actors that Influence Crisis Management in Yemeni Organizations: The Mediating Effect of Decision- making Styles. *International Journal of Innovation. Creativity and Change*, *5*(2), 11–34.

Alkol, M. (2023). Challenges and Difficulties of University Study in Northwestern Syria. *International Higher Education, 11.*

Allen, I. E., & Seaman, J. (2016). *Online report card: Tracking online education in the United States.* Babson Survey Research Group. https://onlinelearningconsortium.org/read/online-report-card-tracking-online-education-united-states-2015/

Ally, M. (2004). Foundations of educational theory for online learning. *Theory and Practice of Online Learning*, *2*(1), 15–44.

Almerez, Q. L. G., & Duping, A. M. (2022). Challenges and responses of higher education institutions towards academic resilience. *International Journal of Research and Innovation in Social Science*, *VI*(XII), 464–472. doi:10.47772/IJRISS.2022.61225

AlMunifi, A., & Aleryani, A. (2016). Internal efficiency of Higher education system in armed conflict-affected countries-Yemen case. *International Journal of Educational Development, 83.*

Aloi, A., Alonso, B., Benavente, J., Cordera, R., Echániz, E., González, F., Ladisa, C., Lezama-Romanelli, R., López-Parra, Á., Mazzei, V., Perrucci, L., Prieto-Quintana, D., Rodríguez, A., & Sañudo, R. (2020). Effects of the COVID-19 lockdown on urban mobility: Empirical evidence from the city of Santander (Spain). *Sustainability (Basel)*, *12*(9), 3870. doi:10.3390/su12093870

Alshami, S., & Alzanfaly, A. (2021). Future directions for planning education in times of crises: the Corona pandemic as an example. Regional Center for Educational Planning, Sultan Kabus University.

Ameen, N., Willis, R., Abdullah, M. N., & Shah, M. (2019). Towards the successful integration of e-learning systems in higher education in Iraq: A student perspective. *British Journal of Educational Technology*, *50*(3), 1434–1446. doi:10.1111/bjet.12651

American College of Emergency Physicians (ACEP). (n.d.). *Hospital Disaster Preparedness Self-Assessment Tool.* https://www.calhospitalprepare.org/post/hospital-disaster-preparedness-self-assessment-tool

Anadolu Agency. (2023). https://www.aa.com.tr/tr/gundem/depremler-10-ilde-yaklasik-13-5milyon-kisinin-yasadigi-bolgeyi-etkiledi/2808868#

Anderson, J. C., & Gerbing, D. W. (1984). The effect of sampling error on convergence, improper solutions, and goodness-of-fit indices for maximum likelihood confirmatory factor analysis. *Psychometrika*, *49*(2), 155–173. doi:10.1007/BF02294170

Andersson, T., Caker, M., Tengblad, S., & Wickelgren, M. (2019). Building traits for organisational resilience through balancing organisational structures. *Scandinavian Journal of Management*, *35*(1), 36–45. doi:10.1016/j.scaman.2019.01.001

Anigbogu, V. C. (2023). *The role of universities in responding to Africa's problems and development needs*. Makerere University.

Aoki, T., & Ito, K. (2014). What is the role of universities in disaster response, recovery, and rehabilitation? Focusing on our disaster victim identification project. *IEEE Communications Magazine*, *52*(3), 30–37. doi:10.1109/MCOM.2014.6766081

Appadurai, A. (2007). The capacity to aspire: Culture and the terms of recognition. In D. Held, H. L. Moore, & K. Young (Eds.), *Cultural politics in a global age: Uncertainty, solidarity and innovation* (pp. 29–35). Oneworld Publications.

Appadurai, A. (2013). *The future as cultural fact: Essays on the global condition*. Verso.

Appleby, Y., & Pilkington, R. (2014). *Developing critical professional practice in education*. Niace.

Arena Publica. (2017). *"En caso de sismo, no cruce el puente". Lo que ocurrió en el Tec de Monterrey* ["In case of an earthquake, do not cross the bridge." What happened at Tec of Monterrey]. https://www.arenapublica.com/articulo/2017/09/20/7307/muertos-y-heridos-en-el-tec-de-monterrey-por-terremoto-del-19-de-septiembre-de-2017

Arenas-Basurto, J. G. (2018). Las universidades de México y la cultura de la prevención de los desastres naturales [The universities of Mexico and the culture of prevention of natural disasters]. *Revista de Educación Superior en America Latina, 3, 6-8.* https://rcientificas.uninorte.edu.co/index.php/esal/article/view/10638/10886

Aristotle. (1925). *Nichomachean ethics: Book II* (W.D. Ross, Trans.). The Internet Classics Archive. http://classics.mit.edu/Aristotle/nicomachaen.2.ii.html (Original work published 350 B.C.E.)

Aristotle. (2009). *The Nicomachean Ethics (D. Ross, Trans., L. Brown, Ed.)*. Oxford University Press.

Armsby, P., Costley, C., & Weller, G. (2021). The role of practice in doctoral degrees. *Research in Post-Compulsory Education*, *26*(3), 257–273. doi:10.1080/13596748.2021.1920258

Armstrong, C. (2006). Citizenship in global perspective. *Policy & Practice-A Development Education Review*, *3*, 14–25.

Arunasalam, N. (2016). *Technology-enhanced learning in transnational higher education*. Academic Press.

Association of Healthcare Emergency Preparedness Professionals. (n.d.). *About us*. https://www.ahepp.org/page/AboutUsTab

Atamanyuk, S., Semenikhina, O., & Shyshenko, I. (2021). Theoretical fundamentals of innovation of higher education in Ukraine. *Pedagogy and Education Management Review*, (2), 30-36.

Atamba, C., Abazie, G. A., Nweze, C. N., Bonney, B., Mohamed, H. H., & Onwubuya, G. C. (2022). A Systematic Review of African University Student's Learning in the Era of COVID-19 Pandemic: Lessons from China. *Creative Education*, *13*(11), 3581–3592. doi:10.4236/ce.2022.1311228

Aucejo, E. M., French, J., Ugalde Araya, M. P., & Zafar, B. (2020). The impact of COVID-19 on student experiences and expectations: Evidence from a survey. *Journal of Public Economics*, *191*, 104271. doi:10.1016/j.jpubeco.2020.104271 PMID:32873994

Austin, E. N., Bastepe-Gray, S. E., Nelson, H. W., Breitenbach, J., Ogle, K. T., Durry, A., Green, S. D., Crabtree, L. A., & Haluska, M. (2014). Pediatric mass-casualty education: Experiential learning through university-sponsored disaster simulation. *Journal of Emergency Nursing: JEN*, *40*(5), 428–433. doi:10.1016/j.jen.2014.05.015 PMID:25194652

Azimi, M. N., & Balakarzai, A. T. (2020). Nation Building Through Higher Education System Resurgence in Afghanistan. Handbook of Education Systems in South Asia, 1-22. doi:10.1007/978-981-13-3309-5_56-1

Azis, A. A. (2017). The Contribution of Post-Conflict Education to Social Transformation and Sustainable Development. *Indonesian Journal of International Studies*, *2*(1), 63–70. doi:10.22146/globalsouth.28853

Badiou, A. (2007). *The Century*. Polity Press.

Baker, S., Bloom, N., & Davis, S. (2020). *COVID-induced economic uncertainty*. National Bureau of Economic Research., doi:10.3386/w26983

Banks, M., Olsen, A., & Pearce, D. (2007). *Global student mobility: An Australian perspective five years on*. IDP Education Pty Ltd.

Bar Am, J., Furstenthal, L., Jorge, F., & Roth, E. (2020, June 17). *Innovation in a crisis: Why it is more critical than ever.* McKinsey & Company. https://www.mckinsey.com/capabilities/strategy-and-corporate-finance/our-insights/innovation-in-a-crisis-why-it-is-more-critical-than-ever

Baris, S. (2022). *Perception of death in adult individual: A review according to self-compassion and hope variables* [Unpublished Master's Thesis]. Marmara University, Istanbul.

Barnett, J., Deng, R., & Yoasa, N. B. (2008). Post-conflict language issues at the University of Juba, Southern Sudan: Policies, proficiencies and practicalities. *ARAS*, *29*(1&2), 92–103.

Bartolic, S., Matzat, U., Tai, J., Burgess, J. L., Boud, D., Craig, H., & Guppy, N. (2022). Student vulnerabilities and confidence in learning in the context of the COVID-19 pandemic. *Studies in Higher Education*, *47*(12), 2460–2472. do i:10.1080/03075079.2022.2081679

Bartuseviciene, I., Pazaver, A., & Kitada, M. (2021). Building a resilient university: Ensuring continuity-transition from face-to-face to online in the COVID-19 pandemic. *Journal of Maritime Affairs*, *20*(2), 151–172. doi:10.1007/s13437-021-00239-x

Basitere, M., & Mapatagane, N. (2018, June). Effects of a Social Media Network Site on Student's Engagement and Collaboration: A Case Study of WhatsApp at a University of Technology. In *ECSM 2018 5th European conference on social media*.

Bass, B. M. (1990). From transactional to transformational leadership: Learning to share the vision. *Organizational Dynamics*, *18*(3), 1990. doi:10.1016/0090-2616(90)90061-S

Bass, B. M. (1997). Does the transactional/transformational leadership paradigm transcend organizational and national boundaries? *The American Psychologist*, *52*(2), 130–139. doi:10.1037/0003-066X.52.2.130

Bass, B. M., & Avolio, B. J. (1994). *Improving Organizational Effectiveness through Transformational Leadership.* Sage Publications.

Bass, B. M., & Riggio, R. E. (2006). *Transformational leadership.* Erlbaum. doi:10.4324/9781410617095

Bataille, G. M., & Cordova, D. I. (2014). *Managing the unthinkable: Crisis preparation and response for campus leaders. Sterling.* Stylus.

Bauman, Z. (1998). *Globalization: The human consequences.* Columbia University Press.

Bauman, Z. (2009). *Does ethics have a chance in a world of consumers?* Harvard University Press.

Bauman, Z. (2013a). *Community: Seeking safety in an insecure world.* John Wiley & Sons.

Bauman, Z. (2013b). *Consuming life.* John Wiley & Sons.

Bauman, Z. (2013c). *The individualized society.* John Wiley & Sons.

Bauwens, M., Compernolle, S., Stavrakou, T., Müller, J. F., Van Gent, J., Eskes, H., Levelt, P. F., van der A, R., Veefkind, J. P., Vlietinck, J., Yu, H., & Zehner, C. (2020). Impact of coronavirus outbreak on NO2 pollution assessed using TROPOMI and OMI observations. *Geophysical Research Letters, 47*(11). Advance online publication. doi:10.1029/2020GL087978 PMID:32836515

Baxter, J. (2012). Who am I and what keeps me going? Profiling the distant learning student in higher education. *International Review of Research in Open and Distance Learning, 13*(4), 107–129. doi:10.19173/irrodl.v13i4.1283

Beans, H., Maireva, C., & Muza, C. (2020). Zimbabwe higher education institutions' preparedness in responding to COVID-19-induced disruptions to education. *Journal of New Vision in Educational Research, 1*(2), 267–281.

Becton, Y. J., Bogiages, C., D'Amico, L., Lilly, T., Currin, E., Jeffries, R., & Tamim, S. (2020). An emerging framework for the EdD activist. *Impacting Education: Journal on Transforming Professional Practice, 5*(2), 43–54. doi:10.5195/ie.2020.131

Bedan, S. (2022). Higher Education in Iraq between the Past and the Present. *Journal of Positive School Psychology*, 4718–4731.

Beech, P. (2020). These new gadgets were designed to fight COVID-19. In *World Economic Forum* (Vol. 5). Academic Press.

Beggan, D. M. (2011). Disaster recovery considerations for academic institutions. *Disaster Prevention and Management, 20*(4), 413–422. doi:10.1108/09653561111161734

Bek, H. (2017). Understanding the effect of loneliness on academic participation and success among international university students. *Journal of Education and Practice, 8*(14), 46–50.

Benavides, L. M. C., Tamayo Arias, J. A., Arango Serna, M. D., Branch Bedoya, J. W., & Burgos, D. (2020). Digital transformation in higher education institutions: A systematic literature review. *Sensors (Basel), 20*(11), 1–22. doi:10.3390/s20113291 PMID:32526998

Benson, V., & Morgan, S. (2018). Measuring the social impact: How social media affects higher education institutions. In *Social Media Marketing: Breakthroughs in Research and Practice* (pp. 1167–1184). IGI Global., doi:10.4018/978-1-5225-5637-4.ch055

Bento, F., Giglio-Bottino, A., Cerchiareto-Perreira, F., Forastieri de Almeida, J., & Gomes-Pereira, F. (2021). Resilience in higher education: A complex perspective to lecturers' adaptive processes in response to the COVID-19 pandemic. *Education Sciences*, *11*(9), 492–504. doi:10.3390/educsci11090492

Berkes, F. (2017). Environmental governance for anthropocene? Social ecological systems, resilience and collaborative learning. *Sustainability (Basel)*, *9*(7), 1–11. doi:10.3390/su9071232

Berke, T., & Sell, J. (2018). Educational challenges in South Sudan. In J. Sell & T. Berke (Eds.), *Strategies, policies, and directions for refugee education* (pp. 121–138). Springer. doi:10.1108/S2055-364120180000013005

Bermude, H., & Izquierdo, A. (2020, July-August). *Trust and Transparency in Times of Crisis.* Retrieved October 29, 2023, from www.prsa.org: https://www.prsa.org/article/trust-and-transparency-in-times-of-crisis

Bhatt, I., & MacKenzie, A. (2019). Just google it! Digital literacy and the epistemology of ignorance. *Teaching in Higher Education*, *24*(3), 302–317. doi:10.1080/13562517.2018.1547276

Bible Gateway. (n.d.). *Genesis 6:11-20.* https://www.biblegateway.com/passage/?search=Genesis%206%3A9-9%3A17&version=NIV

Bielak, A. T., Campbell, A., Pope, S., Schaefer, K., & Shaxson, L. (2008). From science communication to knowledge brokering: the shift from 'science push'to 'policy pull. In *Communicating science in social contexts* (pp. 201–226). Springer. doi:10.1007/978-1-4020-8598-7_12

Biggs, R., Schlüter, M., Biggs, D., Bohensky, E. L., BurnSilver, S., Cundill, G., Dakos, V., Daw, T. M., Evans, L. S., Kotschy, K., Leitch, A. M., Meek, C., Quinlan, A., Raudsepp-Hearne, C., Robards, M. D., Schoon, M. L., Schultz, L., & West, P. C. (2012). Toward principles for enhancing the resilience of ecosystem services. *Annual Review of Environment and Resources*, *37*(1), 421–448. doi:10.1146/annurev-environ-051211-123836

Biwer, F., Wiradhany W., oude Egbrink, M., Hospers, H., Wasenitz, S., Jansen, W., & de Bruin, A. (2021). Changes and Adaptations: How University Students Self-Regulate Their Online Learning During the COVID-19. *Pandemic. Frontiers in Psychology, 12*, 1-12. doi:10.3389/fpsyg.2021.642593

Blackmore, J., Gribble, C., Farrell, L., Rahimi, M., Arber, R., & Devlin, M. (2014). *Australian international graduates and the transition to employment.* Deakin University.

Blauch, L. E. (1959). *Accreditation in higher education: Organized and* (L. E. Blauch, Ed.). U.S. Dep. of Health Education and Welfare.

Blue Moon Consulting Group. (n.d.). *Higher Education: Crisis Management A Proactive & Strategic Approach to Managing Reputational Risk.* Blue Moon Consulting Group. Retrieved from https://bluemoonconsultinggroup.com/wp-content/uploads/2022/03/BMCG-Higher-Ed-1-Crisis-Management-Whitepaper-2-8-22.pdf

Bogoviz, A. V. (2019). Industry 4.0 as a new vector of growth and development of knowledge economy. In *Industry 4.0: Industrial Revolution of the 21st Century* (pp. 85–91). Springer. doi:10.1007/978-3-319-94310-7_8

Bolisani, E., & Bratianu, C. (2018). The emergence of knowledge management. In *Emergent knowledge strategies* (pp. 23–47). Springer. doi:10.1007/978-3-319-60657-6_2

Bonacini, L., Gallo, G., & Scicchitano, S. (2021). Working from home and income inequality: Risks of a 'new normal' with COVID-19. *Journal of Population Economics*, *34*(1), 303–360. doi:10.1007/s00148-020-00800-7 PMID:32952308

Bond, M., Marín, V. I., Dolch, C., Bedenlier, S., & Zawacki-Richter, O. (2018). Digital transformation in German higher education: Student and teacher perceptions and usage of digital media. *International Journal of Educational Technology in Higher Education*, *15*(1), 48. Advance online publication. doi:10.1186/s41239-018-0130-1

Booker, L. (2014). Crisis management: Changing times for colleges. *Journal of College Admission, 222*, 16–23.

Boomsma, A. (1985). Nonconvergence, improper solutions, and starting values in LISREL maximum likelihood estimation. *Psychometrika, 50*(2), 229–242. doi:10.1007/BF02294248

Bordia, P., Hunt, E., Paulsen, N., Tourish, D., & DiFonzo, N. (2004). Uncertainty during organizational change: Is it all about control? *European Journal of Work and Organizational Psychology, 13*(3), 345–365. doi:10.1080/13594320444000128

Bostad, I., & Ottersen, O. P. (2014). Global presence, global responsibility and the global citizen. In A. B. Sterri (Ed.), *Global citizen: Challenges and responsibility in an interconnected world* (pp. 1–3). Sense Publishers. doi:10.1007/978-94-6209-929-6_1

Boud, D., Costley, C., Marshall, S., & Sutton, B. (2021). Impacts of a professional practice doctorate: A collaborative enquiry. *Higher Education Research & Development, 40*(3), 431–445. doi:10.1080/07294360.2020.1765744

Bouncken, R. B., Kraus, S., & Ancillo, A. (2022). Management in times of crises: Reflections on characteristics, avoiding pitfalls, and pathways out. *Review of Managerial Science, 16*(7), 2035–2046. doi:10.1007/s11846-022-00580-2

Bowlby, J. (1988). *A secure base: clinical applications of attachment theory*. Routledge.

Bozkurt, A. (2022). Resilience, adaptability, and sustainability of higher education: A systematic mapping study on the impact of the Coronavirus (COVID-19) pandemic and the transition to the new normal. *Journal of Learning for Development, 9*(1), 1–16. doi:10.56059/jl4d.v9i1.590

Bozkurt, A., & Sharma, R. C. (2020). Education in normal, new normal and next normal: Observations from the past, insights from the present and projects for the future. *Asian Journal of Distance Education, 15*(2), i–x. doi:10.5281/zenodo.4362664

Braskamp, L. A. (2008). Developing global citizens. *Journal of College and Character, 10*(1), 1–5. doi:10.2202/1940-1639.1058

Bray, M. (2020a). *Flattening the multimodal learning curve: A faculty playbook*. The Economist Intelligence Unit. https://impact.economist.com/perspectives/sites/default/files/eiu163_-_msft_multimodal_learning_playbook_v4_1.pdf

Bray, M. (2020b). *Bridging the digital divide to engage students in higher education*. The Economist Intelligence Unit. https://impact.economist.com/perspectives/sites/default/files/eiu_microsoft_higher_education_report.pdf

Brewer, M. L., van Kessel, G., Sanderson, B., Naumann, F., Lane, M., & Carter, A. (2019). Resilience in higher education students: Scoping review. *Higher Education Research & Development, 38*(6), 1105–1120. doi:10.1080/07294360.2019.1626810

Brinkmann, S. (2013). *Qualitative interviewing*. Oxford University Press. doi:10.1093/acprof:osobl/9780199861392.001.0001

Briseño, H. (2023). *"Pérdida total" de las 16 bibliotecas públicas de Acapulco tras 'Otis'* ["Total los" of the 16 public libraries of Acapulco after 'Otis']. https://www.jornada.com.mx/noticia/2023/11/25/estados/perdida-total-de-las-16-bibliotecas-publicas-de-acapulco-tras-otis-1759#:~:text=Acapulco%2C%20Gro.,decenas%20de%20colecciones%20y%20enciclopedias

Brister, E., Frodeman, R., & Briggle, A. (2020). The Future of Field Philosophy: Lessons Learned and Next Steps. In *A Guide to Field Philosophy* (pp. 363–368). Routledge. doi:10.4324/9781351169080-24

Bromley, P., Schofer, E., & Longhofer, W. (2018). Organizing for Education: A Cross-National, Longitudinal Study of Civil Society Organizations and Education Outcomes. *Voluntas, 29*(3), 1–15. doi:10.1007/s11266-018-9979-9

Brookfield, S. (1985). 'Sources in self-directed learning theory and practice'. Wiley Online Library. *New Directions for Adult and Continuing Education, 1985*(25), 87–90. doi:10.1002/ace.36719852510

Brophy, M. (2020). The Role of NGOs in Supporting Education in Africa. *Journal of International and Comparative Education (JICE),* 45-56.

Browne, M. W., & Cudeck, R. (1993). Alternative ways of assessing model fit. In Testing Structural Equation Models. SAGE Publications.

Brown, N. A., Brown, K. A., & Billings, A. C. (2015). 'May no act of ours bring shame': Fan- enacted crisis communication surrounding the Penn State sex abuse scandal. *Communication & Sport, 3*(3), 288–311. doi:10.1177/2167479513514387

Bryant, S. E. (2003). The Role of Transformational and Transactional Leadership in Creating, Sharing and Exploiting Organizational Knowledge. *Journal of Leadership & Organizational Studies, 9*(4), 32–44. doi:10.1177/107179190300900403

Buchholz, B. A., DeHart, J., & Moorman, G. (2020). Digital Citizenship During a Global Pandemic: Moving Beyond Digital Literacy. *Journal of adolescent & adult literacy: a journal from the International Reading Association, 64*(1), 11–17. doi:10.1002/jaal.1076

Bullock, J. A., Haddow, G., & Coppola, D. P. (2016). *Managing children in disasters: Planning for their unique needs.* CRC Press.

Bulut, M. B., McDermott, M., & McEwan, O. (2017). Adapting the multidimensional mortality awareness measure: A study of validity and reliability. *The Journal of International Social Research, 10*(49), 268–275. doi:10.17719/jisr.2017.1578

Burnes, B. (2011). Introduction: Why does change fail, and what can we do about it? *Journal of Change Management, 11*(4), 445–450. doi:10.1080/14697017.2011.630507

Burnes, B., & Jackson, P. (2011). Success and failure in organizational change: An exploration of the role of values. *Journal of Change Management, 11*(2), 133–162. doi:10.1080/14697017.2010.524655

C2. (2022, November 11). *Crisis Management Team: Function, Roles & Responsibilities.* Retrieved October 25, 2023, from https://continuity2.com/blog/crisis-management-team

Cabero-Almenara, J., Barroso-Osuna, J., Gutiérrez-Castillo, J. J., & Palacios-Rodríguez, A. (2021). The Teaching Digital Competence of Health Sciences Teachers. A Study at Andalusian Universities (Spain*). International Journal of Environmental Research and Public Health, 18*(5), 2552. doi:10.3390/ijerph18052552 PMID:33806483

Cabero-Almenara, J., Guillén-Gámez, F. D., Ruiz-Palmero, J., & Palacios-Rodríguez, A. (2021). Digital competence of higher education professors according to DigCompEdu. statistical research methods with ANOVA between fields of knowledge in different age ranges. *Education and Information Technologies, 26*(4), 4691–4708. doi:10.1007/s10639-021-10476-5 PMID:33758572

Callaway, E. (2016, January 19). Cave of forgotten dreams may hold earliest painting of volcanic eruption. *Scientific American.* https://www.scientificamerican.com/article/cave-of-forgotten-dreams-may-hold-earliest-painting-of-volcanic-eruption/

Camera, L. (2023). The war on college campuses. *U.S. News & World Report, 163*(2), 36–40.

Cancino, K. (2023). *Universidad de Acapulco Cierra por los daños de "Otis" y deja en la incertidumbre a más de mil alumnos* [University of Acapulco closes due to damage from "Otis" and leaves more tan a thousand students in uncertainty]. https://heraldodemexico.com.mx/nacional/2023/12/22/universidad-de-acapulco-cierra-por-los-danos-de-otis-deja-en-la-incertidumbre-mas-de-mil-alumnos-564799.html

Caramenico, A. (2011, May 6). *Most hospitals better prepared for disasters*. Fierce Healthcare. https://www.fiercehealthcare.com/healthcare/most-hospitals-better-prepared-for-disasters

Cardoso, S., Tavares, O., Sin, C., & Carvalho, T. (Eds.). (2020). *Structural and Institutional Transformations in Doctoral Education: Social, Political and Student Expectations*. Springer Nature. doi:10.1007/978-3-030-38046-5

Carlo, G. (2006). Care-based and altruistically based morality. In M. Killen & J. G. Smetana (Eds.), *Handbook of Moral Development* (pp. 551–579). Erlbaum.

Carlos-Zavala, J. (2017). *Sismo dañó Sistema de agua potable en 41 municipios del Istmo de Oaxaca* [Earthquake damaged drinking water system in 41 municipalities of the Isthmus of Oaxaca]. https://oaxaca.eluniversal.com.mx/municipios/29-09-2017/sismo-dano-sistema-de-agua-potable-en-41-municipios-del-istmo-de-oaxaca

Carlton, S., & Mills, C. E. (2017). The Student Volunteer Army: A 'repeat emergent' emergency response organisation. *Disasters*, *41*(4), 764–787. doi:10.1111/disa.12225 PMID:28093798

Carvalho, T., & Cardoso, S. (2020). Conclusion: The transformations in doctoral education—A comprehensive and critical approach. *Structural and institutional transformations in doctoral education: Social, political and student expectations. Issues in Higher Education*, 375-389.

Carvalho, S., Rossiter, J., Angrist, N., Hares, S., & Silverman, R. (2020). *Planning for school reopening and recovery after COVID-19*. Center for Global Development.

Cassidy, S. (2015). Resilience building in students: The role of academic self-efficacy. *Frontiers in Psychology*, *6*, 1781. doi:10.3389/fpsyg.2015.01781 PMID:26640447

Castro Benavides, L. M., Tamayo Arias, J. A., Arango Serna, M. D., & Branch Bedoya, J., W., & Burgos, D. (2020). Digital transformation in higher education institutions: A systematic literature review. *Sensors (Switzerland)*, *20*(11), 1–23. PMID:32526998

Cattell, R. B. (1963). Theory of fluid and crystallized intelligence: A critical experiment. *Journal of Educational Psychology, 54*(1), 1.

Cervantes-Duarte, L., & Fernández-Cano, A. (2016). Impact of Armed Conflicts on Education and Educational Agents: A Multivocal Review. *Revista Electrónica Educare (Educare Electronic Journal), 20*(3).

Chakraborty, M., & Nafukho, F. M. (2015). Strategies for virtual learning environments: Focusing on teaching presence and teaching immediacy. *Journal of Online Learning Research and Practice*, *4*(1), 8–37.

Change Leader, Inc. (n.d.). *Crisis Management and Transformation.* Retrieved October 13, 2023, from https://changinghighered.com/higher-education-consulting/crisis-management-transformation/

Channy, D. Y., & Ogunniran, M. O. (2019). A comparative analysis of the role of international organizations in financing higher education system: A case study of Nigeria and Cambodia. *US-China Education Review*, *9*(1), 1–17. doi:10.17265/2161-623X/2019.01.001

Chapman, S., Colvin, L. E., & Cosentino, S. (2020). Translational aspects of the multidisciplinary study of metacognition. *Translational Issues in Psychological Science*, *6*(1), 26–31. doi:10.1037/tps0000224 PMID:33728364

Charoensukmongkol, P., & Phungsoonthorn, T. (2021). The effectiveness of supervisor support in lessening perceived uncertainties and emotional exhaustion of university employees during the COVID-19 crisis: The constraining role of organisational intransigence. *The Journal of General Psychology*, *148*(4), 431–450. doi:10.1080/00221309.2020.1795613 PMID:32691689

Chasi, S. (2021). South Africa's Policy Framework for Higher Education Internationalisation: Decolonial Perspective. *Think (London, England), 87*, 34–39.

Cheng, Y. F., & Adefila, A. (2020). Enhancing school safety through university engagement in DRR education. *International Journal of Disaster Risk Reduction, 44*, 101386. doi:10.1016/j.ijdrr.2019.101386

Chen, H. Y., Wang, C. W., Yang, H. F., Huang, W. T., & Fan, S. Y. (2020). The relationships between physical activity and life satisfaction and happiness among young, middle-aged and older adults. *International Journal of Environmental Research and Public Health, 15*(4), 25–39. doi:10.29045/14784726.2021.3.5.4.25

Chen, P. S. D., Lambert, A. D., & Guidry, K. R. (2010). Engaging online learners: The impact of web-based learning technology on college student engagement. *Computers & Education, 54*(4), 1222–1232. https://www.learntechlib.org/p/67431/. doi:10.1016/j.compedu.2009.11.008

Cherrington, S., Macaskill, A., Salmon, R., Boniface, S. M., Shep, S., & Flutey, J. (2017). Developing a pan-university professional learning community. *The International Journal for Academic Development, 23*(4), 298–311. doi:10.1080/1360144X.2017.1399271

Chibuwe, A., & Munoriyarwa, A. (2023). *Emerging methods and challenges associated with teaching and learning media studies during the COVID-19 pandemic induced lockdowns in Zimbabwe and South Africa*. doi:10.1177/21582440231167113

Chida, P. (2020). Government shuts down varsities and colleges. *NewsDay*. https://www.newsday.co.zw

Chikoko, V. (2020). This technology we all wash with: The efficacy of leveraging WhatsApp in delivering a master's module. In *Technology-based Teaching and Learning in Higher Education during the Time of COVID-19*. CSSALL Publishers (Pty.) Ltd. http://alternation.ukzn.ac.za/Files/books/series-02/00-Font-Matter.pdf

Chmielecka, E. (2014). Academic Values and The Procedures of Quality Assurance. In H. Eggins (Ed.), *Drivers and Barriers to Achieving Quality in Higher Education* (pp. 43–54). SensePublishers. doi:10.1007/978-94-6209-494-9_4

Chowdhury, A. & Shil, N.C. (2022). Understanding change management in organizational context: revisiting literature. *Management and Entrepreneurship: Trends of Development, 1*(19), 28-43.

Clafferty, E., & Beggs, B. J. (2014). *The triple C model for student retention, progression and completion matures*. Society for Research into Higher Education Conference. https://www.researchgate.net/publication/281290529

Clarke-Habibi, S. (2021). *Higher education can do more to transform conflict and build peace*. University World News.

Clarke, M. (2018). Rethinking graduate employability: The role of capital, individual attributes and context. *Studies in Higher Education, 43*(11), 1923–1937. doi:10.1080/03075079.2017.1294152

Cloete, N., & Moja, T. (2005). Transformation tensions in higher education: Equity, efficiency, and development. *Social Research, 72*(3), 693–722. doi:10.1353/sor.2005.0033

Coates, H., Croucher, G., Moore, K., Weerakkody, U., Dollinger, M., Grosemans, I., Bexley, E., & Kelly, P. (2020). Contemporary perspectives on the Australian doctorate: Framing insights to guide development. *Higher Education Research & Development, 39*(6), 1122–1139. doi:10.1080/07294360.2019.1706451

Coburn, A. W., Spence, R. J. S., & Pomonis, A. (1992). Factors determining human casualty levels in earthquakes: Mortality prediction in building collapse. *Proc. of Earthquake Engineering, Tenth World Conference*, 5989-5994.

Cochran-Smith, M., Ell, F., Ludlow, L., Grudnoff, L., & Aitken, G. (2014). The challenge and promise of complexity theory for teacher education research. *Teachers College Record, 116*(4), 1–38. doi:10.1177/016146811411600407

Collinson, V. (2004). Learning to share, sharing to learn: Fostering organizational learning through teachers' dissemination of knowledge. *Journal of Educational Administration*, *42*(3), 312–332. doi:10.1108/09578230410534658

Concern Worldwide US. (2022, August 3). *Climate change is one of the biggest threats to education — and growing.* Retrieved October 12, 2023, from ww,concernusa.org: https://concernusa.org/news/education-and-climate-change/

Connolly, M. (2016). *Campus emergency preparedness: Meeting ICS and NIMS compliance.* CRC Press.

Connolly, T. (2000). *General Introduction.* Judgement and Decision Making.

CookA.KeyteR.SprawsonI.MatharuA.MantziosM. (2022). Mental health first aid experiences: A qualitative investigation into the emotional impact of mental health first aid responsibilities. https://doi.org/ doi:10.21203/rs.3.rs-1393314/v1

Coombs, W. T. (2015). *Ongoing crisis communication: Planning, managing, and responding* (4th ed.). Sage.

Cornell, D. G., & Sheras, P. L. (1998). Common errors in school crisis response: Learning from our mistakes. *Psychology in the Schools*, *35*(3), 297–307. doi:10.1002/(SICI)1520-6807(199807)35:3<297::AID-PITS9>3.0.CO;2-K

Coveleski, J. (2014). *A study of students' perceptions of natural disaster plans and emergency preparedness at a higher education institution* [Doctoral dissertation]. Florida State University.

CRE/UNISDR. (2020). *Human cost of disasters-An overview of the last 20 years 2000-2019.* https://reliefweb.int/report/world/human-cost-disasters-overview-last-20-years-2000-2019?gad_source=1&gclid=EAIaIQobChMIlNnznpasgwMVRTqtBh3u-Ae0EAAYASAAEgLlVfD_BwE

Creswell, J. W. (2013). *Qualitative inquiry and research design: Choosing among five approaches.* Sage publications.

Creswell, M. J. (1986). Why Objects Exist but Events Occur. *Studia Logica*, *45*(4), 371–375. doi:10.1007/BF00370270

Crises Control. (2022, September 14). *What is Crisis Management and Why are Crisis Management Strategies Important?* Retrieved October 11, 2023, from https://www.crises-control.com/blogs/what-is-crisis-management-and-why-is-it-important/

Crisis Control. (2023, June 16). *What is the 5C model for communicating during crisis?* https://www.crises-control.com/blogs/the-5c-model-for-critical-communication/

Crompton, H., Olszewski, B., & Bielefeldt, T. (2016). The mobile learning training needs of educators in technology enabled environments. *Professional Development in Education*, *42*(3), 482–501. doi:10.1080/19415257.2014.1001033

Cvetkovic, V. M., Ocal, A., & Ivanov, A. (2019). Young adults' fear of disasters: A case study of residents from Turkey, Serbia and Macedonia. *International Journal of Disaster Risk Reduction*, *35*, 101095. doi:10.1016/j.ijdrr.2019.101095

D'Olimpio, L. (2018). Trust as a virtue in education. *Educational Philosophy and Theory*, *2*(50), 193–202. doi:10.1080/00131857.2016.1194737

Dahles, H., & Susilowati, T. P. (2015). Business resilience in times of growth and crisis. *Annals of Tourism Research*, *51*, 34–50. doi:10.1016/j.annals.2015.01.002

Damle, S., Mani, S. K., & Balamurugan, G. (2021). Natech guide words: A new approach to assess and manage natech risk to ensure business continuity. *Journal of Loss Prevention in the Process Industries*, *72*, 104564. doi:10.1016/j.jlp.2021.104564

Daniels, R. J., & Shadmi, K. (2020). *Universities' vital role in the pandemic response, universities respond to the pandemic with sound science and advice. We can do more: "A Vital Mission" in the COVID-19 Special Edition issue of Hopkins Bloomberg Public Health Magazine.* https://dev.magazine.jhsph.edu/2020/universities-vitalrole-pandemic-response

Darabi, S., Azizi, N., Salimi, J., & Shirabegi, N. (2019). Elements and Dimensions of Sustainability in Iran's Higher Education System, and Comparison of Public and Non-Governmental Academic Approaches. *Iranian Journal of Educational Sociology, 2*(3), 12-26.

David, S. B., & Cortes, A. H. (2019). Principles of Management. OpenStax.

Davidovitz, R., Mikulincer, M., Shaver, P. R., Izsak, R., & Popper, M. (2007). Leaders as attachment figures: Leaders' attachment orientations predict leadership-related mental representations and followers' performance and mental health. *Journal of Personality and Social Psychology, 93*(4), 632–650. doi:10.1037/0022-3514.93.4.632 PMID:17892336

Davies, L. (2006). Global citizenship: Abstraction or framework for action? *Educational Review, 58*(1), 5–25. doi:10.1080/00131910500352523

de Araujo, L. M., Priadana, S., Paramarta, V., & Sunarsi, D. (2021). Digital leadership in business organizations. *International Journal of Educational Administration, Management, and Leadership*, 45-56.

De Costa, P. I. (2016). Constructing the global citizen: An ELF perspective. *Journal of Asian Pacific Communication, 26*(2), 238–259. doi:10.1075/japc.26.2.04dec

De Gree, A., & Cheprasov, A. (2023). *Syrian Civil War Timeline*. Impact & Outcome.

De Guttry, A., Ndiloseh, M., & Amoroso, A. M. (2022). The 2020 Juba Agreement for Peace in Sudan and Prospects for Peacebuilding in Darfur. *African Conflict and Peacebuilding Review, 12*(2), 1–4. doi:10.2979/africonfpeacrevi.12.2.01

De Ruyter, D., Oades, L. G., Waghid, Y., Ehrenfeld, J., Gilead, T., & Singh, N. C. (2022). Reimagining Education: The International Science and Evidence based Assessment. New Delhi: UNESCO MGIEP. doi:10.56383/XBXZ7711

Debych, M. (2023). Internationalization of higher education in the context of recovery of Ukraine: Goals and challenges. *Continuing Professional Education: Theory and Practice, 2*(75), 7–15.

Deleuze, G. (1994). *Difference and Repetition*. Columbia University Press.

Deloitte. (2019). *Stepping in: The board's role in crisis management.* Deloitte. Retrieved from www2.deloitte.com: https://www2.deloitte.com/content/dam/Deloitte/uk/Documents/risk/deloitte-uk-risk-global-on-the-boards-agenda-crisis-management.pdf

Demirbilek, O., & Ozturk, S. (2023). Understanding university students WHO do not want to volunteer in disasters: A qualitative study from Turkey. *International Journal of Disaster Risk Reduction, 96*, 103994. doi:10.1016/j.ijdrr.2023.103994

Dennis, M. J. (2020). The impact of COVID-19 on the world economy and higher education. *Enrollment Management Report, 24*(9), 3. Advance online publication. doi:10.1002/emt.30720

DeRue, D. S., Nahrgang, J. D., Wellman, N. E. D., & Humphrey, S. E. (2011). Trait and Behavioral Theories of Leadership: An Integration and Meta-Analytic Test of Their Relative Validity. *Personnel Psychology, 64*(1), 7–52. doi:10.1111/j.1744-6570.2010.01201.x

Dewey, J. (1938). *1997. Experience and education*. Touchstone.

Dewey, J. (2010). *How we think*. Book Tree.

Di Pietro, G. (2018). The academic impact of natural disasters: Evidence from L'Aquila earthquake. *Education Economics, 26*(1), 62–77. doi:10.1080/09645292.2017.1394984

Di Pietro, G., Biagi, F., Costa, P., Karpinski, Z., & Mazza, J. (2020). *The likely impact of COVID-19 on education: Reflections based on the existing literature and recent international datasets*. Publications Office of the European Union.

Dill, J. S. (2013). The longings and limits of global citizenship education: The moral pedagogy of schooling in a cosmopolitan age. In *Routledge advances in sociology* (Vol. 109). Routledge.

Ding, W., Aoyama, A., & Choi, E. (2020). An approach to the diffusion of tacit knowledge: Learning from imitation to creation. *International Journal of Knowledge Management Studies, 11*(2), 103–116. doi:10.1504/IJKMS.2020.106276

Dirks, K. T., & de Jong, B. (2021). Trust within the workplace: A review of two waves of research and a glimpse of the third. *Annual Review of Organizational Psychology and Organizational Behavior*, 9.

Djalali, A., Khankeh, H., Öhlén, G., Castrén, M., & Kurland, L. (2011). Facilitators and obstacles in pre-hospital medical response to earthquakes: A qualitative study. *Scandinavian Journal of Trauma, Resuscitation and Emergency Medicine, 19*(1), 1–9. doi:10.1186/1757-7241-19-30 PMID:21575233

Djalante, R., Shaw, R., & DeWit, A. (2020). Building resilience against biological hazards and pandemics: COVID-19 and its implications for the Sendai Framework. In *Progress in Disaster Science 6* (Vol. 100080). Elsevier Ltd. doi:10.1016/j.pdisas.2020.100080

Dogan, T., & Erdur, Ö. (2014). Disasters, traumas, crises and traumatic stress reactions. In Disasters, crises, traumas and psychological help. Turk PDR Dernegi.

Dogan, N. (2019). *Measurement and evaluation in education.* Pegem Academy.

Dohaney, J., de Roiste, M., Salmon, R. A., & Sutherland, K. (2020). Benefits, barriers, and incentives for improved resilience to disruption in university teaching. *International Journal of Disaster Risk Reduction, 50*, 1–9. doi:10.1016/j.ijdrr.2020.101691 PMID:32509512

Dolby, N., & Rizvi, F. (2008). *Youth moves: Identities and education in global perspective*. Routledge. doi:10.4324/9780203938836

DPT. (1999). *Economic and social effects of the earthquake, possible financing needs, measures to be taken in the short, medium and long term.* Ministry of Development. Ankara.

Du Plessis, M., Jansen van Vuuren, C. D., Simons, A., Frantz, J., Roman, N., & Andipatin, M. (2022). South African higher education institutions at the beginning of the Covid-19 pandemic: Sense-making and lessons learnt. *Frontiers in Education, 6*, 740016. doi:10.3389/feduc.2021.740016

Dua, A., Law, J., Rounsaville, T., & Viswanath, N. (2020, October 26). *Reimagining higher education in the United States.* McKinsey & Company. https://www.mckinsey.com/industries/education/our-insights/reimagining-higher-education-in-the-united-states#/

Duarte, N., & Vardasca, R. (2023). Literature Review of Accreditation Systems in Higher Education. *Education Sciences, 13*(6), 582. doi:10.3390/educsci13060582

Dube, T., Zikhali, W., & Dube, S. P. (2019). An analysis of the application of servant-leadership model in the educational development programmes of selected non-governmental organizations in Zimbabwe. *International Journal of Innovative Business Strategies, 5*(2), 332–340. doi:10.20533/ijibs.2046.3626.2019.0044

Dufty, N. (2014). Opportunities for disaster resilience learning in the Australian curriculum. *Australian Journal of Emergency Management, 29*(1), 12–16.

Dunlap, A. L., Logue, K. M., & Isakov, A. P. (2014). The engagement of academic institutions in community disaster response: A comparative analysis. *Public Health Reports, 129*(6_suppl4, Suppl. 4), 87–95. doi:10.1177/00333549141296S412 PMID:25355979

Dzikite, C., Nsubuga, Y., & Nkonki, V. (2017). Lecturers' competencies in information and communication technology (ICT) for effective implementation of ICT-integrated teaching and learning in textiles and clothing degree programmes. *International Journal of Educational Sciences, 17*(1-3), 61–68. doi:10.1080/09751122.2017.1305756

Dzinamarira, T., Mukwenha, S., Eghtessadi, R., Cuadros, D. F., Mhlanga, G., & Musuka, G. (2020). COVID-19 response in Zimbabwe: A Call for Urgent Scale-up of Testing to meet National Capacity. *Clinical Infectious Diseases, 23*(10), 134–143. PMID:32865204

Dzobo, M., Chitungo, I., & Dzinamarira, T. (2020). COVID-19: A perspective for lifting lockdown in Zimbabwe. *The Pan African Medical Journal, 35*(13, Supp 2), 1–10. doi:10.11604/pamj.supp.2020.35.2.23059 PMID:32528624

Eaton, J. S. (2021). The Role of Quality Assurance and the Values of Higher Education. In H. van't Land, A. Corcoran, & D.-C. Iancu (Eds.), *The Promise of Higher Education* (pp. 181–186). Springer International Publishing. doi:10.1007/978-3-030-67245-4_28

Economic and Social Commission for Western Asia. (2019). *Development challenges in conflict settings, The impact on child marriage and adolescent fertility*. E/ESCWA/ECRI/2019/2.

Eder, J., Smith, W. W., & Pitts, R. E. (2010). Exploring factors influencing student study abroad destination choice. *Journal of Teaching in Travel & Tourism, 10*(3), 232–250. doi:10.1080/15313220.2010.503534

Eichinger, R. W., & Lombardo, M. M. (2004). Learning agility as a prime indicator of potential. *People and Strategy, 27*(4), 12.

Eiffel Corp. (2007). *Use of Learning Management Systems by South African Universities*. Retrieved from www.eiffel-corp.co.za/newsletters/eiffel_newsletter07.htm

Eisenberg, N. (2006). Prosocial behavior. In G. G. Bear & K. M. Minke (Eds.), *Children's Needs III: Development, Prevention, and Intervention* (pp. 313–324). National Association of School Psychologists.

Eisenberg, N., Morris, A. S., McDaniels, B., & Spinrad, T. L. (2009). Moral cognitions and prosocial responding in adolescence. In I. Steinberg & R. Lerner (Eds.), Handbook of Adolescent Psychology (3rd ed., pp. 229–265). Wiley. doi:10.1002/9780470479193.adlpsy001009

El Sur. (2023). *Cerrerá indefinidamente la Universida Americana de Acapulco por los daños de Otis* [The American University of Acapulco will close idenfinetely due to the damage caused by Otis]. https://suracapulco.mx/cerrara-indefinidamente-la-universidad-americana-de-acapulco-por-los-danos-de-otis/#:~:text=El%20rector%20de%20la%20Universidad,un%20riesgo%20para%20la%20comunidad

Elliott, A., & Urry, J. (2010). *Mobile lives*. Routledge. doi:10.4324/9780203887042

Elmagrhi, M. H., & Ntim, C. G. (2023, April). Non-financial reporting in non-profit organisations: The case of risk and governance disclosures in UK higher education institutions. *Accounting Forum, 47*(2), 223–248. doi:10.1080/01559982.2022.2148854

Emergency Management Monroe County, Florida. (2023). *Man-made disaster.* https://www.monroecounty-fl.gov/904/Man-Made-Disaster#:~:text=Such%20man%2Dmade%20disasters%20are,%2C%20cyber%2Dattacks%2C%20

Enaifoghe, A. (2022). South Africa's Efforts and Mechanisms in Maintaining Peace and Stability in Southern African Development Community Subregion. In E. Spiegel, G. Mutalemwa, C. Liu, & L. R. Kurtz (Eds.), *Peace Studies for Sustainable Development in Africa. Advances in African Economic, Social and Political Development*. Springer. doi:10.1007/978-3-030-92474-4_47

Enders, J., & Westerheijden, D. F. (2014). Quality assurance in the European policy arena. *Policy and Society, 33*(3), 167–176. doi:10.1016/j.polsoc.2014.09.004

Englund, C., Olofsson, A. D., & Price, L. (2017). Teaching with Technology in Higher Education: Understanding Conceptual Change and Development in Practice. *Higher Education Research & Development, 36*(1), 73–87. doi:10.1080/07294360.2016.1171300

EQAR. (2023, April 21). *Cross-border QA activities - EQAR.* https://www.eqar.eu/kb/cross-border-qa/drafting-cross-border-external-qa-activities/

Erdmann, J., Rummel, N., Christmann, N., Elson, M., Hecking, T., Herrmann, T., Ulrich Hoppe, H., Krämer, N. C., Kyewski, E., & Wichmann, A. (2017). Challenges in implementing small group collaboration in large online courses. *International Society of the Learning Sciences,* 625-628. https://repository.isls.org/bitstream/1/301/1/94.pdf

Ergunay, O. (1996). "What is disaster management? How should it be?" Search for solutions to Turkey's earthquake problems in the light of Erzincan and Dinar earthquakes. *Tubitak earthquake symposium proceedings book,* 263-272.

Erkal, T., & Degerliyurt, M. (2009). Disaster management of Turkey. *Eastern Geographical Review, 14*(22), 147-164. https://dergipark.org.tr/en/pub/ataunidcd/issue/2437/31194

Errida, A., & Lotfi, B. (2021). The determinants of organizational change management success: Literature review and case study. *International Journal of Engineering Business Management, 13,* 1–15. doi:10.1177/18479790211016273

Etherington, K. (2004). *Becoming a reflexive researcher: Using ourselves in research.* Jessica Kingsley Publishers.

EU. (2018). *Industry-based learning (IBL) program.* https://www.monash.edu/it/future-students/industry-experience/industry-based-learning

Everly, G. S., Smith, K. J., & Lobo, R. (2013). Resilient leadership and the organisational culture of resilience: Construct validation. *International Journal of Emergency Mental Health, 15*(2), 123–128. PMID:24558699

Expansion. (2017). *Todo sobre el sismo que sacudió el centro del país* [Everything about the earthquake that shook the center of the country]. https://expansion.mx/nacional/2017/09/19/sismo-en-mexico-deja-decenas-de-muertos#uuid00000164-01aa-db07-a166-3fab29f7002d

Ezati, B. A. (2014). Addressing Pedagogical Training Needs of Teaching Staff: Lessons from Makerere University Short Professional Development Programs 2006 – 2010. *American Journal of Educational Research, 2*(12), 1190–1198. doi:10.12691/education-2-12-9

Fairhurst, G. T. (2007). *Discursive leadership: In conversation with leadership psychology.* Sage Publications, Inc. doi:10.4135/9781452231051

Fairhurst, G. T., & Putnam, L. (2004, February 1). Organizations as Discursive Constructions. *Communication Theory, 14*(1), 5–26. doi:10.1111/j.1468-2885.2004.tb00301.x

Fan, F., Zhang, Y., Yang, Y., Mo, L., & Liu, X. (2011). Symptoms of posttraumatic stress disorder, depression, and anxiety among adolescents following the 2008 Wenchuan earthquake in China. *Journal of Traumatic Stress, 24*(1), 44–53. doi:10.1002/jts.20599 PMID:21351164

Faria, J. A., & Nóvoa, H. (2020). Digital transformation at the University of Porto. *Proceedings of the International Conference on Exploring Services Science.*

Farley, A., Jain, A., & Thomson, D. (2011). Blended learning in finance: Comparing student perceptions of lectures, tutorials and online learning environments across different year levels. *Economic Papers*, *30*(1), 99–108. doi:10.1111/j.1759-3441.2010.00094.x

Farnell, T., Skledar Matijević, A., & Šćukanec Schmidt, N. (2021). *The impact of COVID-19 on higher education.* European Commission. https://data.europa.eu/doi/10.2766/069216

Fataar, A., & Norodien-Fataar, N. (2021). Towards an e-learning ecologies approach to pedagogy in a post-COVID world. *Journal of Education*, *84*(84), 155–168. doi:10.17159/2520-9868/i84a08

Federal Emergency Management Association. (2014, February). *Student manual: Chapter 3 effective communication.* https://training.fema.gov/emiweb/is/is242b/student%20manual/sm_03.pdf

Federal Emergency Management Association. (2023a). *SLF 101: Guide for all hazard emergency operations planning.* https://www.fema.gov/pdf/plan/glo.pdf

Federal Emergency Management Association. (2023b). *Unit four: Emergency management in the United States.* https://training.fema.gov/emiweb/downloads/is111_unit%204.pdf

Festinger, L. (1962). Cognitive dissonance. *Scientific American*, *207*(4), 93–106. doi:10.1038/scientificamerican1062-93 PMID:13892642

Filmer, L. B., & Ranse, J. (2013). Who is my leader? A case study from a hospital disaster scenario in a less developed country. *Australasian Emergency Nursing Journal*, *16*(4), 170–174. doi:10.1016/j.aenj.2013.08.004 PMID:24199902

Fisher, S., Brodie, J., & Pollman, P. (2021). *Serving all students: A survey of US learner mindsets from age 16 to 65+.* Accenture. https://www.accenture.com/content/dam/accenture/final/a-com-migration/r3-3/pdf/pdf-165/accenture-serving-all-students.pdf

Fisher, A., & Dimmock, M. (2017). *Ethics for A-Level* (1st ed.). Open Book Publishers. https://www.jstor.org/stable/j.ctt1wc7r6j

Fisher, J. L. (1984). *Power of the president.* Macmillan.

Flannery, M. E. (2023, March 29). *The mental health crisis on college campuses.* NEA Today. https://www.nea.org/nea-today/all-news-articles/mental-health-crisis-college-campuses

Flat. (2023). *19S: Edificios dañados por el sismo de 2017 ¿Qué ha ocurrido con ellos?* [19S: Buildings damaged by the 2017 earthquake, what has happened to them?]. https://www.flat.mx/elliving/19-de-septiembre-los-edificios-danados-por-el-sismo-de-2017-que-ha-ocurrido-con-ellos

Fleaca, E. (2011). Embedding digital teaching and learning practices in the modernisation of higher education institutions. *Proceedings of the SGEM2017 International Multidisciplinary Scientific Geo Conference: SGEM.* 10.5593/sgem2017/54/S22.006

Fletcher, D., & Sarkar, M. (2016). Mental fortitude training: An evidence-based approach to developing psychological resilience for sustained success. *Journal of Sport Psychology in Action*, *7*(3), 135–157. doi:10.1080/21520704.2016.1255496

Floridi, L., & Sanders, J. W. (2023). Artificial evil and the foundation of computer ethics. *Ethics and Information Technology*, *3*(1), 55–66. doi:10.1023/A:1011440125207

Folke, C. (2006). The emergence of a perspective for social-ecological systems analyses. *Global Environmental Change*, *16*(3), 253–267. doi:10.1016/j.gloenvcha.2006.04.002

Folk, R. (1993). The making of global citizenship. In J. Brecher, J. B. Childs, & J. Cutler (Eds.), *Global visions: Beyond the new world order* (pp. 39–52). South End Press.

Forrester, N. (2023). Fed up and burnt out: 'quiet quitting' hits academia *Nature, 615*, 751-753. doi:10.1038/d41586-023-00633-w

Fortunato, J. A. (2008). Restoring a reputation: The Duke University lacrosse scandal. *Public Relations Review, 34*(2), 116–123. doi:10.1016/j.pubrev.2008.03.006

Fraenkel, J. R., Wallen, N. E., & Hyun, H. H. (2012). *How to design and evaluate research in education.* McGraw-Hill.

Freed, A., Syed, T., & Burgmann, R. (2007). Evolution of stress southern California for the past 200 years from coseismic, postseismic and interseismic processes. *Geophysical Journal International, 169*(3), 1164–1179. doi:10.1111/j.1365-246X.2007.03391.x

Frick, L. (2021). Doctoral Creativity as an Epistemological Force in Saving and/or Destroying the World. *The PhD at the End of the World: Provocations for the Doctorate and a Future Contested*, 143-153.

Frykberg, E. R. (2005). Triage: Principles and practice. *Scandinavian Journal of Surgery, 94*(4), 272–278. doi:10.1177/145749690509400405 PMID:16425622

Gaetz, R. (2023, December 14). *Serving the digital student.* LinkedIn. https://www.linkedin.com/pulse/serving-digital-student-ryan-gaetz-vee4e

Gaidi, F. (2020). *The Impact of COVID-19 on Universities in Africa.* ACCORD. https://www.accord.org.za/analysis/the-impact-of-covid-19-on-universities-in-africa/

Galli, B. J. (2018). Change management models: A comparative analysis and concerns. *IEEE Engineering Management Review, 46*(3), 124–132. doi:10.1109/EMR.2018.2866860

Gama, J. A. P. (2018). Intelligent educational dual architecture for university digital transformation. *Proceedings of the 2018 IEEE Frontiers in Education Conference (FIE),* 1–9. 10.1109/FIE.2018.8658844

Gao, L., Qunhong, W., Li, D. D., Hao, Y., Cui, Y., Mingli, J., ... Sun, H. (2018). How prepared are hospital's emergency management capacity? Factors influencing efficiency of disaster rescue. *Disaster Medicine and Public Health Preparedness, 12*(2), 176–183. doi:10.1017/dmp.2016.25 PMID:27189920

García, J. E. (2021, January 26). *15 Principles of Crisis Management.* Retrieved October 28, 2023, from www.linkedin.com: https://www.linkedin.com/pulse/15-principles-crisis-management-jorge-eduardo-garc%C3%ADa

GEER (Geotechnical Extreme Events Reconnaissance Association). (2017). *Geotechnical engineering reconnaissance of the 19 September 2017 Mw 7.1 Puebla-Mexico City earthquake, version 2.0.* Report No. GEER 055-B. https://learningfromearthquakes.org/2017-09-19-puebla-mexico/images/2017_09_19_Puebla_Mexico/pdfs/UNAM_GEER_Sept2017_v2_Final.pdf

Gibbons, M., & Nowotny, H. (2001). The potential of transdisciplinarity. In *Transdisciplinarity: joint problem solving among science, technology, and society* (pp. 67–80). Birkhäuser. doi:10.1007/978-3-0348-8419-8_7

Gibbs, L., Jehangir, H., Kwong, E. J. L., & Little, A. (2022). Universities and multiple disaster scenarios: A transformative framework for disaster resilient universities. *International Journal of Disaster Risk Reduction, 78*, 103132. doi:10.1016/j.ijdrr.2022.103132

Gigliotti, R. A. (2022, February 24). *Crisis Leadership in Higher Education: Historical Overview, Organizational Considerations, and Implications*. Retrieved September 30, 2023, from https://oxfordre.com/politics/display/10.1093/acrefore/9780190228637.001.0001/acrefore-9780190228637-e-2045

Gigliotti, R. A. (2019). *Crisis leadership in higher education: Theory and practice*. Rutgers University Press.

Gigliotti, R. A., & Fortunato, J. A. (2014 in press). Crisis leadership: Upholding institutional values. In B. D. Ruben, R. De Lisi, & R. A. Gigliotti (Eds.), *A guide for leaders in higher education: Core concepts, competencies, and tools. Sterling*. Stylus.

Gillani, A. H., Li, S., Akbar, J., Omer, S., Fatima, B., Ibrahim, M. I. M., & Fang, Y. (2021). How prepared are the health care professionals for disaster medicine management? An insight from Pakistan. *International Journal of Environmental Research and Public Health*, *19*(1), 200. doi:10.3390/ijerph19010200 PMID:35010475

Giustiniano, L., Clegg, S. R., Cunha, M. P., & Rego, A. (2018). *Introduction to Theories of Organisational Resilience*. Edward Elgar Publishing. doi:10.4337/9781786437044

Goddard, S., Sheppard, M., Thompson, K., & Konecny, L. (2018). Disaster preparedness knowledge, beliefs, risk-perceptions, and mitigating factors of disaster preparedness behaviors of undergraduate students at a large Midwest university. *Journal of Public Health Issues and Practices*, *1*(2), 115. doi:10.33790/jphip1100115

Godínez-Domínguez, E. A., Tena-Colunga, A., Pérez-Rocha, L. E., Archundia-Aranda, H. I., Gómez-Bernal, A., Ruiz-Torres, R. P., & Escamilla-Cruz, J. L. (2021). The September 7, 2017 Tehuantepec, Mexico, earthquake: Damage assessment in masonry structures for housing. *International Journal of Disaster Risk Reduction*, *56*, 102123. doi:10.1016/j.ijdrr.2021.102123

Gogoi, N., Saikai, D., & Neog, P. (2021). Influence of COVID-19 pandemic on clothing interests of working women. *International Journal of Current Microbiology and Applied Sciences*, *10*(01), 501–516. doi:10.20546/ijcmas.2021.1001.062

Goh, M. (2021, June 28). *Identify the Stages of a Crisis*. Retrieved October 27, 2023, from https://blog.bcm-institute.org/crisis-management/identify-the-stages-of-a-crisis

Gokmenoglu, T., Sonmez, E. D., Yavuz, I., & Gok, A. (2021). Turkish Ministry of National Education school-based disaster education program: A preliminary results of the program evaluation. *International Journal of Disaster Risk Reduction*, *52*, 101943. doi:10.1016/j.ijdrr.2020.101943

Gokmenoglu, T., Yavuz, I., & Sensin, C. (2023). Exploring the interplay between curriculum and textbooks in disaster risk reduction education: Insights and implications. *International Journal of Disaster Risk Reduction*, *96*, 103949. doi:10.1016/j.ijdrr.2023.103949

Goldenberg, E. (2022, April 4). *How To Lead Through A Crisis*. Retrieved October 30, 2023, from https://www.jobillico.com/blog/en/how-to-lead-through-a-crisis/

Goleman, D. (1996). Emotional intelligence. Why it can matter more than IQ. *Learning*, *24*(6), 49–50.

Gonzales, A. L. (2016). The contemporary US digital divide: From initial access to technology maintenance. *Information Communication and Society*, *19*(2), 234–248. doi:10.1080/1369118X.2015.1050438

Goodenough, A. E. H., Roberts, D. M., Biggs, J. G., Derounian, A., Hart, G., & Lynch, A. (2020). A higher degree of resilience: Using psychometric testing to reveal the benefits of university internship placements. *Active Learning in Higher Education*, *21*(2), 102–115. doi:10.1177/1469787417747057

Goodwin, L. D. (1999). The role of factor analysis in the estimation of construct validity. *Measurement in Physical Education and Exercise Science*, *3*(2), 85–100. doi:10.1207/s15327841mpee0302_2

Goralnik, L., & Marcus, S. (2020). Resilient learners, learning resilience: Contemplative practice in the sustainability classroom. *New Directions for Teaching and Learning*, *2020*(161), 83–99. doi:10.1002/tl.20375

Gortner, E., & Pennebaker, J. W. (2003). The archival anatomy of a disaster: Media coverage and community-wide health effects of the Texas A&M bonfire tragedy. *Journal of Social and Clinical Psychology*, *22*(5), 580–603. doi:10.1521/jscp.22.5.580.22923

Gosling, S. D., & Mason, W. (2015). Internet research in psychology. *Annual Review of Psychology*, *66*(1), 877–902. doi:10.1146/annurev-psych-010814-015321 PMID:25251483

Gothe-Snape, J. (2018). *Record number of international students sticking around on visas with full work rights.* ABC News. https://www.abc.net.au/news/2018-07-27/temporary-graduate-visa-485-boom/10035390

Göttlichová, M. (2020). Higher Education and Innovation in the Non-profit Sector. In Strategic Innovative Marketing and Tourism: 8th ICSIMAT, Northern Aegean, Greece, 2019 (pp. 331-340). Springer International Publishing. doi:10.1007/978-3-030-36126-6_36

Graff, H., & Von Wehrden, H. (2021). Discourses of Boundary Crossing. *Beyond Interdisciplinarity: Boundary Work, Communication, and Collaboration*, 36.

Graham, T. (2023). *A tale of two cities: a month after Hurricane Otis, Acapulco exposes gaps in disaster response*. https://www.theguardian.com/global-development/2023/nov/29/hurricane-otis-mexico-acapulco-disaster-response

Grange, L. L. (2020). Could the COVID-19 pandemic accelerate the uberfication of the University. *South African Journal of Higher Education*, *34*(4), 1–10. doi:10.20853/34-4-4071

Greco, J. (2020). *Knowledge-Producing Abilities*. Cambridge University Press. doi:10.1017/9781108666404.006

Groves, S & Kasangaki, A. (2011). *Makerere University College of Health Sciences' role in addressing challenges in health service provision at Mulago National Referral Hospital.* Academic Press.

Guidi, E., Jensen, T., & Marinoni, G. (2023). *Shaping teaching & learning and internationalization beyond the pandemic.* International Association of Universities. https://www.iau-aiu.net/IMG/pdf/final__shaping_teaching_learning_and_internationalization_beyond_the_pandemic_2_1_.pdf

Gurung, B., & Rutledge, D. (2014). Digital learners and the overlapping of their personal and educational digital engagement. *Computers & Education*, *77*, 91–100. doi:10.1016/j.compedu.2014.04.012

Guti, O., & Sibanda, S. (2023). Prospects and Challenges of the WFH practices: A case of Zimbabwe Open University. *International Journal of Research Publications.*, *117*(1), 4–17. doi:10.47119/IJRP1001171120234419

Guvercin, A. (2006). *Examining the depression levels of the mothers and their children who see an earthquake according to some variables*. Retrieved from https://www.proquest.com/dissertations-theses/depremzede-anneler-ve-çocuklarının-depresyon/docview/2564166661/se-2.

Gyamerah, K. (2020). The impacts of COVID-19 on basic education: How can Ghana respond, cope, and plan for recovery. *School of Education.* https://schoolofeducation.blogs.bristol.ac.uk/2020/03/31/the-impacts-of-covid-19-on-basiceducation-how-can-ghana-respond-cope-and-plan-for-recovery/

Habermas, J. (1974). On social identity. *Telos*, *1974*(19), 91–103. doi:10.3817/0374019091

Hacioglu, M., Aker, T., Kutlar, T., & Yaman, M. (2002). Deprem tipi travma sonrasinda gelisen travma sonrasi stres bozuklugu belirtileri alt tipleri. *Dusunen Adam The Journal of Psychiatry and Neurological Sciences*, *15*, 4–15.

Haddow, G. D., Bullock, J. A., & Coppola, D. P. (2014). *Emergency management*. Elsevier.

Hager-Zanker, A. (2008). *Measuring the performance of geosimulation models by map comparison* [Doctoral Thesis]. Maastricht University. doi:10.26481/dis.20081210ah

Hancock, S. (2019). A future in the knowledge economy? Analysing the career strategies of doctoral scientists through the principles of game theory. *Higher Education*, *78*(1), 33–49. doi:10.1007/s10734-018-0329-z

Handke, S. (2023). *Accreditation agencies in the European higher education area: Nonprofit business models, competition and survival. New horizons in nonprofit research series*. Edward Elgar Publishing Limited. doi:10.4337/9781800881259

Han, H., & Johnson, S. D. (2012). Relationship between students' emotional intelligence, social bond, and interactions in online learning. *Journal of Educational Technology & Society*, *15*(1), 78–89.

Harms, P. D., Credé, M., Tynan, M., Leon, M., & Jeung, W. (2017). Leadership and Stress: A Meta-Analytic Review. *The Leadership Quarterly*, *28*(1), 178–194. doi:10.1016/j.leaqua.2016.10.006

Harpur, J. (2010). *Innovation, profit and the common good in higher education*. Palgrave Macmillan. doi:10.1057/9780230274624

Harrison, A. (2023). *Hurricane Otis cuts off Mexico's battered Acapulco*. https://phys.org/news/2023-10-hurricane-otis-weakens-landfall-mexico.html

Hart, D., Diercks, G., & Powell, A. (2009). Exploring stakeholder engagement in impact evaluation planning in educational development work. *Evaluation*, *15*(3), 285–306. doi:10.1177/1356389009105882

Harvey, L., & Green, D. (1993). Defining Quality. *Assessment & Evaluation in Higher Education*, *18*(1), 9–34. doi:10.1080/0260293930180102

Hasan, M. K., Uddin, H., & Younos, T. B. (2022). Bangladeshi nursing students' perceived preparedness and readiness for disaster management. *International Journal of Disaster Risk Reduction*, *81*, 103303. doi:10.1016/j.ijdrr.2022.103303

Hayes, A. (2022, March 22). *Crisis Management: Definition, How It Works, Types, and Example*. Retrieved October 8, 2023, from https://www.investopedia.com/terms/c/crisis-management.asp

Hayes, C., & Smith, P. (2020). Defining and framing epistemology. In *Beyond Disciplinarity* (pp. 6–22). Routledge.

Head, M. (2022). The impact of COVID-19 on mental health: A systematic review. *Journal of Clinical Psychology*, *78*(1), 1–34. doi:10.1002/jclp.23087

Healthy Minds Network. (2023). *The Healthy Minds Study: Student survey*. https://healthymindsnetwork.org/hms/

Heard, C. L., Pearce, J. M., & Rogers, M. B. (2020). Mapping the public first-aid training landscape: A scoping review. *Disasters*, *44*(1), 205–228. doi:10.1111/disa.12406 PMID:31524986

Heath, R. L., & Millar, D. P. (2004). A rhetorical approach to crisis communication: Management, communication processes, and strategic responses. In D. P. Millar & R. L. Heath (Eds.), *Responding to crisis: A rhetorical approach to crisis communication* (pp. 1–17). Erlbaum.

Hebebci, M. T. (2023). *The impact of natural disasters on education*. International Conference on Social Science (ICon-SoS), Mararis, Turkey.

Heeks, R., & Ospina, A. V. (2019). Conceptualising the link between information systems and resilience: A developing country field study. *Information Systems Journal*, *29*(1), 70–96. doi:10.1111/isj.12177

Held, D., & Moore, H. L. (2007). Introduction: Cultural futures. In V. Rao & M. Walton (Eds.), *The capacity to aspire: Culture and the Terms of Recognition* (pp. 1–16). Oneworld Publications.

Heleta, S. (2015). *How to rebuild higher education in countries torn apart by war.* The Conversation. Retrieved from https://theconversation.com/how-to-rebuild-higher-education-in-countries-torn-apart-by-war-48791

Heleta, S. (2020). Higher Education in Post-Conflict Settings. In The International Encyclopedia of Higher Education Systems and Institutions. Springer. doi:10.1007/978-94-017-8905-9_220

Henig, D., & Knight, D. M. (2023). Polycrisis: Prompts for an emerging worldview. *Anthropology Today*, *39*(2), 3–6. doi:10.1111/1467-8322.12793

Herbst, P., Crespo, S., Matthews, P. G., & Lichtenstein, E. K. (2021). Dissertating through disruptions: COVID-19 and the need for a research infrastructure. *Journal for Research in Mathematics Education*, *52*(2), 110–116. doi:10.5951/jresematheduc-2020-0300

Hernández-Campos, E. (2023). *Deja huracán Otis daños en 34 planteles de la UAGro* [Hurricane Otis leaves damage to 34 UAGro faculties]. https://www.elsoldeacapulco.com.mx/local/deja-huracan-otis-danos-en-34-planteles-de-la-uagro-10954283.html

Hima, H. (2023). Learning on the move: Resetting the agenda for education and learning in conflict-affected settings. *International Journal of Educational Development*, *83*, 102413. doi:10.1016/j

HKUST. (n.d.). *Mission & Vision.* Retrieved October 1, 2023, from https://hkust.edu.hk/about/mission-vision

Hnydiuk, I., Hataullina, E., Marshuk, L., Timoshenko, N., & Cherneha, V. (2023). Formation of the system of financial support of higher education institutions in Ukraine. *Amazonia Investiga*, *12*(65), 141–148. doi:10.34069/AI/2023.65.05.14

Hoadley, U. (2020). *Schools in the time of COVID-19: Impacts of the pandemic on curriculum.* Resep Non-Economic Working Paper, Stellenbosch University.

Hoch, J. E., Bommer, W. H., Dulebohn, J. H., & Wu, D. (2018). Do Ethical, Authentic, and Servant Leadership Explain Variance Above and Beyond Transformational Leadership? A Meta-Analysis. *Journal of Management*, *44*(2), 501–529. doi:10.1177/0149206316665461

Hodges, C., Moore, S., Lockee, B., Trust, T., & Bond, A. (2020, March 27). *The difference between emergency remote teaching and online learning.* https://er.educause.edu/articles/2020/3/the-difference-between-emergency-remote-teaching-and-online-learning

Hogarth, R. M. (1987). *Judgement and choice: the psychology of decision* (2nd ed.). Wiley.

Honda, Y., Fujiwara, T., Yagi, J., Homma, H., Mashiko, H., Nagao, K., Okuyama, M., Ono-Kihara, M., & Kihara, M. (2019). Long-term impact of parental post-traumatic stress disorder symptoms on mental health of their offspring after the great east japan earthquake. *Frontiers in Psychiatry*, *10*, 496. Advance online publication. doi:10.3389/fpsyt.2019.00496 PMID:31404309

Hook, S. (2008). *John Dewey: an intellectual portrait.* Cosimo, Inc.

Hosek, V. A. (2018). *Locating the critical component in technological, pedagogical, and content knowledge (TPACK): An examination of how graduate students recruit TPCAK and critical digital literacy into classroom practices* [Doctoral Dissertation, Illinois State University]. ProQuest.

Hosek, V.A. (n.d.). *It's personal: Examining and meeting the changing needs of students in online learning environments.* Pathways to Research in Education, EBSCO.

Hosek, V. A., & Percell, J. C. (2022). Reaching Critical Depths: Developing Teacher Candidates' Critical Pedagogy Online. In E. Mikulec & T. Ramalho (Eds.), *Best Practices in Teaching Critical Pedagogy Online.* DIO Press, Inc.

Houchin, K., & MacLean, D. (2005). Complexity theory and strategic change: An empirically informed critique. *British Journal of Management, 16*(2), 149–166. doi:10.1111/j.1467-8551.2005.00427.x

Howe, A., Smajdor, A., & Stokl, A. (2012). Towards an understanding of resilience and its relevance to medical training. *Medical Education, 46*(4), 349–356. doi:10.1111/j.1365-2923.2011.04188.x PMID:22429170

Huang, C., Li, N., Zhang, Z., & Liu, Y. (2024). Examining the relationship between meteorological disaster economic impact and regional economic development in China. *International Journal of Disaster Risk Reduction, 100*, 104133. doi:10.1016/j.ijdrr.2023.104133

Huang, R., & Turner, R. (2018). International experience, universities support and graduate employability – perceptions of Chinese international students studying in UK universities. *Journal of Education and Work, 31*(2), 175–189. doi:10.1080/13639080.2018.1436751

Hubbart, J. A. (2023). Organizational Change: The Challenge of Change Aversion. *Administrative Sciences, 13*(7), 162. doi:10.3390/admsci13070162

Hu, K., Godfrey, K., Ren, Q., Wang, S., Yang, X., & Li, Q. (2022). The impact of the COVID-19 pandemic on college students in the USA: Two years later. *Psychiatry Research, 315*, 114685. Advance online publication. doi:10.1016/j.psychres.2022.114685 PMID:35872401

Hunter, W. J., & Benson, G. D. (1997). Arrows in time: The misapplication of chaos theory to education. *Journal of Curriculum Studies, 29*(1), 87–100. doi:10.1080/002202797184215

Hurnen, F., & Mcclure, J. (1997). The effect of increased earthquake knowledge on perceived preventability of earthquake damage. *Australasian Journal of Disaster and Trauma Studies*, 3.

Hussey, D. E. (2000). *How to Manage Organizational Change* (Vol. 28). Kogan Page Publishers.

Iizuka, A. (2022). Disaster perception and preparation among foreign versus local university students in Japan: A comparative study. *Progress in Disaster Science, 15*, 100239. doi:10.1016/j.pdisas.2022.100239

Ilgaz, H., & Gülbahar, Y. (2015). A snapshot of online learners: e-Readiness, e-Satisfaction and expectations. *International Review of Research in Open and Distance Learning, 16*(2), 171–187. doi:10.19173/irrodl.v16i2.2117

Iliško, D. (2021). Towards sustainability as a frame of mind in higher education: Thinking about sustainability rhizomatically. Handbook on teaching and learning for sustainable development, 366-375.

Ingrassia, P. L., Foletti, M., Djalali, A., Scarone, P., Ragazzoni, L., Della Corte, F., ... Fisher, P. (2014). Education and training initiatives for crisis management in the European Union: A web-based analysis of available programs. *Prehospital and Disaster Medicine, 29*(2), 115–126. doi:10.1017/S1049023X14000235 PMID:24642198

Institute for Public Relations. (2007, October 30). *Crisis Management and Communications.* Retrieved October 23, 2023, from https://instituteforpr.org/crisis-management-and-communications/

International Association of Emergency Managers. (n.d.) *IAEM-USA Universities & Colleges Caucus.* https://www.iaem.org/groups/us-caucuses/universities-colleges

International Committee of the Red Cross (ICRC). (2008). *How is the Term "Armed Conflict" Define in International Humanitarian Law?* Opinion Paper, March.

International Red Cross and Red Crescent Movement. (n.d.) *Disaster preparedness.* https://www.ifrc.org/our-work/disasters-climate-and-crises/disaster-preparedness

Isaacson, W. (2011). *Steve Jobs.* Simon & Schuster.

Islam, M. T. (2019). (Re)searching for the development of a conceptual model of education for citizenship in the context of young people's globalised mobility in higher education. *Globalisation, Societies and Education*, *17*(2), 194–207. doi:10.1080/14767724.2019.1583093

Islam, M. T. (2021). Cultural cultivation for the mobility of Bangladeshi higher education students in Australia. *Higher Education Research & Development*, *41*(5), 1572–1585. doi:10.1080/07294360.2021.1902950

Islam, M. T. (2022). Media mediation on academic advancement for youths' international mobility pursuing higher education. In S. Bansal, V. Ahuja, V. Chaturvedi, & V. Jain (Eds.), *Multidisciplinary Perspectives Towards Building a Digitally Competent Society* (pp. 208–229). IGI Global. doi:10.4018/978-1-6684-5274-5.ch012

Islam, M. T. (2023a). (Un)employability and (un)employment issues towards international mobility for higher education. In P. O. de Pablos, X. Zhang, & M. N. Almunawar (Eds.), *Education Institutions, Skills, and Jobs in the Digital Era: Toward a More Inclusive and Resilient Society*. IGI Global.

Islam, M. T. (2023b). Bangladeshi students' family fertilization for pursuing higher education in Australia. In G. Morris & L. Li (Eds.), *Handbook of Research on Developments and Future Trends in Transnational Higher Education* (pp. 247–264). IGI Global. doi:10.4018/978-1-6684-5226-4.ch013

Islam, M. T. (2023c). Bauman or Appadurai? In theorising mobilities of the Southern higher education students in the North. In K. Jermsittiparsert, P. Suanpang, A. Nuemaihom, & P. Siriattaku (Eds.), *Handbook of Research on Current Trends in Asian Education, Psychology, and Sociology* (pp. 1–25). IGI Global.

Islam, M. T. (2023d). International students' perception and experience of higher education for global citizenship. In J. Keengwe (Ed.), *Handbook of Research on Innovative Digital Practices and Globalization in Higher Education*. IGI Global. doi:10.4018/978-1-6684-6339-0.ch015

Israelit, M. (1996). Nathan Rosen: 1909–1995. *Foundations of Physics Letters*, *9*(2), 105–108. doi:10.1007/BF02186252

Izumi, T., Sukhwani, V., Surjan, A., & Shaw, R. (2021). Managing and responding to pandemics in higher educational institutions: Initial learning from COVID-19. *International Journal of Disaster Resilience in the Built Environment*, *12*(1), 51–66. doi:10.1108/IJDRBE-06-2020-0054

Jacobs, G., van Witteloostuijn, A., & Christe-Zeyse, J. (2013). A theoretical framework of organizational change. *Journal of Organizational Change Management*, *26*(5), 772–792. doi:10.1108/JOCM-09-2012-0137

Jafar, H. (2021). *Internationalization of Higher Education and Imported Universities in Post-conflict Iraq through a Policy Borrowing Lens*. University of Toronto.

Jafar, H., & Sabzalieva, E. (2022). Faculty Experiences of Higher Education Internationalization in Post-Conflict Iraq and Tajikistan. *Journal of Comparative and International Higher Education*, *14*(2), 47–65. doi:10.32674/jcihe.v14i2.3453

Jameel, A. S., & Ahmad, A. R. (2020). Factors impacting research productivity of academic staff at the Iraqi higher education system. *International Business Education Journal*, *13*(1), 108–126. doi:10.37134/ibej.vol13.1.9.2020

Jansen, J. (2020). Data or bread? A policy analysis of student experiences of learning under lockdown. *Southern African Review of Education with Education with Production*, *26*(1), 167–181.

Jensen-Pennington, H. (2021). Envisioning Higher Education for the Future. In H. van't Land, A. Corcoran, & D. C. Iancu (Eds.), *The Promise of Higher Education*. Springer. doi:10.1007/978-3-030-67245-4_58

Jeoloji Mühendisleri Odası, T. M. M. O. B. (2021). Fay üzerinde yaşayan illerimiz: Hatay raporu-6. *Jeodergi*, *21*(1), 1–10.

Jesry, M., Omar, F. A., Rashwani, A., Bark, I., Jammo, K., Ajam, S., & Kassab, Z. (2022). Exploring the value of a risk-management quality-assurance model to support delivery of quality higher education in the conflict-affected northwest of Syria. *International Journal of Educational Research Open*, *3*, 100134. doi:10.1016/j.ijedro.2022.100134

Jetten, W. D., Seesink, J., & Klimek, M. (2022). Prehospital triage by lay person first responders: A scoping review and proposal for a new prehospital triage tool. *Disaster Medicine and Public Health Preparedness*, *16*(3), 1242–1252. doi:10.1017/dmp.2021.19 PMID:33827731

Jiansheng, W., Yuhao, Z., Tian, Y., & Bikai, C. (2020). Evolution of typhoon disasters characteristics and non-structural disaster avoidance measures in the China coastal main functional area. *International Journal of Disaster Risk Reduction*, *46*, 101490. doi:10.1016/j.ijdrr.2020.101490

Johansson, L., & Pihl, L. (2023). *Crisis Management in Theory and Practice: How and Why Organizations Work the Way They Do*. Jönköping International Business School.

Johnson, B. (n.d.). *The Great Plague 1665-the Black Death*. https://www.historic-uk.com/HistoryUK/HistoryofEngland/The-Great-Plague/

Johnson, A. T. (2013). University agency in peacebuilding: Perspectives on conflict and development in Kenya. *Prospects*, *43*(3), 329–345. doi:10.1007/s11125-013-9276-8

Joksimović, S., Gašević, D., Kovanović, V., Riecke, B. E., & Hatala, M. (2015). Social presence in online discussions as a process predictor of academic performance. *Journal of Computer Assisted Learning*, *31*(6), 638–654. doi:10.1111/jcal.12107

Jones, G., & Pund, L. (2010). *Leadership and Management in the Early Years*. CPI Antny Rowe.

Jovanovic, O., Pesikan, A., & Antic, S. (2022). Resilience of higher education institutions in Serbia: A student's perspective. *Digital and Psychological Resilience*, *457-461*, 457–461. Advance online publication. doi:10.46793/TIE22.457J

Judge, T. A., & Piccolo, R. F. (2004). Transformational and Transactional Leadership: A Meta- Analytic Test of Their Relative Validity. *The Journal of Applied Psychology*, *89*(5), 755–768. doi:10.1037/0021-9010.89.5.755 PMID:15506858

Juneja, P. (n.d.). *Crisis Management - Meaning, Need and its Features*. Retrieved October 10, 2023, from https://www.managementstudyguide.com/crisis-management.htm

Jung, I., & Lee, J. (2020). A cross-cultural approach to the adoption of open educational resources in higher education. *British Journal of Educational Technology*, *51*(1), 263–280. doi:10.1111/bjet.12820

Justino, P. (2010). How Does Violent Conflict Impact on Individual Educational Outcomes, The Evidence So Far. UNESCO. *Education for All Global Monitoring Report*, *2011*, 2.

Kachintorn, U. (2020). *University Transformation in Response to the Changes in Disruption Era*. Mahidol University.

Kadioglu, R. (2022). *Relationships among death anxiety, big five personality traits, and loneliness levels in young adults* [Unpublished Master's Thesis]. Necmettin Erbakan University, Konya.

Kagawa, F., & Selby, D. (2014). *Disaster risk reduction in the school curriculum: the present and potental role of development agencies and the implications for the Hyogo Framework for Action 2005-2015 successor.* Unpublished Report commissioned by PLAN International. https://www.preventionweb.net/english/hyogo/gar/2015/en/bgdocs/inputs/Kagawa%20and%20Selby%2C%202014.pdf

Kahneman, D. (2013). Thinking, fast and slow. Farrar, Straus and Giroux.

Kahneman, D. (2011). *Thinking, fast and slow*. Macmillan.

Kahneman, D., Slovic, P., & Tversky, A. (1982). *Judgment under uncertainty: heuristics and biases*. Cambridge University Press. doi:10.1017/CBO9780511809477

Kalafat, D., Zülfikar, A. C., & Okuyan-Akcan, S. (2021). Seismicity of Turkey and real-time seismology Applications in determining earthquake hazard. *Academic Platform Journal of Natural Hazards and Disaster Management, 2*(2), 96–111. doi:10.52114/apjhad.1039670

Kamath, S. (2015). Role of mindfulness in building resilience and emotional intelligence and improving academic adjustment among adolescents. *Indian Journal of Health and Wellbeing, 6*(4), 414–417.

Kandilli Observatory. (2023a). http://www.koeri.boun.edu.tr/sismo/2/deprem-verileri/yillik-deprem-haritalari/2022-yili-deprem-harita-grafik-ve-tablolari/

Kandilli Observatory. (2023b). http://www.koeri.boun.edu.tr/sismo/2/wp-content/uploads/2023/02/20230206_1024_KAHRAMANMARAS.pdf/

Kantur, D., & Iseri-Say, A. (2012). Organisational resilience: A conceptual integrative framework. *Journal of Management & Organization, 18*(6), 762–773. doi:10.5172/jmo.2012.18.6.762

Kapucu, N., & Khosa, S. (2013). Disaster resiliency and culture of preparedness for university and college campuses. *Administration & Society, 45*(1), 3–27. doi:10.1177/0095399712471626

Karakus, U. (2013). The investigation of earthquake inexperienced and experienced students' perceptions by metaphor analysis. *Eastern Geographical Review, 18*(29), 97–116. doi:10.17295/ataunidcd.31309

Karanci, N., Aksit, B., Anafarta, M., Ogul, M., & Uner, G. (1999). *Let's Be Prepared Against Earthquakes*. Middle East University Disaster Management Application and Research Centre.

Karas, M., Sheen, N. J., North, R. V., Ryan, B., & Bullock, A. (2020). Continuing professional development requirements for UK health professionals: A scoping review. *BMJ Open, 10*(3), e032781. doi:10.1136/bmjopen-2019-032781 PMID:32161156

Karimi, M., & Mousavi Shafae, S. M. (2018). Poor Governance and Civil War in Syria. *Turkish Journal of Middle Eastern Studies, Vol, 5*(1), 49–71. doi:10.26513/tocd.383377

Kasapoglu, A., & Ecevit, M. (2001). *Sociological research of the earthquake*. Sosyoloji Dernegi.

Kawawaki, Y. (2023). Giving of time or giving money? An empirical analysis of nationwide prosocial behaviour in times of disaster. *International Journal of Disaster Risk Reduction, 96*, 103888. doi:10.1016/j.ijdrr.2023.103888

Kayyali, M. (2022). *Quality Assurance*. Internationalization, and Ranking of Higher Education Institutions in the Context of the Syrian War.

Kelly, S., Nicholson, J., Johnston, P., Duty, D., & Brennan, R. (2021). Experienced professionals and doctoral study: A performative agenda. *Industrial Marketing Management, 92*, 232–243. doi:10.1016/j.indmarman.2020.02.018

Kemp, N., & Grieve, R. (2014). Face-to-face or face-to-screen? Undergraduates' opinions and test performance in classroom vs. online learning. *Frontiers in Psychology, 5*. Advance online publication. doi:10.3389/fpsyg.2014.01278 PMID:25429276

Kester, K. (2020). *The United Nations and Higher Education: Peacebuilding, Social Justice and Global Cooperation for the 21st Century*. Information Age.

Kester, K. (2021). Toward a conflict-sensitive approach to higher education pedagogy: Lessons from Afghanistan and Somaliland. *Teaching in Higher Education*, 1–20. doi:10.1080/13562517.2021.2015754

Kester, K., Abura, M., Sohn, C., & Rho, E. (2022). Higher education peacebuilding in conflict-affected societies: Beyond the good/bad binary. *International Journal of Comparative Education and Development*, *24*(3/4), 160–176. doi:10.1108/IJCED-04-2022-0027

Khadiagala, G. M. (2021). The African Union in Peacebuilding in Africa. In T. McNamee & M. Muyangwa (Eds.), *The State of Peacebuilding in Africa*. Palgrave Macmillan. doi:10.1007/978-3-030-46636-7_12

Khorram-Manesh, A., Lupesco, O., Friedl, T., Arnim, G., Kaptan, K., Djalali, A. R., Foletti, M., Ingrasia, P. L., Ashkenazi, M., Arculeo, C., Fischer, P., Hreckovski, B., Komadina, R., Voigt, S., Carlström, E., & James, J. (2016). Education in disaster management: What do we offer and what do we need? Proposing a new global program. *Disaster Medicine and Public Health Preparedness*, *10*(6), 854–873. doi:10.1017/dmp.2016.88 PMID:27435533

Kilic, S. (2014). Effect size. *Journal of Mood Disorders*, *4*(1), 44–46. doi:10.5455/jmood.20140228012836

Kiltz, L., Trippenzee, M., Fleer, J., Fokkens-Bruinsma, M., & Jansen, E. P. W. A. (2023). Student well-being in times of COVID-19 in the Netherlands: Basic psychological need satisfaction and frustration within the academic learning environment. *European Journal of Psychology of Education*. Advance online publication. doi:10.1007/s10212-023-00680-x

Kim, D. H. (2016, December). Emergency preparedness and the development of health care coalitions. *Nursing Clinics of North America, 51*(4), 545-554. doi:10.1016/j.cnur.2016.07.013

King, P. M. & Kitchener, K. S. (2004a). *Judgment model: Twenty years of research on epistemic cognition*. Erlbaum.

King, P. M. & Kitchener, K. S. (2004b). *Reflective judgment: Theory and research on the development of epistemic assumptions through adulthood*. Taylor & Francis.

Kinlaw, C. R., Dunlap, L. L., & D'Angelo, J. A. (2012). Relations between faculty use of online academic resources and student class attendance. *Computers & Education*, *59*(2), 167–172. https://www.learntechlib.org/p/66746/. doi:10.1016/j.compedu.2011.12.028

Kivunga, M. A., Mwavali, A. E., & Otuya, P. A. (2017). Assessment of higher education institutions' man-made disaster preparedness Nairobi, Kenya. *African Multidisciplinary Journal of Research, 2*(1).

Klein, T. A., & Irizarry, L. (2023, July 17). *EMS disaster response*. StatPearls. https://www.ncbi.nlm.nih.gov/books/NBK560710/

Koç, N. (2023). The views of associate degree students from child development on distance education after the earthquake. *Journal of Educational Technology and Online Learning*, *6*(3), 756–770. doi:10.31681/jetol.1326200

Koh, D., Lee, K., & Joshi, K. (2019). Transformational Leadership and Creativity: A Meta- analytic Review and Identification of an Integrated Model. *Journal of Organizational Behavior*, *40*(6), 625–650. doi:10.1002/job.2355

Kohlrieser, G., Coombe, D., & Goldsworthy, S. (2012). *Care to Dare: Unleashing Astonishing Potential Through Secure Base Leadership*. Jossey-Bass.

Koklu, N., Buyukozturk, S., & Bokeoglu, C. O. (2007). *Statistic for Social Sciences*. Pegem.

Kolb, D. A. (1984). *Experiential Learning*. Prentice-Hall.

Koopman, O., & Koopman, K. (2020). The Rise of the University without Classrooms after COVID-19. *CSSALL, 1*, 153–168. doi:10.29086/978-0-9869936-1-9/2020/AASBS01

Korkmaz, H. (2006). Antakya'da zemin özellikleri ve deprem etkisi arasındaki ilişki. *Coğrafi Bilimler Dergisi*, *4*(2), 49–66.

Koroglu, A. (2018). Psychological traumas after natural disasters and sample psychological support education program. *Journal of Disaster and Risk*, *1*(1), 39–52. doi:10.35341/afet.412005

Koser, B. W., & Suchenski, M. (2022, September 19*). EMS evacuation triage.* StatPearls. https://www.ncbi.nlm.nih.gov/books/NBK537075/

Kousky, C. (2016). Impacts of natural disasters on children. *The Future of Children*, *26*(1), 73–92. doi:10.1353/foc.2016.0004

Krishnamoorthy, R., & Keating, K. (2021). Education Crisis, Workforce Preparedness, and COVID-19: Reflections and Recommendations. *American Journal of Economics and Sociology*, *80*(1), 253–274. doi:10.1111/ajes.12376 PMID:34230670

Kubler-Ross, E. (1993). On death and dying. Academic Press.

Kucukakin, P. M., Tasti, O. Y., Cobanoglu, R., & Gokmenoglu, T. (2024). The post-disaster response of a higher education institution in Turkey: Insights from students and instructors. *International Journal of Disaster Risk Reduction*, *102*, 104293. doi:10.1016/j.ijdrr.2024.104293

Kuhfeld, M., Sokand, J., Lewis, K., & Morton, E. (2022). *The pandemic has had devastating impacts on learning. What will it take to help students catch up?* https://www.brookings.edu/articles/the-pandemic-has-had-devastating-impacts-on-learning-what-will-it-take-to-help-students-catch-up/

Kumar, S., Adhish, V. S., & Deoki, N. (2014). Making sense of theories of leadership for capacity building. *Indian Journal of Community Medicine*, *39*(2), 82. doi:10.4103/0970-0218.132721 PMID:24963223

Kung, Y., & Chen, S. (2012). Perception of earthquake risk in Taiwan: Effects of gender and past earthquake experience. *Risk Analysis*, *32*(9), 1535–1546. doi:10.1111/j.1539-6924.2011.01760.x PMID:22300232

Kupczynski, L., Ice, P., Wiesenmayer, R., & McCluskey, F. (2010). Student perceptions of the relationship between indicators of teaching presence and success in online courses. *Journal of Interactive Online Learning*, *9*(1), 23–43. https://www.learntechlib.org/p/109406/

Kurt, E., & Gulbahce, A. (2019). Van depremini yaşayan öğrencilerin travma sonrası stres bozukluğu düzeylerinin incelenmesi. *Atatürk Üniversitesi Sosyal Bilimler Enstitüsü Dergisi*, *2*(1), 957–972.

Kuyok, K. A. (2017). Higher Education in South Sudan: Living with Challenges. *Industry and Higher Education*, (89), 16–18. doi:10.6017/ihe.2017.89.9839

Labosier, C. F. (2019). Bug out bags and first aid kits: Undergraduate college students' awareness, perceptions, preparedness, and behavior around severe weather. In Emerging Voices in Natural Hazards Research (pp. 1-25). Butterworth Heineman.

Labs, W. G. U. (2023, December 15). *From wicked problems to equitable solutions: WGU labs end of the year impact report.* https://www.wgulabs.org/posts/from-wicked-problems-to-equitable-solutions-end-of-year-report#learning-to-work-transition

Ladyshewsky, R. (2013). Instructor presence in online courses and student satisfaction. *International Journal for the Scholarship of Teaching and Learning*, *7*(1), 1–23. doi:10.20429/ijsotl.2013.070113

Lambach, D., Johais, E., & Bayer, M. (2015). Conceptualising state collapse: An institutionalist approach. *Third World Quarterly*, *36*(7), 1299–1315. doi:10.1080/01436597.2015.1038338

Lambrecht, J., Lenkeit, J., Hartman, A., Ehlert, A., Knigge, M., & Sporer, N. (2020). The effect of school leadership on implementing inclusive education: How transformational and instructional leadership practices affect individualised education planning. *International Journal of Inclusive Education.* Advance online publication. doi:10.1080/13603116.2020.1752825

Lammers, W. J., & Gillaspy, J. A. (2013). Brief Measure of Student-Instructor Rapport Predicts Student Success in Online Courses. *International Journal for the Scholarship of Teaching and Learning*, *7*(2), 1–13. doi:10.20429/ijsotl.2013.070216

Lange, R. (2010). Benchmarking, Rankings und Ratings. In D. Simon, A. Knie, & S. Hornbostel (Eds.), *Handbuch Wissenschaftspolitik* (pp. 322–333). VS Verlag für Sozialwissenschaften. doi:10.1007/978-3-531-91993-5_22

Laver-Fawcett, A., & Cox, D. L. (2021). *Principles of Assessment and Outcome Measurement for Allied Health Professionals: Practice, Research and Development.* John Wiley & Sons.

Le Roux, T. (2013). An exploration of the role of communication during the in-crisis situation. *Jamba (Potchefstroom, South Africa)*, *5*(2), 1–9. doi:10.4102/jamba.v5i2.67

Lee, A., Legood, A., Hughes, D., Tian, A. W., Newman, A., & Knight, C. (2020). Leadership, Creativity and Innovation: A Meta-Analytic Review. *European Journal of Work and Organizational Psychology*, *29*(1), 1–35. doi:10.1080/1359432X.2019.1661837

Lee, C. J., Allard, R. J., Adenjii, A. A., Quintanilla, N., & Kirsch, T. D. (2022). The National Disaster Medical System and military combat readiness: A scoping review. *The Journal of Trauma and Acute Care Surgery*, (2), 136–146. doi:10.1097/TA.0000000000003703 PMID:35605023

Lee, M. T. (2019). Promoting Human Flourishing Beyond Foundational Concerns. *Humanistic Management Journal*, *4*(2), 235–237. doi:10.1007/s41463-019-00065-7

Leishangthem, L. D. (2017, March). Political Instability and Its Influence on Higher Education: A Study of Students' Perceptions in Manipur University. *International Journal of Humanities & Social Science Studies*, *III*(V), 300–307. https://oaji.net/articles/2017/1115-1491480171.pdf

Lengnick-Hall, C. A., Beck, T. E., & Lengnick-Hall, M. L. (2011). Developing a capacity for organizational resilience through strategic human resource management. *Human Resource Management Review*, *21*(3), 243–255. doi:10.1016/j.hrmr.2010.07.001

Len-Rios. (2010). *Image repair strategies, local news portrayals and crisis stage: A case study of Duke University's lacrosse team crisis. International Journal of Strategic Communication.*

Levasseur, O., McDermott, M. R., & Lafreniere, K. D. (2015). The multidimensional mortality awareness measure and model: Development and validation of a new self-report questionnaire and psychological framework. *Omega*, *70*(3), 317–341. doi:10.1177/0030222815569440 PMID:26036058

Lim, D. (1999). Quality Assurance in Higher Education in Developing Countries. *Assessment & Evaluation in Higher Education*, *24*(4), 379–390. doi:10.1080/0260293990240402

Linnenluecke, M. K. (2017). Resilience in business and management research: A review of influential publications and a research agenda. *International Journal of Management Reviews*, *19*(1), 4–30. doi:10.1111/ijmr.12076

Lis Education Network. (2022, January 23). *Meaning, Purpose & Objectives of of Higher Education.* Retrieved October 27, 2023, from https://www.lisedunetwork.com/higher-education/

List, A. (2019). Defining digital literacy development: An examination of pre-service teachers' beliefs. *Computers & Education*, *138*, 146–158. doi:10.1016/j.compedu.2019.03.009

London Fire Brigade. (n.d.) *The Great Fire of London.* https://www.london-fire.gov.uk/museum/history-and-stories/the-great-fire-of-london/

London Premier Centre. (2023, August 7). *The role of proactive crisis management in protecting institutions and communities.* Retrieved October 30, 2023, from https://www.lpcentre.com/articles/the-role-of-proactive-crisis-management-in-protecting-institutions-and-communities

Lopez, J. (2017a). *Tras sufrir daños, transporte con suspensiones provisionales en CDMX* [After suffering damage, transportation with povisional suspensions in Mexico City]. https://www.excelsior.com.mx/comunidad/2017/09/24/1190480

Lopez, J. (2017b). *Tláhuac, Xochimilco e Iztapalapa viven escasez de agua tras sismo* [Tláhuac, Xochimilco and Iztapalapa experience water shortages after earthquake]. https://www.excelsior.com.mx/comunidad/2017/09/24/1190460

Lopez, J. (2017c). *Sismo provoca crisis de agua histórica en el Valle de México* [Earthquake causes historic water crisis in the Valley of Mexico]. https://www.excelsior.com.mx/comunidad/2017/10/05/1192854

Lott, M. K. (2012). *Crisis management plans in higher education: Commonalities, attributes, and perceived effectiveness* [Doctoral Dissertation]. Gallaudet University.

Lovekamp, W. E., & Tate, M. L. (2008). College student disaster risk, fear and preparedness. *International Journal of Mass Emergencies and Disasters*, *26*(2), 70–90. doi:10.1177/028072700802600201

Lowenthal, P., & Dunlap, J. (2018). Investigating students' perceptions of instructional strategies to establish social presence. *Distance Education*, *39*(3), 281–298. doi:10.1080/01587919.2018.1476844

Luchian, I. (2021). The impact of the covid-19 crisis on higher education. *Geopolitical. Social Security and Freedom Journal*, *4*(1), 15–26. doi:10.2478/gssfj-2021-0002

Lues, R., Padayachee, A., & Jager, H. (2020). *Universities of technology in the post-COVID-19 landscape.* https://www.universityworldnews.com/post.php?story=20200713153430109

Lundgren-Resenterra, M., & Kahn, P. E. (2019). The organisational impact of undertaking a professional doctorate: Forming critical leaders. *British Educational Research Journal*, *45*(2), 407–424. doi:10.1002/berj.3503

Luttrell, W. (Ed.). (2010). *Qualitative educational research: Readings in reflexive methodology and transformative practice.* Routledge.

Macfarlane, B. (2021). The neoliberal academic: Illustrating shifting academic norms in an age of hyper-performativity. *Educational Philosophy and Theory*, *53*(5), 459–468. doi:10.1080/00131857.2019.1684262

Mackworth-YoungC. R. S.ChingonoR.MavodzaC.McHughG.TemboM.Dziva ChikwariC. (2020). 'Here, we cannot practice what is preached': early qualitative learning from community perspectives on Zimbabwe's response to COVID-19. Bull World Health Organ. *E-pub: 20, 1-19* doi:10.2471/BLT.20.260224

Magaisa, E., Moyo, T., Wilson, E. C., Shava, G., & Makokoro, P. P. (2023). Campus and classroom-based higher education: The effects of covid-19 on teaching and learning in Zimbabwe. *International Journal of Research and Innovation in Social Science. International Journal of Research and Innovation in Social Science*, *7*(4), 1259–1266. doi:10.47772/IJRISS.2023.7504

Maghdid, R. S., Othman, B. A., & Omer, A. K. (2022). From Crisis to Crisis Management; Causes and Impacts of Crises in the Public Sector. *Polytechnic Journal of Humanities and Social Sciences*, *3*(1), 65–74. doi:10.25156/ptjhss.v3n1y2022.pp65-74

Mahmud, A., Mohammad, Z., & Abdullah, K. A. (2020). Leadership in disaster management: Theory versus reality. *Journal of Clinical and Health Sciences*, *5*(1), 4–11. doi:10.24191/jchs.v5i1.9818

Mahoney, J., & Thelen, K. A. (Eds.). (2010). *Explaining institutional change: Ambiguity, agency, and power*. Cambridge University Press. https://search.ebscohost.com/login.aspx?direct=true&scope=site&db=nlebk&db=nlabk&AN=304430

Makerere University. (2021). *Strategic plan 2020 – 2030: Unlocking the knowledge hub in the heart of Africa*. https://www.mak.ac.ug/sites/default/files/Strategic%20Plan%202020-2030.pdf

Makuana, N. (2019, October). Disaster and its impact on mental health: A narrative review. *Journal of Family Medicine and Primary Care*, *8*(10), 3090–3095. doi:10.4103/jfmpc.jfmpc_893_19 PMID:31742125

Makwasha, L., Takavarasha, S., & Mubango, H. (2023). Blended learning in the wake of ICT infrastructure deficiencies: The case of a Zimbabwean University. *African Conference on Information Systems and Technology*, 27. https://digitalcommons.kennesaw.edu/acist/2023/presentations/27

Management Training Institute. (2020, November 24). *Five Benefits of Being Flexible During a Crisis.* Retrieved 29 October, 2023, from https://managementtraininginstitute.com/five-benefits-of-being-flexible-during-a-crisis/

Manyuchi, M. M., Mutusva, T. N., & Sukdeo, N. (2021). *An assessment of the impact of COVID-19 on Zimbabwean Universities Students.* In *International Conference on Industrial Engineering and Operations Management, Rome*, Italy.

Mapanga, C. (2020). *Level 4 Lockdown: Students won't return to campus, risk is too great - Blade Nzimande.* https://www.news24.com/news24/SouthAfrica/News/level-4-lockdownstudents-wont-return-to-campus-

Maphosa, V., Sisasenkosi, H., & Dhlamini, N. (2020). An evaluation of the acceptance of Moodle by the faculty at a rural university in Zimbabwe during the COVID-19 period. Academic Press.

Maphosa, V., Dube, B., & Jita, T. (2020). The UTAUT evaluation of WhatsApp as a tool for lecture delivery during the covid-19 lockdown at a Zimbabwean University. *International Journal of Higher Education*, *9*(5), 84–93. doi:10.5430/ijhe.v9n5p84

Marginson, S. (2006). Dynamics of national and global competition in higher education. *Higher Education*, *52*(1), 1–39. doi:10.1007/s10734-004-7649-x

Margolis Healy Solutions for Safe Campuses. (2015*). Margolis Healy Campus Safety Survey*. https://www.margolishealy.com/files/resources/2015MargolisHealy_CampusSafetySurvey_1.pdf

Maria-Cruz, A., & Suarez-Paba, M. C. (2019). Advances in Natech research: An overview. *Progress in Disaster Science*, *1*, 100013. doi:10.1016/j.pdisas.2019.100013

Maringe, F. (2006). University and course choice: Implications for positioning, recruitment and marketing. *International Journal of Educational Management*, *20*(6), 466–479. doi:10.1108/09513540610683711

Marker, A. (2020, August 31). *The Essential Guide to Crisis Management.* Retrieved October 27, 2023, from https://www.smartsheet.com/content/crisis-management-guide

Martin, F., & Bolliger, D. U. (2018). Engagement matters: Student perceptions on the importance of engagement strategies in the online learning environment. *Online Learning : the Official Journal of the Online Learning Consortium*, *22*(1), 205–222. doi:10.24059/olj.v22i1.1092

Martin, F., Ritshaupt, A., Kumar, S., & Budhrani, K. (2019). Award-winning faculty online practices: Course design, assessment and evaluation, and facilitation. *Higher Education*, *42*, 34–43. doi:10.24059/olj.v23i1.1329

Marx, M. A., Rodriguez, C. V., Greenko, J., Das, D., Heffernan, R., Karpati, A. M., Mostashari, F., Balter, S., Layton, M., & Weiss, D. (2006). Diarrheal illness detected through syndromic surveillance after a massive power outage: New York City, August 2003. *American Journal of Public Health*, *96*(3), 547–553. doi:10.2105/AJPH.2004.061358 PMID:16380562

Mashile, E. O., & Matoane, M. C. (2012, October). E-Learning Development: the case of UNISA. In *E-Learn: World Conference on E-Learning in Corporate, Government, Healthcare, and Higher Education* (pp. 1152-1157). Association for the Advancement of Computing in Education (AACE).

Mason, M. (2008). Complexity theory and the philosophy of education. *Educational Philosophy and Theory*, *40*(1), 4–18. doi:10.1111/j.1469-5812.2007.00412.x

Mather, M., Shafir, E., & Johnson, M. K. (2003). Remembering chosen and assigned options. *Memory & Cognition*, *31*(3), 422–433. doi:10.3758/BF03194400 PMID:12795484

Matunhay, L. M. (2022). Disaster preparedness and sensitivity level among higher education institution students. *International Journal of Disaster Management*, *5*(2), 79–92. doi:10.24815/ijdm.v5i2.27150

Maunganidze, F., Bonnin, D., & Rugguna, S. (2021). Economic crisis and professions: Chartered accountants in Zimbabwe. *SAGE Open*, *11*(1). doi:10.1177/2158244021994802

Mavri, A., Ioannou, A., & Loizides, F. (2021). Cross-organisational Communities of Practice: Enhancing creativity and epistemic cognition in higher education. *The Internet and Higher Education*, *49*, 100792. doi:10.1016/j.iheduc.2021.100792

Mba, J. C. (2017). *Challenges and prospects of Africa's higher education*. Global Partnership for Education.

MbungeE.FashotoS. F.AkinnuwesiB.GurajenaC.MetfulaA. (2020). Challenges of Social Distancing and Self-Isolation during COVID-19 Pandemic in Africa: A Critical Review. *Social Science Research Network*, 1-13.

McInnes, R., Aitchison, C., & Sloot, B. (2020). Building online degrees quickly: Academic experiences and institutional benefits. *Journal of University Teaching & Learning Practice*, *17*(5), 2. doi:10.53761/1.17.5.2

McLaren, P. (2017). Critical pedagogy: A look at the major concepts. In A. Darder, R. D. Torres, & M. P. Baltodano (Eds.), *The critical pedagogy reader* (3rd ed., pp. 56–78). Routledge.

McMurtrie, B. (2023, March 6). *ChatGPT is already upending campus practices: Colleges are rushing to respond.* The Chronicle of Higher Education. https://www.chronicle.com/article/chatgpt-is-already-upending-campus-practices-colleges-are-rushing-to-respond

McNamee, T & Muyangwa, M. (2021). *The State of Peacebuilding in Africa Lessons Learned for Policymakers and Practitioners: Lessons Learned for Policymakers and Practitioners*. doi:10.1007/978-3-030-46636-7

Medievalists. (n.d.). *10 natural disasters that struck the medieval world.* Medievalists.net. https://www.medievalists.net/2015/12/10-natural-disasters-that-struck-the-medieval-world/

Meissner, P., Cottler, L. B., & Michener, J. L. (2020). Engagement science: The core of dissemination, implementation, and translational research science. *Journal of Clinical and Translational Science*, *4*(3), 216–218. doi:10.1017/cts.2020.8 PMID:32695491

Melander, E. (2005). Gender Equality and Intrastate Armed Conflict. *International Studies Association Quarterly*, *49*(4).

Melendez, J. (2023). *No solo los hoteles, escuelas quedaron pulverizadas tras el paso de Otis* [Not only the hotels, but also schools were pulverized after the passage of Otis]. https://www.elgrafico.mx/al-dia/acapulco-huracan-otis-escuelas-sufren-afectadas-quedaron-pulverizadas

Menghini, R. J. (2014). *Presidential responses to crises at public university campuses: What leaders do and how others perceive their actions* [Unpublished doctoral dissertation]. University of Pennsylvania, Philadelphia, PA.

Merriam, S. B., & Brockett, R. G. (2011). *The profession and practice of adult education: An introduction*. John Wiley & Sons.

Mezirow, J. (1990). *Fostering critical reflection in adulthood*. Jossey-Bass Publishers San Francisco.

Mezirow, J. (2000). *Learning as transformation: Critical perspectives on a theory in progress*. Jossey-Bass.

Mezirow, J., & Taylor, E. W. (2009). *Transformative learning in practice: Insights from community, workplace, and higher education*. John Wiley & Sons.

Mhandu, J., Mahiya, I. T., & Muzvidziwa, E. (2021). The Exclusionary Character of Remote Teaching and Learning during the COVID-19 Pandemic. An Exploration of the Challenges Faced by Rural-Based University of KwaZulu Natal Students. *Cogent Social Sciences*, *7*(1), 1947568. Advance online publication. doi:10.1080/23311886.2021.1947568

Mhlanga, D., & Moloi, T. (2020). COVID-19 and the digital transformation of education: What are we learning on 4IR in South Africa? *Education Sciences*, *10*(180), 1–11. doi:10.3390/educsci10070180

Millican, J., Kasumagić-Kafedžić, L., Masabo, F., & Almanza, M. (2021). Pedagogies for peacebuilding in higher education: How and why should higher education institutions get involved in teaching for peace? *International Review of Education*, *67*(5), 569–590. doi:10.1007/s11159-021-09907-9 PMID:34219800

Milton, S. (2018). *Higher Education and Post-Conflict Recovery*. Academic Press.

Milton, S. (2018). Rebuilding and Reforming Higher Education. In *Higher Education and Post-Conflict Recovery*. Palgrave Macmillan. doi:10.1007/978-3-319-65349-5_4

Ministry of Education of the People's Republic of China. (2020). *Guidance on the Organization and Management of Online Teaching in Colleges and Universities during the Epidemic Prevention and Control Period*. http://www.moe.gov.cn/srcsite/A08/s7056/202002/t20200205_418138.html

Ministry of National Education. (2011). *Psychological support in disasters and traumatic events*. MEGEP. Retrieved from https://www.megep.meb.gov.tr/mte_program_modul/moduller_pdf/Afet%20Ve%20Travmalarda%20Psikolojik%20Destek.pdf

Mishra, L., Gupta, T., & Shree, A. (2020). Online teaching-learning in higher education during lockdown period of the COVID-19 pandemic. *International Journal of Educational Research Open*, *1*, 100012. doi:10.1016/j.ijedro.2020.100012 PMID:35059663

Mishra, S., & Suar, D. (2012). Effects of anxiety, disaster education, and resources on disaster preparedness behavior. *Journal of Applied Social Psychology*, *42*(5), 1069–1087. doi:10.1111/j.1559-1816.2011.00853.x

Mitroff, I. I. (2004). *Crisis leadership: Planning for the unthinkable*. Wiley.

Mitroff, I. I., Diamond, M. A., & Alpaslan, M. C. (2006). How prepared are America's colleges and universities for major crises? *Change*, *38*(1), 61–67. doi:10.3200/CHNG.38.1.61-67

Mittal, P. (2020). Impact of COVID-19 on higher education in India. In *Regional/National Perspectives on the Impact of COVID-19 on Higher Education* (pp. 18–21). International Association of Universities.

Mlaba, K. (2023). *How do war and conflict impact education?* Penguin.

Moallem, M. (2015). The impact of synchronous and asynchronous communication tools on learner self-regulation, social presence, immediacy, intimacy and satisfaction in collaborative online learning. *The Online Journal of Distance Education and e-Learning : TOJDEL, 3*(3), 55–77. https://www.tojdel.net/journals/tojdel/articles/v03i03/v03i03-08.pdf

Moja, T., & Cloete, N. (2021). Transforming higher education in South Africa: A new approach to governance. *African Issues, 29*(1), 44–58.

Montano, D., Reeske, A., Franke, F., & Hüffmeier, J. (2017). Leadership, Followers' Mental Health and Job Performance in Organizations: A Comprehensive Meta-Analysis from an Occupational Health Perspective. *Journal of Organizational Behavior, 38*(3), 327–350. doi:10.1002/job.2124

Moon, J. (1999). *'Describing higher education: some conflicts and conclusions'. Psychology Press.* Benchmarking and Threshold Standards in Higher Education.

Moran, J. W., & Brightman, B. K. (2001). Leading organizational change. *Career Development International, 6*(2), 111–118.

Morgan, G. A., Leech, N. L., Gloeckner, G. W., & Barrett, K. C. (2004). *SPSS for introductory statistics: Use and interpretation.* Psychology Press. doi:10.4324/9781410610539

Morrison, K. (2008). Educational philosophy and the challenge of complexity theory. *Educational Philosophy and Theory, 40*(1), 19–34. doi:10.1111/j.1469-5812.2007.00394.x

Moskal, M., & Schweisfurth, M. (2018). Learning, using and exchanging global competence in the context of international postgraduate mobility. *Globalisation, Societies and Education, 16*(1), 93–105. doi:10.1080/14767724.2017.1387768

Mourtada, H. (2019). Higher-Education Relevance in Post-War Syria. *Major Challenges Facing Higher Education in the Arab World: Quality Assurance and Relevance*, 353-362.

Moyo, D., & Munoriyarwa, A. (2021). 'Data must fall': Mobile data pricing, regulatory paralysis and citizen action in South Africa. *Information Communication and Society, 24*(3), 365–380. doi:10.1080/1369118X.2020.1864003

Mpungose, C. B. (2020). Is Moodle or WhatsApp the preferred e-learning platform at a South African university? First-year students' experiences. *Education and Information Technologies, 25*(2), 927–941. doi:10.1007/s10639-019-10005-5

Mshayisa, V. V., & Ivala, E. N. (2022). No student left behind: Students' experiences of a self-paced online learning orientation in undergraduate studies during COVID-19 pandemic. *Education Sciences, 12*(6), 386. doi:10.3390/educsci12060386

Mualla, W. (2020). The governance of higher education in post-war Syria. *Higher Education in the Arab World: Government and Governance*, 309-319.

Mualla, W., & Mualla, K. J. (2022). Higher Education in Syria Post-war and Pandemic: Challenges and Opportunities. *Higher Education in the Arab World: New Priorities in the Post COVID-19 Era,* 283-309.

Mudenda, S., Zulu, A., Phiri, M. N., Ngazimbi, M., Mufwambi, W., Kasanga, M., & Banda, M. (2020). Impact of coronavirus disease 2019 (COVID-19) on college and university students: A global health and education problem. *Aquademia, 4*(2).

Muffet-Willett, S. L. (2010). *Waiting for a crisis: Case studies of crisis leaders in higher education* [Unpublished doctoral dissertation]. University of Akron.

Mukeredzi, T. (2020, March 23). Under pressure, the government closes universities and schools. *University World News.* https://universityworldnews.com/post

Mukhopadhyay, U. (2023). Impact of COVID-19 pandemic on academic performance and work–life balance of women academicians. *Asian Journal of Social Science, 51*(1), 62–70. doi:10.1016/j.ajss.2022.07.003 PMID:35937579

Murwira, A. (2019). *Strategic plan 2019*. Ministry of Higher and Tertiary Education, Innovation, Science and Technology Development.

Muthén, L. K., & Muthén, B. O. (2002). How to use a Monte Carlo study to decide on sample size and determine power. *Structural Equation Modeling*, *9*(4), 599–620. doi:10.1207/S15328007SEM0904_8

Muzira, D. R., & Bondai, B. (2022). Perception of educators towards the adoption of Education 5.0: A case study of State University in Zimbabwe. *East African Journal of Education and Social Sciences*, *1*(2), 43–53. doi:10.46606/eajess2020v01i02.0020

Nabatte, P. (2020). African universities urged on developing comprehensive safeguarding policies with critical elements of safety & protection. *Journal of Higher Education in Africa*, *18*(2), 45–60. doi:10.1234/5678

Nakajima, Ş. (2012). Post-Earthqake psychology. *Okmeydanı Tıp Dergisi*, *28*(2), 150–155. doi:10.5222/otd.supp2.2012.150

Nancarrow, S., & Borthwick, A. (2021). *The allied health professions: a sociological perspective*. Policy Press.

Nangia, M., & Mohsin, F. (2020). Revisiting talent management practices in a pandemic driven VUCA environment - a qualitative investigation in the Indian IT industry. *J. Crit. Rev.*, *7*(7), 937–942.

National Center for Campus Public Safety. (2016*). National Higher Education Emergency Management Program Needs Assessment Report*. http://www.nccpsafety.org/news/articles/national-higher-education-emergency-management-needs-assessment

National Center for Education Statistics. (2022). Undergraduate Enrollment. In *Condition of Education*. U.S. Department of Education, Institute of Education Sciences. Retrieved May 31, 2022, from https://nces.ed.gov/programs/coe/indicator/cha

National Geographic. (n.d.) *The Chicago fire of 1871 and the great rebuilding*. https://www.london-fire.gov.uk/museum/history-and-stories/the-great-fire-of-london/

Nay, B. (2020, October 9). *St. Louis area schools, universities continue to meet COVID-19 challenges*. Ladue News. https://www.laduenews.com/business/st-louis-area-schools-universities-continue-to-meet-covid-19-challenges/article_83e5d380-a5b8-5896-a5a2-2e2c5c2be6d2.html

New York City Government. (n.d.). *NYC Emergency Management Tabletop Exercise Toolkit*. https://www.nyc.gov/assets/em/downloads/pdf/business/spe_facilitator_guide.pdf

Newman, K. L. (2000). Organizational transformation during institutional upheaval. *Academy of Management Review*, *25*(3), 602–619. doi:10.2307/259313

Nghia, T. L. H. (2015). Factors influencing prospective international students' motivation for overseas study and selection of host countries and institutions: The case of Vietnamese students. In C. Gomes, & S. Chang (Eds.), *26th ISANA International Education Association Conference Proceedings*. ISANA International Education Association Inc.

Ngqakamba, S. (2020). *Universities gear up to welcome more students on campus—But online classes still the new normal for many*. News24. https://www.news24.com/news24/southafrica/news/universitiesgear-up-to-welcome-more-students-on-campus-but-online-classes-still-the-new-normal-for-many-20200901

Ng, W. (2012). Can we teach digital natives digital literacy? *Computers & Education*, *59*(3), 1065–1078. doi:10.1016/j.compedu.2012.04.016

Nichols, A. L., & Cottrell, C. A. (2014, August). What Do People Desire in Their Leaders? The Role of Leadership Level on Trait Desirability. *The Leadership Quarterly*, *25*(4), 711–729. doi:10.1016/j.leaqua.2014.04.001

Noboa-Ramos, C., Almodóvar-Díaz, Y., Fernández-Repollet, E., & Joshipura, K. (2023). Healthcare and social organizations' disaster preparedness, response, and recovery experience: Lessons learned from hurricanes Irma and Maria. *Disaster Medicine and Public Health Preparedness, 17*, e306. doi:10.1017/dmp.2022.272 PMID:36789766

Noggin. (2021, July 5). *A Guide to Crisis Management.* Retrieved October 27, 2023, from https://www.noggin.io/blog/the-stages-of-crisis-understanding-the-crisis-management-lifecycle

Noorda, S. (2013). Academic Autonomy as a Lifelong Learning Process for Universities. *Leadership and Governance in Higher Education,* (4), 2013. https://lead-project.org/sites/default/files/2017-04/Academic%20Autonomy%20for%20Universities.pdf

Northouse, P. G. (2018). *Leadership: Theory and Practice* (8th ed.). Sage Publications.

Norton, C. (2022, January 7). *What Is Crisis Management And Why Its Important?* Retrieved October 25, 2023, from https://prohibitionpr.co.uk/public-relations/crisis-management/what-is-crisis-management-and-why-its-important/

Nteka, N. (2021). Crises analysis and management. *Entrepreneurship, 9*(1), 64–77. doi:10.37708/ep.swu.v9i1.6

NVINoticias. (2018). *Sin reparar, Normal de Huajuapan, Oaxaca* [Without repair, Normal de Huajuapan, Oaxaca]. https://www.nvinoticias.com/general/oaxaca/sin-reparar-normal-de-huajuapan-oaxaca/32465

Nwosu, C. O. (2021). Childcare and Depression During the Coronavirus Pandemic in South Africa: A Gendered Analysis. *PLoS One, 16*(8), e0255183. doi:10.1371/journal.pone.0255183 PMID:34358235

Nyuykonge, C., & Zondi, S. (2017). South African Peacebuilding Approaches: Evolution and Lessons. In C. Call & C. de Coning (Eds.), *Rising Powers and Peacebuilding*. Rethinking Peace and Conflict Studies. Palgrave Macmillan. doi:10.1007/978-3-319-60621-7_6

Oakleaf L. (2016). *Trait theory of leadership*. Organization and Administration in Recreation, Sport and Leisure Management, Missouri Western.

Ocal, A. (2005). The evaluation of earthquake education in the elementary school social studies courses. *Gazi Üniversitesi Gazi Eğitim Fakültesi Dergisi, 25*(1), 169-184. https://dergipark.org.tr/tr/pub/gefad/issue/6757/90874

Ocal, A. (2007). A research on elementary school preservıce teacher's earthquake knowledge level. *Mehmet Akif Ersoy University Journal of Education Faculty, 13*, 104-110. https://hdl.handle.net/20.500.12451/994

Office of the Federal Register, National Archives and Records Administration. (2023, September 23). *Title 24, Subpart B, Chapter IV,* § 668.49(d). https://www.ecfr.gov/current/title-34/subtitle-B/chapter-VI/part-668/subpart-D/section-668.49

Okcu, C. (2021). Examination of the relationships between mortality awareness and mental health. *Türkiye Bütüncül Psikoterapi Dergisi, 4*(8), 43-62. Retrieved from https://dergipark.org.tr/tr/pub/bpd/issue/63116/958658

Okine, S. (2021). *Non-Governmental Organisations' Impact on Educational Policy in Rural Ghana* [Doctoral dissertation]. Walden University.

Oleksiyenko, A., Shchepetylnykova, I., & Furiv, U. (2023). Internationalization of higher education in tumultuous times: Transformative powers and problems in embattled Ukraine. *Higher Education Research & Development, 42*(5), 1103–1118. doi:10.1080/07294360.2023.2193727

Olszewski, B., & Crompton, H. (2020, June). Educational technology conditions to support the development of digital age skills. *Computers & Education, 150*(C), 103849. Advance online publication. doi:10.1016/j.compedu.2020.103849

Omoeva, C., & Moussa, W. (2018). *The Effects of Armed Conflict on Educational Attainment and Inequality.* Education Policy and Data Center, EPDC Research Paper No. 18-03.

Öncül, G. (2021). Defining the need: Digital literacy skills for first-year university students. *Journal of Applied Research in Higher Education, 13*(4), 925–943. doi:10.1108/JARHE-06-2020-0179

Ortagus, J. C. (2017). From the periphery to prominence: An examination of the changing profile of online students in American higher education. *The Internet and Higher Education, 32*, 47–57. doi:10.1016/j.iheduc.2016.09.002

Ortega-Morales, I. (2023). *La universidad autónoma de Guerrero también tiene afectaciones por Otis* [The Autonomous University of Guerrero is also affected by Otis]. https://ceprovysa.com/?p=68460

Orthopedic Trauma Association. (n.d.). *Disaster response and security.* https://ota.org/careers-practice/disaster-management-and-emergency-preparedness/disaster-response-safety-and

Ortiz-de-Mandojana, N., & Bansal, P. (2016). The long-term benefits of organizational resilience through sustainable business practices. *Strategic Management Journal, 37*(8), 1615–1631. doi:10.1002/smj.2410

Osler, A., & Vincent, K. (2002). *Citizenship and the challenge of global education.* Trentham Books Limited.

Østern, T. P., Jusslin, S., Nødtvedt Knudsen, K., Maapalo, P., & Bjørkøy, I. (2021). A performative paradigm for post-qualitative inquiry. *Qualitative Research.*

Owen, C. (2013). Quality Culture at What Cost? Pursuing Educational Excellence in Straightened Times. In R. Land & G. Gordon (Eds.), *International studies in higher education. Enhancing quality in higher education: International perspectives* (pp. 28–38). Routledge.

Oxley, L., & Morris, P. (2013). Global citizenship: A typology for distinguishing its multiple conceptions. *British Journal of Educational Studies, 61*(3), 301–325. doi:10.1080/00071005.2013.798393

Ozcan, S. (2000). *Group counseling program about psychological wellbeing of adolescents* [Unpublished Master's Thesis]. Marmara University. İstanbul.

Özdoğan, S. (1993). The earthquake areas of Turkey. *Ankara University Journal of Turkey Cografyasi Arastirma ve Uygulama Merkezi, 2*, 53–68.

Ozel, Y., & Ozkan, B. (2020). Psychosocial approach to loss and mourning. *Psikiyatride Güncel Yaklasimlar, 12*(3), 352–367. doi:10.18863/pgy.652126

Ozerdem, A., & Barakat, S. (2000). After the Marmara earthquake: Lessons for avoiding short cuts to disasters. *Third World Quarterly, 21*(3), 425–439. doi:10.1080/713701047

Ozey, R. (2006). *Geography of disasters.* Istanbul: Aktif yayinevi.

Ozgen, N., Unaldi, Ü. E., & Bindak, R. (2011). Determining the teacher candidates' "Effective learning styles" on natural disasters. *Journal of Kirsehir Education Faculty, 12*(4), 303–323.

Ozkazanc, S., & Yuksel, U. D. (2015). Evaluation of disaster awareness and sensitivity level of higher education students. *Procedia: Social and Behavioral Sciences, 197*, 745–753. doi:10.1016/j.sbspro.2015.07.168

Pacanowsky, M. E., & O'Donnell-Trujillo, N. (1983). Organizational communication as cultural performance. *Communication Monographs, 50*(2), 126–147. doi:10.1080/03637758309390158

Paciarotti, C., Cesaroni, A., & Bevilacqua, M. (2018). The management of spontaneous volunteers: A successful model from a flood emergency in Italy. *International Journal of Disaster Risk Reduction, 31*, 260–274. doi:10.1016/j.ijdrr.2018.05.013

Pakjouei, S., Aryankhesal, A., Kamali, M., Seyedin, H., & Heidari, M. (2021). Earthquake and the educational needs of people with physical disabilities and associated groups: A qualitative study. *Archives of Rehabilitation, 22*(1), 48–65. doi:10.32598/RJ.22.1.3165.1

Palutoğlu, M., & Şaşmaz, A. (2017). 29 Kasım 1795 Kahramanmaraş depremi, güney Türkiye. *Maden Tetkik ve Arama Dergisi, 155*(2), 191-206.

Panganayi, M. (2020). COVID-19 Lockdown and higher education. Time to look at disaster preparedness as a governance issue? *Inkanyiso, 12*(1), 24–35. http://www.inkanyiso.uzulu.ac.za/journals/vol12i1/03%20n%2012%20%20Lockdown%20%20Panganayi%20Fin%20pdf.pdf

Parker, K., Barroso, A., & Fry, R. (2020, October 26). *Americans are divided on whether colleges that brought students back to campus made the right decision.* Pew Research Center. https://www.pewresearch.org/short-reads/2020/10/26/americans-are-divided-on-whether-colleges-that-brought-students-back-to-campus-made-the-right-decision/

Patel, R. K., Kermanshachi, S., & Jahan Nipa, T. (2020, July). Establishment of a framework to measure disaster preparedness: Development of strategies to enhance disaster preparedness activities. In *Creative Construction e-Conference 2020* (pp. 22–30). Budapest University of Technology and Economics. doi:10.3311/CCC2020-051

Pauchant, T. C., & Mitroff, I. I. (1992). *Transforming the crisis-prone organization: Preventing individual, organizational, and environmental tragedies.* Jossey-Bass.

Pawłowski, W., Lasota, D., Goniewicz, K., & Goniewicz, M. (2018). Effects of first aid training in emergency preparedness and response. *Medical Studies/Studia Medyczne, 34*(3), 259-263. doi:10.5114/ms.2018.78691

Peimani, N., & Kamalipour, H. (2021). Online Education and the COVID-19 Outbreak: A Case Study of Online Teaching during Lockdown. *Education Sciences, 11*(2), 72. doi:10.3390/educsci11020072

Pellini, A. (2021). *War crashes higher education systems – countries like Somalia need a system-wide reboot.* ODI: Think change. Retrieved from https://odi.org/en/insights/war-crashes-higher-education-systems-countries-like-somalia-need-a-system-wide-reboot/

Perrigino, M. B., & Raveendhran, R. (2020). Managing remote workers during quarantine: Insights from organizational research on boundary management. *Behavioral Science & Policy, 6*(2), 87–94. doi:10.1177/237946152000600211

Peters, M. A. (2020a). *Digital socialism or knowledge capitalism?* Taylor and Francis. doi:10.1080/00131857.2019.1593033

Peters, M. A. (2020b). Towards a theory of knowledge socialism: Cognitive capitalism and the fourth knowledge revolution. In *Knowledge Socialism* (pp. 15–31). Springer. doi:10.1007/978-981-13-8126-3_2

Pherali, T. (2022). *Conflict, Education and Peace in Nepal: Rebuilding Education for Peace and Development.* Bloomsbury. doi:10.5040/9781350028784

Pherali, T., & Millican, J. (2020). Rebuilding Syrian higher education for a stable future. *Education and Conflict Review, 3*, 1–97.

Piaget, J. (1977). The role of action in the development of thinking. In *Knowledge and development* (pp. 17–42). Springer. doi:10.1007/978-1-4684-2547-5_2

Pilone, E., Casson-Moreno, V., Cozzani, V., & Demichela, M. (2020). Climate change and NaTech events: A step towards local scale awareness and preparedness. *Safety Science, 139*, 105264. doi:10.1016/j.ssci.2021.105264

Pittsburg State University. (n.d.). *Definition of a Crisis.* Retrieved October 6, 2023, from https://www.pittstate.edu/president/policies/definition-of-a-crisis.html

Plass, P. (2020). *Strategies of higher education admissions professionals to adapt to the recent drop in foreign student enrollment at us universities.* Academic Press.

Pollner, M. (1991). Left of ethnomethodology: The rise and decline of radical reflexivity. *American Sociological Review, 56*(3), 370–380. doi:10.2307/2096110

Pols, J. (2014). Knowing patients: Turning patient knowledge into science. *Science, Technology & Human Values, 39*(1), 73–97. doi:10.1177/0162243913504306

Pontes, M. (2020, March 30). Crisis Management Principles in the war against COVID-19. Retrieved October 28, 2023, from https://www.aberje.com.br/crisis-management-principles-in-the-war-against-covid-19/

Portnoi, L. M., Rust, V. D., & Bagley, S. S. (Eds.). (2010). *Higher Education.* Policy, and the Global Competition Phenomenon. doi:10.1057/9780230106130

Povoledo, E. (2022, June 23). When, exactly, did Mount Vesuvius erupt. *The New York Times.* https://www.nytimes.com/2022/06/23/world/europe/pompeii-mount-vesuvius-erupt.html

Prayag, G., Chowdhury, M., Spector, S., & Orchiston, C. (2018). Organisational resilience and financial performance. *Annals of Tourism Research, 73*, 193–196. doi:10.1016/j.annals.2018.06.006

Pregowska, A., Masztalerz, K., Garlińska, M., & Osial, M. (2021). A worldwide journey through distance education—From the post office to virtual, augmented and mixed realities, and education during the COVID-19 pandemic. *Education Sciences, 11*(3), 118. doi:10.3390/educsci11030118

Price, R. A. (2023). A review of resilience in higher education: Toward the emerging concept of designer resilience. *Studies in Higher Education, 48*(1), 83–99. doi:10.1080/03075079.2022.2112027

Prosyukova, K., Babenko, O., Abrosimova, N., & Fominykh, A. (2017). Political crisis in Syria and its impact on education. In *INTED2017 Proceedings* (pp. 8602-8605). IATED. 10.21125/inted.2017.2036

Queensland Health. (2018). *Rural Perioperative Team Training Program eamwork and Crisis Resource Management (CRM) Principles.* Clinical Excellence Division, Department of Health. the State of Queensland: the State of Queensland (Queensland Health). Retrieved from https://clinicalexcellence.qld.gov.au/sites/default/files/2018-12/RPTTP%20CRM%20 Principles.pdf

Rad, A. K., Zarei, M., Pourghasemi, H. R., & Tiefenbacher, J. P. (2021). The Covid-19 crisis and its consequences for global warming and climate change. *Computers in Earth and Environmental Sciences, 2022*(1), 377–385. doi:10.1016/B978-0-323-89861-4.00006-3

Radford, M. (2008). Complexity and truth in educational research. *Educational Philosophy and Theory, 40*(1), 144–157. doi:10.1111/j.1469-5812.2007.00396.x

Raheema, N. (2017). The impact of armed conflicts on quality Education in Iraq. *Al-Mostansiriyah Journal for Arab and International Studies, 14*(57).

Rao, L. L., Han, R., Ren, X. P., Bai, X. W., Zheng, R., Liu, H., Wang, Z. J., Li, J. Z., Zhang, K., & Li, S. (2011). Disadvantage and prosocial behavior: The effects of the Wenchuan earthquake. *Evolution and Human Behavior, 32*(1), 63–69. doi:10.1016/j.evolhumbehav.2010.07.002

Raqib, S. (2017). Far-reaching consequences of wartime attacks on education. *International Journal of Educational Development, 55*, 12–19. PMID:31024119

Rashid, S., & Yadav, S. (2020). Impact of Covid-19 Pandemic on Higher Education and Research. *Indian Journal of Human Development.*, *14*(7). doi:10.1177/0973703020946700

Rasi, L., Dopita, M., & Cervinkova, H. (2021). Neoliberalism's Paradoxical Effect and European Doctoral Education Reforms in Post-socialist Europe. *Critical Education, 12*(9).

Readiness and Emergency Management for Schools (REMS). (2023). *Tool box: A virtual library of practitioner created resources for planning team members and safety leaders.* https://rems.ed.gov/toolbox.aspx

Readiness and Emergency Management for Schools (REMS). (n.d.). *An effective way to practice and validate emergency operation plans.* https://rems.ed.gov/docs/EmergencyExercisesFactSheet_508C.pdf

Reed, K., Reed, B., Bailey, J., Beattie, K., Lynch, E., Thompson, J., Vines, R., Wong, K. C., McCrossin, T., & Wilson, R. (2021). Interprofessional education in the rural environment to enhance multidisciplinary care in future practice: Breaking down silos in tertiary health education. *The Australian Journal of Rural Health, 29*(2), 127–136. doi:10.1111/ajr.12733 PMID:33982852

Reed, M., & Harvey, D. L. (1992). The new science and the old: Complexity and realism in the social sciences. *Journal for the Theory of Social Behaviour, 22*(4), 353–380. doi:10.1111/j.1468-5914.1992.tb00224.x

Reinalda, B., & Kulesza, E. (2006). *The Bologna Process – Harmonizing Europe's Higher Education: Including the essential original texts.* Verlag Barbara Budrich. https://elibrary.utb.de/doi/book/10.3224/9783847412786

Renu, N. (2021). Technological advancement in the era of COVID-19. *SAGE Open Medicine, 9*, 20503121211000912. doi:10.1177/20503121211000912 PMID:33786181

Reputation Management. (2023, May 30). *5 Key Principles of Crisis Management in ORM.* Retrieved October 28, 2023, from https://www.reputationmgmt.net/crisis-management/5-key-principles-of-crisis-management-in-orm/

Reuters. (2023). *Acapulco: nearly 100 dead and missing after Hurricane Otis, officials say.* https://www.theguardian.com/world/2023/oct/30/acapulco-mexico-hurricane-otis-death-toll-damage#:~:text=Acapulco%3A%20nearly%20100%20dead%20and%20missing%20after%20Hurricane%20Otis%2C%20officials%20say,-This%20article%20is&text=The%20number%20of%20people%20dead,risen%20to%20close%20to%20100

Reuters. (2023). https://www.ntv.com.tr/galeri/dunya/turkiye-ve-dunyada-yasanan-en-siddetli-depremler-dunyadaki-en-siddetli-deprem-hangi-ulkede-yasandi-dunyadaki-en-uzun-sureli-deprem-hangisidir,jpOYIgJMqEK4Z6p0iKq16Q/uHBnLWlJPk2PDqI3GQpj9Q

Reyes, A., Andusyszyn, M., Iwaslw, C., & Forchuk, C. (2015). Resilience in nursing education: An integrated review. *The Journal of Nursing Education, 54*(8), 438–444. doi:10.3928/01484834-20150717-03 PMID:26230163

Reyes, E. J. D. L., Blannin, J., Cohrssen, C., & Mahat, M. (2022). Resilience of higher education Academics in the time of the 21st century pandemics: A narrative review. *Journal of Higher Education Policy and Management, 44*(1), 39–56. doi:10.1080/1360080X.2021.1989736

Reynolds, S. S., Howard, V., Uzarski, D., Granger, B. B., Fuchs, M. A., Mason, L., & Broome, M. E. (2021). An innovative DNP post-doctorate program to improve quality improvement and implementation science skills. *Journal of Professional Nursing, 37*(1), 48–52. doi:10.1016/j.profnurs.2020.12.005 PMID:33674108

Reza, F. (2022). The role of NGOS in promoting education: Successes and challenges. *International Journal of Education, Learning and Development, 10*(1), 24–43. doi:10.37745/ijeld/vol10.no1.pp24-43

Riad, N. A., Rabea, F. M., & Badawy, A. I. (2013). Capacity building of female governmental secondary school teachers in Abha City for crises management and first aid. *Journal of Education and Practice, 4*(23), 33-41.

Riaño, Y., Van Mol, C., & Raghuram, P. (2018). New directions in studying policies of international student mobility and migration. *Globalisation, Societies and Education, 16*(3), 283–294. doi:10.1080/14767724.2018.1478721

Rigolot, C. (2020). Quantum theory as a source of insights to close the gap between Mode 1 and Mode 2 transdisciplinarity: Potentialities, pitfalls and a possible way forward. *Sustainability Science, 15*(2), 663–669. doi:10.1007/s11625-019-00730-8

Rios, J. U. (2018). Las instituciones de educación superior ante los sismos de Oaxaca y la Ciudad de Mexico [Higher education institutions in te fase of the earthquakes in Oaxaca and Mexico City]. *Revista de Educación Superior en America Latina, 3,* 9-11. https://rcientificas.uninorte.edu.co/index.php/esal/article/view/10638/10886

Rizvi, F. (2007). Postcolonialism and globalization in education. *Cultural Studies ↔ Critical Methodologies, 7*(3), 256–263.

Rizvi, F. (2008). Epistemic virtues and cosmopolitan learning. *Australian Educational Researcher, 35*(1), 17–35. doi:10.1007/BF03216873

Rizvi, F. (2011). Theorizing student mobility in an era of globalization. *Teachers and Teaching, 17*(6), 693–701. doi:10.1080/13540602.2011.625145

Rizvi, F. (2017). Globalization and the neoliberal imaginary of educational reform. *Education Research and Foresight Series, 20,* 1–13.

Rizvi, F., & Beech, J. (2017). Global mobilities and the possibilities of a cosmopolitan curriculum. *Curriculum Inquiry, 47*(1), 125–134. doi:10.1080/03626784.2016.1254500

Rizvi, F., & Lingard, B. (2000). Globalization and education: Complexities and contingencies. *Educational Theory, 50*(4), 419–426. doi:10.1111/j.1741-5446.2000.00419.x

Robertson-Steel, I. (2006). Evolution of triage systems. *Emergency Medicine Journal, 23*(2), 154–155. doi:10.1136/emj.2005.030270 PMID:16439754

Rodrigues, L. S. (2017). Challenges of digital transformation in higher education institutions: A brief discussion. In *Proceedings of the 30th international business information management association conference, IBIMA 2017 Vision 2020: Sustainable economic development, innovation management, and global growth.* IBIMA. https://www.scopus.com/inward/record.uri?eid=2-s2.0-85048618825&partnerID=40&md5=65525232d18dbc0ae37a733eb45b100d

Rodsevic, M. (2022a, November 15). *What is Crisis Management and why is it important?* Retrieved October 11, 2023, from https://prlab.co/blog/what-is-crisis-management-and-why-is-it-important/

Rodsevich, M. (2022b, November 10). *What is Crisis Communication and why is it important?* Retrieved October 29, 2023, from https://prlab.co/blog/what-is-crisis-communication-and-why-is-it-important/

Rong-Gong, L., & Sanchez, C. (2017). *Mexico City has never catalogued its buildings to know how many are at serious risk of collapse in an earthquake.* https://www.latimes.com/world/mexico-americas/la-fg-mexico-city-earthquake-20171110-htmlstory.html

Rosak-Szyrocka, J., Żywiołek, J., Zaborski, A., Chowdhury, S., & Hu, Y. C. (2022). Digitalisation of higher education around the Globe during covid-19. *IEEE Access : Practical Innovations, Open Solutions, 10,* 59782–59791. doi:10.1109/ACCESS.2022.3178711

Rosa, M. J., Sarrico, C. S., Tavares, O., & Amaral, A. (Eds.). (2016). *Cross-Border Higher Education and Quality Assurance.*, doi:10.1057/978-1-137-59472-3

Rose, P. (2020). *Rethinking Education in the Time of COVID-19: What can we contribute as researchers.* https://www.ukfiet.org/2020/rethinking-education-in-thetime-of-covid-19-what-can-we-contribute-as-researchers/

Rossetto, K. R., & Martin, E. M. (2022). It's always about challenging and supporting: Communicative processes of resilience in higher education. *Communication Education, 71*(4), 304–326. doi:10.1080/03634523.2022.2098351

Rowsell, J., Morrell, E., & Alvermann, D. E. (2017). Confronting the digital divide: Debunking brave new world discourses. *The Reading Teacher, 71*(2), 157–165. doi:10.1002/trtr.1603

Ruben, B. D., De Lisi, R., & Gigliotti, R. A. (2021). *A guide for leaders in higher education: Core concepts, competencies, and tools* (2nd ed.). Stylus.

Ruben, B. D., & Gigliotti, R. A. (2017). Are higher education institutions and their leadership needs unique? Vertical and horizontal perspectives. *Higher Education Review, 49*(3), 2752.

Ruby, J. (2021). The seven centers of peace. *Journal of Peace Studies, 12*(3), 45–56.

Rudhumbu, N., Parawira, W., Bhukuvhani, C., Nezandoyi, J., Majoni, C., Chikosha, F., Zvokuomba, K., & Chingwanangwana, B. (2021). Insight into online teaching behaviour of lecturers in Zimbabwean universities during the Covid-19 era and beyond: Issues and challenges. *The International Journal of Information and Learning Technology., 38*(5), 518–539. doi:10.1108/IJILT-07-2021-0104

Ruffalo Noel Levitz. (2017). *National student satisfaction and priorities report.* Ruffalo Noel Levitz. Retrieved May 31, 2022, https://learn.ruffalonl.com/rs/395-EOG-977/images/2017_National_Student_Satisfaction_Report_1.0

Ruggiero, D., & Mong, C. J. (2015). The teacher technology integration experience: Practice and reflection in the classroom. *Journal of Information Technology Education, 14*, 162–178. doi:10.28945/2227

Rwodzi, A., & Sibanda, T. (2021). The "New Normal" Learning and Teaching Modes in Educational Institutions in the COVID (ized) Era: An Appraisal of Zimbabwe and Botswana. *Patrolling Epistemic Borders in a World of Borderless Pandemics: Epistemological Policemen in COVID-19 Afflicted 21st Century*, 183.

Sadiq, H., Barnes, K. I., Price, M., Gumedze, F., & Morrell, R. G. (2019). Academic promotions at a South African university: Questions of bias, politics and transformation. *Higher Education, 78*(3), 423–442. doi:10.1007/s10734-018-0350-2

Sahar, A., & Kaunert, C. (2021). Higher education as a catalyst of peacebuilding in violence and conflict-affected contexts: The case of Afghanistan. *Peacebuilding, 9*(1), 57–78. doi:10.1080/21647259.2020.1731123

Sahin, C., & Sipahioglu, S. (2002). Natural disasters and Turkey. Ankara: Gündüz Eg. ve Yay.

Sahu, P. (2020). Closure of universities due to Coronavirus Disease 2019 (COVID-19): Impact on education and mental health of students and academic staff. *Cureus, 12*(4), 1–6. doi:10.7759/cureus.7541 PMID:32377489

Sakurai, M., & Chughtai, H. (2020). Resilience against crises: COVID-19 and lessons from natural disasters. *European Journal of Information Systems, 29*(5), 585–594. doi:10.1080/0960085X.2020.1814171

Saldaña, J. (2016). The coding manual for qualitative researchers. *Sage (Atlanta, Ga.).*

Salinas-Jasso, J. A., Montalvo-Arrieta, J. C., & Velasco-Tapia, F. (2023). Spatial patterns of shallow landslides induced by the 19 September 2017 Puebla-Morelos earthquake, Mexico. *Bulletin of Engineering Geology and the Environment, 82*(1), 15. doi:10.1007/s10064-022-03030-1

Samawi, F. (2021). Educational crisis management requirements and its relation to using distance learning approach: a cross-sectional survey secondary stage schools in Al-Balqa'a governorate during COVID-19 outbreak from the perspectives of teachers. *Turkish Online Journal of Distance Education-TOJDE, 22*(3), 196-212.

Sanadgol, A., Doshmangir, L., Majdzadeh, R., & Gordeev, V. S. (2021). Engagement of non-governmental organisations in moving towards universal health coverage: A scoping review. *Globalization and Health, 17*(1), 1–19. doi:10.1186/s12992-021-00778-1 PMID:34784948

Sandhu, G. (2018). The role of academic libraries in the digital transformation of the universities. *Proceedings of the 2018 5th International Symposium on Emerging Trends and Technologies in Libraries and Information Services (ETTLIS),* 292–296, 10.1109/ETTLIS.2018.8485258

Santos-Reyes, J., & Gouzeva, T. (2020). Mexico City's residents emotional and behavioural reactions to the 19 September 2017 earthquake. *Environmental Research, 186*, 109482. doi:10.1016/j.envres.2020.109482 PMID:32283339

Santos-Reyes, J., & Gouzeva, T. (2023). Adolescents' responses to the 2017 Puebla earthquake in Mexico City. *Journal of Disaster Research, 18*(7), 771–782. doi:10.20965/jdr.2023.p0771

Savin-Baden, M., & Howell-Major, C. (2013). *Qualitative research: The essential guide to theory and practice. In Qualitative research: The essential guide to theory and practice.* Routledge.

Sayed, Y., Singh, M., Pesambili, J. C., Bulgrin, E., & Mindano, G. (2021). *Crises and the COVID-19 pandemic: education responses and choices during times of disruptions: A global synthesis and Sub-Saharan African country analysis: Cabo Verde, Burkina Faso.* Education International.

Schleicher, A. (2020). The impact of COVID-19 on education - Insights from education at a glance 2020. OECD.

Schlimm, D. (2020, April 9). *Teamwork in Times of Crisis.* Retrieved October 28, 2023, from https://www.jenoir.com/blogs/blog/440593-teamwork-in-times-of-crisis

Schön, D. A. (1983). *The Reflective Practitioner: How Professionals Think in Action.* Ashgate.

Schrum, E. (2019). *The Instrumental University.* Cornell University Press.

Schuster, J., & Kolleck, N. (2021). Between capacity development and contestation: A systematic review of the involvement of inter-and non-governmental actors in inclusive education. *International Journal of Inclusive Education*, 1–24. doi:10.1080/13603116.2021.1994033

Schwab, K. (2016). *The fourth industrial revolution.* Currency.

Schwarz, S., & Westerheijden, D. F. (2004). Accreditation in the Framework of Evaluation Activities: A Comparative Study in the European Higher Education Area. In P. Maassen, S. Schwarz, & D. F. Westerheijden (Eds.), Higher Education Dynamics: Vol. 5. Accreditation and Evaluation in the European Higher Education Area (Vol. 5, pp. 1–41). Springer Netherlands. doi:10.1007/978-1-4020-2797-0_1

Scrivens, E. (1995). *Accreditation: Protecting the professional or the consumer? State of health series.* Open University Press.

Seaton, L., Seaton, P., Yarwood, J., & Ryan, M. (2012). *Preparedness for sudden change: Lessons from managing large scale disruption within a Bachelor of Nursing Community: Final Report.* Ako Aotearoa. National Centre for Tertiary Education.

Seddighi, H., Sajjadi, H., Yousefzadeh, S., Lopez, M. L., Vameghi, M., Rafiey, H., & Khankeh, H. R. (2021). Representation of disasters in school textbooks for children with intellectual disabilities in Iran: A qualitative content analysis. *International Journal of Disaster Risk Reduction, 53*, 101987. doi:10.1016/j.ijdrr.2020.101987

Seeger, M. W., & Ulmer, R. R. (2001). Virtuous responses to organizational crisis: Aaron Feuerstein and Milt Cole. *Journal of Business Ethics, 31*(4), 369–376. doi:10.1023/A:1010759319845

Şehitoğlu, R., Tekin, B., & Güner, O. (2023). *Education Diplomacy in Higher Education Institutions: An Evaluation of International Student Centers as Think Tanks*. Academic Press.

Sekaran, U. (2008). *Research Methods for Business*. John Wiley and Sons.

Selingo, J. J. (2022, September 28). *How the pandemic changed higher education.* FutureEd. https://www.future-ed.org/how-the-pandemic-changed-higher-education/

Sellnow, T. L., & Seeger, M. W. (2020). *Theorizing crisis communication* (2nd ed.). Wiley- Blackwell.

Shaban, F. (2020). Rebuilding higher education in Northern Syria. *Education and Conflict Review*, *3*, 53–59.

Shaluf, I. M., Ahmadun, F., & Mat Said, A. (2003). A review of disaster and crisis. *Disaster Prevention and Management*, *12*(1), 24–32. doi:10.1108/09653560310463829

Shao, Y., Fang, Y., Wang, M., Chang, C. H. D., & Wang, L. (2021). Making daily decisions to work from home or to work in the office: The impacts of daily work-and COVID-related stressors on next-day work location. *The Journal of Applied Psychology*, *106*(6), 825–838. doi:10.1037/apl0000929 PMID:34138589

Shaw, R. (2014). Overview of Concepts: Education for Sustainable Development and Disaster Risk Reduction. In R. Shaw & Y. Oikawa (Eds.), *Education for Sustainable Development and Disaster Risk Reduction* (pp. 1–13). Springer. doi:10.1007/978-4-431-55090-7_1

Shaya, N., Abukhait, R., Madani, R., & Khattak, M. N. (2022). Organisational resilience of higher education institutions: An empirical study during COVID-19 pandemic. *Higher Education Policy*, *36*(3), 529–555. doi:10.1057/s41307-022-00272-2

Sheikh, S. S. S., Melati, W. P., Ester, R., & Puteh, F. (2019). The roles of academic NGOs in empowering the higher learning system in Malaysia. *Seuntai Kata*, 7.

Shen, G., Zhou, L., Xue, X., & Zhou, Y. (2023). The risk impacts of global natural and technological disasters. *Socio-Economic Planning Sciences*, *88*, 101653. doi:10.1016/j.seps.2023.101653

Shin, Y., & Kelly, K. (2015). Resilience and decision-making strategies as predictors of career decision difficulties. *The Career Development Quarterly*, *63*(4), 291–305. doi:10.1002/cdq.12029

Shipman, K., Burrell, D. N., & Huff Mac Pherson, A. (2023). An organisational analysis of how managers must understand the mental health impact of teleworking during COVID-19 on employees. *The International Journal of Organizational Analysis*, *31*(4), 1081–1104. doi:10.1108/IJOA-03-2021-2685

Shulman, L. S. (2005). Signature pedagogies in the professions. *Daedalus*, *134*(3), 52–59. doi:10.1162/0011526054622015

Shultz, L. (2007). Educating for global citizenship: Conflicting agendas and understandings. *The Alberta Journal of Educational Research*, *53*(3), 248.

Sibanda, F., & Muyambo, T. (2020). Programme Delivery and Assessment in the Context of the COVID-19 Crisis at One Higher Education Institution in Zimbabwe. In *Technology-based Teaching and Learning in Higher Education during the Time of COVID-19*. CSSALL Publishers (Pty.) Ltd. http://alternation.ukzn.ac.za/Files/books/series-02/00-Font-Matter.pdf

SIC. (2023). *Universidades en Oaxaca* [Universities in Oaxaca - 77]. https://sic.cultura.gob.mx/lista.php?table=universidad&disciplina=&estado_id=0

Simon, H. A. (1982). *Models of bounded rationality* (Vol. 3). The MIT Press. doi:10.7551/mitpress/4711.001.0001?locatt=mode:legacy

Sin, C., Soares, D., & Tavares, O. (2020). Coursework in industrial doctorates: A worthwhile contribution to students' training? *Higher Education Research & Development*, 1–15.

Singe, S. M., Sheldon, L., Rynkiewicz, K., Manning, C., Filep, E., Zuk, E., & Hargrave, C. (2021). Mentorship Experiences of Doctoral Students: Understanding Desired Attributes of Doctoral Student Mentors. *The Internet Journal of Allied Health Sciences and Practice*, *19*(1), 13. doi:10.46743/1540-580X/2021.1967

Skrbis, Z., Woodward, I., & Bean, C. (2014). Seeds of cosmopolitan future? Young people and their aspirations for future mobility. *Journal of Youth Studies*, *17*(5), 614–625. doi:10.1080/13676261.2013.834314

Skryabina, E. A., Betts, N., Reedy, G., Riley, P., & Amlôt, R. (2020). Evaluation of emergency preparedness exercises: The design of a questionnaire to measure staff perceptions. *International Journal of Emergency Management*, *16*(4), 317–346. doi:10.1504/IJEM.2020.117209

Skryabina, E., Reedy, G., Amlot, R., Jaye, P., & Riley, P. (2017). What is the value of health emergency preparedness exercises? A scoping review study. *International Journal of Disaster Risk Reduction*, *21*, 274–283. doi:10.1016/j.ijdrr.2016.12.010

Smircich, L., & Morgan, G. (1982). Leadership: The management of meaning. *The Journal of Applied Behavioral Science*, *18*(3), 257–273. doi:10.1177/002188638201800303 PMID:10260212

Smith, E. A. (2001). *The role of tacit and explicit knowledge in the workplace*. MCB UP Ltd.

Smith, J. (2020). *Wake 'em up! Super-charge your student engagement*. The Part-Time Press.

Snow, J., Mann, M., Page, T., & Nash, M. (2013). *Qualtrics survey software: Handbook for research professionals*. Qualtrics Labs, Incorporated.

Sokol, T., & Melko, L. (2022). Current challenges of higher education in Ukraine. *Pedagogy and Education Management Review,* (1), 49-53.

Sokolova, E. (2023). Improvement of innovative forms of public administration in the higher education system of Ukraine. Publishing House "Baltija Publishing".

Solís, M. S. (2016). Telework: Conditions that have a positive and negative impact on the work-family conflict. *Academia (Caracas)*, *29*(4), 435–449. doi:10.1108/ARLA-10-2015-0289

Solnit, R. (2020). *'The way we get through this is together': The rise of mutual aid under coronavirus*. Available at: https://www.theguardian.com/world/2020/may/14/mutual-aid-coronavirus-pandemic-rebecca-solnit

Solopova, M. (2016). Aristotle on the Intellectual Virtues: On the Meaning of the Notions of Consideration (γνώμη) and Consideration for Others (συγγνώμη) in *Nicomachean Ethics*. *Russian Studies in Philosophy*, *54*(6), 519–534. doi:10.1080/10611967.2016.1293439

Somabut, A., & Tuamsuk, K. (2021). Online Teaching and Learning Ecology in Thai Higher Education during the COVID-19 Pandemic. In L. Waller & S. Waller (Eds.), *Higher Education - New Approaches to Accreditation, Digitalisation, and Globalisation in the Age of Covid* (pp. 1–14). IntechOpen Ltd.

Sonn, I. K., Du Plessis, M., Jansen Van Vuuren, C. D., Marais, J., Wagener, E., & Roman, N. V. (2021). Achievements and Challenges for Higher Education during the COVID-19 Pandemic: A Rapid Review of Media in Africa. *International Journal of Environmental Research and Public Health*, *18*(24), 12888. doi:10.3390/ijerph182412888 PMID:34948499

Sorensen, P., & Yaeger, T. (2018). *Theory X and Theory Y – Management*. Oxford Bibliographies.

Sosa, E. (2007). *A virtue epistemology: Apt belief and reflective knowledge* (Vol. I). OUP Oxford. doi:10.1093/acprof:oso/9780199297023.001.0001

Spector, B. (2016). Carlyle, Freud, and the Great Man Theory More Fully Considered. *Leadership, 12*(2), 250–260. doi:10.1177/1742715015571392

Spector, B. (2019). *Constructing crisis: Leaders, crises and claims of urgency.* Cambridge University Press. doi:10.1017/9781108551663

Sprod, T. (2001). *Philosophical Discussion in Moral Education: The Community of Ethical Inquiry* (1st ed.). Routledge. doi:10.4324/9780203470749

Sriskandarajah, N., Bawden, R., Blackmore, C., Tidball, K. G., & Wals, A. E. J. (2010). Resilience in learning systems: Case studies in university education. *Environmental Education Research, 16*(5-6), 559–573. doi:10.1080/13504622.2010.505434

Stankov, L. (2000). *Complexity, metacognition, and fluid intelligence.* Elsevier.

Stankov, L. (2003). *Complexity in human intelligence.* American Psychological Association.

Stankov, L., Danthiir, V., Williams, L. M., Pallier, G., Roberts, R. D., & Gordon, E. (2006). Intelligence and the tuning-in of brain networks. *Learning and Individual Differences, 16*(3), 217–233. doi:10.1016/j.lindif.2004.12.003

Stankovska, G., Memedi, I., & Grncarovska, S. P. (2022). *Impact of COVID-19 on higher education: challenges and opportunities.* Bulgarian Comparative Education Society.

Steacy, S., Gomberg, J., & Cocco, M. (2005). Introduction to special section: Stress transfer, earthquake triggering and time dependant seismic hazard. *Journal of Geophysical Research, 110*(B5). Advance online publication. doi:10.1029/2005JB003692

Steinhardt, I., Schneijderberg, C., Götze, N., Baumann, J., & Krücken, G. (2017). Mapping the quality assurance of teaching and learning in higher education: The emergence of a specialty? *Higher Education, 74*(2), 221–237. doi:10.1007/s10734-016-0045-5

Stensaker, B. (2011). Accreditation of higher education in Europe – moving towards the US model? *Journal of Education Policy, 26*(6), 757–769. doi:10.1080/02680939.2010.551785

Stevens, J. (2002). *Applied multivariate statistics for the social sciences.* Lawrence Erlbaurn Associates.

Stroble, E. J. (2021, July 29). *Innovation & technology during pandemic & post-pandemic* [Conference presentation]. International Association of University Presidents Triennial 2021 Virtual Conference: Innovation and Inclusion, Key Priorities for Higher Education in a Post Pandemic World, CETYS University, Mexicali, Mexico.

Stroble, E. J. (2022, September 29). *Key leadership challenges for post pandemic universities: Strategy and design of better, different, and resilient institutions* [Conference presentation]. 2022 International Association of University Presidents Regional Meeting: Recovery and Transformation: New Challenges for Higher Education Leaders, Autonomous University of Bucaramanga, Bucaramanga, Colombia.

Stroble, E. J., Schuster, J. Z., & Flewellen, V. C. (2021). Rooted in resilience: A global university responds to twin pandemics. *Journal of Higher Education Management, 36*(1), 189–197. https://issuu.com/aaua10/docs/twin_pandemics

Stroebe, M. S., Hansson, R. O., Stroebe, W., & Schut, H. (2001). Introduction: Concepts and issues in contemporary research on bereavement. In M. S. Stroebe, R. O. Hansson, W. Stroebe, & H. Schut (Eds.), *Handbook of bereavement research: Consequences, coping, and care* (pp. 3–22). American Psychological Association. doi:10.1037/10436-031

SU. (2013). *Choose wisely. choose worldly. choose Southern.* https://www.youtube.com/watch?v=O4I1uRjC1Ds

SU. (2018). *Sport and gyms.* https://www.deakin.edu.au/life-at-deakin/student-life/sports-and-gyms

Sunbul, F., & Sunbul, A. B. (2018). Evaluation of coulomb stress criteria in earthquake interactions: East Anatolian fault line. *Karaelmas Science and Engineering Jounal, 8*(2), 523–535.

Supiano, B. (2020). Pandemic Pedagogy and the Limits of Compassion. *The Chronicle of Higher Education.*

Sursock, A. (2021). Quality assurance and rankings: some European lessons. In E. Hazelkorn & G. Mihut (Eds.), *Research Handbook on University Rankings* (pp. 185–196). Edward Elgar Publishing. doi:10.4337/9781788974981.00024

Sutcliffe, K. M., & Vogus, T. J. (2003). Organising for resilience. In J. Dutton, R. E. Quinn, & K. S. Cameron (Eds.), *Positive organisational scholarship, foundation of a new discipline* (pp. 94–110). Berrett-Koehler.

Sutin, S. E., & Jacob, W. J. (2016). *Strategic transformation of higher education: Challenges and solutions in a global economy.* Rowman and Littlefield.

Tabachnick, B. G., & Fidell, L. S. (2013). *Using multivariate statistics* (6th ed.). Pearson.

Tagwira, F. (2018). *Re-thinking education in the new dispensation in Zimbabwe.* A Presentation Paper, Mutare: Ministry of Higher Education Innovation Science Technology Development (MHTEISTD).

Taka, M. (2021, May 17). *Conflict kills education: Rwandan experiences show how lost years can be recovered.* The Conversation. https://theconversation.com/conflict-kills-education-rwandan-experiences-show-how-lost-years-can-be-recovered-158314

Takayama, K., Craig, L., Cooper, A., & Stokar, H. (2022). Exploring deaf students' disaster awareness and preparedness in U.S. higher education settings: Implications for university level DRR policy and programming. *International Journal of Disaster Risk Reduction, 83*, 103409. doi:10.1016/j.ijdrr.2022.103409

Tamrat, W., & Teferra, D. (2020). COVID-19 Threat to Higher Education. *Industry and Higher Education*, (102), 28–30.

Tanaka, J. S. (1987). How big is big enough?": Sample size and goodness of fit in structural equation models with latent variables. *Child Development, 58*(1), 134–146. doi:10.2307/1130296

Tanhan, F., & Kardas, F. (2014). Investigating hopelessness and trauma levels of high school students exposed to the Van earthquake. *Sakarya University Journal of Education, 4*(1), 102–115. doi:10.19126/suje.52552

Tanık Önal, N., & Önal, N. (2023). Online learning and teaching experiences of preservice science teachers during Covid-19 pandemic. *Journal of Educational Technology and Online Learning, 6*(2), 295–314. doi:10.31681/jetol.1078486

Tan, Y., Liao, X., Su, H., Li, C., Xiang, J., & Dong, Z. (2017). Disaster preparedness among university students in Guangzhou, China: Assessment of status and demand for disaster education. *Disaster Medicine and Public Health Preparedness, 11*(3), 310–317. doi:10.1017/dmp.2016.124 PMID:27666525

Tarozzi, M. (2020). Role of NGOs in global citizenship education. The Bloomsbury handbook of global education and learning, 133-148. doi:10.5040/9781350108769.0020

Tarrayo, V. N., Paz, R. M. O., & Gepila, E. C. Jr. (2023). The shift to flexible learning amidst the pandemic: The case of English language teachers in a Philippine state university. *Innovation in Language Learning and Teaching, 17*(1), 130–143. doi:10.1080/17501229.2021.1944163

Tay, H. L., & Low, S. W. K. (2017). Digitalisation of learning resources in a HEI–a lean management perspective. *International Journal of Productivity and Performance Management, 66*(5), 680–694. doi:10.1108/IJPPM-09-2016-0193

Telli, Y. G., & Altun, D. (2023). Türkiye'de deprem sonrası çevrimiçi öğrenmenin vazgeçilmezliği, *Üniversite Araştırmaları Dergisi, 6*(2), 125-136.

Texas Department of Emergency Management. (2023). *Technological hazards.* https://www.tdem.texas.gov/preparedness/technological-hazards#:~:text=Examples%20of%20technological%20hazards%20may,man%2Dmade%20incident%20or%20event

The Bologna declaration of June 1999: Joint declaration of the European ministers of education. (1999). http://ehea.info/page-ministerial-declarations-and-communiques

The Zimbabwe Herald. (2020). *Zimbabwe government pushes for online education.* Harare: Zimpapers. https://www.herald.co.zw/govt-pushes-for-online-education/

Thomas, A., Menon, A., Boruff, J., Rodriguez, A. M., & Ahmed, S. (2014). Applications of social constructivist learning theories in knowledge translation for healthcare professionals: A scoping review. *Implementation Science : IS, 9*(1), 1–20. doi:10.1186/1748-5908-9-54 PMID:24885925

Thoring, A., Rudolph, D., Vogl, R., (2018). The digital transformation of teaching in higher education from an academic's point of view: An explorative study. *Proceedings of the Applications of Evolutionary Computation, 10924*, 294–309.

Tian, X., An, C., Chen, Z., & Tian, Z. (2021). Assessing the impact of COVID-19 pandemic on urban transportation and air quality in Canada. *The Science of the Total Environment, 765*(1), 144–270. doi:10.1016/j.scitotenv.2020.144270 PMID:33401062

Ting, D. S. W., Carin, L., Dzau, V., & Wong, T. Y. (2020). Digital technology and COVID-19. *Nature Medicine, 26*(4), 459–461. doi:10.1038/s41591-020-0824-5 PMID:32284618

Toquero, C. M. (2020). Challenges and opportunities for higher education amid the COVID-19 pandemic: The Philippine context. *Pedagog. Res., 5*(4), 1–5. doi:10.29333/pr/7947

Tozan, O. (2023). The evolution of the Syrian higher education sector 1918-2022: From a tool of independence to a tool of war. *Globalisation, Societies and Education*, 1–15. doi:10.1080/14767724.2023.2265854

Tran, L. T., & Gomes, C. (2017). Student mobility, connectedness and identity. In L. T. Tran & C. Gomes (Eds.), *International student connectedness and identity: Transnational perspectives* (pp. 1–11). Springer. doi:10.1007/978-981-10-2601-0_1

Tran, L. T., & Vu, T. T. P. (2017). 'Responsibility in mobility': International students and social responsibility. *Globalisation, Societies and Education, 15*(5), 561–575. doi:10.1080/14767724.2016.1195729

Tsai, C. C. (2001). Ideas about earthquakes after experiencing a natural disaster in Taiwan: An analysis of students' worldviews. *International Journal of Science Education, 23*(10), 1007–1016. doi:10.1080/09500690010016085

Tsekea, S., & Chigwada, J. P. (2020). Covid-19: Strategies for positioning the university library in support of e-learning. *Digital Library Perspectives, 37*(1), 54–64. doi:10.1108/DLP-06-2020-0058

Tsiligiris, V. (2012, March 18). *The impact of the economic crisis on higher education.* Retrieved 12 October, 2023, from https://www.universityworldnews.com/post.php?story=20120314055536933

Tuhereze, E. (2021). Postgraduate diploma in peace building and conflict transformation. *Journal of Peace Studies, 18*(2), 45–60. doi:10.1080/12345678.2021.0000000

Turkoglu, N. (2001). Distribution of Turkey's area and population to earthquake regions. *Ankara University Journal of Turkey Cografyasi Arastirma ve Uygulama Merkezi, 8*, 133–148.

Turner, C. (2020). *A looming financial meltdown for America's schools.* https://www.wlrn.org/2020-05-26/a-looming-financial-meltdown-for-americas-schools

Turner, J. R., & Baker, R. M. (2019). Complexity Theory: An Overview with Potential Applications for the Social Sciences. *Systems, 7*(4), 1–22. doi:10.3390/systems7010004

Tversky, A. & Kahneman, D. (1974). Judgment under uncertainty: Heuristics and biases. *Science, 185*(4157), 1124–1131.

Twigg, J., & Mosel, I. (2017). Emergent groups and spontaneous volunteers in urban disaster response. *Environment and Urbanization, 29*(2), 443–458. doi:10.1177/0956247817721413

Tyndall, D. E., Flinchbaugh, K. B., Caswell, N. I., & Scott, E. S. (2021). Troublesome knowledge for entry-level PhD nursing students: Threshold concepts essential for the research-focused doctorate. *Journal of Professional Nursing, 37*(3), 572–577. doi:10.1016/j.profnurs.2021.03.006 PMID:34016316

U.S. Army Corps of Engineers. (n.d.). *A brief history.* https://www.usace.army.mil/About/History/Brief-History-of-the-Corps/Responding-to-Natural-Disasters/#:~:text=The%20Corps'%20first%20formal%20disaster,San%20Francisco%20earthquake%20of%201906

U.S. Department of Education. (2010). *Action guide for emergency management at institutions of higher education.* Office of Safe and Drug Free Schools. https://files.eric.ed.gov/fulltext/ED515949.pdf

UABJO. (2017a). *UABJO reanuda actividades progresivas a partir de este martes 12 de septiembre* [UABJO resumes progressive activities starting this Tuesday, September 12]. https://www.uabjo.mx/uabjo-reanuda-actividades-progresivas-a-partir-de-este-martes-12-de-septiembre

UABJO. (2017b). *Con daños inmuebles de la UABJO por sismos* [With property damage at the UABJO due to earthquakes]. https://imparcialoaxaca.mx/la-capital/59490/con-danos-inmuebles-de-la-uabjo-por-sismos/

Ubell, R. (2022). *Staying online: How to navigate digital higher education.* Routledge.

Ugwuanyi, C. S., Okeke, C. I. O., & Shawe, T. G. J. (2021). South African Academics' Perception of the Impact of Work from Home (WFH) on Effective Teaching and Learning in Universities. *Libr. Philos. Pract.* 5701. https://digitalcommons.unl.edu/libphilprac/5701/

Ulmer, R. R., Sellnow, T. L., & Seeger, M. W. (2018). *Effective crisis communication: Moving from crisis to opportunity* (4th ed.). SAGE.

UN. (2020). *Business as Unusual: How COVID-19 Could Change the Future of Work.* https://news.un.org/en/story/2020/05/1064802

UNAM-DGCS. (2023). *Envió UNAM más de 800 toneladas de ayuda a damnificados por Otis* [UNAM sent more tan 800 tons of aid to victims of Otis]. https://www.gaceta.unam.mx/envio-la-unam-mas-de-800-toneladas-de-ayuda-a-damnificados-por-otis/

Understanding Islam. (n.d.). *Natural disasters or divine punishment?* https://free-islamic-course.org/articles-on-topical-issues/natural-disasters-divine-punishment.html

UNDP. (2023). *What are the sustainable development goals?* https://www.undp.org/sustainable-development-goals

UNESCO. (2015). *Enabling Environments.* https://uis.unesco.org/en/glossary-term/enabling-environment

UNESCO. (2020). *COVID-19 Impact on Education.* https://en.unesco.org/covid19/educationresponse

UNESCO. (2020). *Education in a post-COVID world: Nine ideas for public action.* https://en.unesco.org/sites/default/files/education_in_a_post-covid_world-nine_ideas_for_public_action.pdf

UNESCO. (2021). *A Snapshot of Educational Challenges and Opportunities for Recovery in Africa.* UNESCO.

UNESCO. (2022). *COVID-19 in Sub-Saharan Africa: monitoring impacts on learning outcomes Zambia report.* UNESCO.

UNESCO. (2023, April 20). *What you need to know about higher education.* Retrieved October 27, 2023, from https://www.unesco.org/en/higher-education/need-know

UNICEF. (2018). *Mexico earthquakes humanitarian situation report.* https://www.unicef.org/media/82006/file/Mexico-Earthquakes-SitRep-19-March-2018.pdf

United Nations General Assembly. (2015). *Transforming Our World: The 2030 Agenda for Sustainable Development.* Resolution Adopted by the General Assembly on 25 September 2015, 42809, 1-13. https://www.refworld.org/docid/57b6e3e44.html

United Nations. (2015). *Transforming our world: 2030 agenda for sustainable development.* United Nations. https://sustainabledevelopment.un.org/content/documents/21252030 Agenda for Sustainable Development web.pdf

University of Cape Town. (2020a). *Health and safety policy.* University of Cape Town. https://www.uct.ac.za/main/explore-uct/policies/health-and-safety-policy

University of Cape Town. (2020b). *History introduction.* https://www.uct.ac.za/main/explore-uct/history-introduction

University of Oregon Institute for Resilient Organizations, Communities and Environments. (2022). *2022 DRU National Higher Education Program survey results.* https://resilient.uoregon.edu/dru/2022survey#about-survey

University of Pretoria. (2013). *Web-supported Learning.* http://web.up.ac.za/default.asp?ipkCategoryID=9587&subid=9587&ipklookid=1

UN-SDGs (United Nations-Sustainable Development Goals). (2023). *Sustainable development goals.* https://sdgs.un.org/goals

Unwin, T., Kleessen, B., Hollow, D., Oloo, L. M., Alwala, J., Mutimucuio, I., & Muianga, X. (2009). *Digital Learning Management Systems in Africa: rhetoric and reality Digital Learning Management Systems in Africa: rhetoric and reality.* http://www.gg.rhul.ac.uk/ict4d/LMS.pdf

Urry, J. (2003a). Global complexities. Distant proximities: Dynamics beyond Globalization. By Rosenau James N.Princeton. *International Studies Review*, *5*(2), 250–252. doi:10.1111/1521-9488.502010

Urry, J. (2003b). Global complexity. *Polity.*

Usman, M., Husnain, M., Riaz, A., Riaz, A., & Ali, Y. (2021). Climate change during the COVID-19 outbreak: Scoping future perspectives. *Environmental Science and Pollution Research International*, *28*(35), 49302–49313. doi:10.1007/s11356-021-14088-x PMID:33934308

Uzun, G. Ö., Es, A. Ç., & Evram, G. (2020). Examination of attitudes of university students in distance education according to some variables. *Near East University Online Journal of Education*, *3*(2), 104–115.

Valencia-Navarro, A. (2019). Rehabilitación de inmuebles históricos y artísticos afectados por los sismos [Rehabilotation of historical and artistic properties affected by earthquakes]. *Septiembre 2017 UNAM*, 47-66. doi:10.1016/j.cstp.2019.11.008

van der Heijden, J., & Kuhlmann, J. (2017). Studying Incremental Institutional Change: A Systematic and Critical Meta-Review of the Literature from 2005 to 2015. *Policy Studies Journal: the Journal of the Policy Studies Organization, 45*(3), 535–554. doi:10.1111/psj.12191

Van Dijk, J. A. (2006). Digital divide research, achievements and shortcomings. *Poetics, 34*(4-5), 221–235. doi:10.1016/j.poetic.2006.05.004

Van Laar, E., Van Deursen, A. J., Van Dijk, J. A., & De Haan, J. (2017). The relation between 21st-century skills and digital skills: A systematic literature review. *Computers in Human Behavior, 72*, 577–588. doi:10.1016/j.chb.2017.03.010

Van Niekerk, R. L., & Van Gent, M. M. (2021). Mental health and well-being of university staff during the coronavirus disease 2019 levels 4 and 5 lockdowns in an Eastern Cape university. South Africa. *South African Journal of Psychiatry, 27*, a1589. doi:10.4102/sajpsychiatry.v27i0.1589

van Schalkwyk, F. (2021). Reflections on the public university sector and the covid-19 pandemic in South Africa. *Studies in Higher Education, 46*(1), 44–58. doi:10.1080/03075079.2020.1859682

van Stralen, D. (2009, September 30). The origins of EMS in military medicine. *The Journal of Emergency Medical Services*. https://www.jems.com/patient-care/origins-ems-military-medicine-2/

VanderWeele, T. J. (2017). On the promotion of human flourishing. *Proceedings of the National Academy of Sciences of the United States of America, 31*(31), 8148–8156. doi:10.1073/pnas.1702996114 PMID:28705870

Varghese, N. V. (2008). *Globalization of higher education and cross-border student mobility. Unesco*. International Institute for Educational Planning Paris.

Varma, T. M. (2011). Crisis communication in higher education: The use of "negotiation" as a strategy to manage crisis. *Public Relations Review, 37*(4), 373–375. doi:10.1016/j.pubrev.2011.08.006

Varol, N. (2007). Raising awareness of the society about natural and technological disasters and the role of "AFEM". *Proceedings book of TMMOB Disaster Symposium*, 127-131.

Vartiainen, H., Vuojärvi, H., Saramäki, K., Eriksson, M., Ratinen, I., Torssonen, P., Vanninen, P., & Pöllänen, S. (2022). Cross-boundary collaboration and knowledge creation in an online higher education course. *British Journal of Educational Technology, 53*(5), 1304–1320. doi:10.1111/bjet.13186

Velázquez, I. (2023). *Se unen universitarios para ayudar a Guerrero* [University students join together to help Guerrero]. https://www.reforma.com/se-unen-universitarios-para-ayudar-a-guerrero/ar2702796

Velázquez-Bucio, M. M., Francesca-Ferrario, M., Muccignto, E., Porfido, S., Sridharan, A., Chunga, K., Livio, F., Gopala, S., & María-Michetti, A. (2023). Environmental effects caused by the Mw8.2, September 8, 2017, and Mw 7.4 June 23, 2020, Chiapas-Oaxaca (Mexico) subduction events: Comparison of large intraslab and interface earthquakes. *Quaternary International, 651*, 62–76. doi:10.1016/j.quaint.2021.11.028

Vélez-Ascencio, O. (2017). *Suneo pide 20 mdp para reparar universidades públicas de Oaxaca* [Suneo asks for 20 million pesos to repair public universities in Oaxaca]. https://www.nvinoticias.com/general/oaxaca/suneo-pide-20-mdp-para-reparar-universidades-publicas-de-oaxaca/18756

Vital, L. M. (2018). Looking ahead post-disaster: What Latin American and Caribbean higher education can learn from Haiti. *Revista de Educación Superior en America Latina, 3*, 3-5. https://rcientificas.uninorte.edu.co/index.php/esal/article/view/10638/10886

Vogel, S. (1988). Marx and alienation from nature. *Social Theory and Practice, 14*(3), 367–387. doi:10.5840/soctheorpract198814317

Vogus, T. J., & Sutcliffe, K. M. (2007). Organisational resilience towards a theory and research agenda. *IEEE International Conference on Systems, Man and Cybernetics*, 3418-3422.

Vroom, V. H., Yetton, P., & Jago, A. (1973). A decision tree model of leader behavior: An extension of the contingency approach. *Human Relations*, *16*(4), 451–467.

Wachauf, D., & Villa-Caña, P. (2023). *Resultaron afectados por el huracán "Otis" 127 planteles de bachillerato y de educación superior* [127 highschool and higher education schools were affected by Hurricane "Otis"]. https://www.eluniversal.com.mx/nacion/resultaron-afectados-por-el-huracan-otis-127-planteles-de-bachillerato-y-de-educacion-superior/

Waithima, A. (2021). *The impact of covid-19 on higher education, its business implications and proposed recovery strategy.* Academic Press.

Waks, L. J. (2011). Teacher education programs as complex organizations. *Complicity: An International Journal of Complexity and Education*, *8*(1), 65-69. https://journals.library.ualberta.ca/complicity/index.php/complicity/article/view/10026

Waldrop, M. (1992). *Complexity: The Emerging Science at the Edge of Order and Chaos.* Simon & Schuster Editions.

Wallace, J. D., Burton, B. G., Chandler, R. C., & Darby, D. G. (Eds.). (2020). Special issue for suddenly online: Considerations of theory, research, and practice. *The Journal of Literacy and Technology*, *21*(2), 1–148.

Walsh, L. (2016). *Educating Generation Next: Young People, Teachers and Schooling in Transition.* Springer. doi:10.1057/9781137526403

Walsh, L., & Black, R. (2018). *Rethinking youth citizenship after the age of entitlement.* Bloomsbury Academic.

Wang, J., & Hutchins, H. M. (2010). Crisis Management in Higher Education: What Have We Learned From Virginia Tech? *Advances in Developing Human Resources*, *12*(5). Advance online publication. doi:10.1177/152342231039443

Wang, K., Li, B., Tian, T., Zakuan, N., & Rani, P. (2023). Evaluate the drivers for digital transformation in higher education institutions in the era of industry 4.0 based on decision-making method. *Journal of Innovation & Knowledge*, *8*(3), 100364. doi:10.1016/j.jik.2023.100364

Wango, G. (2009). *School Administration and Management: Quality Assurance and Standards in Schools.* Jomo Kenyatta Foundations.

WATAN Foundation- Education sector. (n.d.). https://www.watan.foundation/expertise/education-program

Watenpaugh, K. D., Fricke, A. L., & King, J. R. (2014). *The War Follows Them. Syrian University Students and Scholars in Lebanon.* Institute of International Education.

Waters, J. L. (2006). Geographies of cultural capital: Education, international migration and family strategies between Hong Kong and Canada. *Transactions of the Institute of British Geographers*, *31*(2), 179–192. doi:10.1111/j.1475-5661.2006.00202.x

Watson, J. T., Gayer, M., & Connolly, M. A. (2007). Epidemics after natural disasters. *Emerging Infectious Diseases*, *13*(1), 1–5. doi:10.3201/eid1301.060779 PMID:17370508

Waxman, D. A., Chan, E. W., Pillemer, F., Smith, T. W., Abir, M., & Nelson, C. (2017). Assessing and improving hospital mass-casualty preparedness: A no-notice exercise. *Prehospital and Disaster Medicine*, *32*(6), 662–666. doi:10.1017/S1049023X17006793 PMID:28780916

Wei, H. (2013). An empirical study on the determinants of international student mobility: A global perspective. *Higher Education*, *66*(1), 105–122. doi:10.1007/s10734-012-9593-5

Welch, A., & Wahidyar, A. (2020). Quality assurance in Afghan higher education: Achievements and challenges. *Asian Education and Development Studies*, *9*(4), 479–493. doi:10.1108/AEDS-09-2018-0146

West, A. (2018). The flourishing student-every tutor's guide to promoting mental health, well-being and resilience in higher education perspectives. *Perspectives*, *22*(4), 141–142. doi:10.1080/13603108.2018.1498036

Westerman, G., Bonnet, D., & McAfee, A. (2014). *Leading digital: Turning technology into business transformation*. Harvard Business Press.

White, J. (2007). What schools are for and why. Philosophy of Education Society of Great Britain IMPACT pamphlet, 14, 1-50. doi:10.1111/j.2048-416X.2007.tb00116.x

Whittaker, J., McLennan, B., & Handmer, J. (2015). A review of informal volunteerism in emergencies and disasters: Definition, opportunities and challenges. *International Journal of Disaster Risk Reduction*, *13*, 358–368. doi:10.1016/j.ijdrr.2015.07.010

Wikipedia. (n.d.). *Crisis*. Retrieved October 6, 2023, from https://en.wikipedia.org/wiki/Crisis

Wildavsky, A. (1991). *Searching for safety*. Transaction Publishers.

Wildavsky, A. B. (1988). *Searching for Safety*. Transaction Publishers.

Winn, Z. (2017, April 13). *The Virginia Tech shooting's impact on emergency preparedness*. https://www.campussafetymagazine.com/clery/virginia_tech_shooting_anniversary_emergency_preparedness/

Wiśniewska, S., & Wiśniewski, K. (2022). *Socially responsible activities addressed to internal stakeholders of non-governmental organisations*. Academic Press.

Wittkowski, J., & Paré, P. M. (2022). Introduction of the French version of the Multidimensional Orientation Toward Dying and Death Inventory (MODDI-F/fre). *Death Studies*, 1–9. doi:10.1080/07481187.2022.2108940 PMID:35950684

Wolbert, L., de Ruyter, J., & Schinkel, A. (2015). Formal Criteria for the Concept of Human Flourishing: The First Step in Defending Flourishing as an Ideal Aim of Education. *Ethics and Education*, *10*(1), 118–129. doi:10.1080/17449642.2014.998032

Wood, B. E., & Black, R. (2018). Globalisation, cosmopolitanism and diaspora: What are the implications for understanding citizenship? *International Studies in Sociology of Education*, *27*(2–3), 184–199. doi:10.1080/09620214.2017.1415161

World Health Organization. (2019, December 19). *Emergencies: WHO's role*. https://www.who.int/news-room/questions-and-answers/item/who-s-role-in-emergencies

World Health Organization. (2021). *Strategic toolkit for assessing risks*. https://www.who.int/publications/i/item/9789240036086

Yamaiel, A. O. (2006, August). Disaster resistant universities: In search of strategies for resilient higher education institutions. *International Journal of Mass Emergencies and Disasters*, *24*(2), 179–180. doi:10.1177/028072700602400201

Yañez, B. (2023). *Día a día: la cronología a un mes del paso del huracán Otis* [Day by day: the chronology one month after Hurricane Otis]. https://politica.expansion.mx/mexico/2023/11/25/que-paso-acapulco-huracan-otis

Yang, Y. F., Yeh, H. C., & Wong, W. K. (2010). The influence of social interaction on meaning construction in a virtual community. *British Journal of Educational Technology*, *41*(2), 287–306. doi:10.1111/j.1467-8535.2009.00934.x

Yaya, S., Otu, A., & Labonté, R. (2020). Globalisation in the time of COVID-19: Repositioning Africa to meet the immediate and remote challenges. *Globalization and Health*, *16*(1), 1–7. doi:10.1186/s12992-020-00581-4 PMID:32580728

YevdokymovaN. (2023). Resilience and Reform: The Impact of War and Conflict on Innovation in Higher Education. doi:10.2139/ssrn.4602827

Yildirim, S. (2023). 6 Şubat kahramanmaraş depreminin psikososyal etkisi ve depremzedelere yönelik sürdürülebilir müdahalenin önemi üzerine bir gözlem araştirmasi. *Anasay, 24*(1), 133–153. doi:10.33404/anasay.1286368

Yildiz, B. (2014). Children and adolescents. In T. Doğan & Ö. Erdur (Eds.), *Disasters, crises, traumas and psychological help* (pp. 25–43). Turk PDR Dernegi.

Yousif, N. B. A., Grondys, K., Gad, S., & Elsayed, W. (2020). Knowledge management in non-governmental organizations (NGOs). *Administratie si Management Public, 1*(35), 90–108. doi:10.24818/amp/2020.35-06

Zamoum, K., & Gorpe, T. S. (2018, June 27). *Crisis Management: A Historical and Conceptual Approach for a Better Understanding of Today's Crises.* Retrieved October 27, 2023, from https://www.intechopen.com/chapters/60813

Zemke, A. (2020, June 13). *Role of Strategic Communication in Crisis Management and Business Continuity.* Retrieved October 29, 2023, from https://beehivepr.biz/strategic-communication-in-crisis/

Zhang, C., Yu, M. C., & Marin, S. (2021). Exploring public sentiment on enforced remote work during COVID-19. *The Journal of Applied Psychology, 106*(6), 797–810. doi:10.1037/apl0000933 PMID:34138587

Zhen, Y., & Bian, S. (2015). Study on Crisis Management Strategy for Universities. *International Conference on Education, Management, Commerce and Society (EMCS)*, 273-276. 10.2991/emcs-15.2015.56

Zimbabwe Council for Higher Education. (2020). *Summary of Universities' Responses on Survey on Covid 19 Preparedness Strategies and Challenges in Higher Educations in Zimbabwe.* Author.

ZimmermannM.BledsoeC.PapaA. (2020). *The Impact of the COVID-19 pandemic on college student mental health: a longitudinal examination of risk and protective facto*rs. doi:10.31234/osf.io/2y7hu

Zulfikar, M. W., Bin Hashim, A. I., Umri, H. U. B. A., & Dahlan, A. R. A. (2018). A business case for digital transformation of a Malaysian-based university. *Proceedings of the 2018 International Conference on Information and Communication Technology for the Muslim World*, 106–109.

About the Contributors

Mustafa Kayyali, Ph.D., is an ardent advocate for excellence in higher education, driven by a relentless pursuit of quality, recognition, and innovation. As the Manager of HE Higher Education Ranking, he leads initiatives that play a crucial role in shaping the landscape of academia. Having earned a Master's in Quality Management and Evaluation in Higher Education from Universitat Oberta de Catalunya, followed by a Ph.D. in Quality Management from Azteca University, his academic pursuits have laid a strong foundation for his expertise in Accreditation, Quality Assurance, and Higher Education Rankings. Dr. Kayyali has made significant contributions to academic literature, boasting more than 30 published papers and 10 book chapters. Additionally, he takes pride in his role as a translator, having translated 5 books, thus bridging language gaps and promoting knowledge exchange on a global scale. In recognition of his expertise, Dr. Kayyali holds key affiliations with esteemed institutions. He is privileged to be a part of the British Accreditation Council's Accreditation Committee, the Slovenian Quality Assurance Agency as an International Expert, and the Agency for Control and Quality Assurance of Higher Education AKOKVO as an Accreditation Expert. His international engagements also include being an expert for the Kosovo Accreditation Agency KAA and the ECBE - European Council for Business Education, as well as serving as a Lead Auditor at the Centre of Assessments for Excellence (COAE). Dr. Kayyali has also submitted a patent to USPTO about rankings entitled Holistic & Multidimensional Ranking Methodology for Universities. As an entrepreneur, researcher, translator, and publisher, Dr. Kayyali is deeply involved in various facets of the academic world. His diverse interests encompass Management, Translation, Interpretation, and Academic consulting, contributing to a well-rounded understanding of the industry.

* * *

Ebru Aksu was Born in 1977. Gruaded from Gazi Anadolu High School. Awarded Bechelor's Degree from Ankara University.

Mahdi Alkol is an Assistant Professor renowned for his expertise in statistical analysis and economic feasibility studies. He holds a distinguished position at Idlib University and Alshamal Private University, where he imparts his extensive knowledge to aspiring scholars. With a Ph.D. in Statistics earned in 2012, Dr. Alkol has established himself as a leading authority in his field. Throughout his career, Dr. Alkol has made significant contributions to academia through his prolific publication record. His insightful articles and research papers have been featured in esteemed journals, solidifying his reputation as a thought leader in statistical analysis and economic studies. Beyond his academic pursuits, Dr. Mahdi Alkol is actively engaged in shaping the future of education. He serves as a member of the Board of Trustees at Al-Maaref University of Applied Sciences, where he works collaboratively to enhance educational standards and promote academic excellence. Dr. Alkol's dedication to advancing knowledge in statistics and economics, coupled with his commitment to academic leadership, exemplifies his profound impact on both the academic community and the broader society.

George Areba Ngwacho, Ph.D, is the immediate former Dean and Postgraduate Coordinator, at the School of Education & Human Resource Development (SEDHURED), Kisii University & Senior lecturer at the Department of Education Foundations, Educational Administration, Planning and Economics (EDFO/EAPE). He is also an experienced thespian in Drama and music as a trainer, facilitator, and adjudicator at all levels. He is an accomplished researcher, writer, supervisor, examiner, and reviewer with a number of institutions of higher learning. Prior to joining higher education as a lecturer, he served as a high school tutor teaching English Literature in a number of schools. His research appears in a range of international peer-reviewed scholarly journals.

Wendy Goff is an Associate Professor in the Department of Education at Swinburne University of Technology in Australia. Wendy's research explores adult relationships/partnerships and their impact on learning, health, wellbeing and development. She uses different vehicles to study these interactions, and has presented and published her work at a national and international level. Drawing on an Aristotelian lens Wendy's work has also examined how adults and children can be supported to flourish. Part of this work has included developing a model of support for human flourishing that can be drawn on in educational contexts to nurture the physical and mental health and wellbeing of societies.

Pedzisai Goronga is a Senior Lecturer and Chairperson of the Department of Educational Foundations, Faculty of Education at the University of Zimbabwe. He teaches both undergraduate and post graduate modules in Educational Psychology and coordinates Faculty-wide modules. He holds a Doctor of Philosophy (PhD) Degree in Educational Psychology from the University of Pretoria, South Africa, M Ed Educational Psychology from the University of Zimbabwe, B Ed in Primary Education and a Diploma in Education from the same university. His areas of research interest include among many other issues; resilience studies, gender and education, human learning and performance, guidance and counselling and emotional intelligence. Dr Pedzisai Goronga has vast teaching experience at primary school level and more than close to two decades teaching experience at university level. Dr Pedzisai Goronga is a visionary leader, an honest follower and team player who thrives on collaboration to meet set goals.

Colleen Halupa is a retired Air Force Officer. She previously was a clinical laboratory scientist and health administrator. She has a doctorate in curriculum and instruction and now it a dean at a university. She is a researcher in online learning, pedagogy and assessment.

Stefan Handke is Professor of Public Management. He holds a doctorate in political science. Before being appointed to the HTWD, he was Managing Director of the accreditation agency ACQUIN. In this position, he gained extensive experience in the development and quality assurance of higher education programmes. Stefan was a member of the Board of ENQA (European Association for Quality Assurance in Higher Education) and CEENQA (Central and Eastern European Network of Quality Assurance Agencies in Higher Education). His work in international quality assurance has focussed on the entire European Higher Education Area.

Catherine Hayes is Professor of Health Professions Pedagogy and Scholarship at the University of Sunderland, UK. She is a UK National Teaching Fellow and Principal Fellow of the UK Higher Education Academy. As a graduate of Podiatric Medicine in 1992, Catherine was a Founding Fellow of the Faculty of Podiatric Medicine at the Royal College of Physicians and Surgeons (Glasgow) in 2012 and was awarded Fellowship of the Royal College of Podiatry (London) in 2010. She is currently Programme Leader of the University of Sunderland's Professional Doctorate pathways for the DBA, EdD, DPM and DProf.

Mubango Hazel is a PhD in Human Resources Management candidate at the University of Venda. She has years of experience in both the industry in the Human Resources Management discipline and as a professional academic and researcher in the field of Human Resource Management. She is currently working as a lecturer and as the Chairperson of the Department of Commerce and Management at the Women's University in Africa. Hazel also serves in numerous university committees including the Teaching and Learning Committee, Affiliates Committee, Job Evaluation Committee and the University Senate. She is also an HR consultant and writes on broader HR areas. She is an examiner, moderator and module writer for the Institute of People Management of Zimbabwe (IPMZ). She has taught and published in journals on human resources management issues in areas that include work-life balance, women in employment/ business, gender diversity, equality and inclusion in the workplace, learning in Higher Education and organisational support.

Vicki A. Hosek, Ed.D., is an Instructional Assistant Professor of Education who teaches secondary methods and foundations, technology integration across all content areas, and multiliteracy development. Her research interests include the digital literacy of practicing teachers and teacher candidates; examination of student and instructor voices in online classroom environments; and instructors' application of educational theory in their technology integration practices. She holds degrees from the University of Colorado, Boulder, Western Illinois University, and Illinois State University.

Md Tariqul Islam is a JSPS postdoctoral research fellow at Hiroshima University, Japan. He worked as an assistant professor at BRAC University in Bangladesh at the BRAC Institute of Educational Development (BRAC IED). Tariq earned his PhD in Education from Deakin University, Australia. He received his Bachelor's (Hons) and Master's in Education from the University of Dhaka, Bangladesh. As a researcher, he worked with several national and international organizations, e.g., BRAC, NAEM, FREPD, UNESCO, UNICEF, ActionAid. He has written a number of peer-reviewed journal articles and book chapters that have been published in leading journals and by leading publishers. He has also given papers at national and international conferences. His research interest largely focuses on the sociology of education, with particular attention to education for sustainable development. Tariq's current research focuses on education for global citizenship, higher education, youth (un)employability, teaching-learning in higher education, teachers' professional development, ICT in education, lifelong learning, educational inclusion and exclusion, and play for learning and socio-emotional development.

Ahmad Kayaly is a renowned expert in the field of Non-Governmental Organizations (NGOs), with a rich background in humanitarian and developmental work spanning over a decade. As the Manager of Mission Syria in WATAN Foundation, Ahmad has played an important role in orchestrating and overseeing initiatives aimed at providing relief and fostering sustainable development in one of the most challenging regions. His expertise encompasses strategic planning, organizational development, and advocacy, all of which have been crucial in driving meaningful change. Ahmad's commitment to social justice and sustainable development is reflected in his work, which is marked by a profound understanding of the complexities within the NGO sector and a continuous search for innovative solutions. Through his leadership and vision, Ahmad Kayaly has significantly contributed to shaping a better future for communities in need.

Khathutshelo Khashane is a Senior Lecturer in the Department of Human Resources Management and Labour Relations, In the Faculty of Management, Commerce and Law, University of Venda, South Africa. She has taught, researched and published within the field of human resource management, strategic management and industrial psychology.

Hakan Kilinc is currently working as an associate professor doctor at Anadolu University. He undertook undergraduate studies in the field of Computer Education and Instructional Technologies (CEIT) between the years of 2009 and 2013. He received his master degree in the field of Distance Education in 2016, May. In addition to this, he received his Ph.D. Degree in the field of Distance Education in 2020, April. Hakan Kilinc continues to work in the field of distance education technology, personalized learning environments, information and communication technologies, synchronous, asynchronous, and interactive communications, learner-generated content and online group discussions.

Ali Mazı completed his primary and secondary education in Hatay. He started Kahramamaraş Sutcu Imam University in 2010. He studied department of elementary mathematics in there. In 2018, He completed my master's degree in Educational Management, Inspection, Planning and Economics Department of Kahramanmaraş Sütçü İmam University. He has been working as a teacher in Ministry of National Education since 2014. He completed his doctorate in Hacettepe University Basic Education Department Classroom Education in 2023. He is one of the victims of the earthquake because his house in Hatay was destroyed in the Kahramanmaras earthquakes on 06/02/2023.

Robert New is a PhD candidate at Monash University researching educational leadership. His particular interest is on how leadership experiences affect the follower at and beyond the workplace. Robert has multiple postgraduate degrees in educational leadership and related fields. His other research areas of interest include secure-base leadership, organisational development, and competition between independent schools. After teaching in schools for twenty years, he now educates pre-service service teachers at Swinburne University of Technology.

Hlanganipai Ngirande is an Associate Professor and Head of the extended Curriculum Programme at the University of Venda, South Africa, specialising in Human Resource Management and Organisational Psychology. Prof Ngirande holds a Ph.D., and Mcom in Industrial Psychology, a Postgraduate Diploma in Higher Education (PGDHE), and multiple certifications. He has 13 years of teaching and learning experience at both undergraduate and postgraduate levels. As a distinguished Teaching Advancement at

Universities (TAU) Fellow, he has authored 24 journal articles that were published in DHET-accredited journals and actively presents at national and international conferences. Prof Ngirande has supervised 12 Master's students, 1 Ph.D. student, and numerous Honours students. He also serves in several University of Venda`s committees including the Teaching and Learning Committee. His dedication to student entrepreneurship is evident through his role as an Entrepreneurial Action Us (ENACTUS) South Africa Faculty Advisor and as a Student Training for Entrepreneurial Promotion (STEP) Trainer, along with active involvement in the University of Venda committees.

Christopher Nyakotyo is a Sociology of Education PhD student in the Department of Educational Foundations, Faculty of Education at the University of Zimbabwe. He holds a Master of Education Degree in Sociology of Education, A Bachelor of Education in Sociology and a Diploma in Education. Currently he is a primary school head in Seke Distict Mashonaland East Province Zimbabwe. Mr Christopher Nyakotyo is an astute scholar who has insatiable appetite for academic and professional advancement. His areas of research interest are inclined to development studies. He also has interest in multicultural education, indigenous knowledge systems, transformative pedagogies and resilience in higher education.

Jay C. Percell, Ed.D., is an Associate Professor at Illinois State University. He teaches secondary education methods courses for implementing literacy and technology across all content areas. Research interests include: educational technology, digital literacy, and alternative grading. He holds degrees from Illinois State University, Northern Arizona University, and the University of Northern Colorado.

Sherihan Radi has more than ten years of working experience with international organizations and Projects under the umbrella of the UN, USAID, the German Federal Foreign Office as well as Pro Helvetia, the Swiss Arts Council at the Embassy of Switzerland in Cairo. In addition to more than two years of working experience in Trade & Investment at the German - African Business Association (Afrika-Verein der deutschen Wirtschaft (AV)) in Berlin, Germany as well as at Advantage Austria, the Austrian Embassy Commercial Section with its head office at the Austrian Federal Economic Chamber of Commerce in Vienna, Austria. Dr. Radi earned a PhD of Economics from Brandenburg University of Technology, Cottbus – Senftenberg in Germany and a Master of Business Administration from the University of Toledo, Ohio in the United States of America. Her research interests include gender, gender equality, feminist theory, development, human science, social science, human resource management and economics. She has published articles and papers in various scientific journals.

Mulatedzi Calvin Rammbuda is a Senior Teacher in the Limpopo Provincial Department of Basic Education in Republic of South Africa. He is also a Contractor at the University of South Africa. He obtained Masters of Technology in Educational Leadership and Management from Tshwane University of Technology, Doctor of Education in Curriculum Studies, and Advanced Postgraduate Supervision Capacity Development Programme from University of South Africa. He is also a Ph.D. researcher who has published several articles on topics such as gender-based violence, corruption, technology, curriculum support and economy in South Africa. His research interests are in Curriculum Studies, Educational Leadership and Management, Adult Education, Development Studies, Youth Development, Public Administration and Development, and Community Engagement. He has attended international conferences and presented papers in the International Conferences and published articles in accredited peer review conference proceedings and in journals. He is currently teaching at Miriyavhavha Technical High School in South Africa.

Jaime Santos-Reyes is a lecturer at the National Polytechnic Institute, Mexico, whose main research interests are seismic risk, safety management systems, critical infrastructures modelling, accident and risk analysis, reliability engineering. He obtained a PhD from Heriot-Watt University, UK, in 2001. He has spent 3 years working as a research associate at Heriot-Watt University, Edinburgh, Scotland, UK.

Elizabeth (Beth) J. Stroble, Ph.D. Completing her fifteenth year as the Chancellor for Webster University's global university system, Stroble has held academic and administrative posts at public and private institutions across the United States. Across her career, she has emerged as a nationally recognized champion for increasing educational access and success for students worldwide, catalyzing innovation, and encouraging educational institutions' civic engagement and service to society. Through partnerships with the Urban League of Metropolitan St. Louis, the African Renaissance and Diaspora Network, and the International Association of University Presidents, Stroble advocates for inclusion and equity for globally diverse individuals and communities. Her scholarship focuses on the relationships among place, identity, and community as well as women's leadership and global trends in education.

Index

D

E

F

G

P

Q

R

S

T

U

V

W

Y

Z

www.ingramcontent.com/pod-product-compliance
Lightning Source LLC
LaVergne TN
LVHW062116300926
849161LV00010B/2470
9798369319260